BLUEPRINT READING BASICS

Gulf Publishing Company
Book Division
Houston, London, Paris, Tokyo

BLUEPRINT READING BASICS

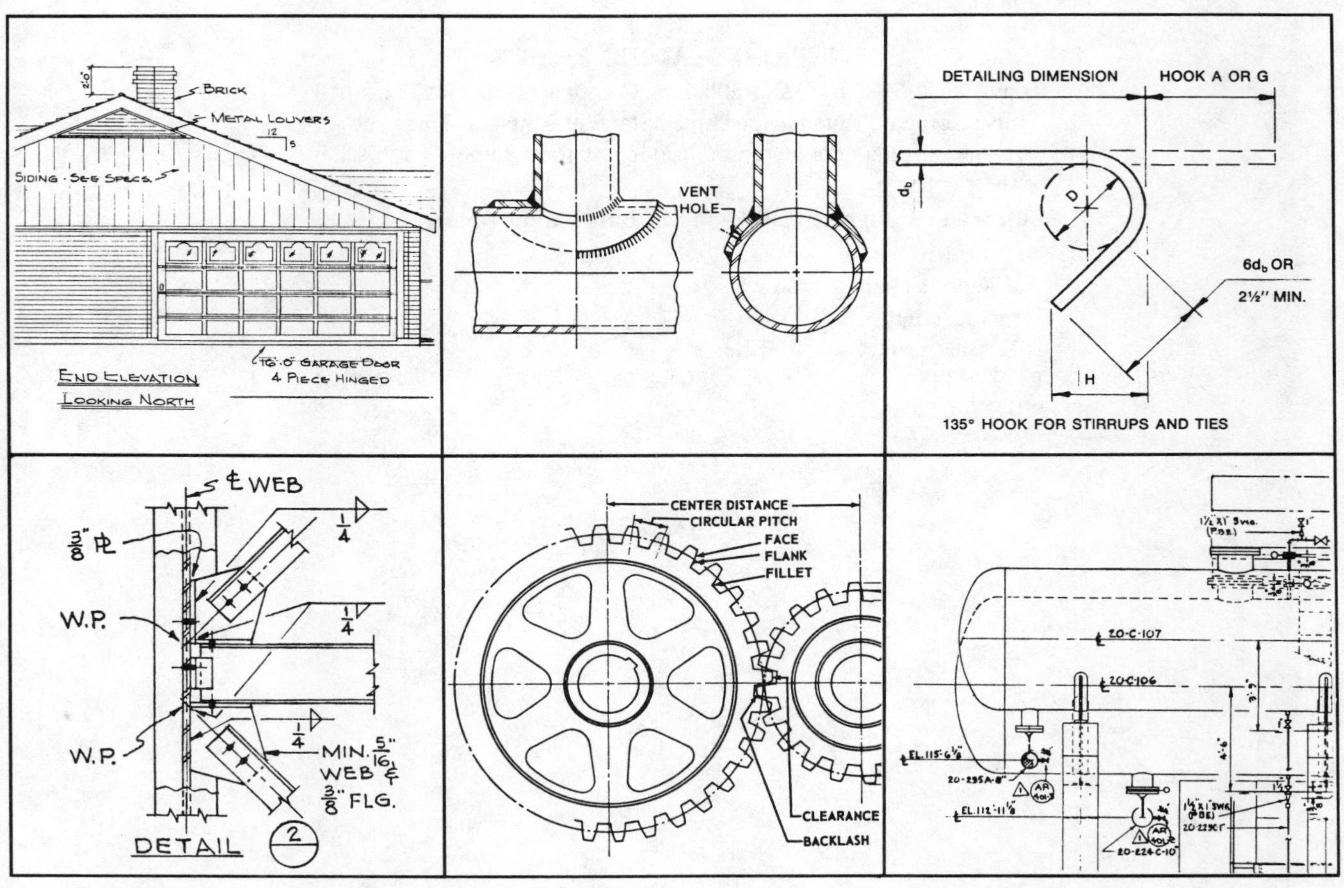

Rip Weaver

BLUEPRINT READING BASICS

Library of Congress Cataloging in Publication Data

Weaver, Rip.
Blueprint reading basics.
Includes index.
1. Blue-prints. I. Title.

T379.W38 604.2'5 82-6104
ISBN 0-87201-075-9 AACR2

Contents

Preface

In the past year I visited more than fifty colleges and universities. Almost all of the instructors I met asked me to do a modern blueprint reading book that would encompass the fields of piping, structural steel, concrete foundations, and other industrial disciplines. Of course, they also wanted the standard disciplines covered, such as machine and architectural.

This book covers all these items and more, including metrics. It is designed to be beneficial to a wide range of people who must be able to read and understand blueprints, which in this technological age includes sales personnel and secretaries, as well as craftsmen, technicians, and students.

This is the only blueprint reading book I know of that explains plant coordinate systems which are used by almost every type of industrial complex. And for those who have forgotten how to handle fraction addition, subtraction, multiplication, and division, there is a chapter that explains traditional methods and presents some new and novel methods, as well.

I hope you enjoy this book and that it will aid you in your career.

Rip Weaver
June 1982

CHAPTER 1

Blueprints and Working Drawings

Webster's dictionary defines blueprint as a photographic reproduction in white on a blue background, usually of architectural or engineering plans. However, most drawing reproductions today are not actually blueprints, and the term used, "print," is usually a diazo reproduction on white paper. The lines on this print are usually black or blue but, because of modern chemicals, may be any color.

Print paper is treated with a chemical that is light sensitive. The type of chemical determines the color of the lines on the print. The treated paper is placed beneath the original drawing, run through a diazo machine, which exposes it to intense light, and the print is made. The light "burns off" the portion of the chemical that is unprotected by the dark lines of the original. As it runs through a developer, the chemical left on the treated paper darkens to form lines on the print. The print can be made in less than a minute on modern reproduction machines.

Print paper is opaque and comes in many weights. A reproducible, transparent copy can be made. This is called a sepia because the color of the lines are sepia, or brownish. The sepia can be sent to locations that have reproduction machines and prints may be run from it.

Another type of print is the electrostatic copy. This is run on a copy machine that makes black lines on a white background and is usually limited to smaller sizes, such as A, B, or C sizes (see Table 1-1). Electrostatic copies are sometimes made from a microfilm. In this case, a microfilm copy is mailed to a location, usually foreign, which has the electrostatic copy machine. The microfilm, fastened on a card, is fed into the machine and a reproduction, usually C size, is made. This saves postage as the microfilm is much lighter than several prints.

Table 1-1. Drawing Paper Sizes

Size Designation	Normal Size, Inches	Alternate Size, Inches
A or 1	8½ x 11	9 x 12
B or 2	11 x 17	12 x 18
C or 3	17 x 22	18 x 24
D or 4	22 x 34	24 x 36
E or 5	34 x 44	36 x 48

Whatever the type of reproduction, some distortion occurs during the print-making process. Usually, the print is slightly larger than the original. This enlargement is not visible with the naked eye, but is pronounced enough that scaled drawings should not be measured on prints. Only the given dimensions, should be used for any construction. Prints may be scaled to ascertain approximate dimensions, but can not be scaled for any fabrication dimensions.

Working Drawings

Working drawings are the original drawings prepared by the draftsman, or drafter if you prefer. These are made on very high quality, high rag content, transparent vellum (a drafting paper), or

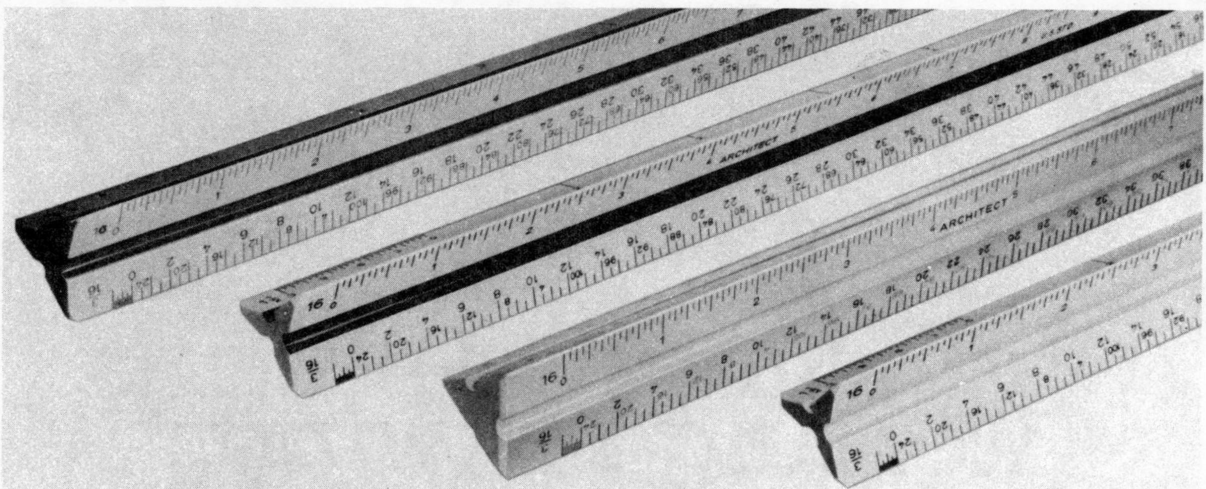

Figure 1-1. Triangular scales. (Courtesy of Bruning Division, AM Corp.)

on a plastic film. If made on vellum, sometimes called tracing paper, a graphite pencil is normally used. When working drawings are made on plastic film, a plastic leaded pencil or a special ink that fuses with the plastic may be used. Most drafting rooms prefer to use the plastic leaded pencil.

Working drawings are normally made to scale, although some views are to proportion. The drafting scale is a precision instrument that has two main purposes. It measures distances and enables a scaled replica of an item to be drawn on the drawing paper. Ordinarily, this replica drawing is made to a scale smaller than full size. If the object to be drawn is small enough, full scale drawings are made. Occasionally, for very small items, drawings are made to larger than full-size scale. So scales are used to make drawings to a reduced size, full size, and enlarged size.

The Architect's Scale

The most common scales are the architect's scale, mechanical engineer's scale, civil engineer's scale, and the metric scale. These measuring scales come in varied sizes and shapes but the most popular one is the triangular shape pictured in Figure 1-1. This is the architect's scale, which is used not only for drawing buildings and homes but for drawing machine parts, concrete, steel, piping, vessels, exchangers, and most other items that require fabrication.

The architect's triangular scale is divided into fractions of an inch that equal one foot. The scales shown in Figure 1-1 have a 16 on the top scale. This means that each inch is divided into sixteenths on this scale. So 1/16″ will represent one foot on the scaled drawing. Below that on the triangular scale, note the 3/16. On this scale, 3/16″ will equal one foot on the scaled drawing. Other scales on the triangular architect's scale are ⅛, 3/32, ¼, ⅜, ½, ¾, 1½, and 3. These are all normally used to equal one foot on the scale drawing such as ½″ = 1′–0″. These may also be used for proportional drawings. For an item drawn a fourth size, then 12″ ÷ 4 = 3″, so the 3″ = 1′–0″ scale would be used. For an eighth size, 12″ ÷ 8 = 1½″ so the 1½″ = 1′–0″ scale would be used.

Table 1-2 shows the scales for making full-size and reduced drawings with the architect's scale.

Table 1-2. Scales for Use with Architects' Scale

Size	Scale				
Full Size	1 ″	=	1″		
¾ Size	¾ ″	=	1″		
½ Size	½ ″	=	1″		
¼ Size	¼ ″	=	1″ or 3″	=	1′-0″
⅛ Size	⅛ ″	=	1″ or 1½″	=	1′-0″
1/12 Size	1 ″	=	12″ or 1″	=	1′-0″
1/16 Size	¾ ″	=	1′-0″		
1/24 Size	½ ″	=	1′-0″		
1/32 Size	⅜ ″	=	1′-0″		
1/48 Size	¼ ″	=	1′-0″		
1/64 Size	3/16″	=	1′-0″		
1/96 Size	⅛ ″	=	1′-0″		

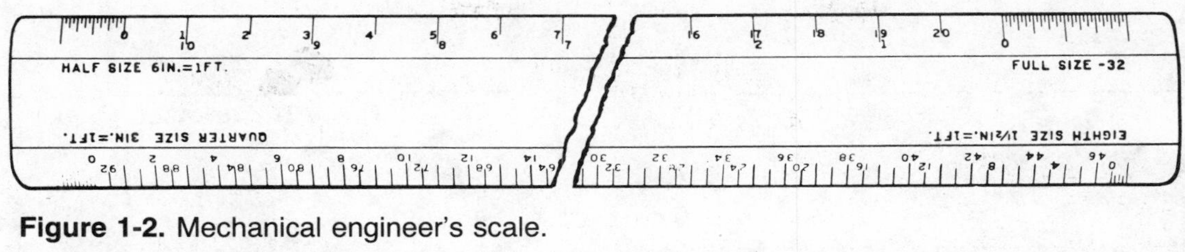

Figure 1-2. Mechanical engineer's scale.

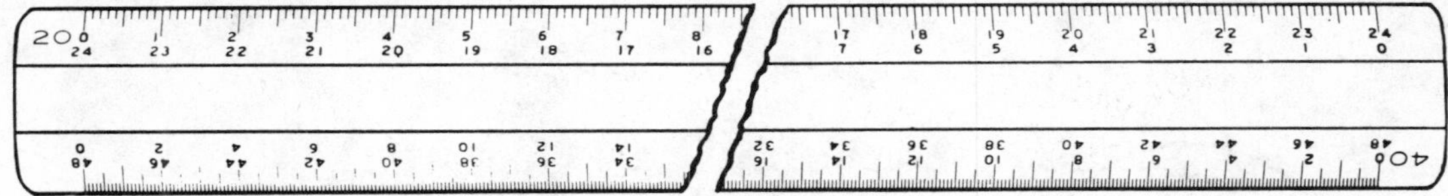

Figure 1-3. Civil engineer's scale.

The Mechanical Engineer's Scale

This scale is divided to make drawings ⅛, ¼, ½, and full size. Figure 1-2 pictures this scale. Since the architect's scale also allows ⅛, ¼, ½, and full-size drawings (see Table 1-2), very few mechanical engineer's scales are used.

The Civil Engineer's Scale

This scale is divided into decimal units. These units are 10, 20, 30, 40, 50, and 60 parts to the inch. This scale is also called the chain scale or the decimal scale, but most often it is referred to as just the engineer's scale. It is usually used when making very large reductions such as mapping and is used for some machine shop drawings where dimensions must be shown in decimals. Figure 1-3 shows a flat engineer's scale but triangular models are also available.

The Metric Scale

This scale is used to measure items in metric units, usually millimeters. It is divided into decimeters (tenth of a meter), centimeters (hundredths of a meter), and millimeters (thousandths of a meter). The metric scale is also divided into ratios. For instance, 1:100, 1:75, 1:50, 1:25, 1:20, and 1:33⅓, which is very close to the U.S. ⅜″ = 1′-0″ scale. Table 1-2 shows that a scale of ⅜″ = 1′-0″ is actually 1⁄32 size or on a ratio of 1:32. The metric system is covered in Chapter 4.

Other Scale Shapes

Professionals in drawing rooms usually purchase the triangular shaped scale, which at over a foot long is a little large for nondrafting people to carry around and use in estimating lengths from plans. So, a handy scale to have is the 6″ long flat scale shown in Figure 1-4, which will easily fit in pocket or purse. These are available at any blueprint supply company and come in several types. The two-bevel type has two scales at each end for a total of four scales. The opposite bevel has two scales on each side. The four-bevel has two scales on each of its four edges.

Terminology

The first drawing made is the *layout* or *layout sketch*. This is a nonproduction drawing, which

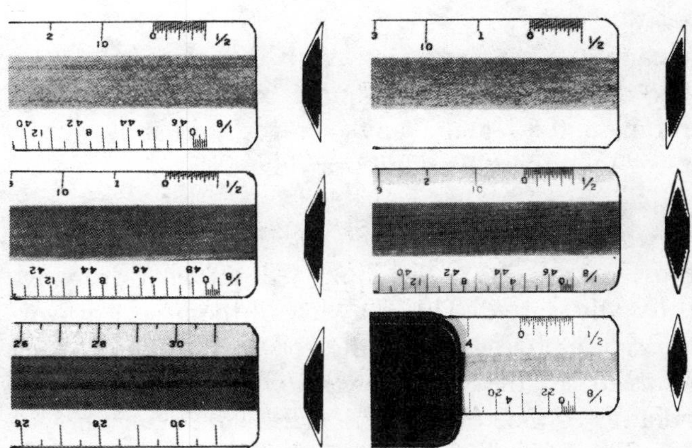

Figure 1-4. Flat scales.(Courtesy of Bruning Division, AM Corp.)

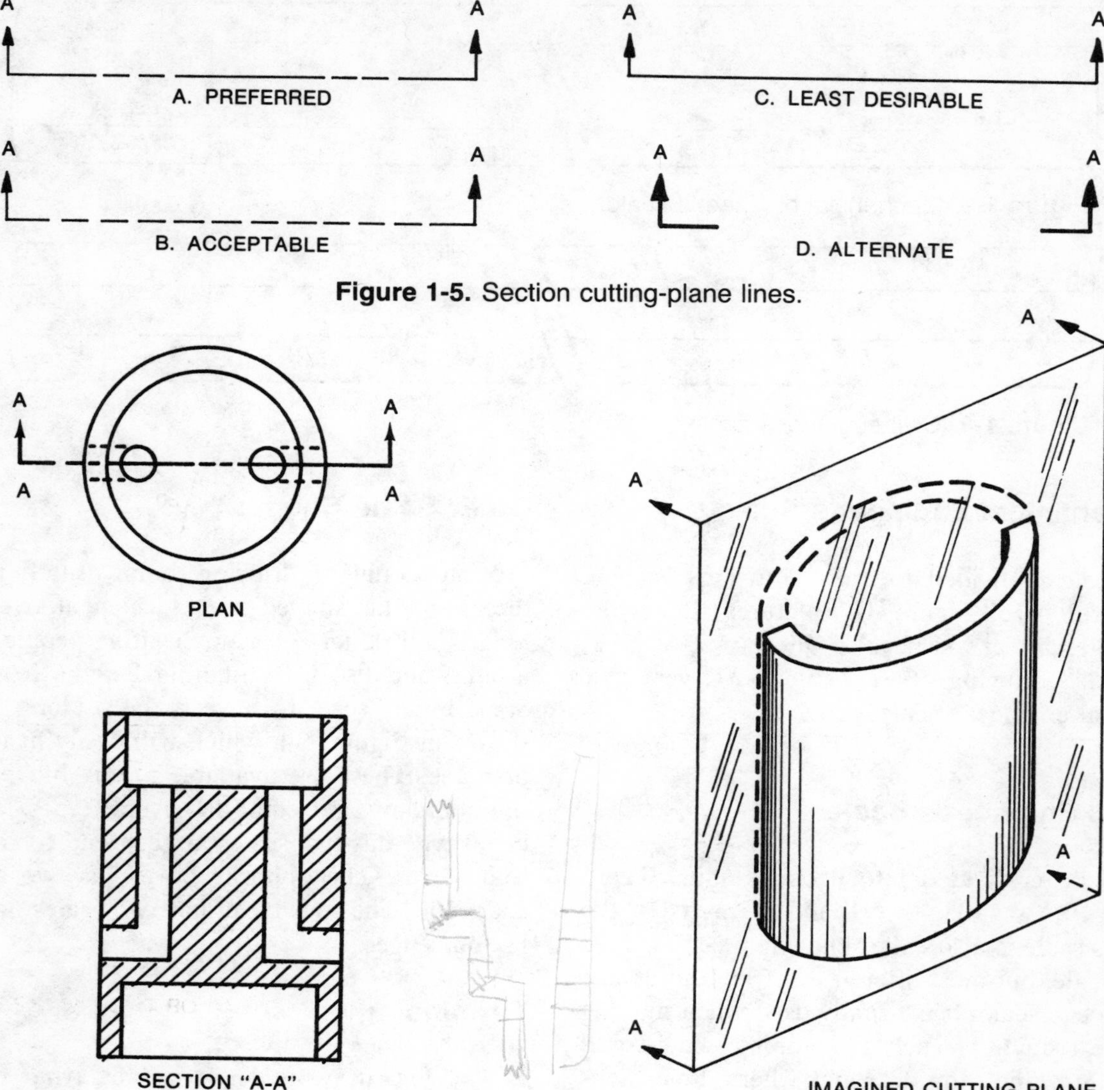

Figure 1-5. Section cutting-plane lines.

Figure 1-6. The cutting-plane sectioning theory.

means it will never be issued. It is a planning drawing made to assure adequate scale, dimensioning room, and to make any changes that may occur prior to making the final drawing.

When the layout drawing has been reviewed by those responsible and is approved, the *plan* drawing is usually started. This is a view of the object drawn to scale as if seen from the top of the item. Most plans orient the viewer by showing a north arrow pointing to the north. North arrows are never shown on any other types of drawings, only plans. So, anytime a drawing has a north arrow on it, it is a plan of the object.

Sections are drawn along a section line cut across the plan drawing at some point. The plan will show just where this section line is cut and the section drawing will be drawn vertically as if standing and looking at the object from that point. Figure 1-5 shows commonly used section cutting plane lines. Figure 1-6 shows the cutting plane sectioning theory. Solid lines cut by the section cutting line will have some type of sectioning lines shown in the section view. Figure 1-6 has slanted sectional cutting lines where the object is cut across a solid surface. Note that the drilled holes do not have these lines, showing that the cutting line does not cut through a solid portion here.

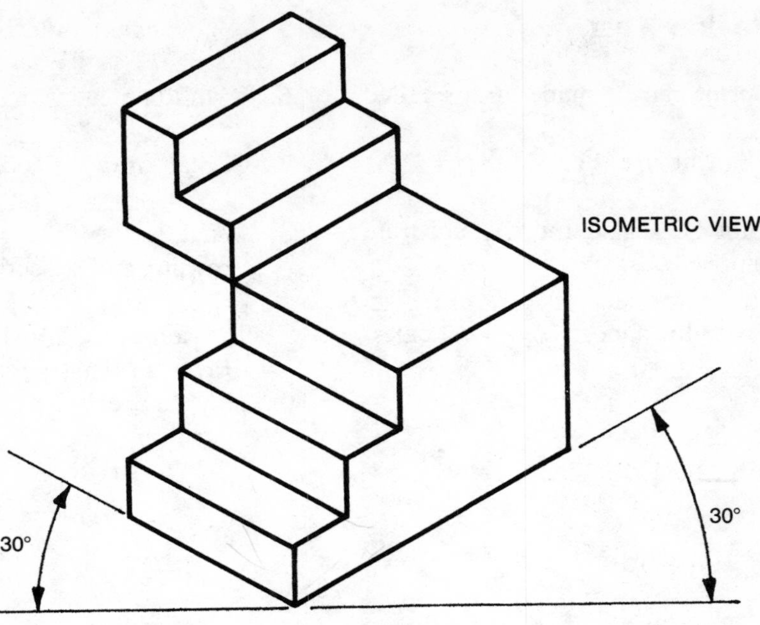

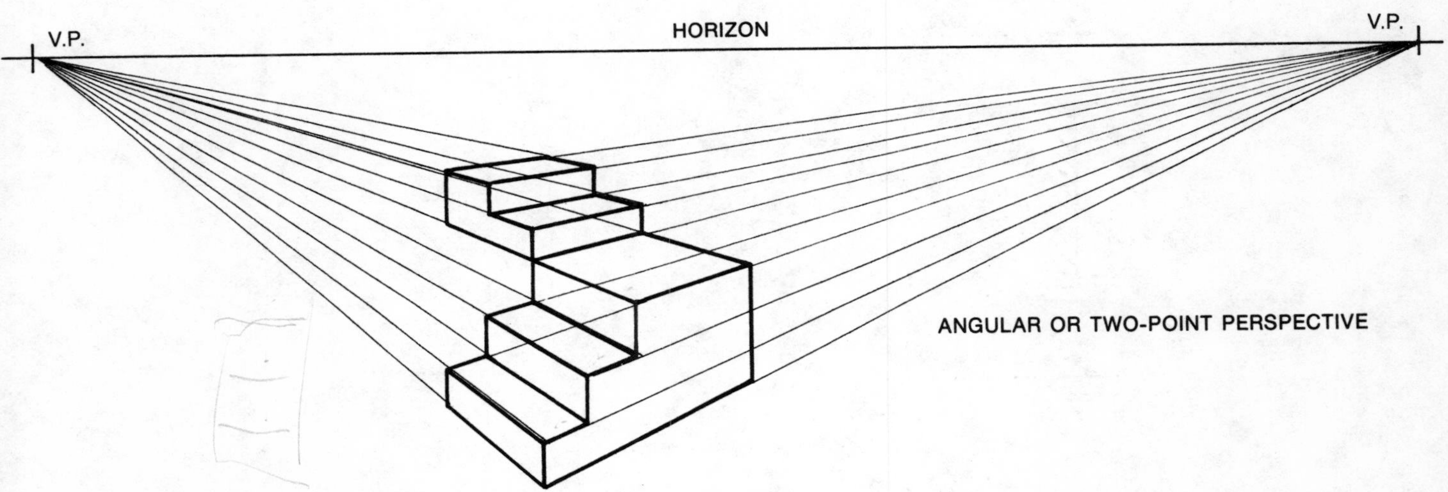

Figure 1-7. Two-direction steps drawn as isometric and angular perspective views.

There are many types of sections and others will be covered in Chapter 9.

Elevation drawings are drawn vertically but, unlike sections, do not cut through an object. A good example of an elevation is a view of the side of a house. Builders want to have a plan and elevation drawings of each side of the house they are building.

Detail drawings are usually drawn to a scale larger than the plan drawing to enable more detail to be shown. Enlarged sections are very common and are called detail drawings.

Isometric drawings are made to enable the viewer to see the object in three dimensions. These are usually drawn to scale, but not always. They are normally the easiest to read for the nonprofessional. Many architects prepare a *perspective* drawing, also a three-dimensional type drawing. Isometric drawings have horizonal lines drawn on a line 30° from the horizontal, while perspective drawings are drawn to a vanishing point, much as the natural eye sees objects. Figure 1-7 shows the isometric and the two-point perspective drawings.

Review Test

1. A transparent print can be made and is called a ________________.
2. An 8½″ × 11″ print size is called the ______ size.
3. Prints should not be scaled for any accurate dimension because ____.
4. Name three of the four most common scales.
   a. ________________
   b. ________________
   c. ________________
5. What scale would be used to make a drawing ¼ size? ________________
6. A millimeter is what part of a meter? ____________________
7. A plan drawing is one viewed from ____.
8. What is the difference between an elevation drawing and a section drawing? ____________________________________
9. In isometric drawings, horizontal lines are drawn on what angle? ____________
10. How do perspective drawings differ from isometric drawings? ____________________

CHAPTER 2
Dimensioning

Dimensioning is the process of applying lines and numbers to a pictorial representation of an object, so it can be properly fabricated. While most drawings are made to some scale, the drawing can not be measured with an appropriate scale (scaled) to determine fabrication dimensions. While the blueprint reader will not be required to ascertain proper dimensions, there are several dimensioning functions that must be known to properly read the print.

The two basic types of dimensioning are *machine shop* and *architectural*. For machine shop dimensioning, dimension lines are broken and dimensions are located within this break. Most other drafting fields use architectural dimensioning, which has an unbroken dimension line running between extension lines, with dimensions located above the line. Most computer-plotter drawings reflect this style. Figure 2-1 shows these two types of dimensioning.

Arrowheads

Arrowheads terminate dimension and leader lines, usually at a line extended from the point being dimensioned to. In rare cases, the dimension may go to the actual visible line, not to an extension line, but this is not normal drafting room practice. Arrowheads may be one of several types but types should not be mixed on the same drawing. Figure 2-2 shows several arrowhead types.

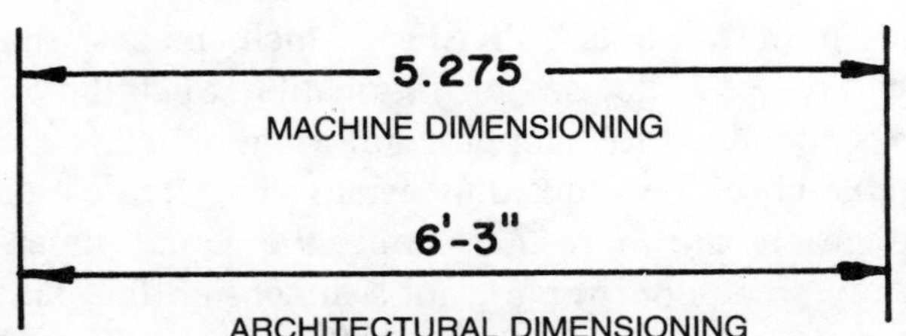

Figure 2-1. The two basic types of dimensioning.

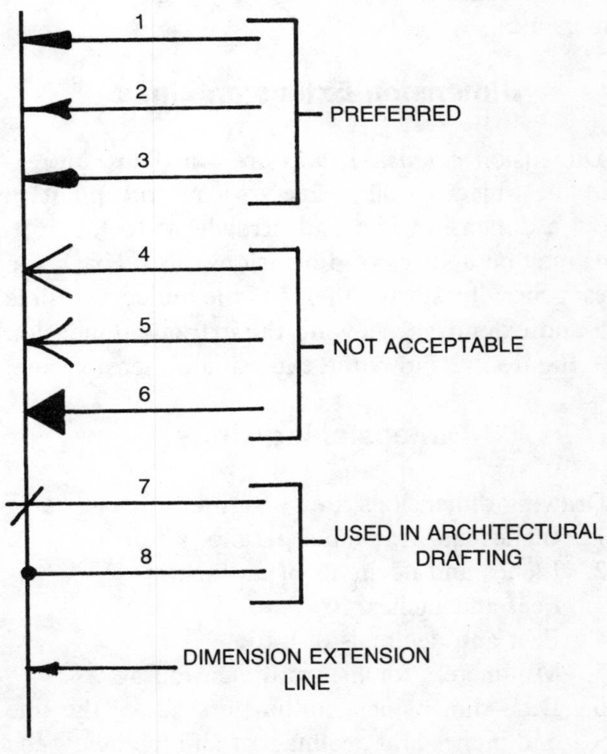

Figure 2-2. Arrowhead types.

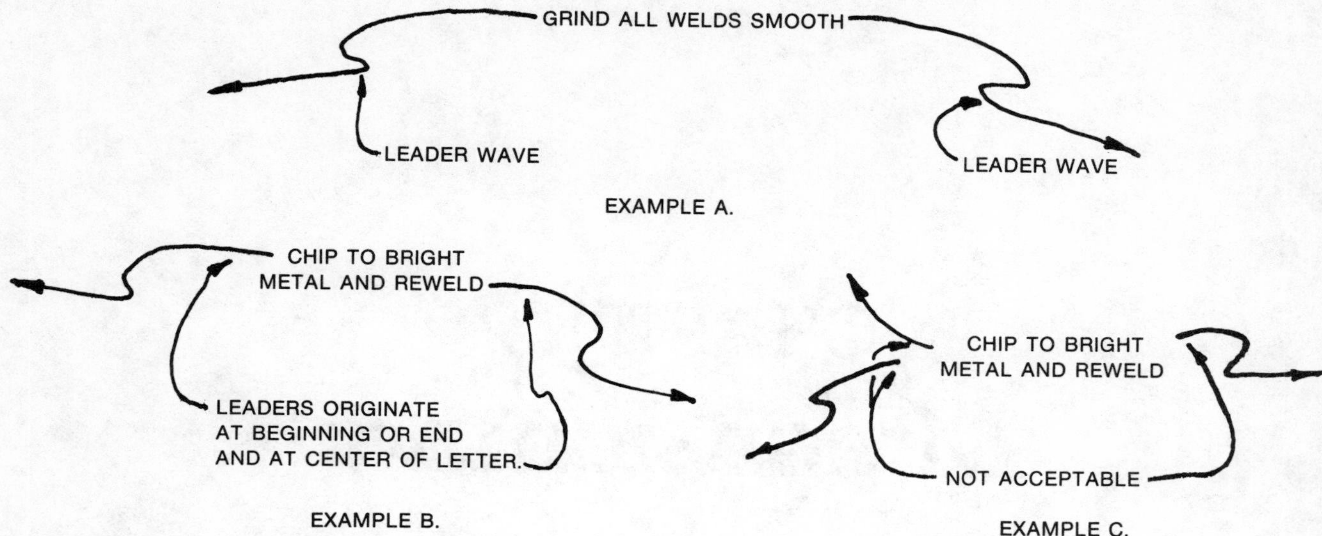

Figure 2-3. Freehand leader lines; preferred for architectural drafting.

Leaders

Leader lines are used to go from special notes to the point being noted. Leader lines begin at the first word of a note or at the last word of the note. Leader lines may be drawn as straight lines or, if made freehand, should be wavy. Figure 2-3 shows freehand leader lines and Figure 2-4 shows straight leader lines.

Dimension Extension Lines

Dimension *extension lines* are thin lines and extend the object's visible line to a remote point to allow a dimension line and arrowhead to touch it. This may be a series of dimensions, also. Extension lines generally start ⅛″ from the object's visible line and extend ⅛″ beyond the extreme dimension line. Figure 2-2 shows the extension dimension line.

Dimensioning Units

Drawing dimensions are given in:

1. Inches and fractional parts of an inch.
2. Inches and decimals of an inch.
3. Feet and inches.
4. Feet and decimals of a foot.
5. Millimeters for metric dimensioning.
6. Dual dimensions; millimeters above the line and inches and decimals of an inch below the line.

Each drawing will note the scale used, either in the title block or on the drawing. It is quite common to note "none" in the scale box on the title block but the drawing is actually drawn to scale. This is done to keep people from scaling the drawing when fabricating. The fabricator must utilize the given dimensions. If some dimension is not shown, the fabricator must contact the office that prepared the drawing and obtain the dimension.

Some firms will place "As Noted" in the title box instead of noting the scale used. This is done when more than one scale is used. The actual scale is provided underneath each view shown.

A common term used on dimensions is "*NTS*". This note may be next to a dimension or beneath the dimension line. This means "*not to scale*" or "*not true scale,*" and occurs when one or more dimensions are changed and the drafter decides not to redraw the object to scale. A wavy line under a dimension also indicates the dimension is not to scale.

Overall dimensions are given for references by many firms. These are *reference dimensions* only and may be noted "REF," which means that they are not to be used for fabrication. Such dimensions are used for shipping information and as a double check on other dimensions. The fabrication dimensions are those that supply the actual dimension of a component part, not a dimension that may total several other dimensions. Sometimes this reference dimension may be shown in brackets.

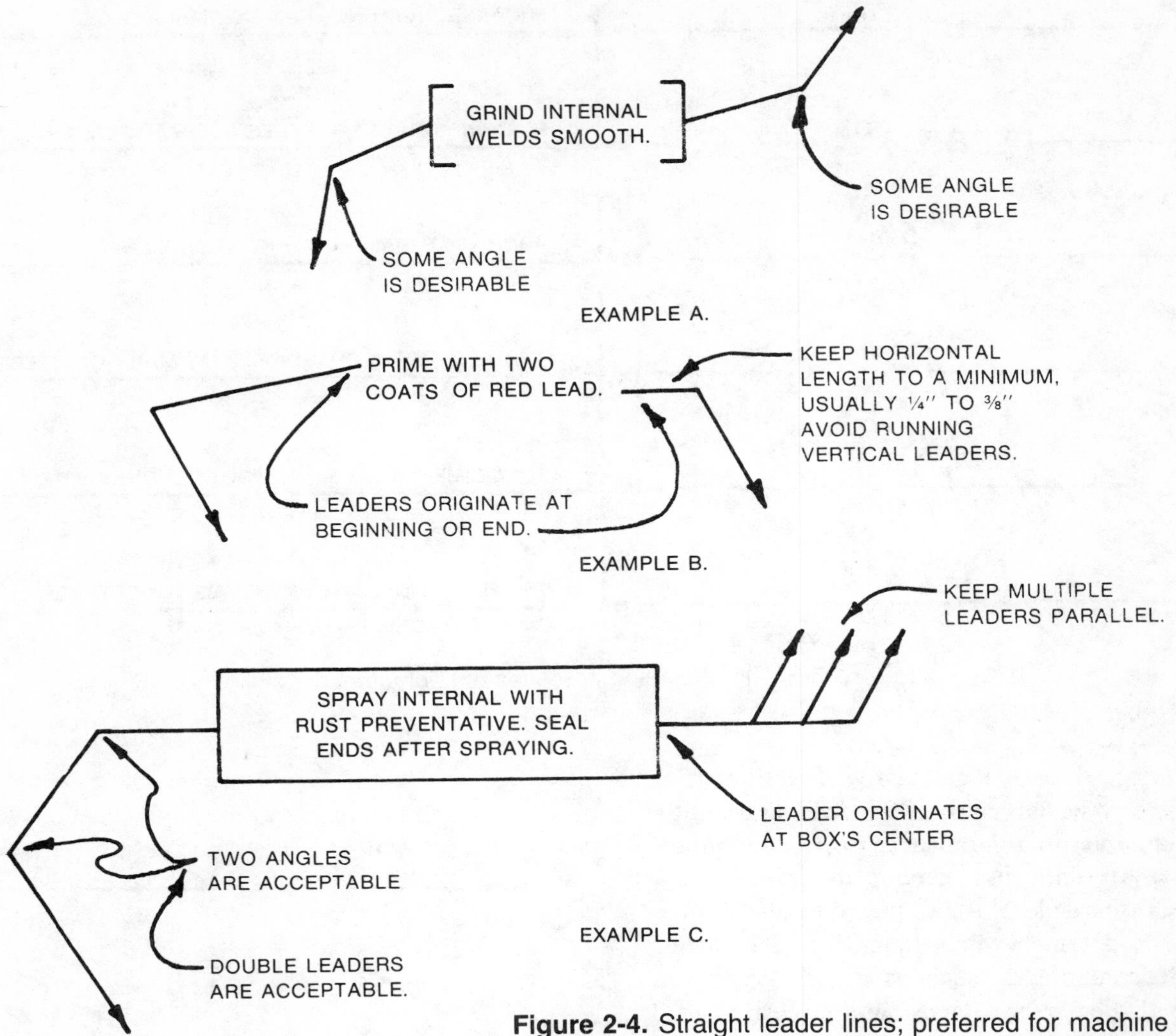

Figure 2-4. Straight leader lines; preferred for machine drafting.

Machine shop drafting practice is not to show the inch symbol (") when all dimensions are expressed in inches. Exceptions are made to this rule when there is the possibility of confusion. For example, diameters usually have the inch marks shown.

When dimensions are given in feet and inches, both symbols are to be provided, however some firms will omit them. Most firms express all dimensions 12" and above as feet and inches but there are exceptions to this rule. A few firms will make this break at 24". In any event, all diameters are expressed in inches, as 36" pipe, 109½" bolt circle and 2¼" hole. Extremely large diameters are expressed in feet, such as 100′ storage tank.

Coordinate dimensioning is used where close tolerances are not necessary. This dimensioning type is explained in Chapter 5. Coordinates may be expressed in feet and inches, feet and decimals of a foot, and in millimeters for metric dimensioning.

When special fabrication accuracy is not required, fractions are dimensioned to the closest 1/16" and in 1/16" units. However, 8/16" is shown as ½" and all fractions are reduced to their lowest form. This is called *general dimensioning*.

Precision dimensioning uses inches and decimals of an inch or millimeters and decimals of a millimeter. Precision dimensioning is used for parts of items that must be interchangable, such as parts of an engine that are made by thousands and assembled at a factory or plant. Each part must fit the engine so tolerances must be very small. When parts must be fabricated to a very close tolerance, the tolerance is shown after the dimension as:

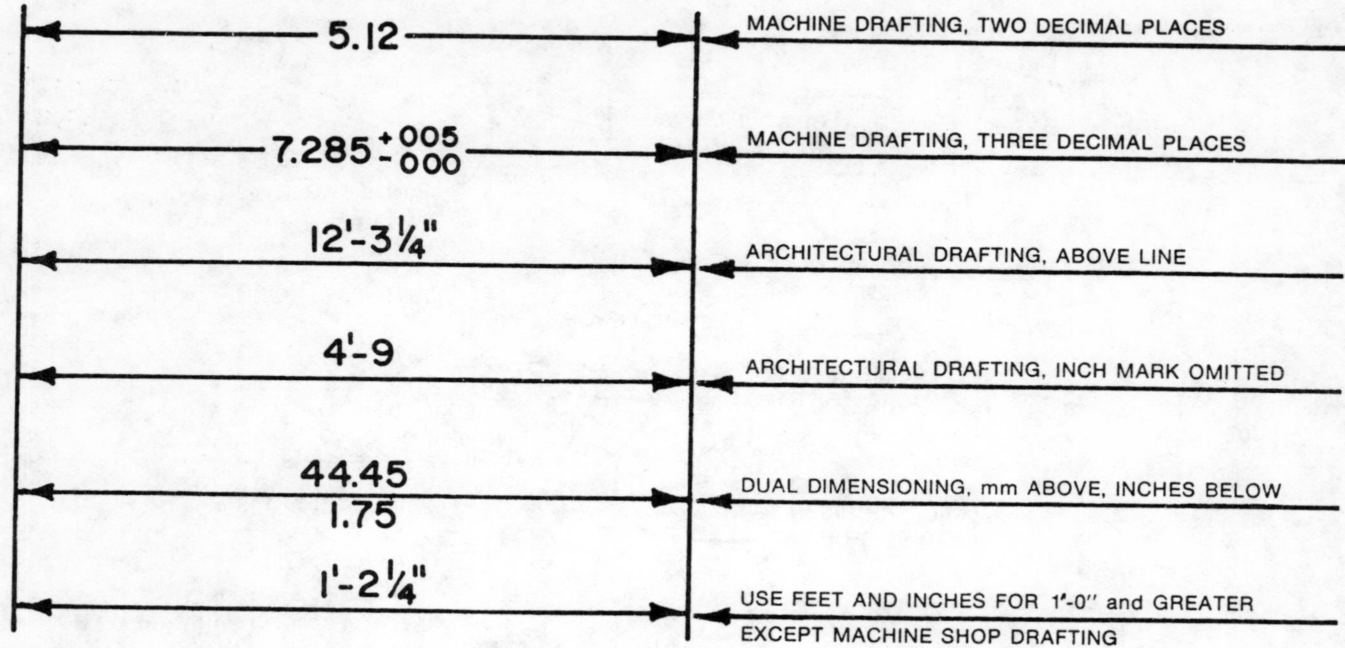

Figure 2-5. Dimensioning practices, straight lines.

21.468 $^{+.005}_{-.000}$, which means the specified dimension (21.468) may be larger by .005, five thousandths of an inch, but no tolerance is permitted on the small side. Any item not meeting the specified tolerance is discarded. Normal machine shop tolerances are ± .01 inch when dimensions are carried to two decimal places, such as 2.34". When this tolerance is too large, dimensions are carried to 3 decimal places and the tolerance is shown. Figure 2-5 shows straight line dimensioning practices.

Angles

Angles are dimensioned by an arc using the angle vertex as the radius point of the arc. The dimension arc line may be broken with the angle inserted within the break or the arc line may be solid with the angle located above or to the side of the arc line. Figure 2-6 shows angle dimensioning. All eight types are used by industry. Note example G shows degrees (°) and minutes (′) of the angle. Example H goes one step further and shows seconds (″) of the angle. Only in precision dimensioning are the angles' seconds shown or considered. Machine shops require dimensioning to seconds for precision dimensioning. An angle's tolerance may be expressed in several ways as shown by Figure 2-7.

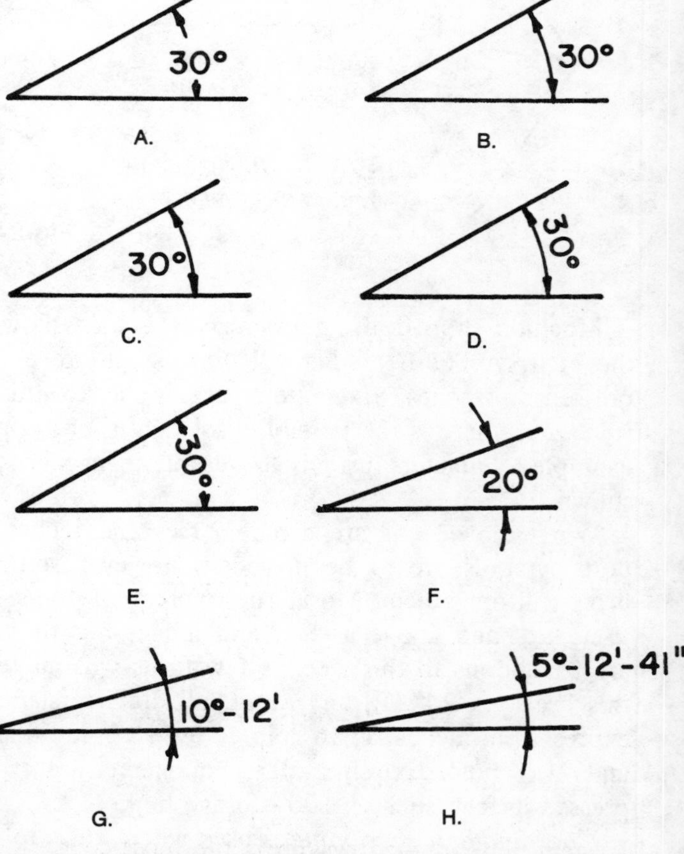

Figure 2-6. Methods of angle dimensioning.

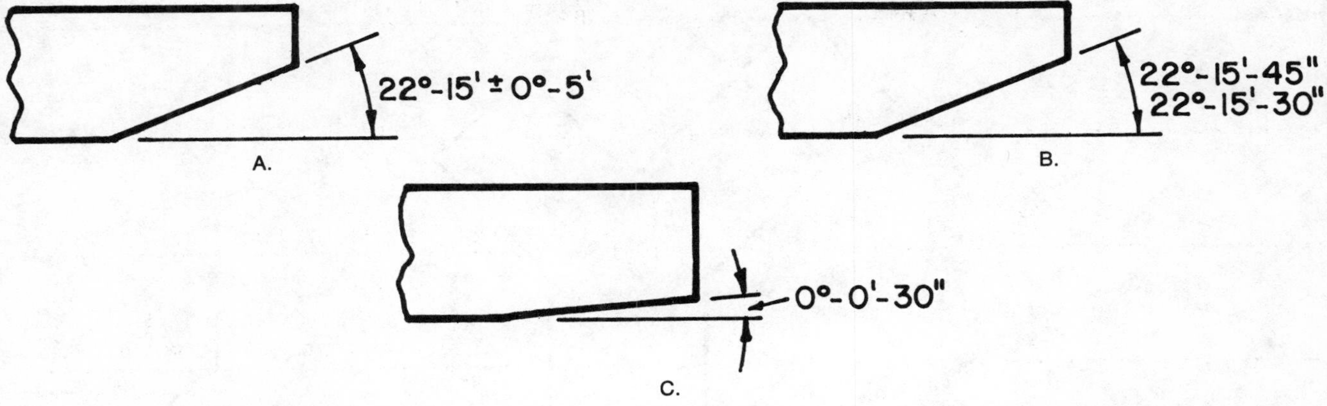

Figure 2-7. Machine shop drafting procedure for angle dimensioning with tolerances.

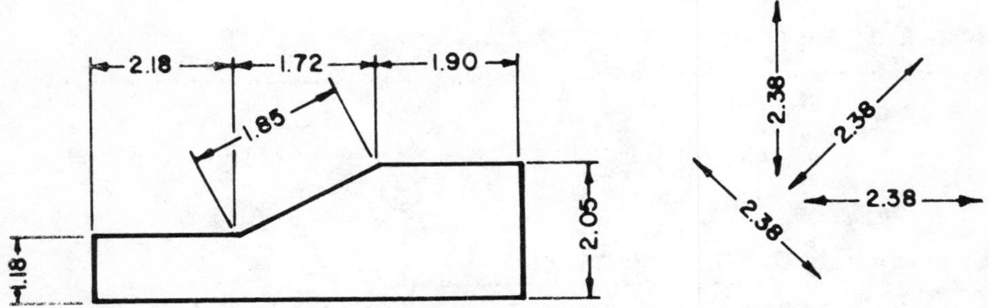

Figure 2-8. Aligned dimensioning.

Example A of Figure 2-7 shows the desired angle as 22°-15′ and permits a tolerance of five minutes, so the acceptable angle would be from 22°-10′ to 22°-20′. This is considered a large tolerance.

Example B allows a maximum angle of 22°-15′ 45″ and a minimum angle of 22°-15′-30″, which is a tolerance of only 15 seconds of a degree—a very tight tolerance. There are 60 seconds to a minute and 60 minutes to a degree. One degree of the angle contains 3,600 seconds. The tolerance of 15 seconds is four thousandths (0.004) of one degree!

Example C shows that with no degrees and no minutes, zeros are used to communicate those facts. In every case, a hyphen is shown between the degrees, seconds, and minutes of the angle.

Dimensioning Systems

There are two basic systems for locating dimensions, the *aligned* system and the *unidirectional* system. The aligned system is the most commonly used system and calls for all dimensions to be read from the bottom or right side of the drawing. Figure 2-8 shows aligned dimensioning utilizing broken dimension lines as practiced by many machine shops.

The unidirectional system has all dimensions read from the bottom of the drawing, regardless of the angle of the dimension line. When this system is used with fractions, the fraction bar will be horizontal. The unidirectional system is gaining industry popularity because many firms use a typing machine for all dimensions and the horizontal dimensions are easy to type. Figure 2-9 shows unidirectional dimensioning.

When dimensioning space is limited, the drafter may deviate from normal dimensioning practices. The notes in Figure 2-10 show how to dimension in crowded places:

1. While extension lines may be inclined at any angle, the dimension line will parallel the line of the item being dimensioned.
2. Arrowheads are located outside the dimensional length.

(text continued on page 13)

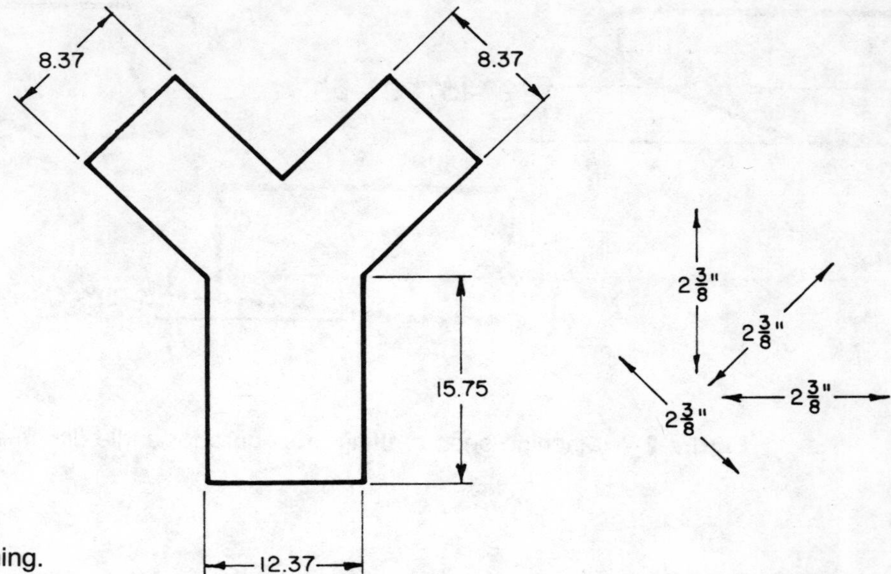

Figure 2-9. Unidirectional dimensioning.

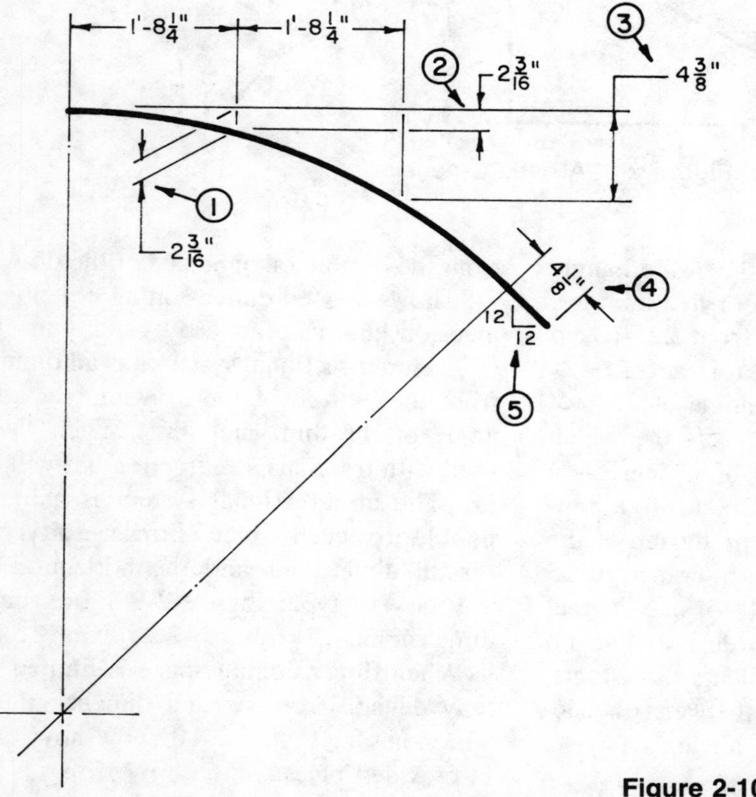

Figure 2-10. Dimensioning in crowded places.

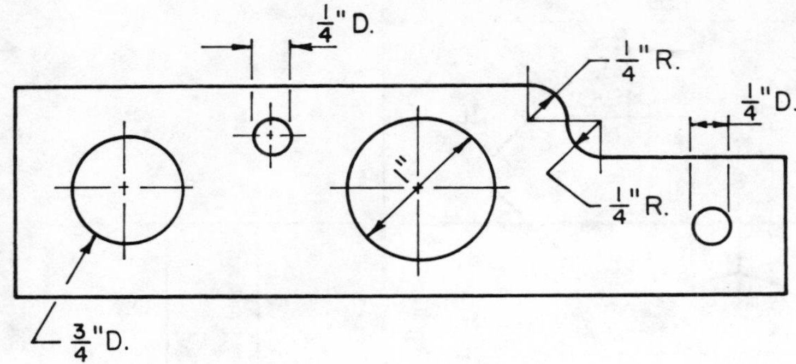

Figure 2-11. Dimensioning circles and arcs.

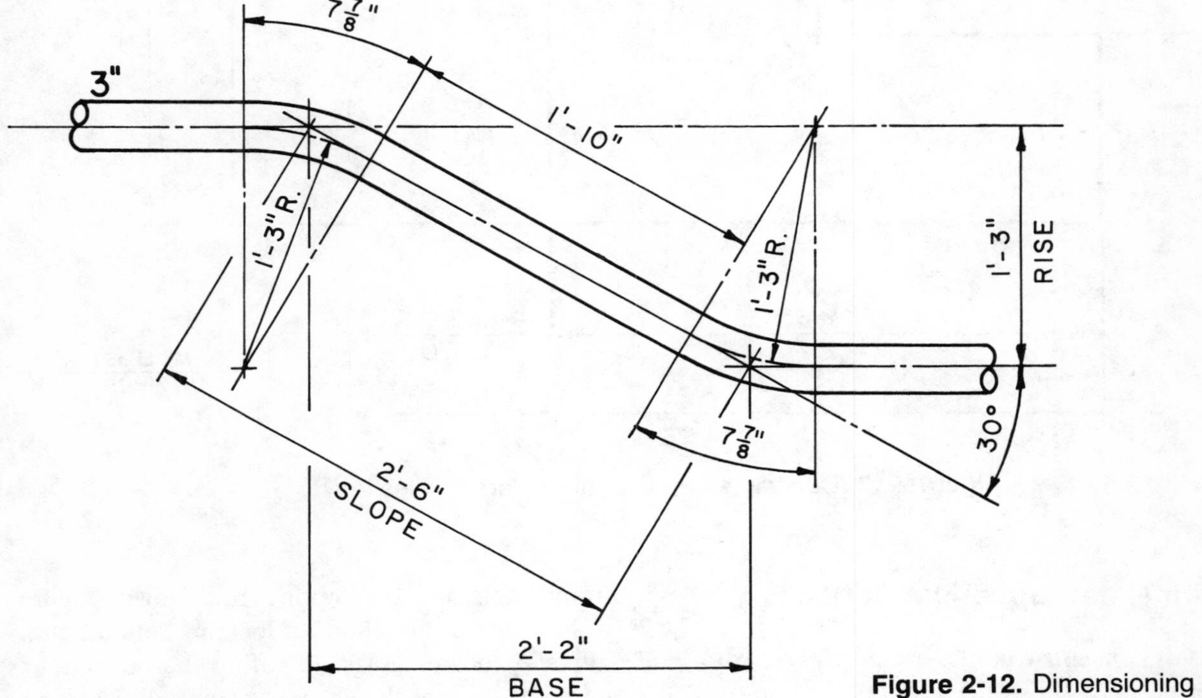

Figure 2-12. Dimensioning pipe bends.

3. Arrowheads and dimension line are inside the extension lines, but the dimension is placed outside.

4. Arrowheads are outside the extension lines and the dimension is inside.

5. This is a slope symbol. A 12:12 slope shows that both sides of the triangle's short legs are equal, making it a 45° slope.

Dimensioning Curved Surfaces

A circle will have its *diameter* dimensioned, not the radius. A radius is given for arcs or partial circles. Figure 2-11 shows methods for circle and arc dimensioning. Figure 2-12 shows how 3″ pipe bends are dimensioned. Pipe is dimensioned to center lines so the 7⅞″ dimension is the amount of arc at the pipe's center line. Dimensions are also given to center line intersections. The rise, base, and slope form the three sides of a triangle formed by center line intersections.

Dual Dimensioning

Figure 2-13 shows dual dimensioning. Many drawings will have dual dimensioning and the reader must remember that millimeter units are located on top the dimension line. Inches are given below the dimension line.

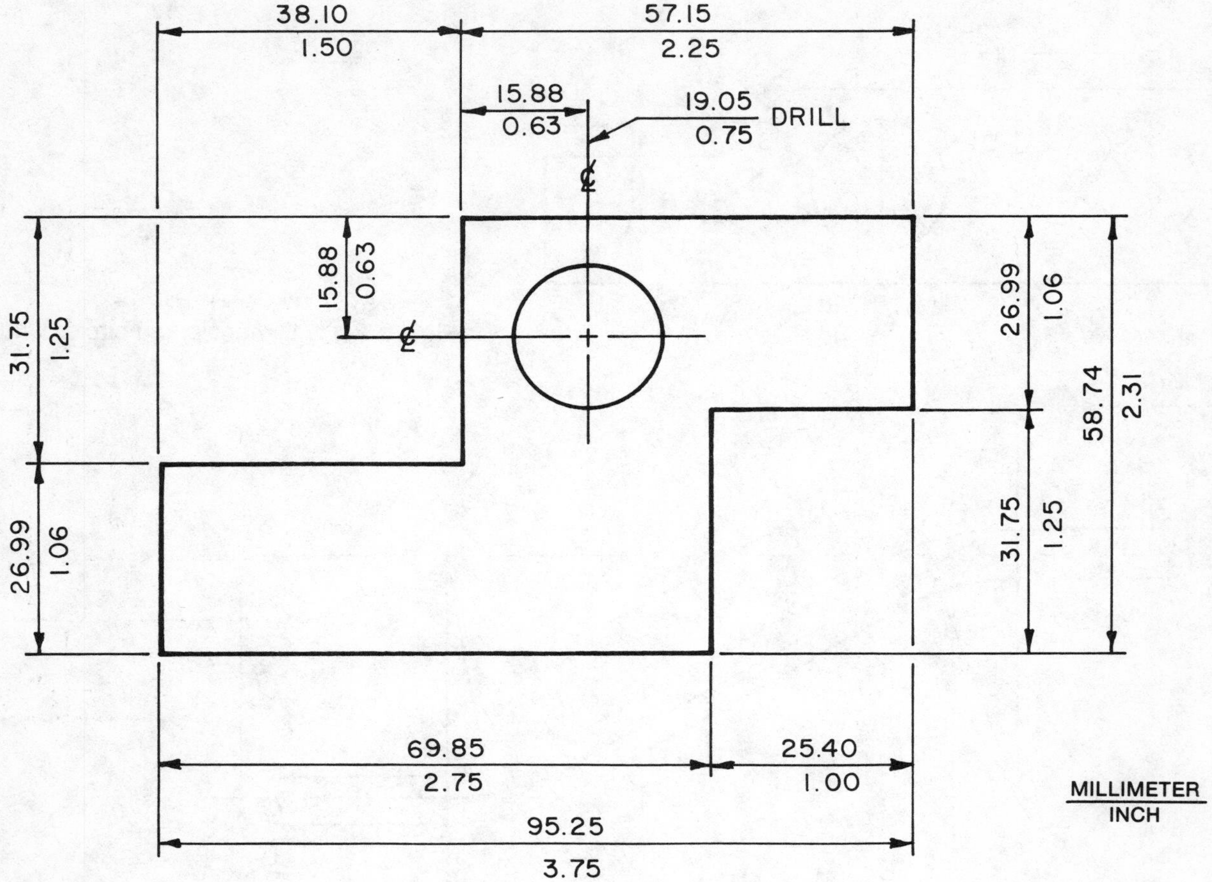

Figure 2-13. Dual dimensioning, millimeters and inches.

Cylindrical and Spherical Surfaces

Diameters of concentric circles are always shown, but the drafter may show these diameters in the side view, leaving the front view less congested. Figure 2-14 shows how the side view is dimensioned with diameters. Spheres and spherical surfaces are dimensioned by a radius, as shown by Figure 2-15.

Special Features

Machine drafting has many special features that require special notes or dimensions to fully communicate manufacturing data. At times these features are used in other drafting fields, so the blueprint reader must be familiar with them.

Round holes are located by dimension, sometimes by angle, and their diameters, depths, and number must be given by dimension or note. If the drawing doesn't show otherwise, it is assumed to be a through hole and the depth is not shown. Figure 2-16 shows several methods of locating round holes. Example A locates holes by noting the bolt circle. Example B locates holes by straight dimensions. Example C is a structural steel member with holes located by a spacing note. Here, the architectural style dimensioning is used with dimensions above the line and hole size given below the line.

Blind holes must have their depth shown. The depth noted is only the full diameter length and does not include the drill point. Figure 2-17 shows two ways to dimension blind holes. Example A shows depth by notation while Example B provides dimensioned depth. Either way is correct.

Counterbored holes are shown in Figure 2-18. The notation method shown is used most often, however dimensions are also acceptable.

Figure 2-19 pictures countersunk hole dimensioning. Normally, countersunk holes are dimensioned by notation.

(text continued on page 17)

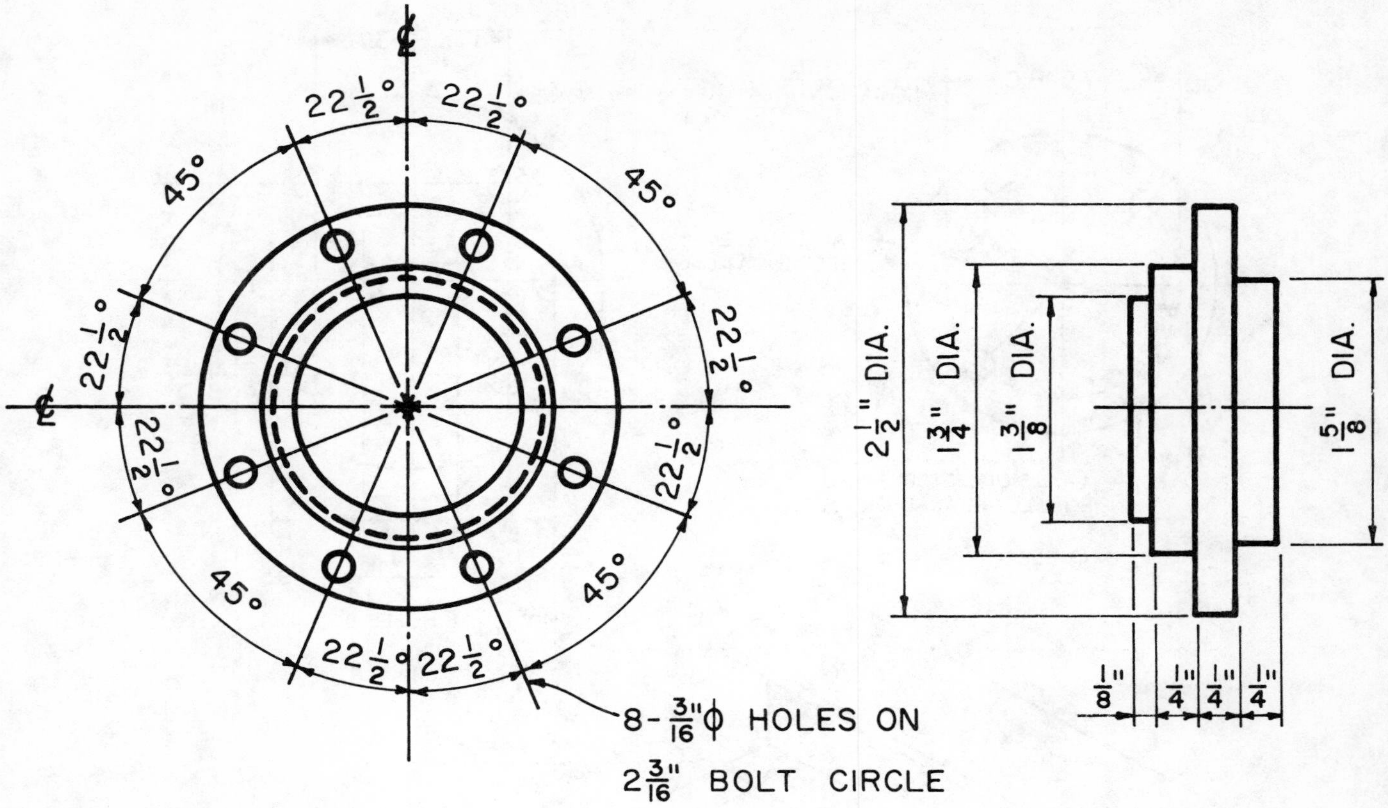

8- $\frac{3}{16}$" ϕ HOLES ON
2 $\frac{3}{16}$" BOLT CIRCLE

FRONT VIEW

SIDE VIEW

Figure 2-14. Sometimes it is better to show the diameter of a concentric circle in side views.

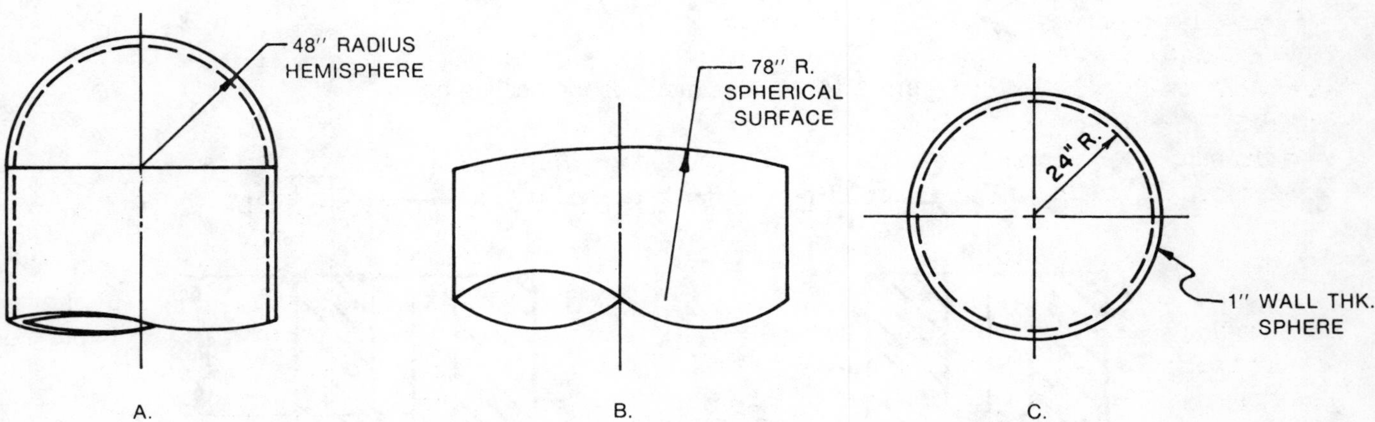

Figure 2-15. Dimensioning spherical surfaces.

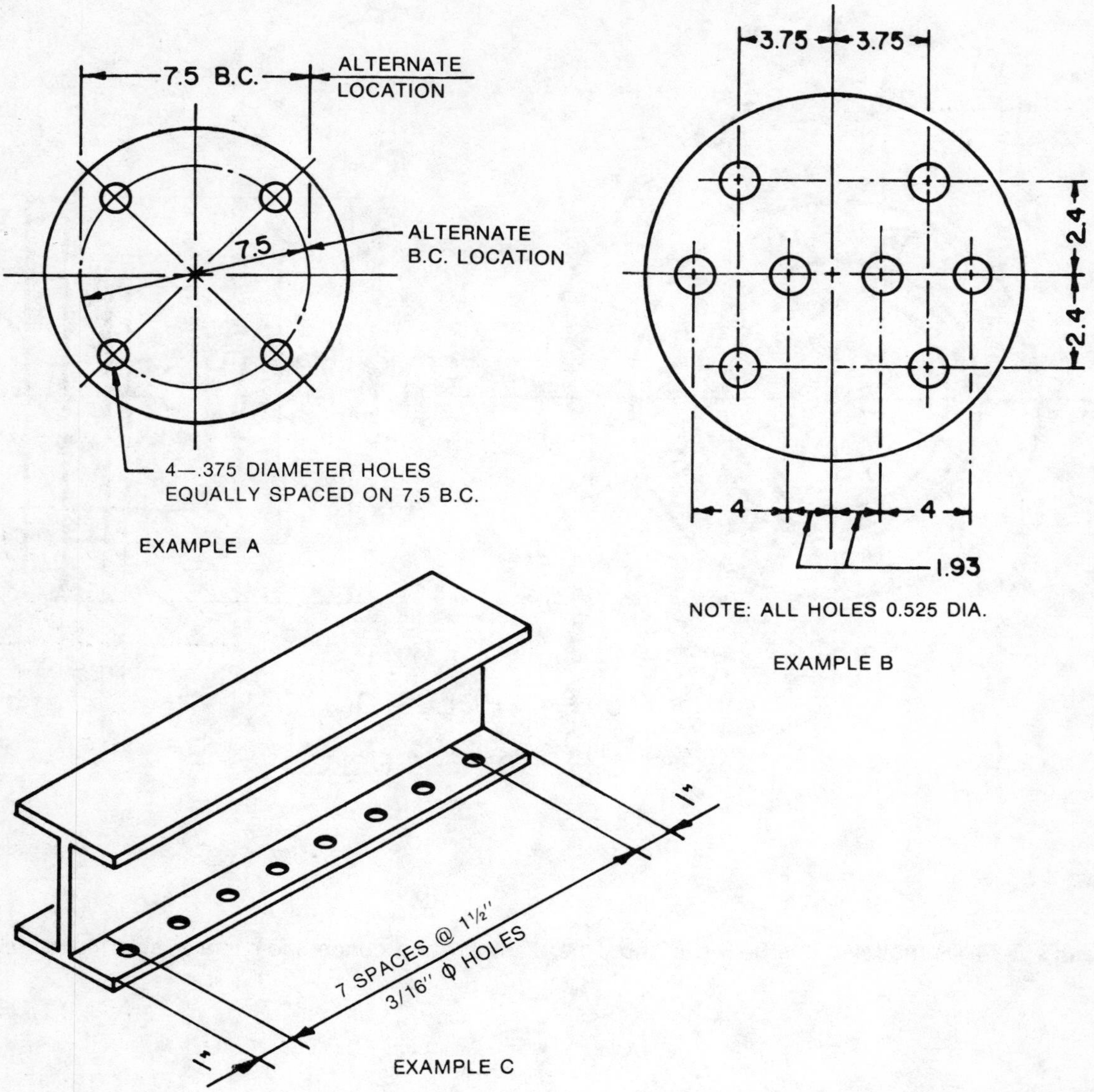

7.5 B.C.
ALTERNATE
LOCATION
7.5
ALTERNATE
B.C. LOCATION
4—.375 DIAMETER HOLES
EQUALLY SPACED ON 7.5 B.C.
EXAMPLE A
3.75
3.75
2.4
2.4
4
4
1.93
NOTE: ALL HOLES 0.525 DIA.
EXAMPLE B
1"
7 SPACES @ 1½"
3/16" ϕ HOLES
1"
EXAMPLE C

Figure 2-16. Various methods for locating holes.

.437
.435
DIA.- .75 DEEP
EXAMPLE A
.437
.435
DIA.
.75
EXAMPLE B

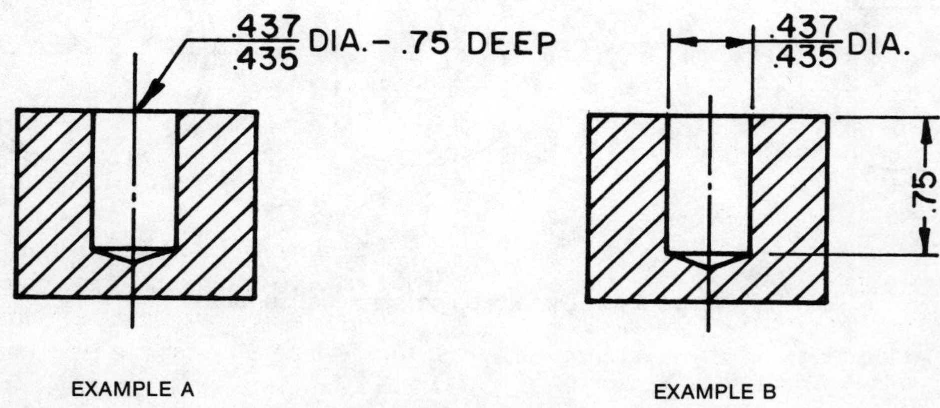

Figure 2-17. Blind hole dimensioning.

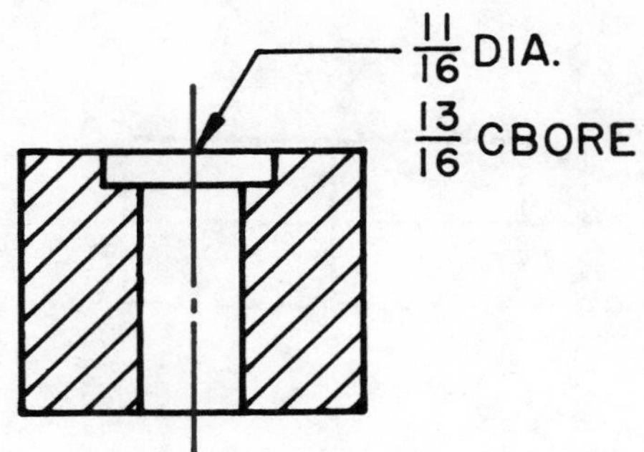

Figure 2-18. The counterbore.

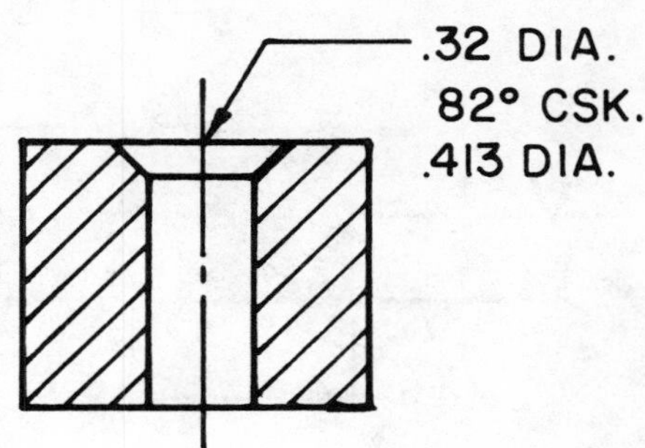

Figure 2-19. Countersunk hole.

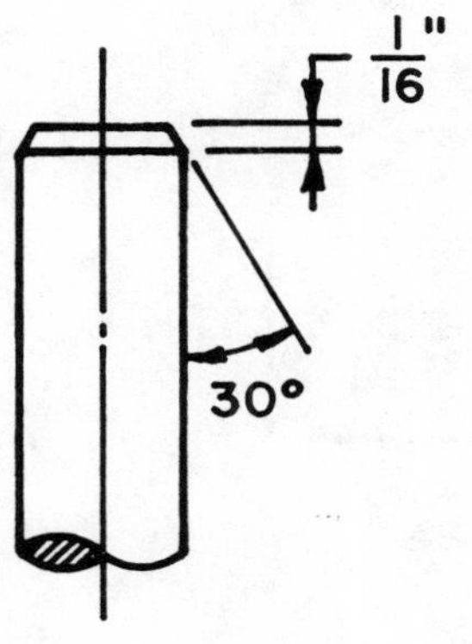

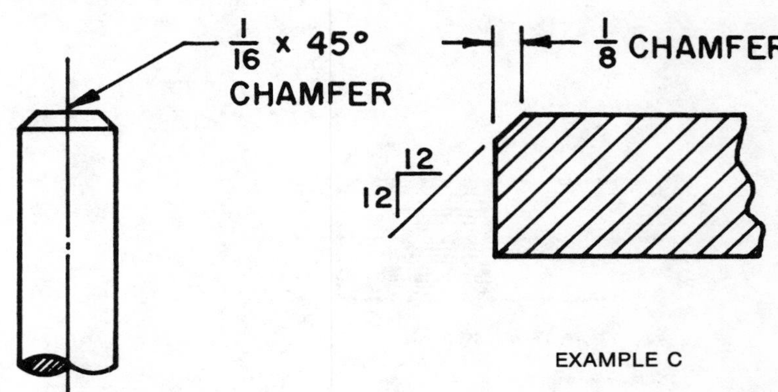

EXAMPLE A

EXAMPLE B

Figure 2-20. Chamfer dimensioning.

continued from page 14

Chamfers, or beveled edges, are dimensioned by the angle and length as shown by Figure 2-20. Three styles of dimensioning are standard. Example A provides the angle by a dimension. Example B shows the angle by notation. Example C shows the angle by a 12:12 bevel note. This is a 45° angle. Only when the angle is 45° should example B be used.

Knurls are used to roughen a surface. This gives the surface a better grip. Knurled surfaces may have straight or diamond shaped patterns. Figure 2-21 shows knurled surfaces with ANSI (American National Standards Institute) symbols.

Surface Finish Designation

There are two basic types of surface finishing—*controlled* and *noncontrolled*. The controlled surface finish has numerous types and tolerances that are used in machine drafting. The ANSI B46.1 specification shows these symbols and tolerances. If the reader is doing machine shop work, this specification should be part of your library and may be purchased from a local book store.

The more common type of surface finishing is the noncontrolled finish. This is machine finished, but little, if any, control is specified as to the finish quality. When a surface is to be machine finished, a 60° V is shown pointing to the surface. Occasionally, the letters R or G are placed in the V to designate "rough" or "grind." A slanted "f" was the finish symbol for many years and is still used occasionally, but it has generally been replaced by the V. Figure 2-22 shows these finish symbols.

Dimensioning Terms

Other dimensions and terms one might encounter when reading blueprints include the following:

Datum point is a reference point, usually a centerline, which establishes a base line. It is desirable to show dimensions from a datum point, thereby eliminating cumulative errors which may occur in a string of intermediate dimensions.

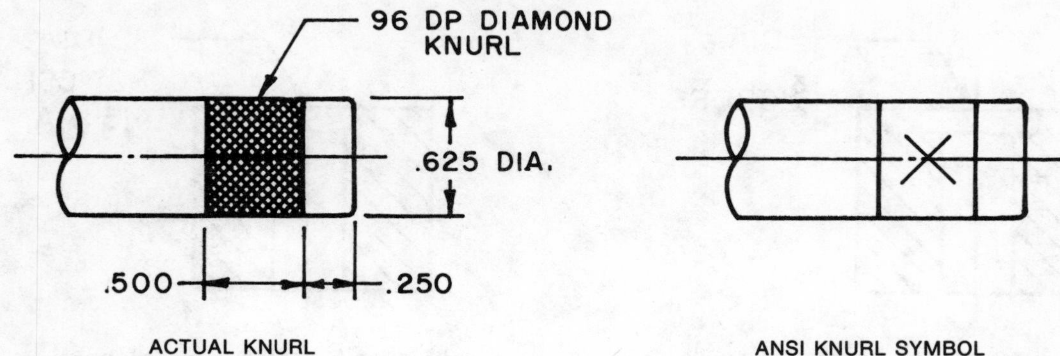

EXAMPLE A. DIAMOND KNURL

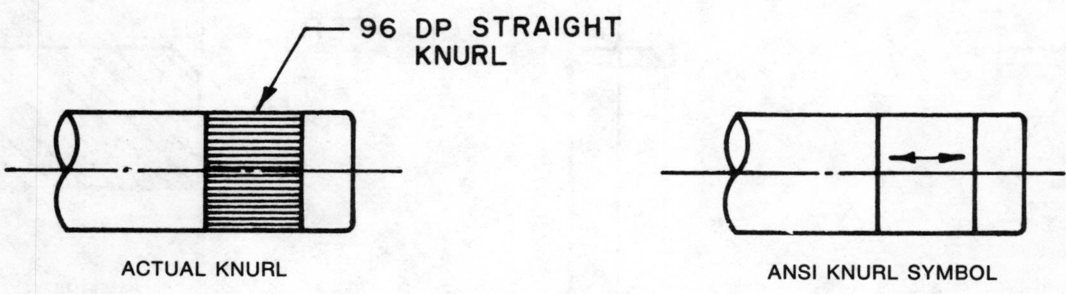

EXAMPLE B. STRAIGHT KNURL

Figure 2-21. Knurled surfaces.

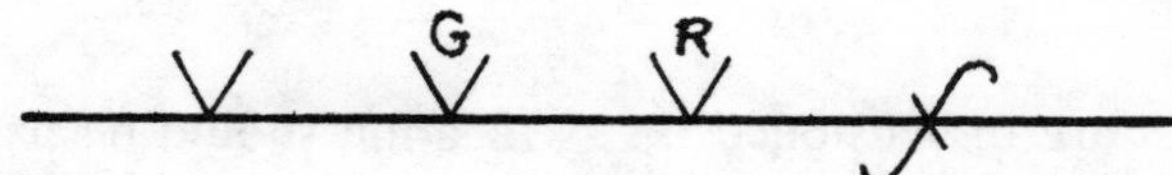

Figure 2-22. Surface finish designation. Surface to be machined.

Hidden line dimensioning is to be avoided. A special section may be needed, but do not originate or terminate dimensions at hidden lines.

Hold dimensions have two meanings, depending on their use. If just the word "Hold" appears beneath or beside the dimension, it means to manufacture it as close as possible (no tolerance) to that dimension. If both the dimension and "Hold" appear in a cloud, it means the dimension may change in a later revision. Manufacture is held until the hold is released.

Intermediate dimensions, also called breakdown dimensions, are the smaller, point-to-point dimensions used for fabrication or manufacture. Tolerance, when needed, should be noted on these dimensions.

Limit dimensions show the maximum size permitted.

Location dimensions locate an item within the object.

Size has many meanings in drafting. Basic size is the size of the object within tolerance limits. If no tolerance limits are shown, the basic size is referred to as the design size. Actual size is measured from the manufactured item. It is also called the as-built size. Nominal size is used for general identification, such as "3″ pipe" which actually has a 3½″ outside diameter.

Symmetrical dimensions are those with identical dimensions on the other half of the object. A symmetrical centerline divides two identical parts. Usually only one part is drawn with a note "Symmetrical About This Centerline" shown.

CHAPTER 3

Math Review and Shortcuts

Anyone reading prints must be able to comfortably add, subtract, divide, and multiply numbers and fractions. Because many students have lost their efficiency in performing these operations, this chapter provides a review to enable the student to be more comfortable while performing simple math. For those who must constantly work with basic math, the last portion of this chapter will offer several shortcuts to basic math. If one masters these shortcuts, math problems can be done in 20% of the time it would take to solve them using traditional methods. Also, the shortcut methods will result in fewer errors.

Common Denominator

Fractions cause the greatest problems. Since fractions must always be finally expressed in their lowest possible form, it is often necessary to convert the fraction to another fraction with a common denominator. For instance, to add 3/8 and 1/4 is impossible because the denominators (8 and 4) are unlike. With these two fractions the common denominator, *a number which is divisible by all denominators,* is 8. Eight is divisible by 4 (equals 2) and by 8 (equals 1).

To convert 1/4 to eighths the numerator (the top number, here 1) and the denominator (the bottom number, 4) must be multiplied by the same number, in this case 2 since 4 goes into 8 two times. Two times 1 equals 2, and 2 times 4 equals 8, so 1/4 = 2/8. Now 3/8 + 1/4 is converted to 3/8 + 2/8 = 5/8″.

Table 3-1. Fraction Breakdown by Sixteenths

1/16	9/16
2/16 = 1/8	10/16 = 5/8
3/16	11/16
4/16 = 2/8 = 1/4	12/16 = 6/8 = 3/4
5/16	13/16
6/16 = 3/8	14/16 = 7/8
7/16	15/16
8/16 = 4/8 = 2/4 = 1/2	16/16 = 8/8 = 4/4 = 2/2 = 1

Table 3-1 provides a breakdown of fractions to the lowest possible form by sixteenths of an inch. Most drawings utilizing fractions are dimensioned to the closest 1/16″.

Adding Common Denominator Fractions

When fractions have common denominators their addition is relatively easy.

1.
```
   9′-6 3/8″
 + 3′-2 5/8″
  12′-8 8/8″ = 12′-9″
```

2.
```
   3′-3 1/4″
 + 6′-4 1/4″
   9′-7 2/4″ = 9′- 7 1/2″
```

3.
```
   4′-7 1/16″
   2′-0 3/16″
   1′-1 5/16″
   7′-8 9/16″
```

4.
```
   2′-3 7/8″
   2′-4 3/8″
   1′-2 1/8″
   5′-9 11/8″ = 5′-10 3/8″
```

In the first problem 8/8 must be reduced to its lowest form, which is one. The sum (answer) then becomes 12′-8″ + 8/8″ or 12′-8″ + 1″ = 12′-9″.

The second problem has a sum of 9′-7 2/4″ which must be expressed in its lowest form which is 9′-7 1/2″. The fourth problem has 5′-9 11/8″ as the first sum. Reducing 11/8″ to its lowest form brings 8/8″ + 3/8″ or 1″ + 3/8″. So 1″ + 3/8″ is added to 5′-9″ to arrive at 5′-10 3/8″.

In their final form, fractions are always expressed as less than one and are reduced to their lowest form. The numerator must be smaller than the denominator.

Adding Fractions with Uncommon Denominators

When adding fractions that have unlike denominators a common denominator must be found.

5. 4′-2 1/16″ = 4′-2 1/16″
   1′-3 1/8″ = 1′-3 2/16″
   2′-1 1/4″ = 2′-1 4/16″
   7′-6 7/16″

6. 1 1/2″ = 1 4/8″
   2 1/8″ = 2 1/8″
   3 3/4″ = 3 6/8″
   6 11/8″ = 7 3/8″

In the fifth problem the common denominator is 16 so 1/8″ and 1/4″ are converted to 2/16″ and 4/16″ and added to 1/16″ to total 7/16″. Problem 6 has 8 as the common denominator and the original sum is 6 11/8″ which must be reduced to its lowest form where the numerator is less than the denominator. So 11/8″ becomes 8/8″ + 3/8″ or 1″ + 3/8″ which is added to the original sum, 6″, to equal 7 3/8″.

Find the final sum for problems 7 through 14. Show all work and fraction reduction.

7. 1′-0 3/8″
   2′-2 1/4″
   3′-1 3/16″

8. 2′-3 5/8″
   9′-1 1/4″
   5′-2 5/8″

9. 7′-2 11/16″
   2′-2 7/8″
   5′-4 1/2″

10. 3 1/4″
   6′-2 1/2″
   9′-4 3/4″

11. 9′-0 1/4″
   6′-3 1/8″
   5′-2 1/2″
   9′-1 11/16″

12. 1′-3 3/8″
   2′-0 7/8″
   5′-2 3/4″
   7′-1 7/8″

13. 4′-3 11/16″
   9′-0 15/16″
   4′-2 7/8″
   5′-1 3/4″

14. 8′-0 5/8″
   1′-3 9/16″
   2′-2 13/16″
   1′-1 1/2″

Adding Feet and Inches

When adding feet and inches remember that in their final form, inches must be expressed as less than 12″. So inches, too, must be reduced to their lowest possible form. For instance, a final answer of 7′-15″ would be incorrect and counted wrong. Since there are 12 inches to a foot, 15″ must be converted to 1′-3″ making the final correct answer 8′-3″.

15. 7′- 8″
   4′-11″
   5′- 6″
   16′-25″ = 16′ plus 2′-1″ = 18′-1″

In problem 15 the inches column totaled 25. To reduce to the lowest possible form 25 is divided by 12 which equals 2 plus one left over, or 2′-1″ which is added to 16′ to arrive at the final answer, 18′-1″.

16. 7′- 5 1/4″ = 7′- 5 2/8″
   3′- 8 3/4″ = 3′- 8 6/8″
   4′-11 7/8″ = 4′-11 7/8″
   14′-24 15/8″ = 14′-25 7/8″ = 16′-1 7/8″

Problem 16 combines the reduction of fractions and the reduction of inches. The common denominator is 8 and the first step is to convert all fractions to eighths. Then the answer of 14′-24 15/8″ must be reduced so 15/8″ is selected first. This becomes 1 7/8″ so the 1″ is added to 24″ becoming 25″. Now this must be converted to feet and inches since it is over 12″. Dividing 25″ by 12″ gives 2′-1″ which is added to 14′ to arrive at 16′-1 7/8″ for the final answer. Any other answer is incorrect.

Adding Exercise

For practice, work problems 17 through 31.

17. 4′-9 3/8″
   +7′-6 1/4″

18. 14′-6 3/16″
   +11′-9 1/2″

19. $6'\text{-}11\frac{3}{4}''$ $+5'\text{-}11\frac{5}{8}''$

20. $7'\text{-}4\frac{1}{4}''$ $8'\text{-}8\frac{7}{8}''$ $5'\text{-}11\frac{1}{2}''$

21. $9'\text{-}7\frac{7}{8}''$ $5'\text{-}5\frac{3}{4}''$ $4'\text{-}9\frac{13}{16}''$

22. $12'\text{-}11\frac{11}{16}''$ $8'\text{-}0\frac{1}{8}''$ $9'\text{-}10\frac{15}{16}''$

23. $108'\text{-}5\frac{3}{8}''$ $96'\text{-}9\frac{9}{16}''$ $72'\text{-}11\frac{3}{8}''$

24. $5\frac{1}{4}''$ $9'\text{-}9\frac{1}{2}''$ $11'\text{-}10\frac{3}{8}''$

25. $19'\text{-}11\frac{9}{16}''$ $27'\text{-}8\frac{5}{8}''$ $19'\text{-}9\frac{1}{4}''$

26. $1'\text{-}8\frac{3}{4}''$ $2'\text{-}9\frac{5}{8}''$ $5'\text{-}2\frac{1}{2}''$

27. $10\frac{9}{16}''$ $11\frac{3}{8}''$ $9\frac{7}{8}''$

28. $117'\text{-}9\frac{5}{16}''$ $117'\text{-}11\frac{7}{8}''$ $117'\text{-}7\frac{7}{16}''$

29. $42'\text{-}9\frac{1}{4}''$ $76'\text{-}10\frac{5}{8}''$ $14'\text{-}11\frac{1}{2}''$ $22'\text{-}7\frac{7}{8}''$

30. $18'\text{-}5\frac{7}{8}''$ $12'\text{-}7\frac{15}{16}''$ $6'\text{-}9\frac{13}{16}''$ $5'\text{-}7\frac{1}{2}''$

31. $22'\text{-}7\frac{3}{4}''$ $44'\text{-}8\frac{5}{16}''$ $37'\text{-}9\frac{15}{16}''$ $2'\text{-}11\frac{1}{8}''$

Subtracting Fractions

Subtracting fractions is easy when fractions have common denominators.

32. $8'\text{-}6\frac{5}{8}'' - 4'\text{-}3\frac{3}{8}'' = 4'\text{-}3\frac{2}{8}'' = 4'\text{-}3\frac{1}{4}''$

33. $9'\text{-}7\frac{3}{4}'' - 3'\text{-}3\frac{1}{4}'' = 6'\text{-}4\frac{2}{4}'' = 6'\text{-}4\frac{1}{2}''$

Note that in each case the fraction is expressed in its lowest possible form in the final answer.

To subtract fractions with uncommon denominators a common denominator must be found.

34. $9'\text{-}11\frac{11}{16}'' - 5'\text{-}3\frac{3}{8}'' = 9'\text{-}11\frac{11}{16}'' - 5'\text{-}3\frac{6}{16}'' = 4'\text{-}8\frac{5}{16}''$

35. $26'\text{-}9\frac{3}{4}'' - 21'\text{-}6\frac{1}{8}'' = 26'\text{-}9\frac{6}{8}'' - 21'\text{-}6\frac{1}{8}'' = 5'\text{-}3\frac{5}{8}''$

Problem 34 has a common denominator of 16. To convert $\frac{3}{8}''$ to sixteenths both the numerator (3) and denominator (8) are multiplied by 2 to arrive at $\frac{6}{16}''$. Then $\frac{6}{16}''$ can be subtracted from $\frac{11}{16}''$ leaving $\frac{5}{16}''$.

The common denominator in Problem 35 is 8. By multiplying 3 and 4 by 2, $\frac{3}{4}''$ is converted to $\frac{6}{8}''$, and $\frac{1}{8}''$ can be subtracted from $\frac{6}{8}''$ leaving $\frac{5}{8}''$.

But what if the fractions were reversed?

		Step 1		Step 2
36.	$26'\text{-}9\frac{1}{8}'' =$	$26'\text{-}9\frac{1}{8}''$	=	$26'\text{-}8\frac{9}{8}''$
	$-21'\text{-}6\frac{3}{4}'' =$	$-21'\text{-}6\frac{6}{8}''$	=	$-21'\text{-}6\frac{6}{8}''$
				$5'\text{-}2\frac{3}{8}''$

The first step is to find the common denominator, which is known to be 8, and convert $\frac{3}{4}''$ to $\frac{6}{8}''$. But since $\frac{6}{8}''$ cannot be subtracted from $\frac{1}{8}''$, a whole unit must be borrowed. Step 2 shows that $9\frac{1}{8}''$ is converted to $8\frac{9}{8}''$. One has been borrowed from 9 to equal $8\frac{8}{8}''$. The original $\frac{1}{8}''$ is added to $8\frac{8}{8}''$ to total $8\frac{9}{8}''$. Now $6\frac{6}{8}''$ can be subtracted from $8\frac{9}{8}''$.

		Step 1		Step 2
37.	$14'\text{-}5\frac{3}{16}'' =$	$14'\text{-}5\frac{3}{16}''$	=	$14'\text{-}4\frac{19}{16}''$
	$-6'\text{-}2\frac{5}{8}'' =$	$-6'\text{-}2\frac{10}{16}''$	=	$-6'\text{-}2\frac{10}{16}$
				$8'\text{-}2\frac{9}{16}''$

Problem 37 is another two-step problem. Sixteen is the common denominator and $\frac{5}{8}''$ is converted to $\frac{10}{16}''$ in Step 1. Since $\frac{10}{16}''$ cannot be subtracted from $\frac{3}{16}''$, one unit is borrowed from the 5 and converted to $\frac{16}{16}''$. So we have $5\frac{3}{16}'' = 4'' + 1'' + \frac{3}{16}'' = 4'' + \frac{16}{16}'' + \frac{3}{16}'' = 4\frac{19}{16}''$. Now $2\frac{10}{16}''$ can be subtracted from $4\frac{19}{16}''$ leaving $2\frac{9}{16}''$.

38. 17′-3 3/16″ − 6′-8 7/8″

Step 1: = 17′-3 3/16″ = − 6′-8 14/16″

Step 2: = 17′-2 19/16″ = − 6′-8 14/16″

Step 3: = 16′-14 19/16″ = − 6′- 8 14/16″ ; 10′- 6 5/16″

Problem 38 shows inches and fractions must both be converted before subtraction can be accomplished. Step 1 is to establish a common denominator but, again, 14/16″ cannot be subtracted from 3/16″ so in step 2, 3 3/16″ is converted to 2″ + 1″ + 3/16″ = 2″ + 16/16″ + 3/16″ = 2 19/16″. Now 14/16″ may be subtracted from 19/16″ but 8″ cannot be subtracted from 2″. So 1′ must be borrowed from 17′. Now 17′ becomes 16′-12″. So 16′-12″ is added to the 2 19/16″ to total 16′-14 19/16″ in step 3. To put it another way 17′-2 19/16″ = 16′-12″ + 2 19/16″ = 16′-14 19/16″. Now 6′-8 14/16″ can be subtracted.

Review Exercise

Work problems 39-58 on separate paper. Show only the final answer reduced to the lowest possible form.

39. 7′- 9 3/8″ +5′- 4 5/8″

40. 2′- 6 5/16″ − 4 15/16″

41. 27′- 9 9/16″ + 8′- 9 15/16″

42. 6′- 3 7/8″ − 5′- 4 1/4″

43. 19′- 0 3/16″ − 7′- 5 1/2″

44. 21′- 9 1/4″ − 4′-11 7/8″

45. 9 15/16″ +11 13/16″

46. 1′- 0 1/8″ − 11 13/16″

47. 108′- 2 5/16″ + 71′-11 3/8″

48. 19′- 4 11/16″ − 7 13/16″

49. 4 7/16″ −2 3/4″

50. 8′- 3 15/16″ +8′- 3 15/16″

51. 22′- 9 7/8″ +22′- 9 7/8″

52. 16′- 3 1/16″ −12′- 7 1/2″

53. 11/16″ − 3/8″

54. 13′-11 9/16″ +26′-10 7/8″

55. 27′-11 3/8″ ; 42′-10 3/4″ ; 17′-11 13/16″ ; 19′-11 1/2″

56. 28′- 7 5/8″ ; 4′-10 1/2″ ; 26′-11 3/4″ ; 17′-10 7/8″

57. 19′-10 1/4″ ; 12′- 8 1/2″ ; 72′-11 13/16″ ; 4′- 8 5/16″

58. 46′- 8 1/8″ ; 12′-11 1/2″ ; 19′-10 1/2″ ; 42′- 9 7/8″

Fraction Multiplication

To multiply fractions the rule is: numerator x numerator = numerator, and denominator x denominator = denominator.

59. 1/2 x 1/2 = ______. Answer 1 x 1 = 1; 2 x 2 = 4, so 1/2 x 1/2 = 1/4.
60. 1/4 x 1/4 = ______. Answer 1 x 1 = 1; 4 x 4 = 16, so 1/4 x 1/4 = 1/16.

When whole numbers are combined with fractions to be multiplied, they must be converted to fractions before multiplication.

61. 1 1/2 x 3/4 = ______. Answer 3/2 x 3/4 = 9/8 = 1 1/8.

In problem 61, 1 1/2 is converted to 3/2 so it can be multiplied by 3/4. Then, 3 x 3 = 9; 2 x 4 = 8. 9/8 is reduced to 1 1/8 for the final answer.

Review Exercise

Multiply problems 62-71. Reduce all answers to lowest possible form.

62. $\frac{3}{4} \times \frac{1}{2} =$

63. $\frac{1}{8} \times \frac{1}{2} =$

64. $\frac{9}{16} \times \frac{3}{4} =$

65. $\frac{5}{8} \times \frac{1}{4} =$

66. $3\frac{1}{4} \times \frac{1}{4} =$

67. $2\frac{1}{8} \times 5\frac{1}{2} =$

68. $1\frac{1}{8} \times 7\frac{3}{8} =$

69. $4\frac{3}{16} \times 1\frac{1}{2} =$

70. $3\frac{5}{8} \times \frac{1}{4} =$

71. $\frac{3}{16} \times \frac{3}{16} =$

Dividing Fractions

To divide fractions, invert one fraction and multiply.

72. $\frac{1}{4} \div \frac{1}{4} = \frac{1}{4} \times \frac{4}{1} = \frac{4}{4} = 1.$

Note the second fraction $\frac{1}{4}$ has been inverted and becomes $\frac{4}{1}$ before multiplication.

73. $\frac{3}{8} \div \frac{1}{4} = \frac{3}{8} \times \frac{4}{1} = \frac{12}{8} = 1\frac{4}{8} = 1\frac{1}{2}.$

74. $5\frac{1}{4} \div \frac{1}{2} = \frac{21}{4} \div \frac{1}{2} = \frac{21}{4} \times \frac{2}{1} = \frac{42}{4} = 10\frac{2}{4} = 10\frac{1}{2}.$

Problem 74 can also be worked as two separate problems: $5 \div \frac{1}{2} = 5 \times \frac{2}{1} = \frac{10}{1} = 10$ and $\frac{1}{4} \div \frac{1}{2} = \frac{1}{4} \times \frac{2}{1} = \frac{2}{4} = \frac{1}{2}$. Then the 10 and the $\frac{1}{2}$ are added to arrive at $10\frac{1}{2}$, the same answer as problem 74.

Another way of dividing fractions is the up-down-up and down-up-down method. This is quicker and done visually.

75. $\frac{3}{8} \div \frac{5}{8}'' = \frac{24}{40} = \frac{3}{5}.$

In problem 75 the first top number (3) is visually multiplied by the second bottom number (8), and the answer (24) becomes the top number of the answer fraction. Then the first bottom number (8) is multiplied by the second top number (5), and the answer (40) becomes the bottom number of the answer fraction. Then, of course, $\frac{24}{40}$ is reduced to $\frac{3}{5}$ by dividing 24 and 40 by 8.

Review Exercise

Divide problems 76-85. Reduce all answers to lowest possible form.

76. $\frac{3}{8} \div \frac{3}{8} =$

77. $\frac{5}{16} \div \frac{1}{2} =$

78. $\frac{1}{4} \div \frac{3}{4} =$

79. $\frac{5}{8} \div \frac{3}{4} =$

80. $2\frac{1}{4} \div 2 =$

81. $\frac{3}{16} \div \frac{11}{16} =$

82. $\frac{1}{2} \div \frac{1}{8} =$

83. $3\frac{1}{2} \div 3\frac{1}{2} =$

84. $\frac{5}{8} \div \frac{9}{16} =$

85. $1\frac{5}{8} \div \frac{5}{8} =$

Dividing Fractions and Numbers by 2

To divide $\frac{5}{8}$ by 2, conventional methods dictate that the 8 be multiplied by 2 to arrive at $\frac{5}{16}$ as half of $\frac{5}{8}$. Adding an even number in front of a fraction does not make this problem any more difficult since $12\frac{5}{8} \div 2 = 6\frac{5}{16}$. The 12 is divided by 2 and the 8 is multiplied by 2. When the number added in front of a fraction is an odd number, the problem becomes more difficult by conventional methods.

To divide $11\frac{5}{8}$ by 2, the shortcut method is to divide 2 into 11, resulting in 5 (disregard any remainder). For the fraction's numerator add the two numbers in the fraction $(5 + 8 = 13)$ and for the denominator, double the given denominator $(2 \times 8 = 16)$. So, the shortcut answer is $5\frac{13}{16}$.

This method works only when the whole number is an odd number. After learning this method, any odd number with any fraction can be easily divided by 2 by observation. For practice, $7\frac{3}{8} \div 2 = 3\frac{11}{16}$, $17\frac{3}{4} \div 2 = 8\frac{7}{8}$, and $21\frac{1}{2} \div 2 = 10\frac{3}{4}$. Now do the following math problems:

1. $5\frac{5}{8} \div 2 =$
2. $7\frac{3}{8} \div 2 =$
3. $19\frac{3}{4} \div 2 =$
4. $33\frac{1}{2} \div 2 =$
5. $17\frac{1}{8} \div 2 =$
6. $13\frac{1}{4} \div 2 =$
7. $3\frac{7}{8} \div 2 =$
8. $11\frac{1}{2} \div 2 =$
9. $23\frac{3}{4} \div 2 =$

Whole Number Subtraction

Subtraction of whole numbers is simple enough when the top numbers are larger than the bottom numbers.

```
Example:4,582,796
       —2,370,663
        2,212,133
```

When the number to be subtracted has larger numbers, conventional methods require that numbers be borrowed, added to the next number, and then subtraction is done. To simplify this procedure, do not do any marking out of numbers and adding to top numbers. Just add one to the left lower number.

```
Example:   3,456
          —  567
           2,889
```

To work this problem, say 7 from 16 is 9, then because one was borrowed, *add 1 to the next left number.* Then the 6 becomes a 7 and 7 from 15 is 8. Again, because 1 was borrowed, add 1 to the 5 and subtract 6 from 14, which is 8. Now 1 must be added to the left lower number, which is zero, so it becomes 1. Then 1 from 3 is 2. So the answer is 2,899.

Do not add to the bottom number when standard subtraction can be done. The one is added only when the previous subtraction required borrowing.

```
Example: 382,745
        —197,326
         185,419
```

To solve: 6 from 15 is 9, 3 from 4 is 1 (no borrowing), 3 from 7 is 4 (no borrowing), 7 from 12 is 5 (borrowing here), 10 from 18 is 8, and 2 from 3 is 1.

Do the following subtraction problems:

```
1.  375,234          4.  382,751
   —286,547             —  95,956

2.  257,322          5.  632,040
   — 88,445             —523,050

3.  571,063          6.  271,012
   —280,557             —189,138
```

Subtracting Feet and Inches

When subtracting feet and inches, subtraction is easy when no borrowing is necessary.

```
Example:   4′ - 3″
          —2′ - 2″
           2′ - 1″
```

The problem occurs when a number such as 2′ - 8″ must be subtracted from, say, 4′ - 3″.

```
Example:   4′ - 3″
          —2′ - 8″
           1′ - 7″
```

Because 8″ is larger than 3″, conventional methods require several steps of addition and subtraction, all leading to possible errors.

The shortcut method is to always subtract from 12″, the amount being borrowed. Here, subtract 8″ from 12″ (the amount being borrowed), leaving 4″. Then add the upper number to it. So, 4″ + 3″ = 7″ and this goes in the answer. Then, because 1′ was borrowed (the 12″), add 1 to the lower number (2 + 1 = 3) and subtract 3′ from 4′ = 1′. So the final answer is 1′ - 7″.

To work another problem—

```
Example:   131′ -  8″
          —  58′ - 10″
             72′ - 10″
```

To solve: subtract 10″ from 12″ and add the answer to the 8″. Then add 1′ to the 8′ making 9′. Now, 9′ from 11′ = 2′. Because 1 was borrowed, add 1′ to the 5 and subtract 6 from 13 to get 7.

Practice this method by working the following problems:

```
1.  73′ - 3″          4.  26′ -  6″
   —26′ - 5″             —16′ - 10″

2.  42′ - 8″          5.  47′ -  7″
   —17′ - 9″             —39′ -  9″

3.  16′ - 2″          6.  53′ -  1″
   —15′ - 7″             —44′ -  6″
```

Subtracting Feet, Inches, and Fractions

The shortcut method becomes more desirable when fractions are added and the fraction to be subtracted is larger than the one above.

Example: $\begin{array}{r} 10\ 1/8 \\ -\ \underline{6\ 5/8} \\ 3\ 1/2 \end{array}$

Since 5/8 cannot be subtracted from 1/8, borrowing must be done. But in the shortcut method, 8/8 is borrowed. Now, subtract 5 (5/8) from 8 (8/8) leaving 3 (3/8). Note that to this point, the top fraction has been ignored. Now the top fraction, 1/8, is added to the 3/8 to total 4/8 or 1/2. Since 1 (8/8) was borrowed, 1 is added to the 6 to equal 7. Then, 7 is subtracted from 10 leaving 3.

The same rules apply when the fractions have uncommon denominators.

Example: $\begin{array}{r} 11\ 3/16 \\ -\ \underline{8\ 3/4} \\ 2 \end{array}$

To subtract fractions, 16 is the common denominator and 3/4 is 12/16. Since 12 cannot be subtracted from 3, borrowing must be done. Then 16/16 is borrowed and 12/16 is subtracted from 16/16, leaving 4/16. Then, the 3/16 is added to 4/16, totaling 7/16. Since borrowing was done, 1 is added to the 8 to make 9, and 9 is subtracted from 11, leaving 2.

Putting the feet, inches, and fractions together and doing the subtraction by the shortcut method:

Example: 27′ - 4 3/8″
—16′ - 8 7/8″

To solve: subtract 7/8 from 8/8, leaving 1/8. Then, add the 3/8 to 1/8 to total 4/8 or 1/2. Then subtract 9″ from 12″ = 3″, and add the 4″ to equal 7″. Then, 17 from 27 leaves 10. So, the answer is 10′ — 7½″.

With uncommon denominators, the shortcut method has a greater advantage.

Example: 14′—3 5/8″
— 8′—9 11/16″

Subtract 11/16″ from 16/16″ leaving 5/16″. Then, add 10/16″ (5/8″) to total 15/16″. Subtract 10 (9 + 1) from 12″, which is 2″, add the 3″ to total 5″. Then, 9′ (8′ + 1′) from 14′ = 5′. So, the final answer is 5′ — 5 15/16″.

Practice the shortcut method by doing the following problems:

1. 10′ - 4 1/4″
— 6′ - 5 3/4″

2. 62′ - 3 3/8″
—14′ - 7 7/8″

3. 24′ - 0 3/16″
—18′ - 2 5/16″

4. 19′ - 1 3/16″
— 6′ - 3 3/8″

5. 32′ - 2 5/16″
—18′ - 4 5/8″

6. 13′ - 6 1/4″
— 8′ - 7 11/16″

7. 11′ - 8 1/2″
—10′ - 8 3/4″

8. 63′ - 0 11/16″
—27′ - 10 7/8″

9. 32′ - 5 3/4″
—17′ - 10 7/8″

How to Square Numbers Ending in 5

Many times the draftsman must work certain calculations that often have numbers ending in 5. Often these numbers must be squared. There is a method that draftsmen use that allows them to square numbers ending in 5 without doing any but the easiest math.

The first rule is that when squaring a number ending in 5, the answer will always end in 25.

How does one know what the first number or numbers will be? The second rule is to add one to the number and multiply it by itself. To square 25, add 1 to the 2 and multiply 2 × 3 which is 6. Then add 25 after the 6 to total 625. So. $25^2 = 625$. To square 35, add 1 to the first number (3) to make 4. Then multiply 3 × 4 = 12 and tack on the 25 to make the answer 1,225. To square 65, multiply 7 × 6 = 42 and add on the 25 after the 42 to make 4,225.

Practice with the following problems:

1. $45^2 =$
2. $55^2 =$
3. $65^2 =$
4. $75^2 =$
5. $85^2 =$
6. $95^2 =$

Multiplying Numbers Ending in 5

When two numbers ending in 5 are multiplied, the first thing to consider is the sum of the first two digits of the two numbers.

Example: $\begin{array}{r} 45 \\ \times\ \underline{25} \end{array}$

The first digits of the two numbers are 4 and 2. Since the sum of these two numbers is 6 (an even number), the answer will end in 25. Any time the sum of the first two digits is an even number, the answer will end in 25.

To find the first number or numbers, the two digits are multiplied (2×4) and half the sum is added to the answer. So, $2 \times 4 = 8$ plus half the sum (6) or 3 is added to total 11. So, $45 \times 25 = 1{,}125$.

To work another problem—

Example:
$$\begin{array}{r} 55 \\ \times\ 75 \\ \hline 4{,}125 \end{array}$$

To solve: $7 + 5 = 12$ (an even number), so the answer will end in 25. Half of 12 is 6. Then $5 \times 7 = 35$. Adding 6 to $35 = 41$. The answer is 4,125.

The following are some problems for practice:

1. 25×65
2. 45×65
3. 36×75
4. 65×85
5. 55×35
6. 95×35

When the sum of the first digits of the two numbers is an odd number, the answer will end in 75, not 25.

Example:
$$\begin{array}{r} 35 \\ \times\ 65 \\ \hline \end{array}$$

The first digits of the two numbers are 3 and 6. The sum of these two numbers is 9, an odd number, so the answer will end in 75.

To find the first number or numbers, the two digits are multiplied (3×6) and the answer is 18. Then half the sum, without any remainder, of 6 and 3 (9) is 4. Add the 4 to 18 to total 22. The answer is 2,275.

To work another problem—

Example:
$$\begin{array}{r} 55 \\ \times\ 65 \\ \hline \end{array}$$

The first digits of the two numbers (5 and 6) total 11 (an odd number), so the answer will end in 75. To find the first part of the answer, multiply $6 \times 5 = 30$ and add 5 (half of $5 + 6$ and ignoring the decimal remainder) to total 35. Then, the answer is known to be 3,575.

Now do the following problems using the methods described above as quickly as you can:

1. 15×25
2. 25×35
3. 35×75
4. 45×95
5. 45×55
6. 75×25
7. 35×15
8. 45×35
9. 75×25
10. 85×35
11. 45×45
12. 55×95
13. 75×85
14. 95×15
15. 55×85

CHAPTER 4

The Metric System

The United States is taking steps to convert to the metric system of weights and measures. While this will be a complicated and very costly conversion the end result will be an easier understood system. For instance, there are no fractions to be learned in the metric system. Every unit is expressed in whole and/or decimal numbers.

Why Change?

The United States is the last country in the world to make the conversion. Maintaining the English measurement units is hurting United States foreign trade as the rest of the world is on the metric system. Anyone that has worked on a foreign-made car or motorcycle knows they must use metric tools. This is because other nations build all their products to metric dimensions. English units simply do not match.

The United States military has been using metric units for years because NATO allies all use them. All Olympic sports use metric units. Many United States companies are changing their designs to metric units to enable them to better compete on the foreign market and because they know metric units are here to stay. Physicians, scientists and astronauts use exclusively the metric system. All the pharmaceutical and photographic and many data processing firms now utilize metric units.

The Metric System is Established

The forerunner of the present metric system was established in France after the French Revolution and was adopted in 1799. Since then numerous changes have been made. Today the world is adopting the newest system, Systeme International D'Unites (International System of Units), which is commonly referred to as SI.

In 1866 the Congress of the United States passed a law making it lawful to employ the weights and measures of the metric system. Not many people paid much attention to this law. On July 2, 1971, the Department of Commerce recommended a 10-year changeover to the metric system similar to the one Britain is now using. The Commerce Department estimated the United States was losing up to $25 billion in foreign trade annually because our products, manufactured in English units, did not match those manufactured in foreign countries.

Besides the commercial advantages there are many other advantages to the metric system. Schools have estimated they will save $700 million each year just because it is easier to teach metric units. For a few years both systems will have to be taught, so initially conversion will be costly to schools. But, business will bear the greatest conversion costs. It is estimated their costs will be over $1 billion a year for 10 years! However, the experts agree that this cost is small compared to the gain in international trade alone.

The SI System

The SI is based on decimal arithmetic where units of different sizes are formed by multiplying or dividing a single base number by powers of ten. This eliminates the use of unwieldy fractions. In the past fractions had to be converted to decimals for computer input, and our adding machines wouldn't take fractions.

The SI system has seven base-units. These base-units and their abbreviations are

meter (sometimes metre), m
kilogram, kg
second, s
ampere, A
kelvin, K
mole, mol
candela, cd

While these terms may sound strange to one used to the English units, constant use of the metric system will prove them easier to learn and use. In the English units, we speak of tons, pounds and ounces where the metric system uses grams. For miles, yards, feet and inches the metric system uses meters. And for barrels, gallons, quarts, pints and cups the metric system uses liters (sometimes litres). So for 12 English units one must learn only three metric units.

The metric prefixes must be remembered. These prefixes are used for all metric units. A kilo is 1000, so a kilogram is 1000 grams and a kilometer is 1000 meters. A hecto is 100, so a hectogram is 100 grams. Table 4-1 supplies decimal multiples and sub-multiples for SI units. However the most commonly used ones are

kilo = 1000
hecto = 100
deka = 10
deci = 0.1
centi = 0.01
milli = 0.001

Table 4-1. Decimal Multiples and Sub-multiples for SI Units

Unit Factor	Prefix	Symbol
10^{12}	tera	T
10^{9}	giga	G
10^{6}	mega	M
10^{3}	kilo	k
10	deka	da
10^{-1}	deci	d
10^{-2}	centi	c
10^{-3}	milli	m
10^{-6}	micro	μ
10^{-9}	nano	n
10^{-12}	pico	p
10^{-15}	femto	f
10^{-18}	atto	a

Metric vs. English

Length

To establish a relationship between metric and English units, the meter is 39.37 inches or slightly longer than an English yard. Instead of a mile, which is 5,280 feet or 1,760 yards, the metric length would be expressed as a kilometer, 1000 meters, which is 0.6214 of a mile. Miles per hour is expressed as kilometers per hour, so 80 kilometers per hour is 50 miles per hour. A quick conversion method is to multiply the kilometers per hour by ⅝ to get miles per hour. Eighty times 5 equals 400 which is divided by 8 to get 50 mph. The same formula applies to miles. A 500 kilometer trip becomes 500 x ⅝ = 2500 ÷ 8 = 312.5 miles. To convert miles to kilometers multiply miles by 8/5. A 300 mile trip becomes 300 x 8/5 = 2400 ÷ 5 = 480 kilometers.

Smaller metric units are the centimeter, 1/100 of a meter, and the millimeter, 1/1000 of a meter. There are 30.48 centimeters in one foot. How many millimeters are in one foot? The answer is simple using the decimal system. Centi is 1/100 and milli is 1/1000 so to convert from centi to milli move the decimal point to the right one place, and the answer is 304.8 millimeters in one foot. One inch equals 25.4 millimeters, 2.54 centimeters or 0.0254 meter. Figure 4-1 shows a guide for converting metric lengths to English units.

In the metric system most drawings are dimensioned in millimeters, but the mm is not noted. In this text some metric dimensions are shown next to the English units and will be in millimeters unless otherwise noted.

Weight

Figure 4-2 explains how the metric gram replaces many English units. Since the kilogram is

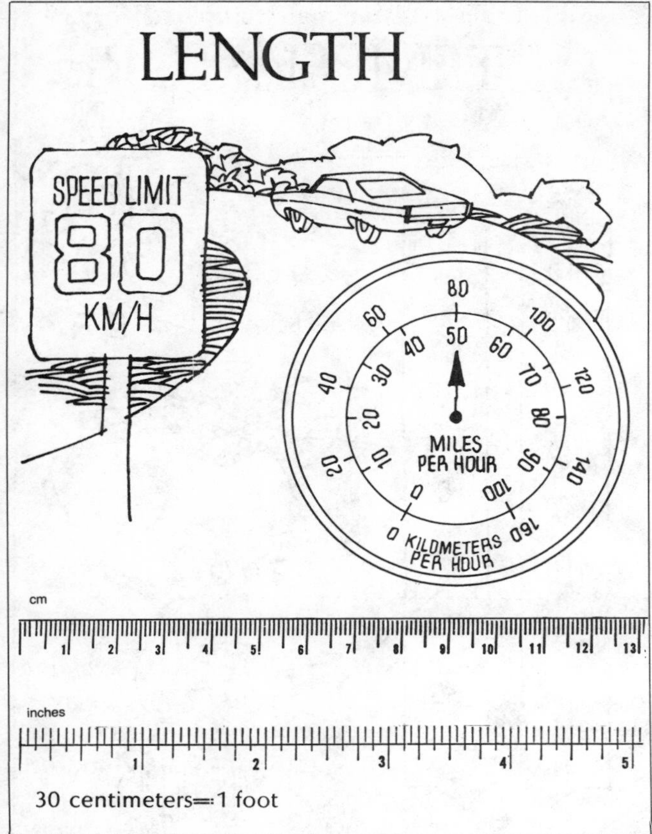

Figure 4-1. Relationship of meters to miles and feet.

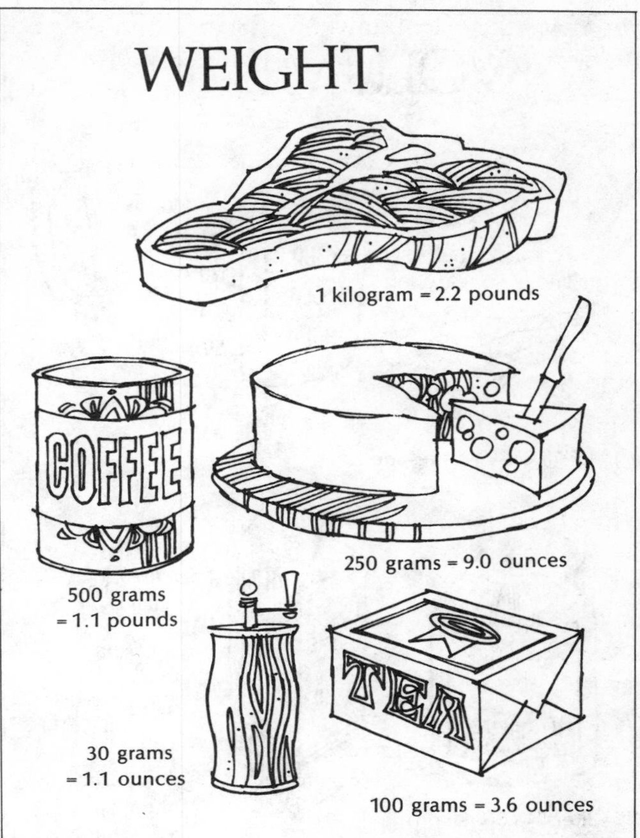

Figure 4-2. Relationship of grams to pounds and ounces.

2.2 pounds a person weighing 140 pounds would weigh only 63.6 kilograms or kilos (pronounced kee′ los), as they are commonly called. A six-pound roast would be how many kilos?

Volume

Figure 4-3 shows the relationship of the metric liter to the English volume units. Note that the liter is slightly larger than the quart. In the metric system all volume is measured in liters. Things normally expressed in gallons, such as gasoline, milk, ice cream and water are purchased by the liter. For instance a car would contain 80 liters of gasoline instead of 21 gallons. A liter is 0.2642 of a gallon. If a chef at a hotel bought 50 gallons of milk every day, how many liters must the chef buy to have the same volume of milk?

Temperature

The SI unit for temperature is Celsius. Figure 4-4 shows the Celsius-Fahrenheit relationship. Celsius is the same measurement as the old term Centigrade.

The ASTM Guide

The American Society for Testing and Materials (ASTM) has published a complete booklet on metric practices. It is the *Metric Practice Guide* and it is numbered E380. It may be ordered by writing ASTM, 1916 Race Street, Philadelphia, Pa. 19103.

Metric vs. English Solutions

To prove the metric system simplicity the three problems below are solved using both English and metric units.

Problem 1

What is the area of the floor of a room with the following dimensions?

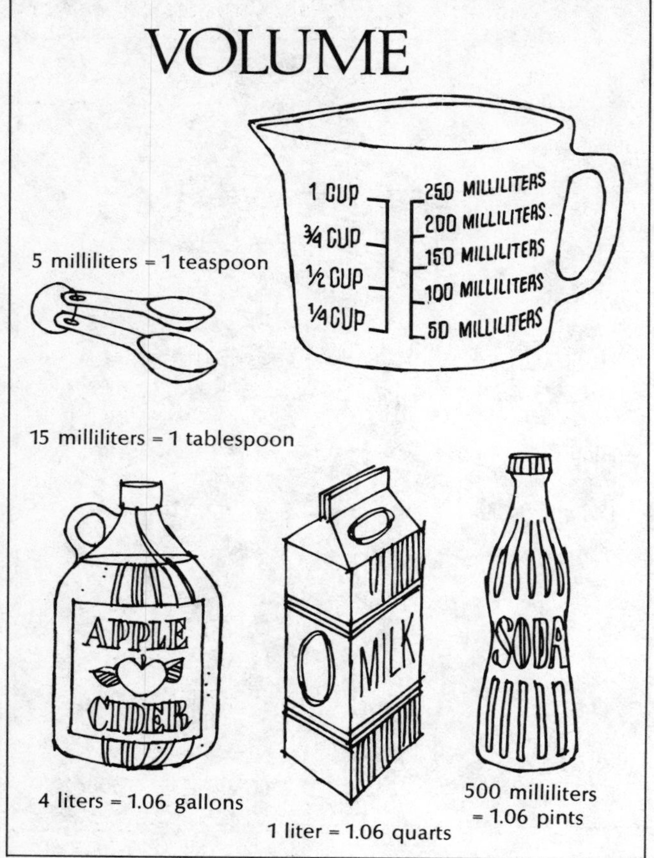

Figure 4-3. Relationship of liters to English units of volume.

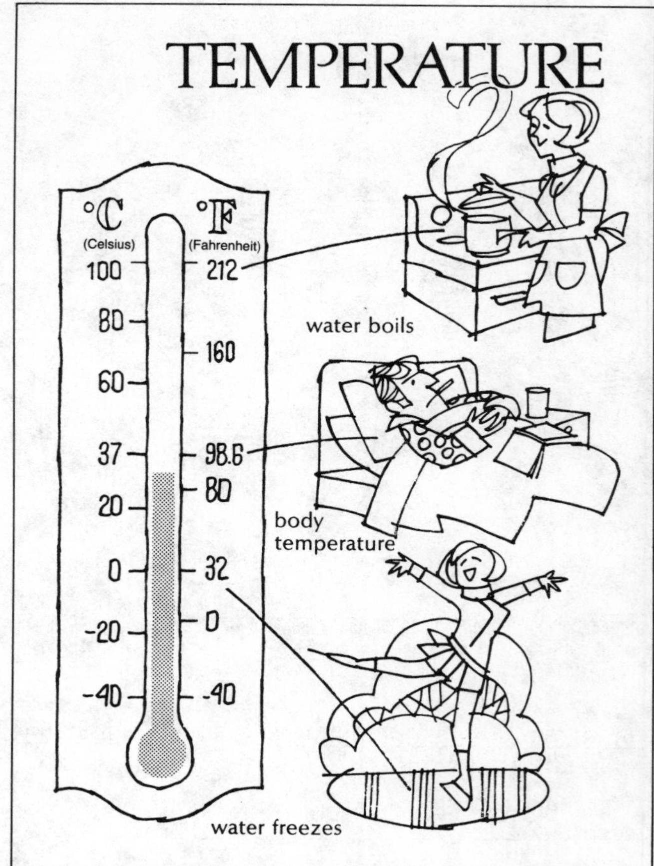

Figure 4-4. Celsius-Fahrenheit relationship.

	English Units
Length	15′-7″
Width	12′-6″

Metric Units
475 centimeters (4.75 meters)
380 centimeters (3.80 meters)

Solution. The area is determined by multiplying the length of the room by its width. However, the English units must be converted into a common unit expression which, in this case, may be either feet or inches.

English Unit Solution

Multiply feet by 12 to convert to inches. Length 15′-7″ = (15 x 12)+7 = 187″, width 12′-6″ = (12 x 12)+6 = 150″. Length x width = 187″x 150″ = 28,050 square inches. Since there are 144 square inches in a square foot, 28,050 ÷ 144 = 194.79, rounded to 195 square feet. There are nine square feet in a square yard, so 195 ÷ 9 = 21.67, rounded to 22 square yards.

Metric Unit Solution

Since length and width are given in common units no conversion is necessary. Length x width = 475 cm x 380 cm = 180,500 square cm or 4.75 m x 3.80 m = 18.05 square meters.

Problem 2

What is the approximate total weight of the contents of a basket that contains the following items:

	English Units	Metric Units
Meat	4 lb 9 oz	2.07 kilograms
Potatoes	3 lb 4 oz	1.47 kilograms
Tomatoes	2 lb 15 oz	1.33 kilograms
Cereal	1 lb 7 oz	650 grams

Solution of Problem

	English Weight in Ounces	Metric Weight in Grams
Weight in pounds multiplied by 16 gives weight in ounces		
Meat	(4 x 16) + 9 = 73	2070
Potatoes	(3 x 16) + 4 = 52	1470
Tomatoes	(2 x 16) + 15 = 47	1330
Cereal	(1 x 16) + 7 = 23	650
	195	5520

195 divided by 16 = 12 lb (approx.) or 5.5 kilograms (approx.)

Problem 3

What is the volume of the following two comparable but not equal mixtures:

	English Units	Metric Units
Milk	1 gal 2 qt 1 pt	6.15 liters
Water	3 qt 1 pt	3.31 liters
Flavoring	½ pt	236 milliliters

Solution of Problem

English Volume in Pints		Metric Volume in Milliliters	
Multiply gallons by 8, and quarts by 2 to convert to pints		6150	multiply liters by 1000 to convert to milliliters
		3310	
Milk		236	
(1 x 8) + (2 x 2) + 1 =	13	9696	
Water			
(3 x 2) + 1 =	7		
Flavoring	½		
	20½		

20½ ÷ 2 = 10 qt (approx.) or 10.25 liters
10 ÷ 4 = 2½ gal (approx.) or 10 liters (approx.)

Reference Tables

Table 4-2 supplies conversion factors from metric to English units. Table 4-3 converts feet and inches to meters and millimeters. Table 4-4 converts millimeters and meters to inches and feet.

Table 4-2. Metric to English Conversion Factors

Length			
Centimeter	=	0.3937	inch
Meter	=	39.37	inches
Meter	=	3.281	feet
Meter	=	1.094	yards
Kilometer	=	39,372	inches
Kilometer	=	3,281	feet
Kilometer	=	1,094	yards
Kilometer	=	0.6214	statute mile
Kilometer	=	0.5396	nautical mile
Inch	=	25.4	millimeters
Inch	=	2.54	centimeters
Inch	=	0.0254	meter
Foot	=	304.8	millimeters
Foot	=	30.48	centimeters
Foot	=	3.048	dekameters
Foot	=	0.3048	meters
Yard	=	914.402	millimeters
Yard	=	91.4402	centimeters
Yard	=	0.9144	meter
Statute mile	=	1.60935	kilometers
Nautical mile	=	1.853	kilometers
Area			
Square millimeter	=	0.00155	square inch
Square centimeter	=	0.155	square inch
Square meter	=	1,550	square inches
Square meter	=	10.764	square feet
Square meter	=	1.196	square yards
Hectare	=	2.471	acres
Square kilometer	=	0.3861	square mile
Square inch	=	645.2	square millimeters
Square inch	=	6.452	square centimeters
Square foot	=	929.034	square centimeters
Square foot	=	0.0929	square meter
Square yard	=	8,361	square centimeters
Square yard	=	0.8361	square meters
Acre	=	4,047	square meters
Acre	=	0.4047	hectare
Square mile	=	259	hectares
Square mile	=	2.59	square kilometers
Volume			
Cubic centimeter	=	0.06102	cubic inch
Cubic centimeter	=	0.001	liter
Cubic meter	=	61,023	cubic inches
Cubic meter	=	35.314	cubic feet
Cubic meter	=	1.308	cubic yards
Cubic meter	=	1,000	liters
Cubic inch	=	16.39	cubic centimeters

(table continued on next page)

Table 4-2 continued

Cubic foot	=	28,320	cubic centimeters
Cubic foot	=	0.02832	cubic meter
Cubic yard	=	0.7646	cubic meter

Capacity

Milliliter	=	0.0338	U.S. fluid ounce
Liter	=	1.057	U.S. liquid quarts
U.S. fluid ounce	=	29.57	milliliters
U.S. liquid quart	=	0.946	liter

Mass or Weight

Gram	=	15.4324	grains
Gram	=	0.03527	ounces, avoirdupois
Gram	=	0.03215	ounces, troy
Kilogram	=	2.205	pounds, avoirdupois
Ton, metric	=	1.1023	tons
Grain	=	0.0648	gram
Pound	=	0.4536	kilogram
Ton	=	0.907	metric ton

Table 4-3. Inches to Millimeters and Feet to Meters

Inches								Inches							
1/2s	1/4s	1/8s	1/16s	1/32s	1/64s	Decimals[a]	Milli-meters	1/2s	1/4s	1/8s	1/16s	1/32s	1/64s	Decimals[a]	Milli-meters
					1	0.0156 25	0.397						33	0.515 625	13.097
				1	2	0.031 25	0.794					17	34	0.531 25	13.494
					3	0.046 875	1.191						35	0.546 875	13.891
			1	2	4	0.062 5	1.588				9	18	36	0.562 5	14.288
					5	0.078 125	1.984						37	0.578 125	14.684
				3	6	0.093 75	2.381					19	38	0.593 75	15.081
					7	0.109 375	2.778						39	0.609 375	15.478
		1	2	4	8	0.125 0	3.175[a]			5	10	20	40	0.625 0	15.875[a]
					9	0.140 625	3.572						41	0.640 625	16.272
				5	10	0.156 25	3.969					21	42	0.656 25	16.669
					11	0.171 875	4.366						43	0.671 875	17.066
			3	6	12	0.187 5	4.762				11	22	44	0.687 5	17.462
					13	0.203 125	5.159						45	0.703 125	17.859
				7	14	0.218 75	5.556					23	46	0.718 75	18.256
					15	0.234 375	5.953						47	0.734 375	18.653
	1	2	4	8	16	0.250 0	6.350[a]		3	6	12	24	48	0.750 0	19.050[a]
					17	0.265 625	6.747						49	0.765 625	19.447
				9	18	0.281 25	7.144					25	50	0.781 25	19.844
					19	0.296 875	7.541						51	0.796 875	20.241
			5	10	20	0.312 5	7.938				13	26	52	0.812 5	20.638
					21	0.328 125	8.334						53	0.828 125	21.034
				11	22	0.343 75	8.731					27	54	0.843 75	21.431
					23	0.359 375	9.128						55	0.859 375	21.828
		3	6	12	24	0.3750	9.525[a]			7	14	28	56	0.875 0	22.225[a]
					25	0.390 625	9.922						57	0.890 625	22.622
				13	26	0.406 25	10.319					29	58	0.906 25	23.019
					27	0.421 875	10.716						59	0.921 875	23.416
			7	14	28	0.437 5	11.112				15	30	60	0.937 5	23.812
					29	0.453 125	11.509						61	0.953 125	24.209
				15	30	0.468 75	11.906					31	62	0.968 75	24.606
					31	0.484 375	12.303						63	0.984 375	25.003
1	2	4	8	16	32	0.500 0	12.700[a]	2	4	8	16	32	64	1.000 0	25.400[a]

(table continued on next page)

Table 4-3 continued

Inches ↓ / →	0	1	2	3	4	5	6	7	8	9
	Millimeters[a]									
0	–	25.4	50.8	76.2	101.6	127.0	152.4	177.8	203.2	228.6
10	254.0	279.4	304.8	330.2	355.6	381.0	406.4	431.8	457.2	482.6
20	508.0	533.4	558.8	584.2	609.6	635.0	660.4	685.8	711.2	736.6
30	762.0	787.4	812.8	838.2	863.6	889.0	914.4	939.8	965.2	990.6
40	1016.0	1041.4	1066.8	1092.2	1117.6	1143.0	1168.4	1193.8	1219.2	1244.6
50	1270.0	1295.4	1320.8	1346.2	1371.6	1397.0	1422.4	1447.8	1473.2	1498.6
60	1524.0	1549.4	1574.8	1600.2	1625.6	1651.0	1676.4	1701.8	1727.2	1752.6
70	1778.0	1803.4	1828.8	1854.2	1879.6	1905.0	1930.4	1955.8	1981.2	2006.6
80	2032.0	2057.4	2082.8	2108.2	2133.6	2159.0	2184.4	2209.8	2235.2	2260.6
90	2286.0	2311.4	2336.8	2362.2	2387.6	2413.0	2438.4	2463.8	2489.2	2514.6
100	2540.0	–	–	–	–	–	–	–	–	–

Feet ↓ / →	0	1	2	3	4	5	6	7	8	9
	Meters[a]									
0	–	0.3048	0.6096	0.9144	1.2192	1.5240	1.8288	2.1336	2.4384	2.7432
10	3.0480	3.3528	3.6576	3.9624	4.2672	4.5720	4.8768	5.1816	5.4864	5.7912
20	6.0960	6.4008	6.7056	7.0104	7.3152	7.6200	7.9248	8.2296	8.5344	8.8392
30	9.1440	9.4488	9.7536	10.0584	10.3632	10.6680	10.9728	11.2776	11.5824	11.8872
40	12.1920	12.4968	12.8016	13.1064	13.4112	13.7160	14.0208	14.3256	14.6304	14.9352
50	15.2400	15.5448	15.8496	16.1544	16.4592	16.7640	17.0688	17.3736	17.6784	17.9832
60	18.2880	18.5928	18.8976	19.2024	19.5072	19.8120	20.1168	20.4216	20.7264	21.0312
70	21.3360	21.6408	21.9456	22.2504	22.5552	22.8600	23.1648	23.4696	23.7744	24.0792
80	24.3840	24.6888	24.9936	25.2984	25.6032	25.9080	26.2128	26.5176	26.8224	27.1272
90	27.4320	27.7368	28.0416	28.3464	28.6512	28.9560	29.2608	29.5656	29.8704	30.1752
100	30.4800	–	–	–	–	–	–	–	–	–

[a]Exact figure

Table 4-4. Millimeters to Inches and Meters to Feet

Milli- meters	Inches Nearest 1/16"	Inches Nearest 1/64"	Inches Decimals	Milli- meters	Inches Nearest 1/16"	Inches Nearest 1/64"	Inches Decimals
1	1/16	3/64	0.03937	51	2	2 1/64	2.00787
2	1/16	5/64	0.07874	52	2 1/16	2 3/64	2.04724
3	1/8	1/8	0.11811	53	2 1/16	2 3/32	2.08661
4	3/16	5/32	0.15748	54	2 1/8	2 1/8	2.12598
5	3/16	13/64	0.19685	55	2 3/16	2 11/64	2.16535
6	1/4	15/64	0.23622	56	2 3/16	2 13/64	2.20472
7	1/4	9/32	0.27559	57	2 1/4	2 1/4	2.24409
8	5/16	5/16	0.31496	58	2 5/16	2 9/32	2.28346
9	3/8	23/64	0.35433	59	2 5/16	2 21/64	2.32283
10	3/8	25/64	0.39370	60	2 3/8	2 23/64	2.36220
11	7/16	7/16	0.43307	61	2 3/8	2 13/32	2.40157
12	1/2	15/32	0.47244	62	2 7/16	2 7/16	2.44094
13	1/2	33/64	0.51181	63	2 1/2	2 31/64	2.48031
14	9/16	35/64	0.55118	64	2 1/2	2 33/64	2.51969
15	9/16	19/32	0.59055	65	2 9/16	2 9/16	2.55906
16	5/8	5/8	062992	66	2 5/8	2 19/32	2.59843
17	11/16	43/64	0.66929	67	2 5/8	2 41/64	2.63780
18	11/16	45/64	0.70866	68	2 11/16	2 43/64	2.67717
19	3/4	3/4	0.74803	69	2 11/16	2 23/32	2.71654
20	13/16	25/32	0.78740	70	2 3/4	2 3/4	2.75591
21	13/16	53/64	0.82677	71	2 13/16	2 51/64	2.79528
22	7/8	55/64	0.86614	72	2 13/16	2 53/64	2.83465
23	7/8	29/32	0.90551	73	2 7/8	2 7/8	2.87402
24	15/16	15/16	0.94488	74	2 15/16	2 29/32	2.91339
25	1	63/64	0.98425	75	2 15/16	2 61/64	2.95276
26	1	1 1/32	1.02362	76	3	2 63/64	2.99213
27	1 1/16	1 1/16	1.06299	77	3 1/16	3 1/32	3.03150
28	1 1/8	1 7/64	1.10236	78	3 1/16	3 5/64	3.07087
29	1 1/8	1 9/64	1.14173	79	3 1/8	3 7/64	3.11024
30	1 3/16	1 3/16	1.18110	80	3 1/8	3 5/32	3.14961
31	1 1/4	1 7/32	1.22047	81	3 3/16	3 3/16	3.18898
32	1 1/4	1 17/64	1.25984	82	3 1/4	3 15/64	3.22835
33	1 5/16	1 19/64	1.29921	83	3 1/4	3 17/64	3.26772
34	1 5/16	1 11/32	1.33858	84	3 5/16	3 5/16	3.30709
35	1 3/8	1 3/8	1.37795	85	3 3/8	3 11/32	3.34646
36	1 7/16	1 27/64	1.41732	86	3 3/8	3 25/64	3.38583
37	1 7/16	1 29/64	1.45669	87	3 7/16	3 27/64	3.42520
38	1 1/2	1 1/2	1.49606	88	3 7/16	3 15/32	3.46457
39	1 9/16	1 17/32	1.53543	89	3 1/2	3 1/2	3.50394
40	1 9/16	1 37/64	1.57480	90	3 9/16	3 35/64	3.54331
41	1 5/8	1 39/64	1.61417	91	3 9/16	3 37/64	3.58268
42	1 5/8	1 21/32	1.65354	92	3 5/8	3 5/8	3.62205
43	1 11/16	1 11/16	1.69291	93	3 11/16	3 21/32	3.66142
44	1 3/4	1 47/64	1.73228	94	3 11/16	3 45/64	3.70079
45	1 3/4	1 49/64	1.77165	95	3 3/4	3 47/64	3.74016
46	1 13/16	1 13/16	1.81102	96	3 3/4	3 25/32	3.77953
47	1 7/8	1 27/32	1.85039	97	3 13/16	3 13/16	3.81890
48	1 7/8	1 56/64	1.88976	98	3 7/8	3 55/64	3.85827
49	1 15/16	1 59/64	1.92913	99	3 7/8	3 57/64	3.89764
50	1 15/16	1 31/32	1.96850	100	3 15/16	3 15/16	3.93701

Milli- meters	Feet-Inches Nearest 1/16"	Feet-Inches Nearest 1/64"	Decimal Inches
100	0-3 15/16	0-3 15/16	3.93701
200	0-7 7/8	0-7 7/8	7.87402
300	0-11 13/16	0-11 13/16	11.81102
400	1-3 3/4	1-3 3/4	15.74803
500	1-7 11/16	1-7 11/16	19.68504
600	1-11 5/8	1-11 5/8	23.62205
700	2-3 9/16	2-3 9/16	27.55906
800	2-7 1/2	2-7 1/2	31.49606
900	2-11 7/16	2-11 7/16	35.43307
1000	3-3 3/8	3-3 3/8	39.37008

Meters	Feet-Inches Nearest 1/16"	Decimal Feet
1	3- 3 3/8	3.2808
2	6- 6 3/4	6.5617
3	9-10 1/8	9.8425
4	13- 1 1/2	13.1234
5	16- 4 7/8	16.4042
6	19- 8 1/4	19.6850
7	22-11 9/16	22.9659
8	26- 2 15/16	26.2467
9	29- 6 5/16	29.5276
10	32- 9 11/16	32.8084
11	36- 1 1/16	36.0892
12	39- 4 7/16	39.3701
13	42- 7 13/16	42.6509
14	45-11 3/16	45.9318
15	49- 2 9/16	49.2126
16	52- 5 15/16	52.4934
17	55- 9 5/16	55.7743
18	59- 0 11/16	59.0551
19	62- 4 1/16	62.3360
20	65- 7 3/8	65.6168
21	68-10 3/4	68.8976
22	72- 2 1/8	72.1785
23	75- 5 1/2	75.4593
24	78- 8 7/8	78.7402
25	82- 0 1/4	82.0210
26	85- 3 5/8	85.3018
27	88- 7	88.5827
28	91-10 3/8	91.8635
29	95- 1 3/4	95.1444
30	98- 5 1/8	98.4252
31	101- 8 1/2	101.7060
32	104-11 13/16	104.9869
33	108- 3 3/16	108.2677
34	111- 6 9/16	111.5486
35	114- 9 15/16	114.8294

CHAPTER 5
Plant Coordinate Systems

Plant Coordinate Systems

Just as the earth has latitudes and longitudes to enable locations to be expressed mathematically, almost all industrial installations utilize a grid system which is called a **plant coordinate system.** To define a particular spot on the earth, two coordinates are given, such as latitude X and longitude Y. This establishes two straight lines, 90 degrees apart, referred to as **crosshairs.** The two coordinate lines meet at a point which is unlike any other point on the earth and so becomes a precise location.

Plant coordinate systems use north-south and east-west lines as coordinates. A **bench mark,** or base point, is established and coordinate lines are assigned to this point. The bench mark must be something that will remain stable and is usually a concrete marker firmly planted in the ground. As an example, this bench mark may be assigned the coordinates of north 0.0′ and east 0.0′. An item assigned coordinates of east 10.0′ and north 150.0′ would have it crosshairs located 10′ east of the bench mark and 150′ north of the bench mark.

The bench mark is also assigned a plant elevation. This elevation might be called "Elevation 100.00′". In some plants the actual elevation above sea level is used at the bench mark. Usually the mythical elevation of 100 is given to bench marks. Why don't plants call the bench mark as elevation 0.0′? This would make items located below grade have a *minus* elevation, such as −6.50′, and this would be confusing and could cause errors. So minus elevations are avoided in industrial plants.

In overall plants plans, coordinates are usually shown in the decimal system, such as N.1438.63′. Plant drawings are dimensioned in feet and inches. The draftsman must convert these decimal coordinates to feet and inches on his drawings. A few plant coordinate systems are in feet and inches, however since most are in decimals, the draftsman must know how to convert decimal coordinates to feet and inches.

Decimal Conversion

The coordinate N.150.00′ is expressed in feet and decimals of a foot. This is easily converted to N.150′-0″ since the decimal shown was 0′. If this had been N.150.50′, it is apparent that .50′ is one-half of a foot or 6 inches, so N.150.50′ becomes N.150′-6″. With a coordinate of N.150.76′, conversion to feet and inches is not visually apparent. To convert decimals to feet and inches, draftsmen refer to a conversion table. Table 5-1 supplies these conversions.

To convert N.150.76′, it is known that we have 150′. The unknown is how many inches, plus fractions of an inch, are in the .76′? Referring to Table 5-1, .76 is located under the 9″ column, across from ⅛″ on the left. So, .76′ is converted to 9⅛″. The fully converted coordinate then becomes N.150′-9⅛″.

Table 5-1. Decimals of a Foot by 1/16th's

	0"	1"	2"	3"	4"	5"
0"	.0000	.0833	.1667	.2500	.3333	.4167
1/16"	.0052	.0855	.1719	.2552	.3385	.4219
1/8"	.0104	.0937	.1771	.2604	.3437	.4271
3/16"	.0156	.0990	.1823	.2656	.3490	.4323
1/4"	.0208	.1042	.1875	.2708	.3542	.4375
5/16"	.0260	.1094	.1927	.2760	.3594	.4427
3/8"	.0312	.1146	.1979	.2812	.3646	.4479
7/16"	.0365	.1198	.2031	.2865	.3698	.4531
1/2"	.0417	.1250	.2083	.2917	.3750	.4583
9/16"	.0469	.1302	.2135	.2969	.3802	.4635
5/8"	.0521	.1354	.2188	.3021	.3854	.4688
11/16"	.0573	.1406	.2240	.3073	.3906	.4740
3/4"	.0625	.1458	.2292	.3125	.3958	.4792
13/16"	.0677	.1510	.2344	.3177	.4010	.4844
7/8"	.0729	.1562	.2396	.3229	.4062	.4896
15/16"	.0781	.1615	.2448	.3281	.4115	.4948

	6"	7"	8"	9"	10"	11"
0"	.5000	.5833	.6667	.7500	.8333	.9167
1/16"	.5052	.5885	.6719	.7552	.8385	.9219
1/8"	.5104	.5937	.6771	.7604	.8437	.9271
3/16"	.5156	.5990	.6823	.7656	.8490	.9323
1/4"	.5208	.6042	.6875	.7708	.8542	.9375
5/16"	.5260	.6094	.6927	.7760	.8594	.9427
3/8"	.5312	.6146	.6979	.7812	.8646	.9479
7/16"	.5365	.6198	.7031	.7865	.8698	.9531
1/2"	.5417	.6250	.7083	.7917	.8750	.9583
9/16"	.5469	.6302	.7135	.7969	.8802	.9635
5/8"	.5521	.6354	.7188	.8021	.8854	.9688
11/16"	.5573	.6406	.7240	.8073	.8906	.9740
3/4"	.5625	.6458	.7292	.8125	.8958	.9792
13/16"	.5677	.6510	.7344	.8177	.9010	.9844
7/8"	.5729	.6562	.7396	.8229	.9062	.9896
15/16"	.5781	.6615	.7448	.8281	.9115	.9948

To practice using the decimal conversion table, supply the coordinate, in feet and inches, for problems 1 through 10. Correct answers are found in Table 5-2.

1.	N.123.75′	6.	E.10073.3843′
2.	E.736.27′	7.	N.593.8671′
3.	N.428.79′	8.	S.4328.9313′
4.	W.177.43′	9.	S.5277.7433.′
5.	S.377.12′	10.	W.384.7992′

Often, the draftsman may know two coordinates and need to know the distance between them. If the smaller coordinate is subtracted from the larger one, the difference is the distance between them. It must be noted that this applies only to like coordinates, such as subtracting a north coordinate from another north coordinate. See examples 11 and 12.

11. N.1824.73′
−N.1638.92′
185.81′
= 185′-9¾″

12. E.278.3724′
−E.123.7269′
154.6455′
= 154′-7¾″

The reverse of this method is used to establish a coordinate when one coordinate is known and the distance between the known and unknown coordinate is known. If the known coordinate is N.183.5622′ and the known dimension is 13′-2⅜″ north of this coordinate, then

1. Convert 13′-2⅜″ to feet and decimals of a foot. Using Table 5-1, 13′-2⅜″ equals 13.1979′.
2. Since 13′-2⅜″ is north of the known coordinate, the answer must be larger than the N.183.5622′. In other words, the coordinate is farther north from the bench mark. So 13.1979 must be added to the known coordinate.

13. N.183.5622′
\+ 13.1979′
N.196.7601′ is the unknown coordinate

Had the known dimension, 13′-2⅜″, been south of the known coordinate, N.183.5622′, we would be getting closer to the bench mark, so the dimension, 13.1979′, would have to be subtracted from the known coordinate.

14. N.183.5622′
− 13.1979′
N.170.3643′ is the unknown coordinate south of the known one.

Normally a plant will not have both north and south coordinates. Occasionally they will, so the student must remember that each coordinate is a dimension a certain distance from the bench mark. If N.18′-0″ is 18′-0″ north of the bench mark then S.18′-0″ is 18′-0″ south of the same bench mark. So the dimension between N.18′-0″ and S.18′-0″ is 36″-0″. Here, the rule is, *to find the distance between north and south coordinates, add the two coordinates.*

15. N.138.6623′
\+S. 36.5277′
175.1900′=175′-2¼″, the dimension between coordinates.

The same rule applies when both east and west coordinates are given in the same plant.

North and south coordinates have been used in this chapter but they could have just as easily been east and west coordinates. The same rules apply. A dimension 10′-6″ east of E.18.5′ would make the unknown coordinate

16. E.18.5′
\+ 10.5′ (10′-6″ converted to decimals)
\+E.29.0′

Since the 10′-6″ dimension was east of the known, coordinate, the unknown coordinate was further east of the bench mark, so 10′-6″ is added to get a larger east coordinate than E.18.5′. Had the dimension 10′-6″ been west of E.18.5′, it would be closer to the 0.0′ bench mark so it would have to be subtracted from E.18.5′.

17. E.18.5′ known coordinate
− 10.5′ known dimension
E. 8.0′ unknown coordinate

Table 5-2. Answers To Problems 1-10

1.	N.123′-9″	6.	E.10073′-4⅝″
2.	E.736′-4 7/16″	7.	N.593′-10⅜″
3.	N.428′-9½″	8.	S.4328′-11 3/16″
4.	W.177′-5 3/16″	9.	S.5277′-8 5/16″
5.	S.377′-1 7/16″	10.	W.384′-9 9/16″

Coordinate Review Test

Supply the proper answer for problems 18 through 37.

18. Convert E.483.2768′ to feet and inches.
19. Convert N.128′-4 15/16″ to feet and decimals of a foot.
20. Supply the coordinate for a line 10′-11¼″ north of N.3658.2373′. Express the answer in feet and decimals of a foot.
21. Supply the coordinate for a line 26′-3⅜″ east of E.2742.5763′. Express the answer in feet and decimals of a foot.
22. Supply the coordinate for a line 7′-5 11/16″ west of E.127.36′. Express the answer in feet and decimals of a foot.
23. Supply the coordinate for a line 32′-7¾″ south of N.1375.3822′. Express the answer in feet and decimals of a foot.
24. In feet and inches, supply the dimension between coordinates N.132.3718′ and N.163.27′.
25. In feet and inches, supply the dimension between coordinates E.372.37′ and E.480.00′.
26. In feet and inches, supply the dimension between coordinates N.372.213 and S.32.397′.
27. In feet and inches, supply the dimension between coordinates E.37.375′ and W.1.225′
28. Supply the dimension between coordinates E.138′-5⅝″ and E.117′-6 15/16″. Express the answer in feet and inches.
29. Supply the dimension, in feet and inches, between coordinates N.128′-3⅛″ and S.31′-5 15/16″.
30. Supply the coordinate, in feet and decimals of a foot, for a line that is 62′-8 3/16″ south of N.12.3722′.
31. Supply the coordinate for a line 68′-4″ east of W.42.7878′. Express the answer in feet and decimals of a foot.
32. Supply the coordinate for a line 17′-3 5/16″ west of W.127.2727′. Express the answer in feet and decimals of a foot.
33. In feet and inches, supply the dimension between coordinates S.373.2852′ and S.528.3773′.
34. In feet and inches, supply the dimension between coordinates E.27.3233′ and W.11.3321′.
35. Supply the coordinate for a line that is 63′-10⅜″ west of E.27.3333′. Express the answer in feet and decimals of a foot.
36. Supply the coordinate for a line that is 12′-3 13/16″ east of E.37.5212′. Express the answer in feet and decimals of a foot.
37. Supply the coordinate for a line 67′-3¼″ north of S.18.1755′. Express the answer in feet and decimals of a foot.

CHAPTER 6

Welding Details and Drawings

Welding Defined

When two metal pieces are fused together by heat the complete process is called welding. Most welding is accomplished with a *filler metal,* called *welding rod,* or an *electrode.* Welds have many uses. *Tack welds* are preliminary welds, very small and short, used for aligning two or more metal pieces prior to making the strength weld. Tack welds do not show on drawings. *Strength welds* are the type commonly seen. These are made by various methods, but their purpose is to transfer stresses, shear forces and tension and compression loads from one metal piece to another. In theory, the weld is stronger than the parent metal, pieces being joined. When subjected to controlled destruction testing the parent metal gives before the weld metal.

Weld Testing

Destructive Tests

Welds are tested—thus, testing the welders proficiency—by two basic means: *destructive* and *non-destructive* testing. As the name implies, the destructive test actually destroys the metal. Destructive testing is conducted on samples of like material and welding procedures, not on the actual strength weld.

Welders new on a jobsite must pass qualification tests to make welds of various materials. They must weld two *coupons,* small pieces of steel, which are then bent until the metal or weld material actually cracks or gives. If the weld cracks before the metal, the welder doesn't get the job. If the metal is first to go, the welder "qualifies" and is issued a paper of approval for welding that type of metal. The welder is qualified to perform this weld on this job. The welder must requalify for each material specification welded. For instance, someone may be qualified to weld carbon steel, Type 304 stainless steel and Type 316 stainless steel, but not qualified to weld 5% chrome-½% molybdenum steel.

Non-destructive Tests

Non-destructive testing is performed on the actual weld and doesn't harm it in any manner. One non-destructive weld test commonly used is X-ray. An X-ray is made on film, much as people have lung and bone X-rays made. Any *voids* will show up. Voids appear as white lines or lighter areas and indicate incomplete weld areas where air was entrained. Under stress these voids would give and the weld would shear. An incomplete weld would fail the test and, if the voids were small, could be chipped out, rewelded and retested. Pipe buttwelds usually are tested by X-ray, shooting a picture of one out of ten welds. Sometimes 10% of each weld is X-rayed. There are many other non-destructive tests used, such as dye penetrant and magnetic particle, but this chapter cannot cover them all.

Pressure Tests

The destructive and non-destructive tests discussed are non-pressure tests which can be conducted on welds which cannot be pressurized. Pressure tests are conducted on vessels which can contain pressure, such as closed piping systems, pressure vessels, exchangers and heater tubes. Pressure tests conducted with water are called *hydrostatic* tests. In hydrostatic testing the item is filled with water or another liquid (an anti-freeze solution in freezing climates) and pressure of at least one and one-half times the design pressure is applied. The item must contain this pressure for a set length of time to pass the test. Since liquid doesn't compress, a very slight weld leak will quickly cause a loss in system pressure and be noted. Should a rupture occur, water will almost immediately attain normal atmospheric pressure and not cause an explosion.

Pneumatic pressure testing uses air, which does compress. Because air will compress, large volumes must be induced into the item to elevate pressure. Should a rupture occur the explosion sjze will be on a ratio of the amount of air contained in the tested item. Because of this hazard, pneumatic testing is usually limited to 100 pounds per square inch gage (psig) pressure, containers of relatively small volume and 1.1 times design pressure is used as the test pressure.

When a system operates at a vacuum, less than atmospheric pressure, pressure testing is still used. Hydrostatic or pneumatic internal pressure tests are still used to test weld quality, but test pressure is usually limited to 15 psig.

Welding Processes

Manual Shielded Metal-Arc

Manual shielded metal-arc welding is the most commonly used welding process in fabrication shops and at job sites where automatic welding is prohibited by the size or shape of items. This welding process is most often referred to as *manual, hand* or *stick* welding. The *stick* refers to the electrode, or welding rod, a small piece of filler metal with a diameter smaller than a pencil and about 5-6 inches long. The electrode is chemically coated. During the welding process the coating melts, forming a flux and a gas shield. The flux in the coating forms a slag covering over the deposited metal and purifies it. The portion of the coating that vaporizes forms a gas shield over the area being welded, preventing the absorption of impurities from the air which would reduce weld strength.

Welding is accomplished when an electric arc is produced between the electrode end and the steel components being welded. This arc heats the base metal and the electrode to the point where they both melt, forming a metal molten pool on the surface. As the arc is moved along the surface, the molten pool solidifies behind it as it cools, forming a weld and actually fusing the two metal pieces together. Effectively the two metal pieces have become one piece.

So the term "manual shielded metal-arc" welding means a weld done by hand, shielded by a gas formed by the electrode cover vaporizing, using a metal filler material and forming an electric arc. Figure 6-1 pictures this common welding process.

Submerged-Arc

Submerged-arc welding, a process similar to manual shielded metal-arc welding, is performed mostly in fabrication shops by automatic or semi-automatic welding machines. Submerged-arc welding employs a bare electrode, and flux is introduced separately. In the automatic operation a machine feeds both welding rod and flux to the joint by separate nozzles.

The automatic machine and the items to be welded are controlled by instumentation. The welding point remains constant, and the metal pieces being welded slowly move as the weld is made. Modern pipeline companies now use a traveling submerged-arc welder. The ends of the two pieces of pipe are butted together and held in place. The submerged-arc welding machine is attached; it welds and travels at the same time until the complete circle is made and the joint is complete. The machine is removed and transported to the next joint to be welded, usually 40 feet down the line.

The semi-automatic process needs a welder-operator to operate a hand-guided device which automatically feeds weld rod and flux through one nozzle.

Submerged-arc welding makes a deep weld penetration and does it faster, consuming less

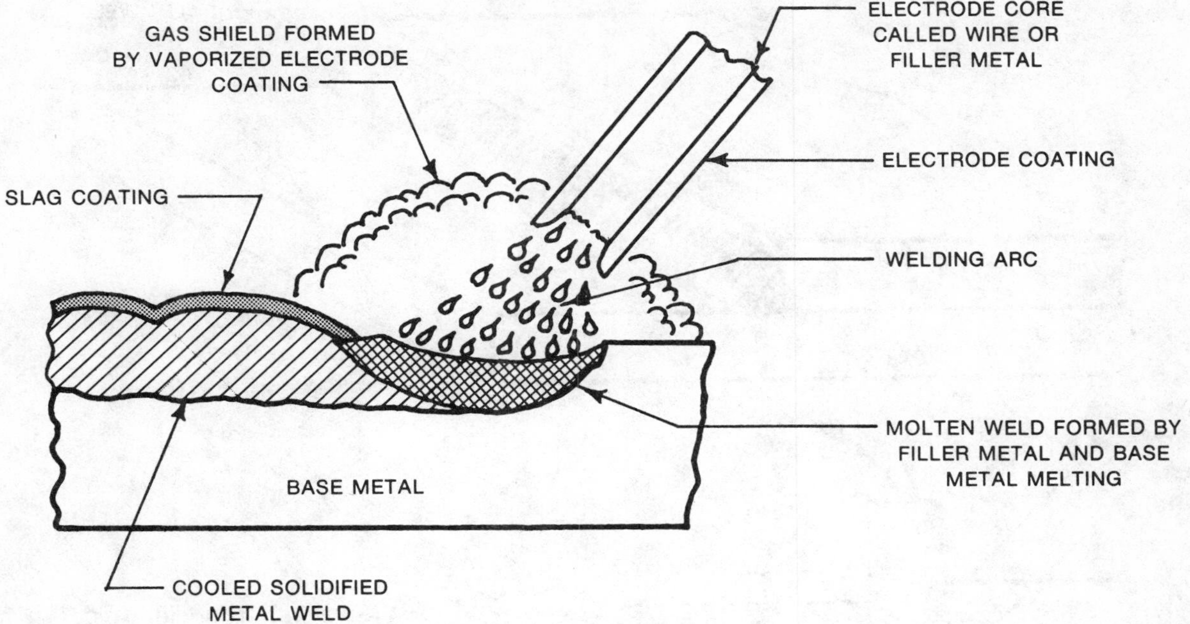

Figure 6-1. Manual shielded metal-arc weld.

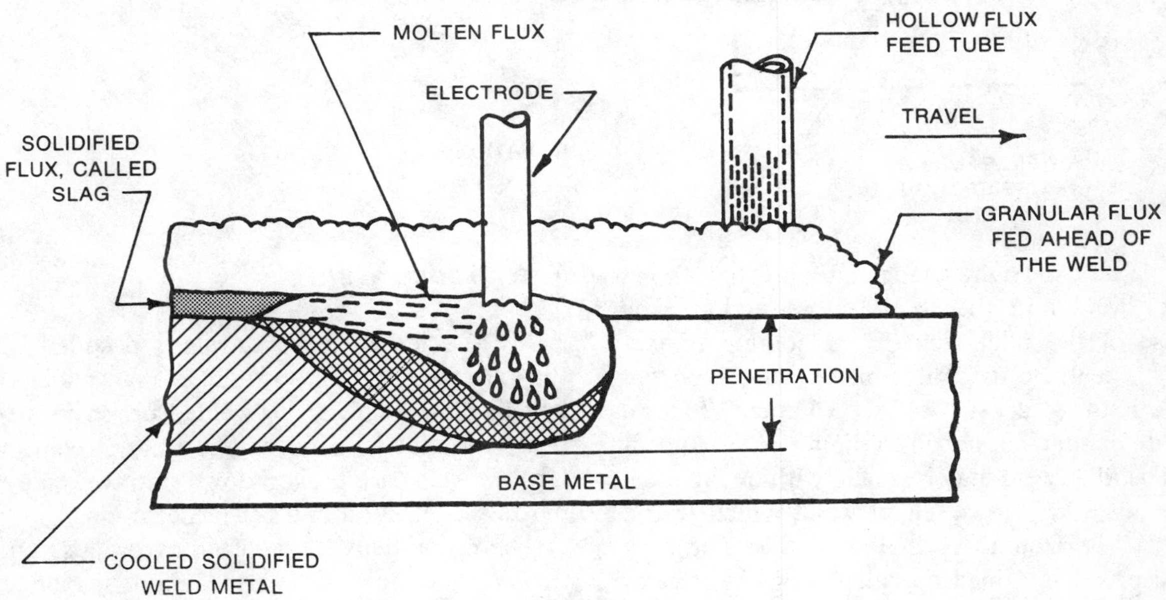

Figure 6-2. Submerged-arc welding.

manhours, than the manual shielded metal-arc process. Consequently, the submerged-arc weld is less expensive. Why, then, aren't more or all welds mady by this process? The welding machine with its fully instrumented controls is just not as mobile as the human who makes manual welds. Figure 6-2 shows the submerged-arc welding process. Note that the penetration is very deep.

There are many other processes used in welding, each having their own particular use. As a weld gets thicker a different process is needed. Some special material specifications require special processes.

Strength Weld Types

The Fillet Weld

A weld's cross-section identifies its type. One common type is *fillet weld*, which has a triangular cross-section. A ½″ fillet weld will have two ½″

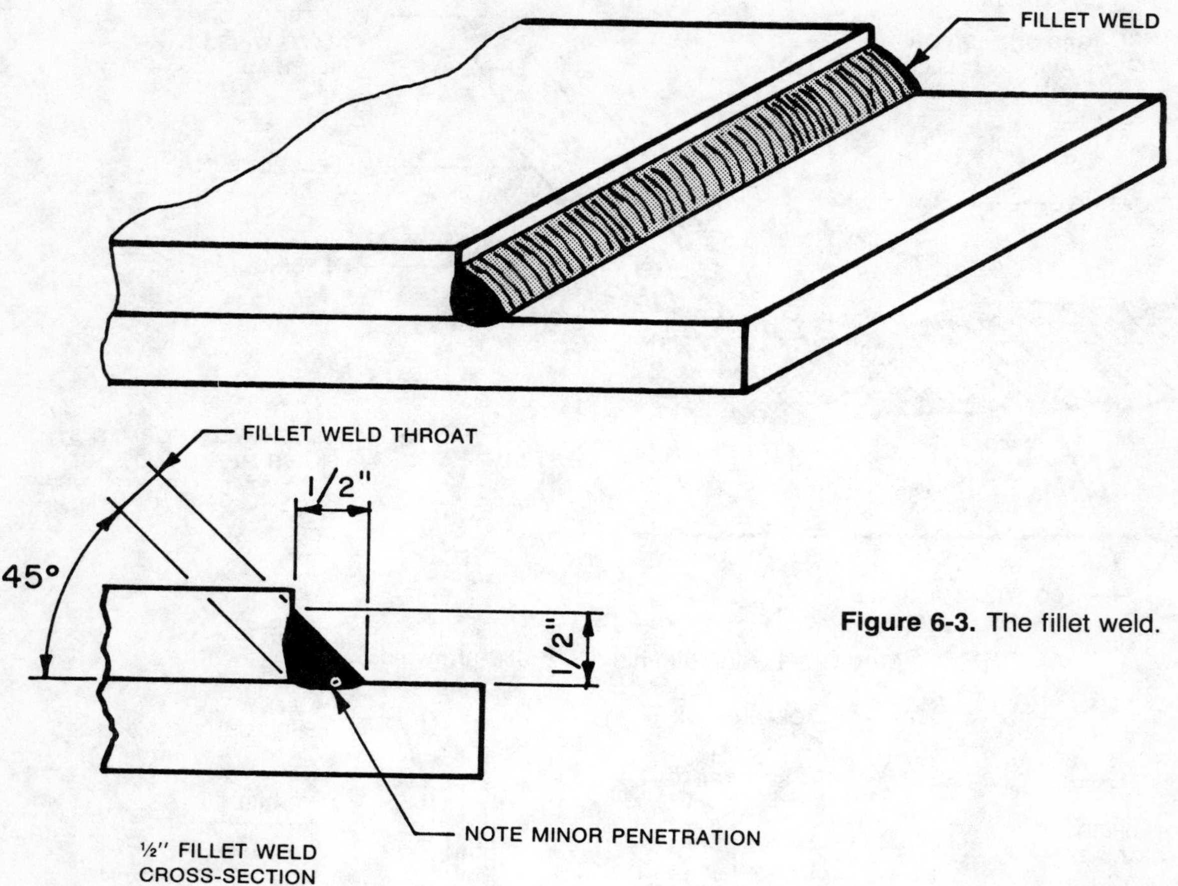

Figure 6-3. The fillet weld.

legs as part of its right triangle. Figure 6-3 shows the fillet weld and the weld's cross-section. The main use of the fillet weld is the joining of two structural steel pieces. These pieces might be flat plate, structural steel shapes (see Chapter 11) or a steel plate or shape attached to piping or equipment items. The fillet weld may be made with unequal leg lengths, say a ½″ × ¾″ fillet weld, which would have a ¾″ horizontal leg in Figure 6-3. The fillet weld's size is determined by calculating the stresses and forces applied to the joint and supplying a weld with a cross-sectional area large enough to withstand them.

The Plug Weld

Plug welds are used to join overlapping steel members and resist shear loads. Plug welds are round and fill a hole cut in one of the members being joined. Figure 6-4 pictures the plug weld and its cousin the slot weld, an elongated plug weld specified for high shear forces.

The Groove Weld

Most welds are groove welds, so called because they are made in a groove either man-made by prebeveling the edges to be welded or, as in a square butt weld, cut by torch as the weld is being made. Groove welds are broken down into two categories: partial penetration and full penetration.

There are many kinds of groove welds. The particular kind depends on the configuration of the pieces being welded. Figure 6-5 pictures several kinds of full penetration groove welds.

Weld Prequalification

The AWS (American Welding Society) publishes complete welding specifications, procedures and technical drawing weld symbols. Joints which conform to all AWS procedures and specifications are called *prequalified joints*. These joints are used in all steel fabrication in the United States and in many other countries.

(text continued on page 44)

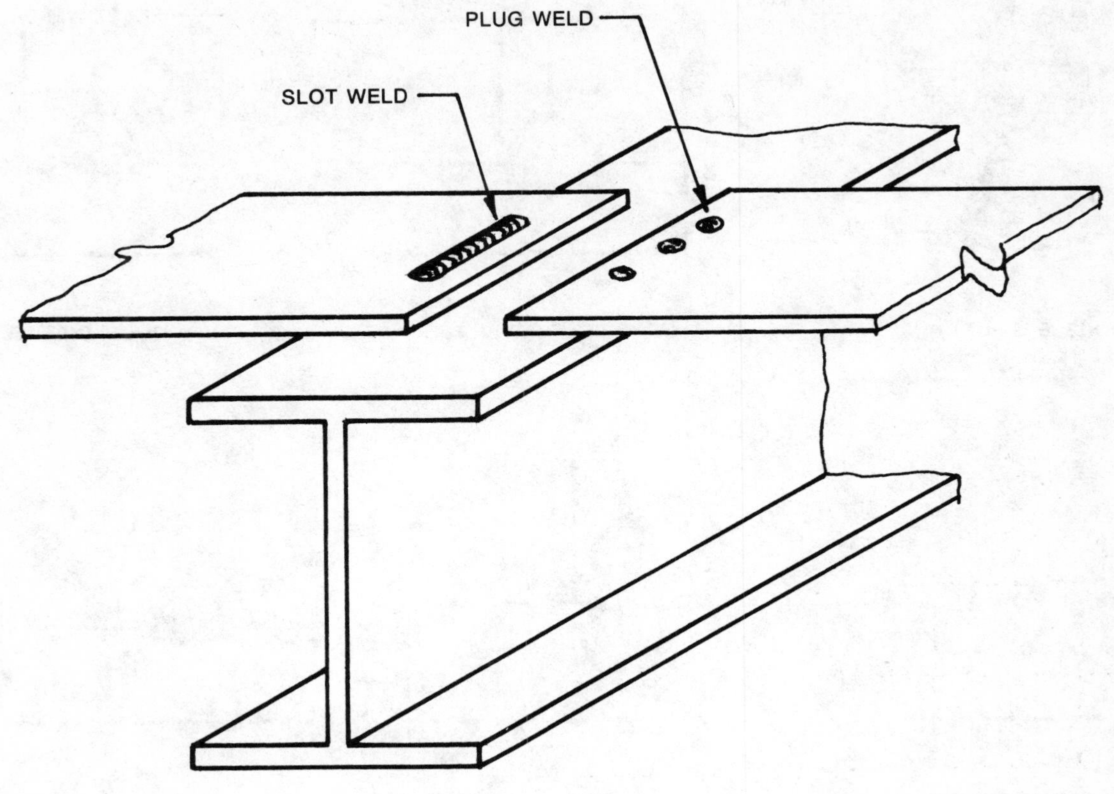

Figure 6-4. The plug and slot weld.

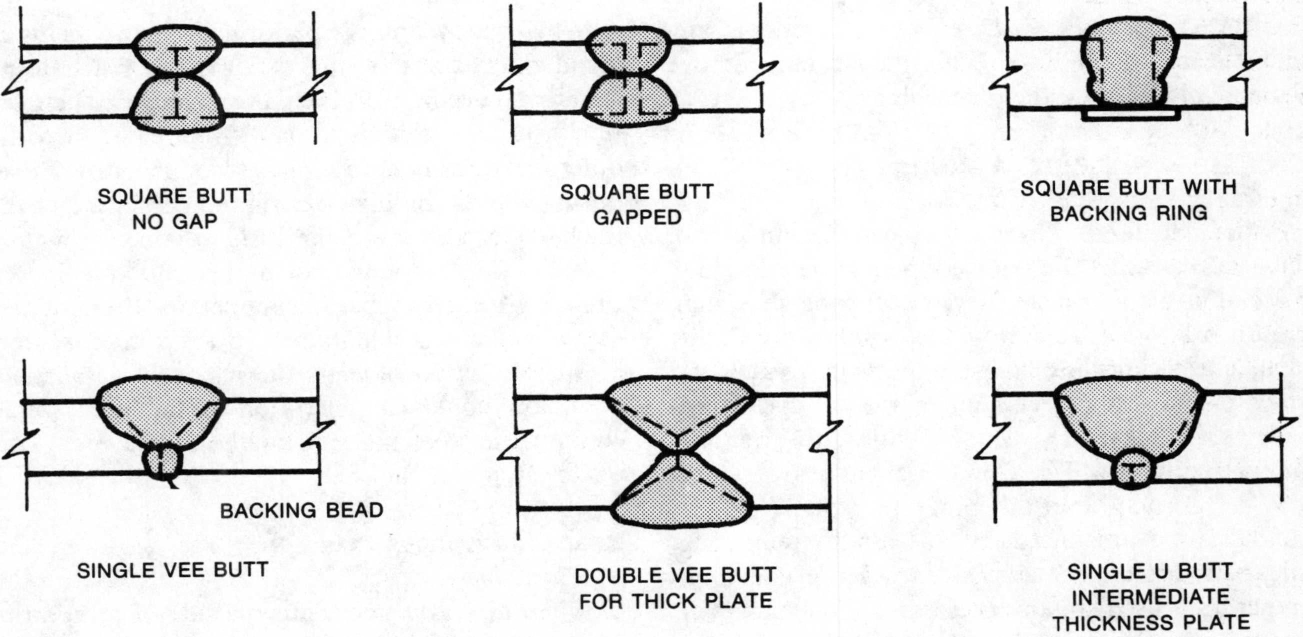

Figure 6-5. Full penetration groove welds.

(table continued on next page)

Figure 6-5 continued

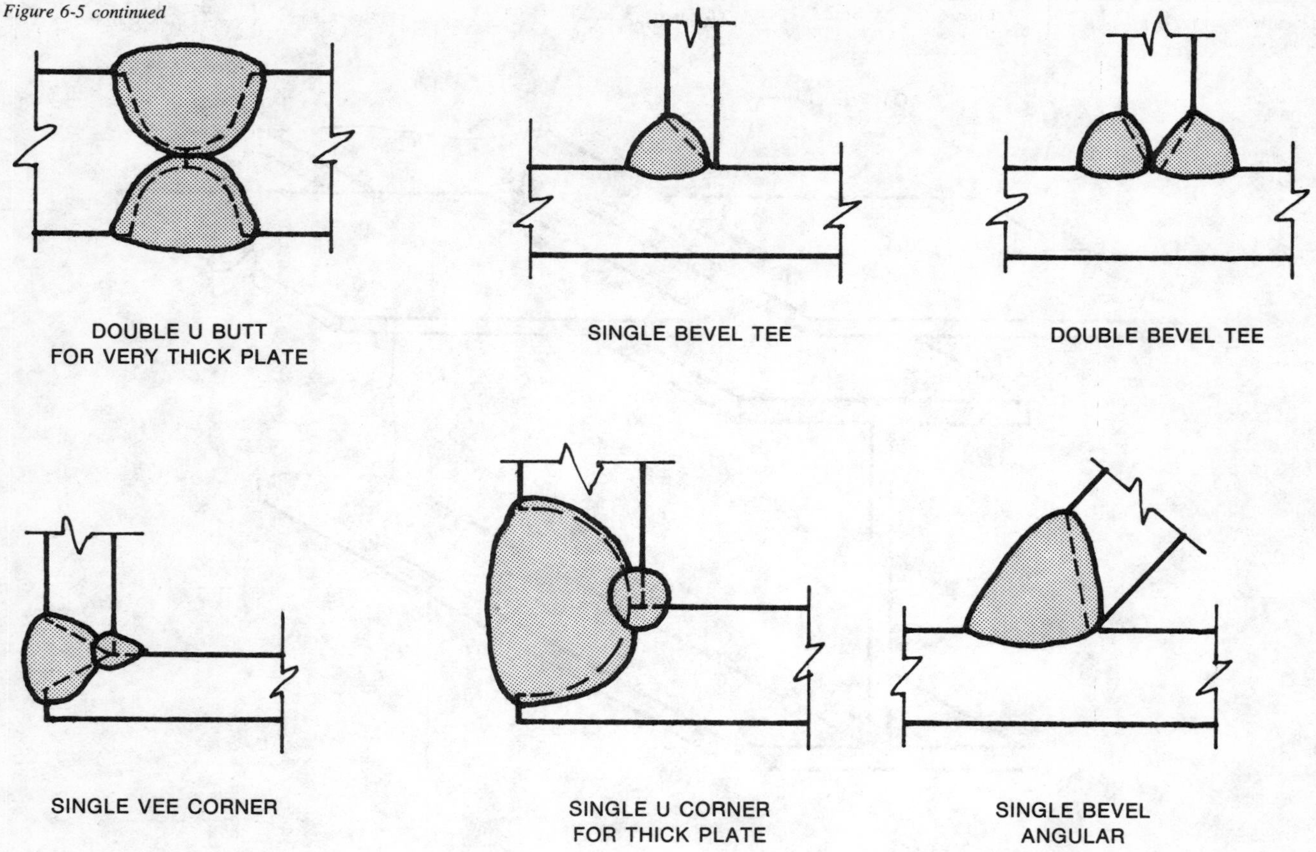

DOUBLE U BUTT
FOR VERY THICK PLATE

SINGLE BEVEL TEE

DOUBLE BEVEL TEE

SINGLE VEE CORNER

SINGLE U CORNER
FOR THICK PLATE

SINGLE BEVEL
ANGULAR

The AWS technical drawing weld symbols are pictured in Figure 6-6. The student must know these symbols and their meanings.

Pipe Welding

When welding cylindrical objects together "on the run" (axially) the groove type buttweld is used. Normally this is the single vee butt type shown in Figure 6-5. This weld would be used for welding pipe and fittings, exchangers, pressure vessels or other cylindrical objects. When the weld is made with very thick plate, over ¾″ thick, the welding end preparation will be changed slightly. Instead of a 37½° normal bevel the joint will have 37½° for the inside ¾″ thickness and 10° for the remaining thickness. Figure 6-7 shows buttwelding end joint preparation used for thickness of ¾″ and for ¾″ and greater. Several end preparations are used for thicknesses greater than ¾″. Refer to Figure 6-5 for other types.

When buttwelding is performed on two items of equal thickness a normal buttweld is made using standard beveling for joint preparation. When the thickness of one item is over 1/16″ greater than the other thickness *taper boring* is recommended. Taper boring is the grinding or boring of the inside of the thicker piece to make the inside diameters match. Taper boring is done on an angle of 30° maximum. Figure 6-8 shows taper boring details. Inside diameter alignment is maintained to keep internal flow as smooth as possible. If the inside is misaligned, fluid flow could cause erosion at the joint which could cause wear and eventually induce stress corrosion attack at the joint.

Branch Connections

When pipe is welded into the side of a pipe run this pipe is called a branch or sometimes a nozzle. Hence the name *branch weld* or *nozzle weld*. Figure 6-9 shows a typical branch or nozzle weld. This is

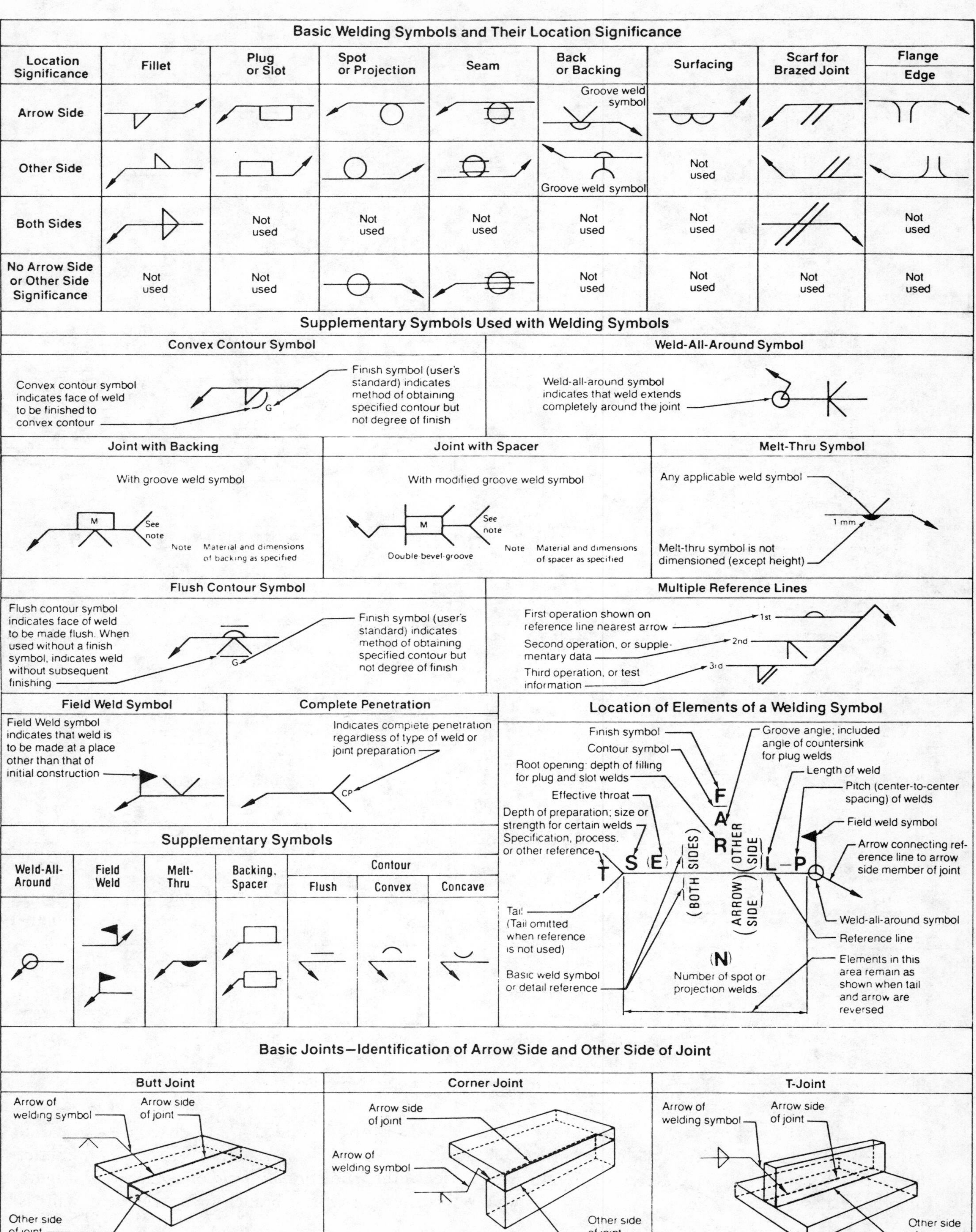

Figure 6-6. Welding symbols. (Courtesy of Texas Pipe Bending Co., Inc.)

Figure 6-6 continued

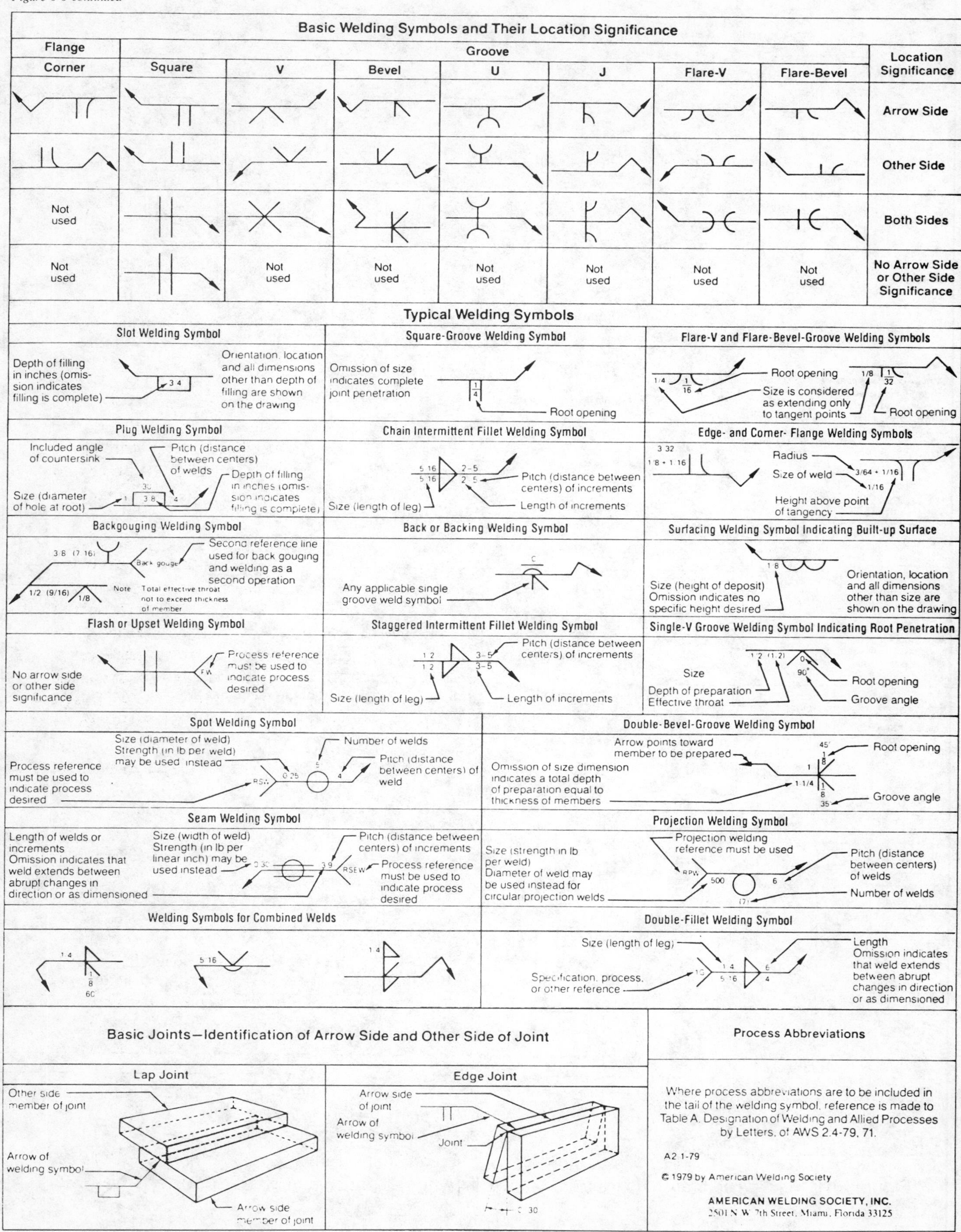

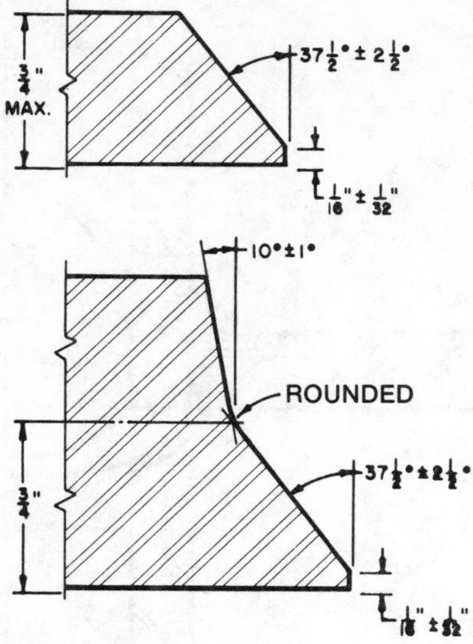

Figure 6-7. Joint preparation.

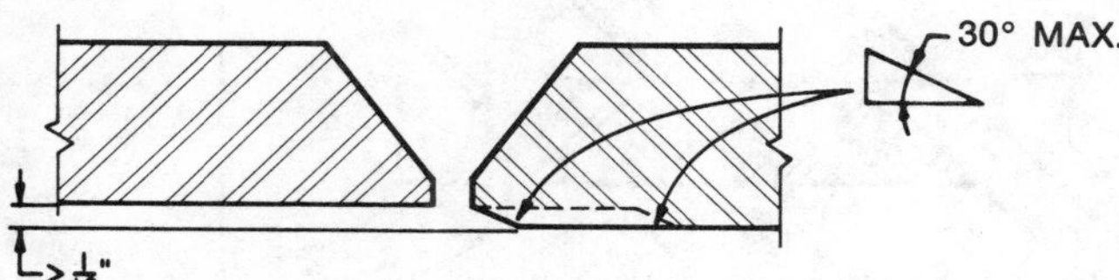

Figure 6-8. Taper boring detail.

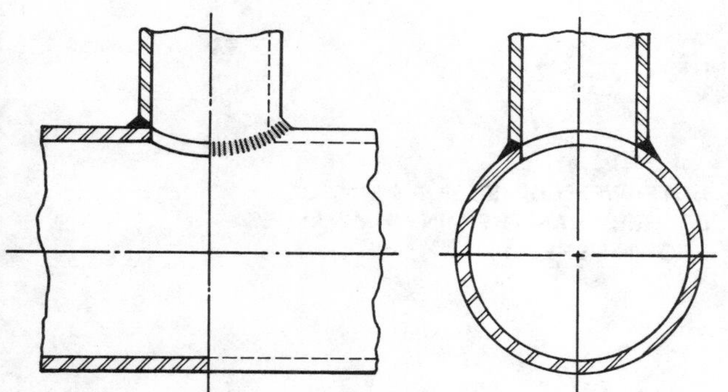

Figure 6-9. Typical branch connection.

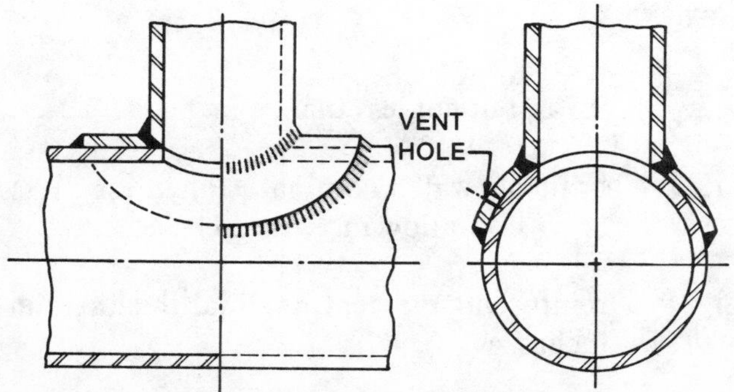

Figure 6-10. Reinforced branch connection.

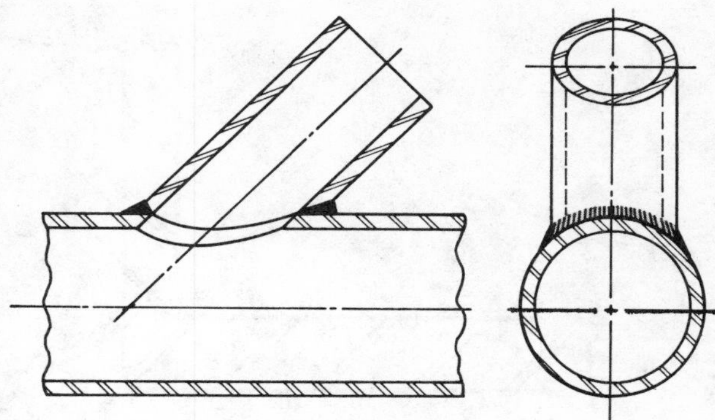

Figure 6-11. Angular branch connection.

(continued from page 44)

called an ID to OD type where the inside diameter of the branch rests on the outside diameter of the header. In this example the branch is pictured as a smaller size than the header. Branches may be equal to or smaller than the header size, but cannot be larger than the header size.

Figure 6-9 pictures an unreinforced branch connection. When pressures are moderate to high many larger branch connections must be reinforced. Figure 6-10 shows an ID to OD branch connection which has been reinforced. The connections are reinforced by the addition of a *reinforcing pad*, a piece of shaped metal cut from the header material. Also available is the *weld saddle*, a commercially manufactured reinforcement fitting which is similar in shape to the reinforcing pad, but has a neck that rises up on the branch pipe a few inches. Both of these reinforcement methods have a diameter of 2D, twice the branch nominal diameter. When actual engineering calculations are performed, the reinforcing pad may be less diameter than 2D, but usually it is more expensive to calculate the exact cross-sectional area needed than to use 2D as a diameter.

When branch or nozzle connections are not 90° to the header axis the welding is more difficult and more expensive. The angular branch connection is shown in Figure 6-11. Any angle may be required, but the usual angle is 30° or 45° from the axis of the header.

Minimum Branch Weld Sizes

Figure 6-12 shows ID to OD and OD to ID branch weld minimum sizes. Detail *a* is an ID to

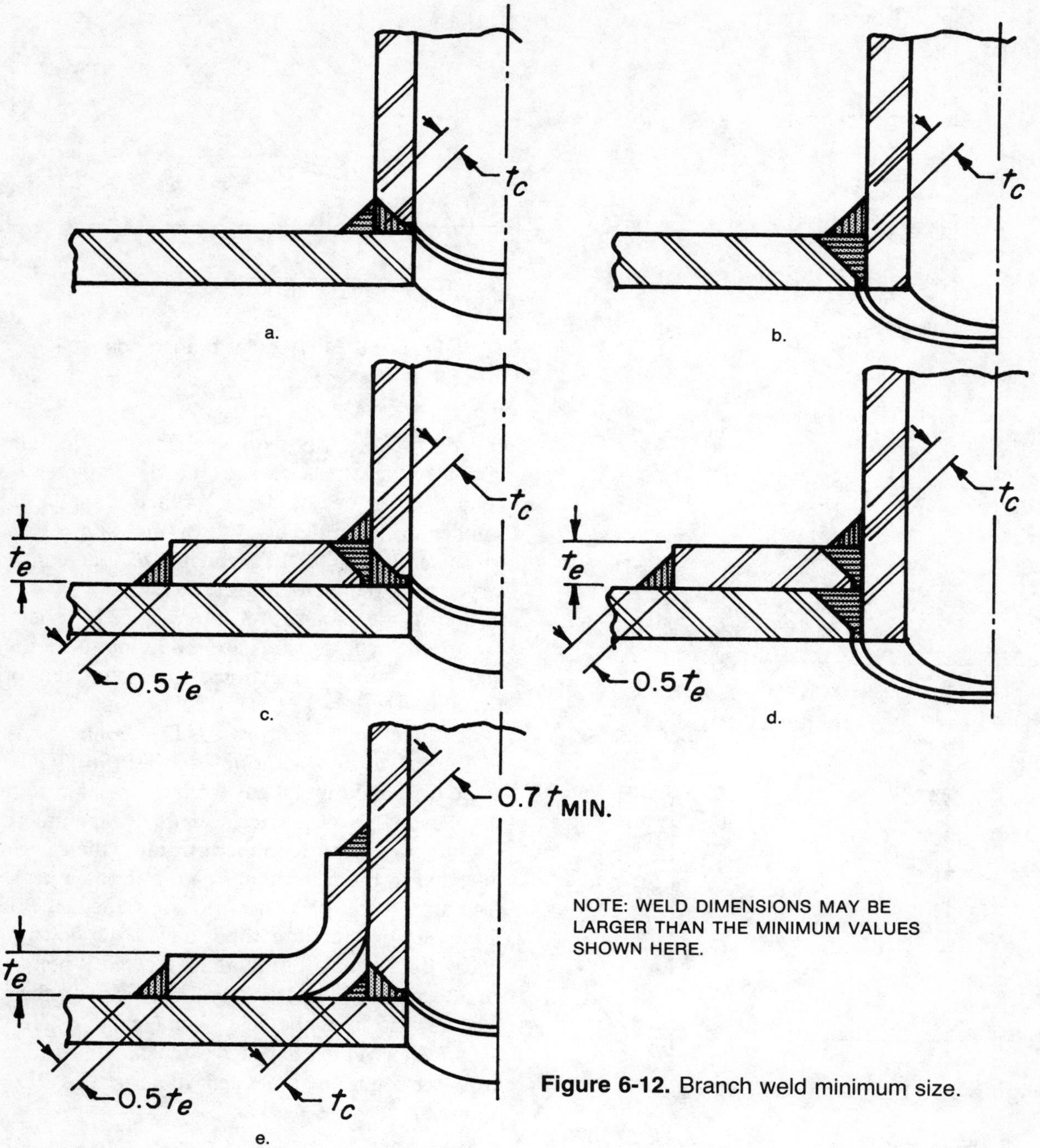

NOTE: WELD DIMENSIONS MAY BE LARGER THAN THE MINIMUM VALUES SHOWN HERE.

Figure 6-12. Branch weld minimum size.

OD unreinforced branch connection. Detail *b* is an OD to ID unreinforced branch connection. Detail *c* is an ID to OD branch connection reinforced with a reinforcing pad cut from the header material or from steel plate of similar material. Detail *d* is an OD to ID branch connection with a reinforcing pad. Detail *e* is an ID to OD branch connection reinforced with a weld saddle. The notations used in Figure 6-12 are:

t_c = 0.7 t_n, but not less than ¼ inch.

t_n = branch wall nominal thickness less corrosion allowance, in inches.

t_e = reinforcing element nominal thickness in inches.

t_{min} = the smaller of t_n or t_e.

Review Test

1. Welding is the fusing together of two ____________.

2. A welding rod is also called ____________.

3. Define the purpose of tack welds.

4. Define the purpose of strength welds.

5. Name the two basic types of weld testing.

6. Define a weld coupon.

7. Define a void in a weld.

8. Define hydrostatic testing.

9. Why is hydrostatic testing preferred over pneumatic testing?

10. Define "stick".

11. What is the purpose of flux?

12. Define manual shielded metal-arc welding.

13. A fillet weld is one that fills a hole with weld metal. (True of false).

14. Define AWS.

Define the following weld symbols:

15.

16.  ½

17. 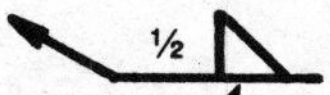 ½

18.  ¼

19.  ¼

20.

21.

22. 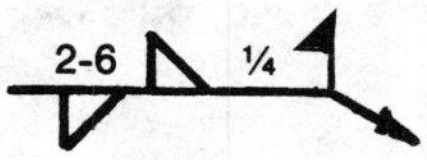 2-6 ¼

23. 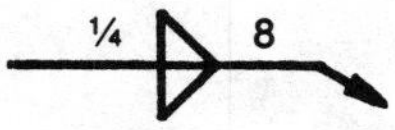 ¼ 8

24.

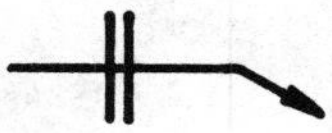

25.  12

26. Define taper boring.

27. Define OD to ID branch connection.

28. Define ID to OD branch connection.

29. Name 2 items used to reinforce branch connections.

30. Normally branch reinforcement has an OD of 2D. Define 2D.

CHAPTER 7

Reading Perspective and Isometric Drawings

Of the many types of drawings produced, the easiest type of print to read is the pictorial drawing. The pictorial drawing encompasses several methods of presentation. Real estate people are used to seeing perspective drawings of houses, shopping centers and tall buildings.

Perspective Drawings

These drawings are based on the fact that, as the observer views an object, all receding lines extended far enough, appear to converge at some distant point. A railroad track is a good example. If one stands in the center of a straight railroad track and looks at the rails going to the horizon, the rails appear to converge at some far off point. This point, where all lines seem to meet, is called the vanishing point (V.P.). The vanishing point is located at a distance but at eye level. Lines below eye level appear to run upward to the vanishing point and lines above eye level appear to go downward to the vanishing point on the horizon. Perspectives may be drawn so the vanishing point is either to the right, left, or directly in the center of vision. Figure 7-1 illustrates the six usual perspective views converging to a single vanishing point. This is called a *parallel perspective* drawing.

Perspective drawings are also made with two vanishing points. These are called *angular* or *two-point perspective* drawings. This is because lines in such drawings recede to two vanishing points. The front face is not drawn flat as in the parallel perspective. Figure 7-2 shows how the basic angular perspective is constructed. Figure 7-3 is a residential drawing made using the angular perspective method. This type of drawing is very much like a picture and is easily read by the public. The main purpose of these residential type perspectives is for a sales tool, a picture to show prospects how the finished house will look. These drawings are drawn from a vantage point that shows the most attractive features of the house. Trees and shrubs are usually added to enhance sales.

Figure 7-4 shows the house plan used with the perspective drawing. This, also, is a sales drawing. Only room sizes are given and even some of them are not provided. No overall dimensions are shown and windows are not located or defined. This is not a working drawing. Basically, it is a picture of the house plan.

Isometric Objects

Another pictorial drawing is the *isometric*. Isometric drawings, often sent to shops for construction drawings, are constructed by using *three axes*, one vertical and two inclined 30° to the right and left. Vertical lines appear on the vertical axis but horizontal lines will appear on the 30° axes. Figure 7-5 shows isometric construction lines. Isometric objects can be drawn to any scale, but the lines are all

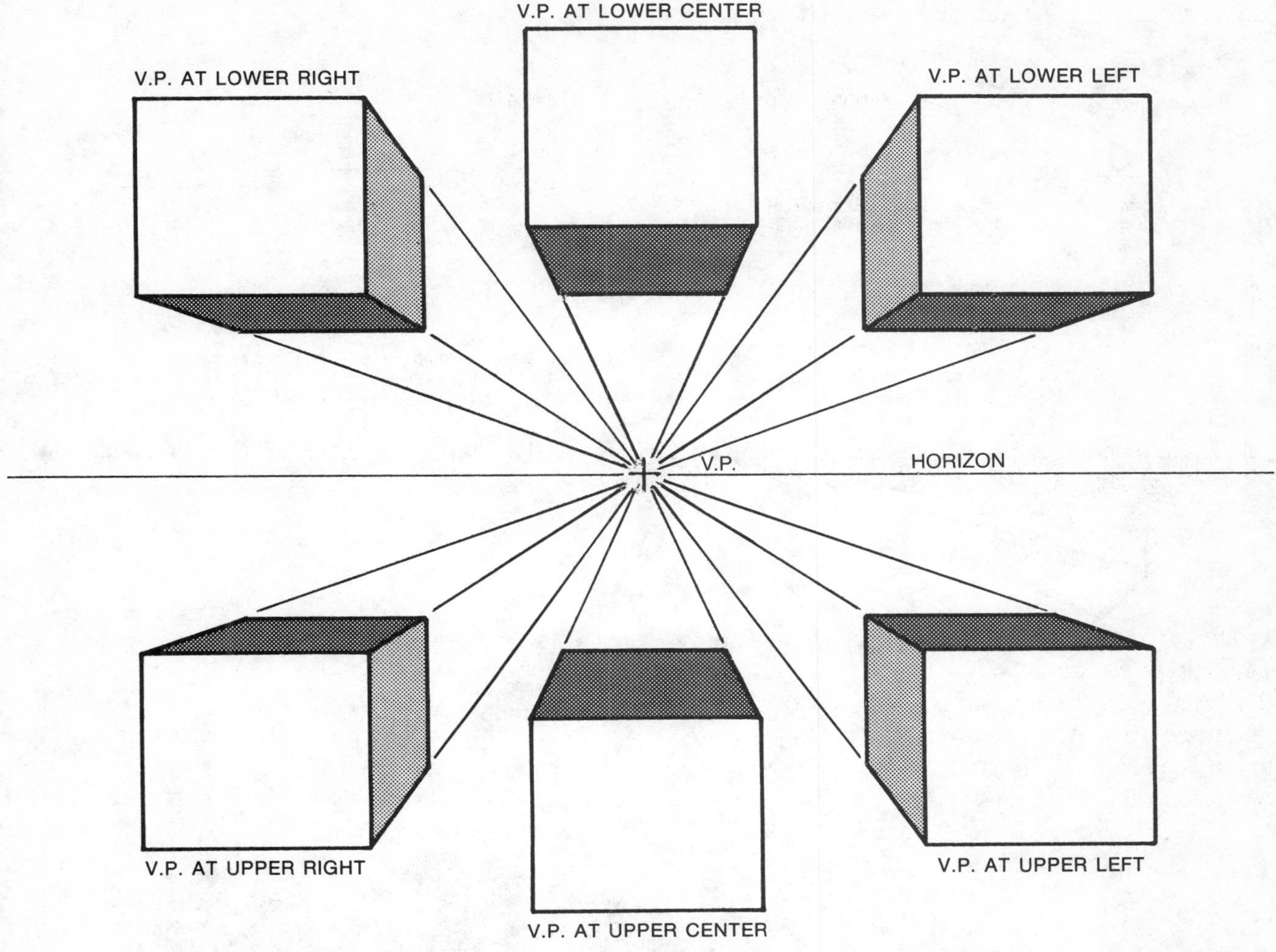

Figure 7-1. The six positions of the vanishing point in parallel perspective sketching.

drawn to the same scale. Figure 7-6 shows how a filing cabinet is drawn in isometric form and to scale.

Do not confuse the isometric with the two-point perspective. The isometric is made with horizontal lines on a 30° inclined line, while the two-point perspective has horizontal lines running to a vanishing point. Figure 7-7 shows the difference in these two drawing types.

Circles and Arcs in Isometric Drawings

In isometric presentation circles and arcs will appear as ellipses and sections of ellipses. Professional isometricians have templates that draw these ellipses and partial ellipses. All horizontal circles become horizontal ellipses with the major axis remaining horizontal. Vertical ellipses depict vertical circles, but their major axes are not drawn on the 30° line. It is actually 35°-16′ from the horizontal. Figure 7-8 shows the isometric ellipses used for drawing circles in isometric drawings. Figure 7-9 shows arcs in isometric drawings.

Nonisometric Lines

Lines that are not horizontal or vertical are called *nonisometric lines* and cannot be drawn at 30° or vertical in an isometric, but they are still drawn to actual scale. Figure 7-10 shows how these

(text continued on page 56)

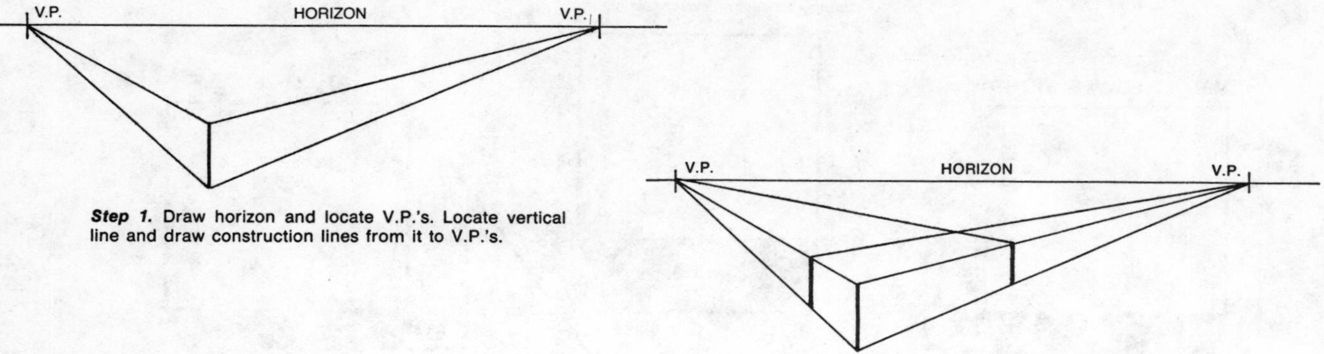

Step 1. Draw horizon and locate V.P.'s. Locate vertical line and draw construction lines from it to V.P.'s.

Step 2. Establish depth vertical lines and draw construction lines to vanishing points from them.

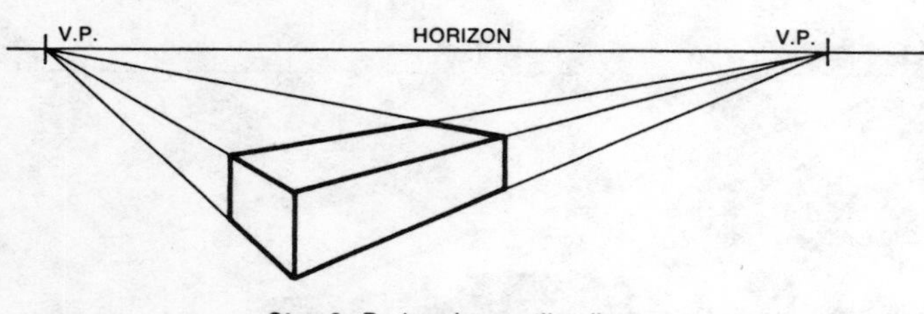

Step 3. Darken box outline lines.

Figure 7-2. How to construct the angular perspective drawing.

Figure 7-3. Residential drawing, angular perspective type.

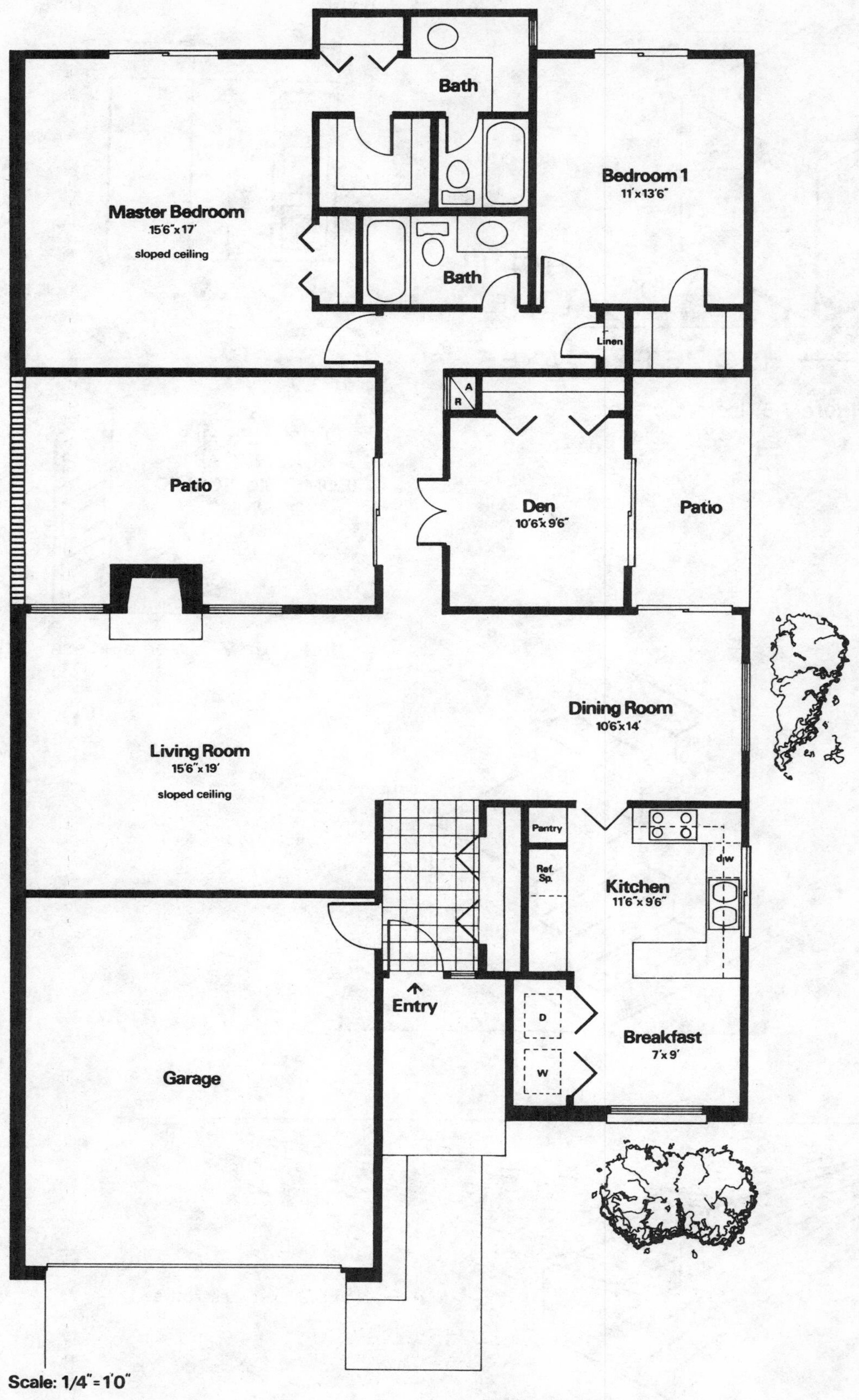

Figure 7-4. Plan of house shown in Figure 7-3.

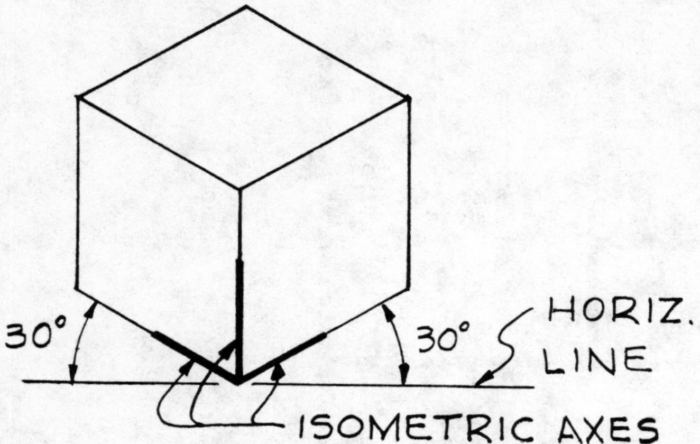

Figure 7-5. Isometric axes.

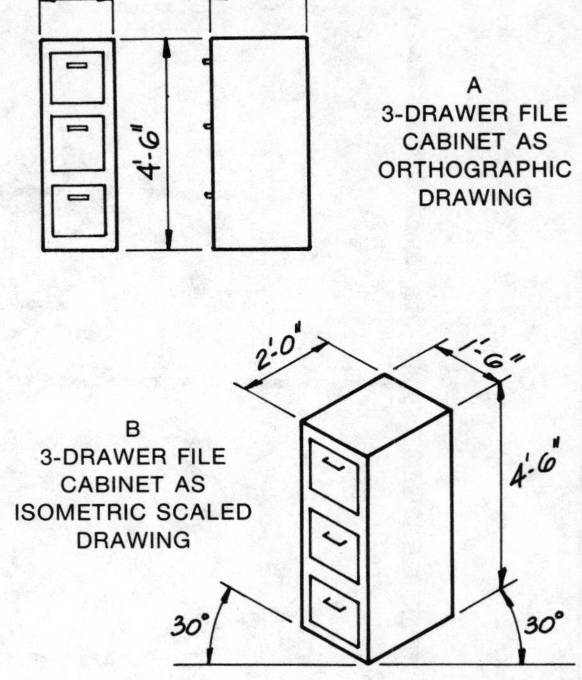

Figure 7-6. Isometric scaled drawing.

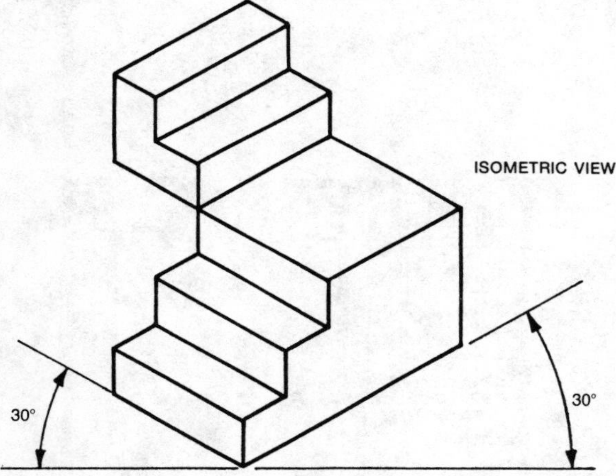

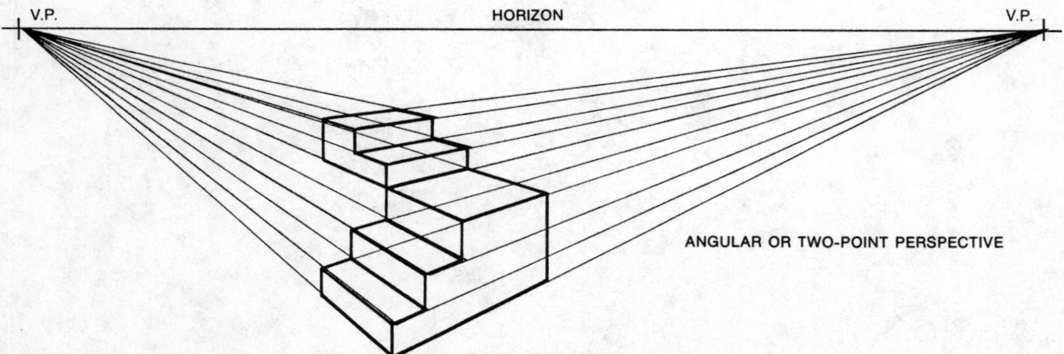

Figure 7-7. Two-direction steps drawn as isometric and angular perspective views.

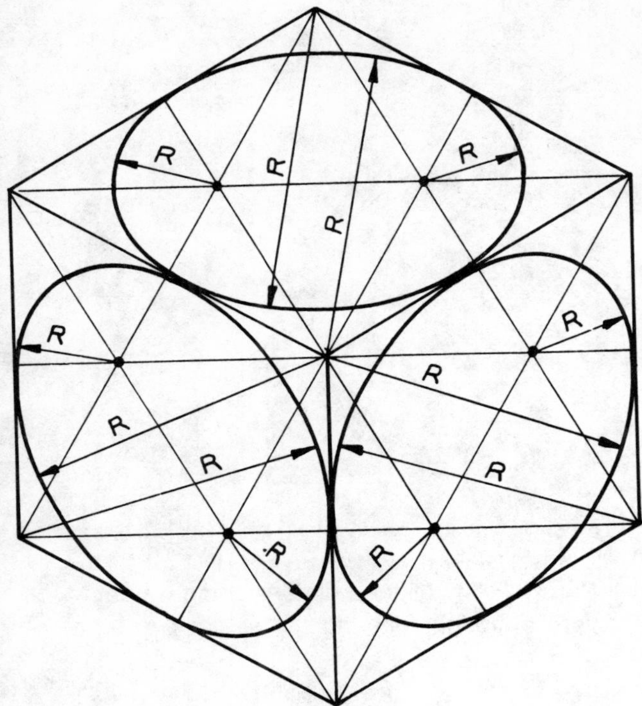

Figure 7-8. Four-center ellipsis.

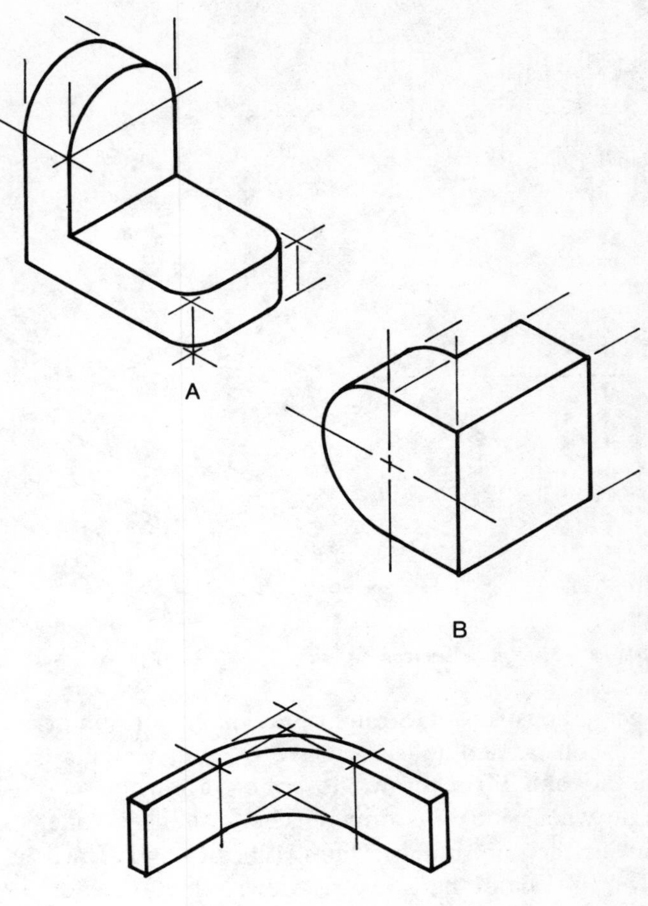

Figure 7-9. Use of isometric arcs.

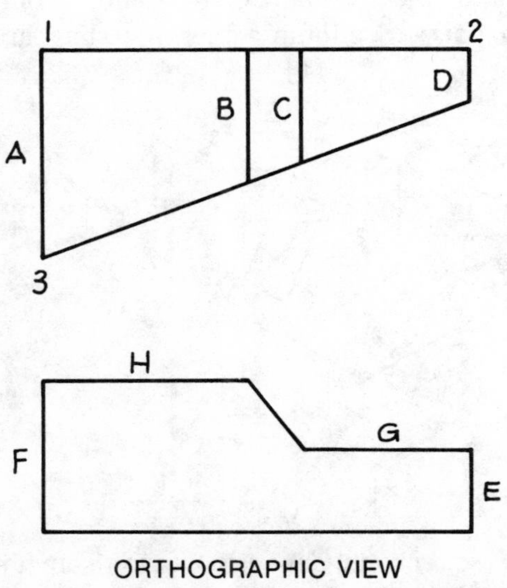

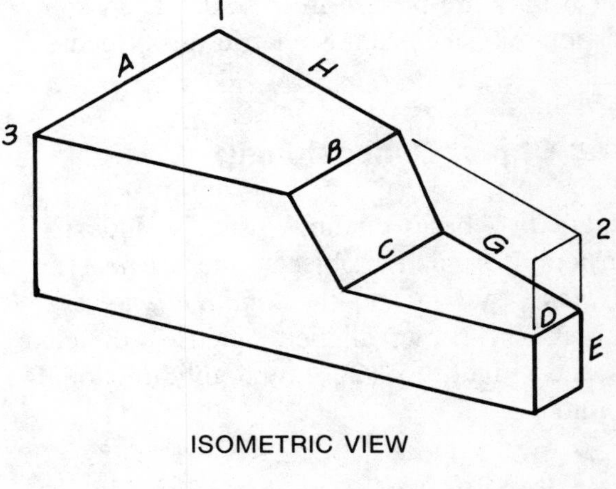

Figure 7-10. How to construct non-isometric lines.

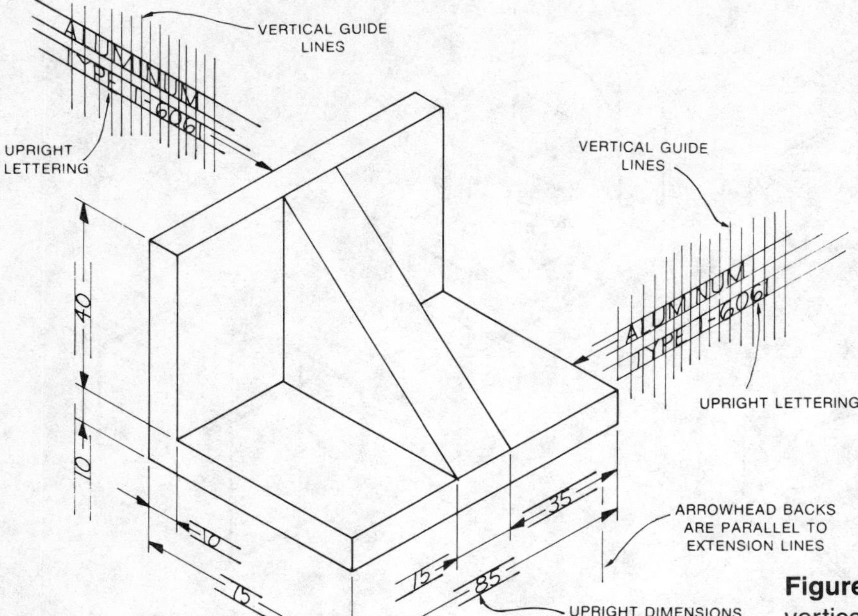

Figure 7-11. Aligned dimensioning with all vertical upright lettering.

continued from page 51

lines are constructed. Straight line (point 1 to point 2) is constructed to scale on 30° isometric line. Lines E and F are drawn to scale forming a rectangle when side D is drawn. This establishes one point of the nonisometric lines. Lines G and H are drawn to scale in the isometric view. When the ends of these lines are connected, the nonisometric line is drawn to scale. Then lines A, B, and C are drawn to scale in the isometric view and plotted on the 30° line. When these end points are connected, the nonisometric lines are formed to scale.

So, to construct nonisometric lines, all horizontal and vertical lines are plotted in to scale. Then connect end points to form nonisometric lines, including angles.

Isometric Object Dimensioning

There are two basic methods used by industry of dimensioning isometric objects: the *aligned*, or *pictorial plane*, system and the *unidirectional* system. Figure 7-11 shows aligned (metric) dimensioning, and Figure 7-12 shows unidirectional dimensioning.

Isometric Sectioning

At times it is desirable to show isometric objects in partial views to better show details not seen in the full isometric view. This is accomplished by isometric sections. There are three basic sectioning types: the *full section*, the *half section*, and the *partial*, or *broken-out section*.

Figure 7-13 shows the full section. In the complete view it is not known whether the hole at the top of the conical piece is continuous or conical, or whether the base plate is solid or has a hole in it.

The isometric full section has an imaginery plane cutting through the center of the object, dividing it in two equal parts.

The section lining symbol shown in Figure 7-13 indicates that the object is made of steel and is composed of two parts. Sectioning lines for steel are

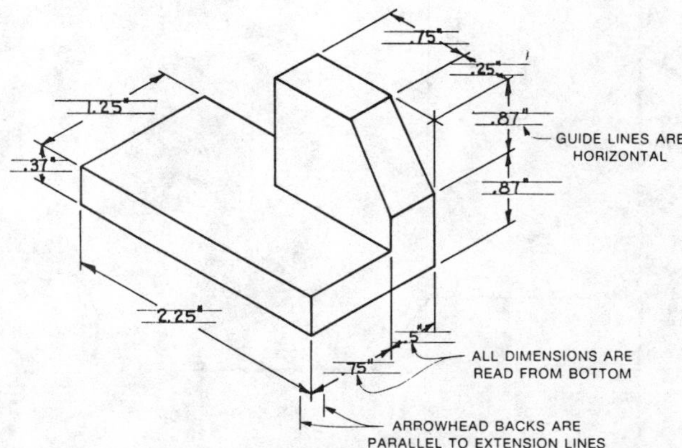

Figure 7-12. Unidirectional dimensioning for isometric objects.

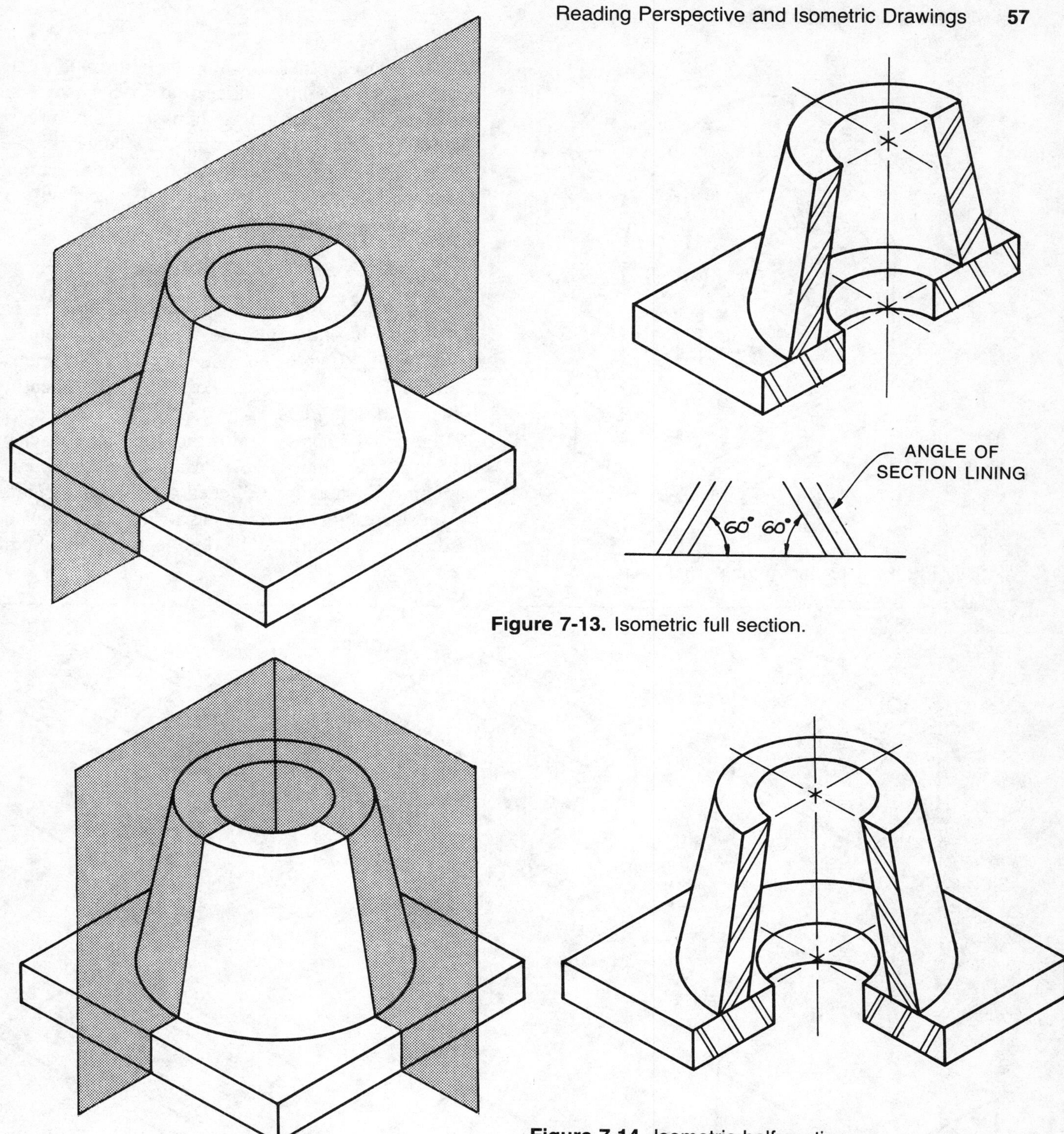

Figure 7-13. Isometric full section.

Figure 7-14. Isometric half section.

two lines close together and a wide space before two more lines together. While sectioning lines are drawn at 60° in isometric sections, they are drawn at the same direction for each piece of metal. When two separate pieces of metal make contact, the sectioning lines are turned the opposite way to designate this. In Figure 7-13 the base piece is one piece of steel while the conical piece is another piece. How they are joined is not shown in this detail, but they are probably welded.

Figure 7-14 shows the isometric half section. Here the isometric has two cutting planes, perpen-

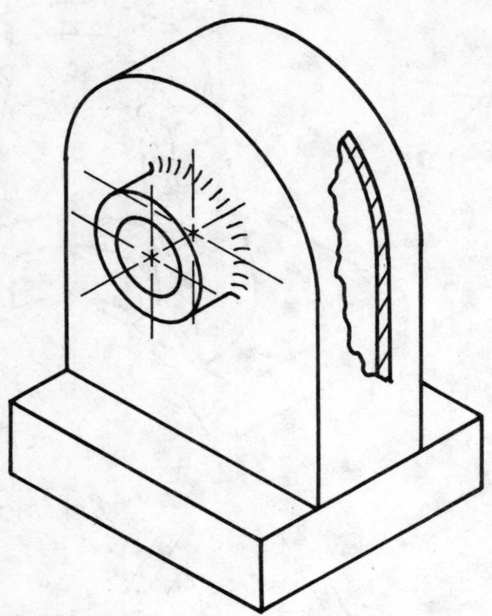

Figure 7-15. Isometric partial section.

dicular to each other, forming a right angle at the object's center point. Notice that in the isometric half section the sectioning line slant direction is reversed on each side of the object's center line.

Figure 7-15 shows the isometric partial section. This type is used when only a small piece must be broken out to show necessary detail.

Student Problems

Figure 7-16 has several isometric objects for student study and reference. Isometrics are much better understood when one can make isometric sketches. Practice isometric sketching by duplicating some of these figures.

Figure 7-17 has several orthographic (multiview) drawings with dimensions. Practice scale isometric sketching by making sketches of objects assigned by the instructor. In most cases full scale may be used. If this is too small, double scale should be tried.

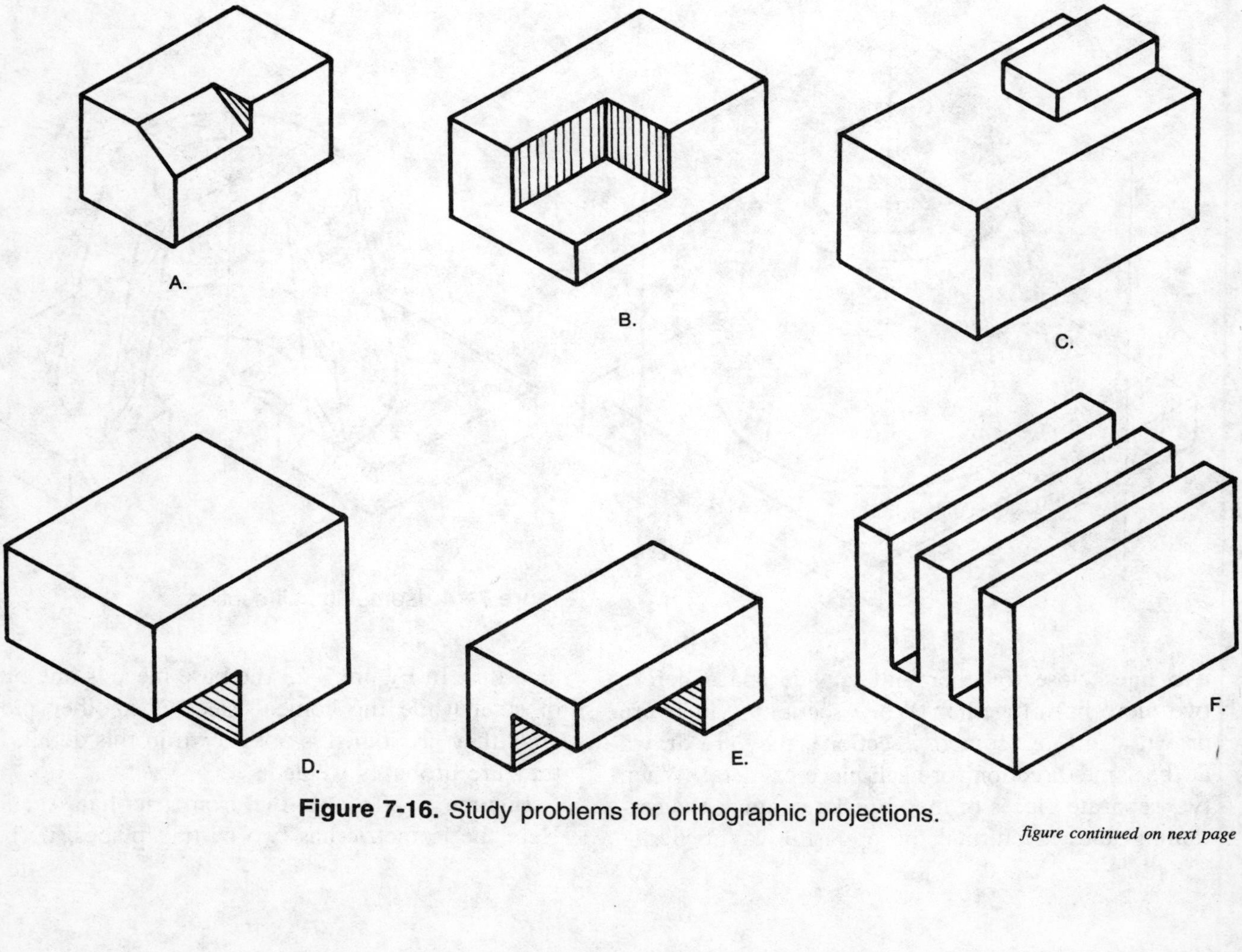

A.

B.

C.

D.

E.

F.

Figure 7-16. Study problems for orthographic projections.

figure continued on next page

Figure 7-16 continued

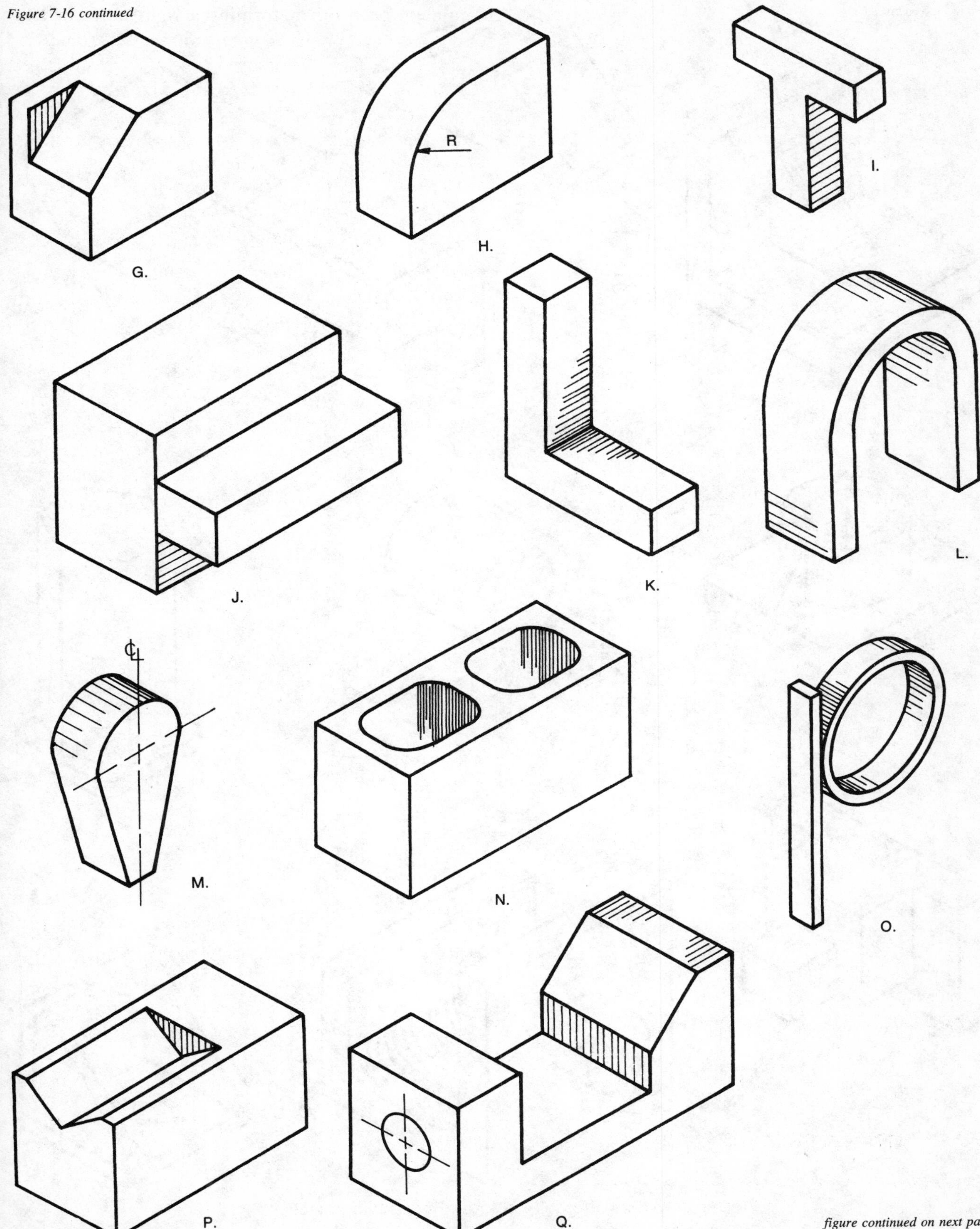

figure continued on next page

Figure 7-16 continued

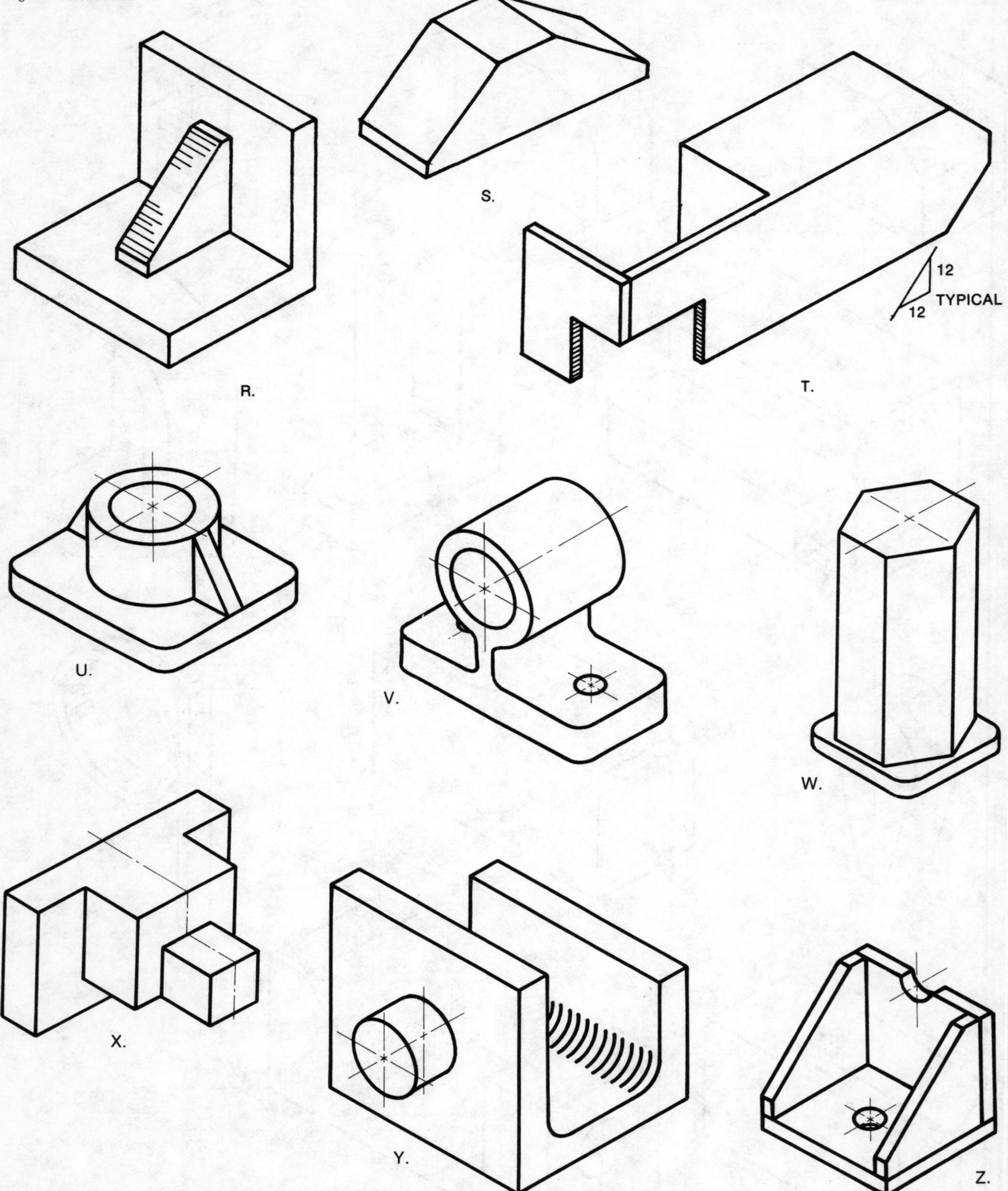

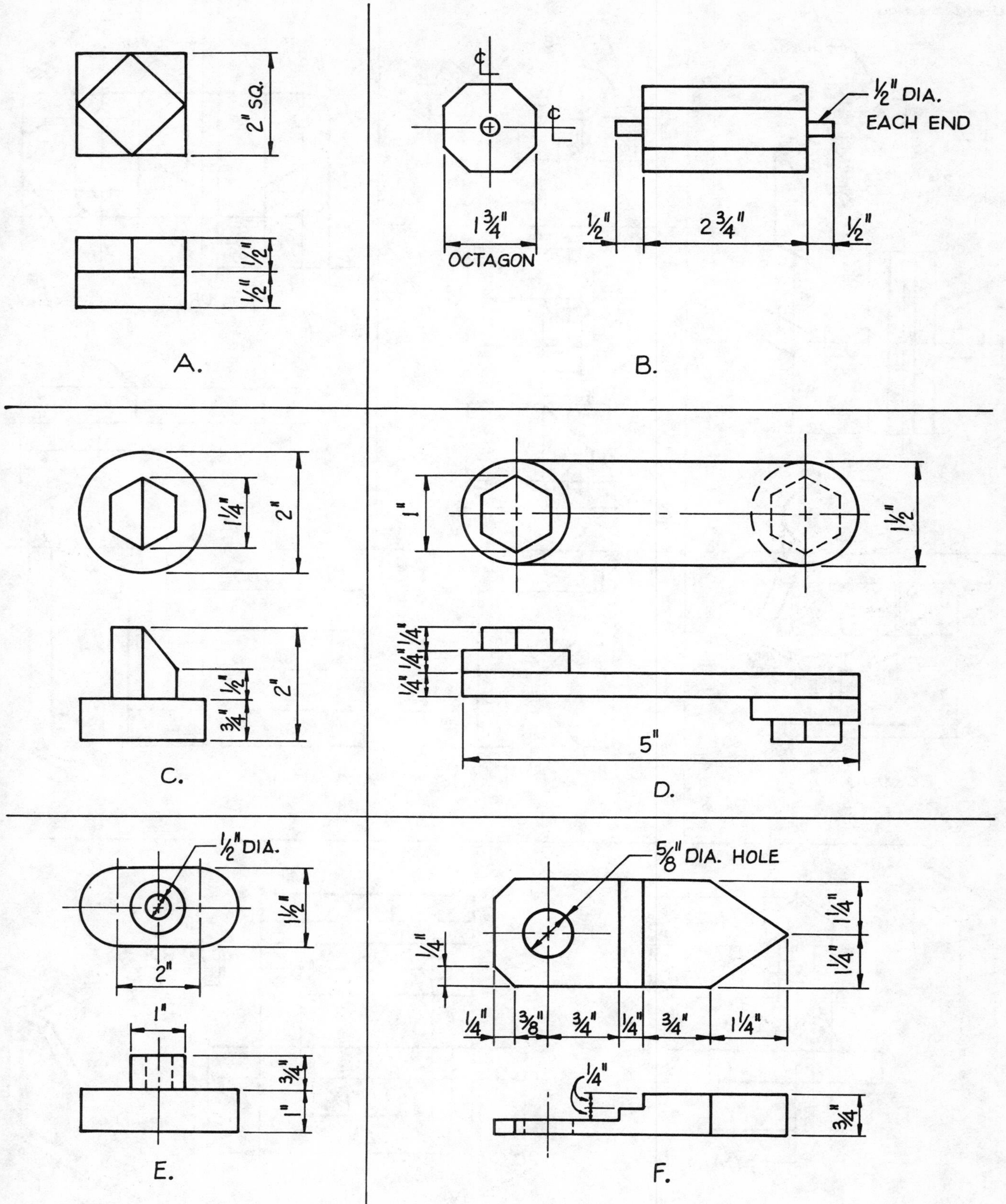

Figure 7-17. Orthographic drawings for isometric exercises.

figure continued on next page

Figure 7-17 continued

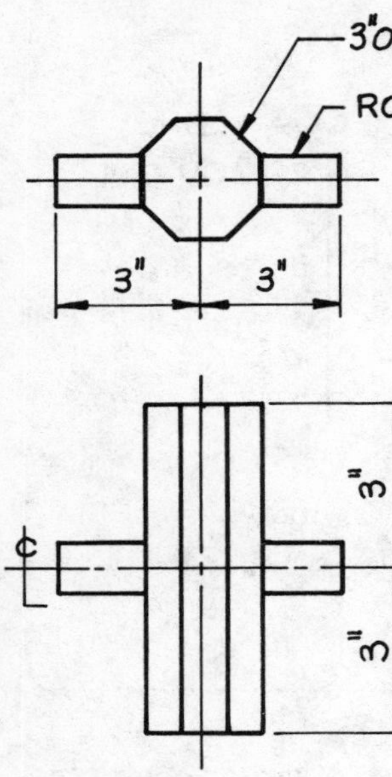

3" OCTAGON
ROUND ROD
3"
3"
℄
3"
3"

G.

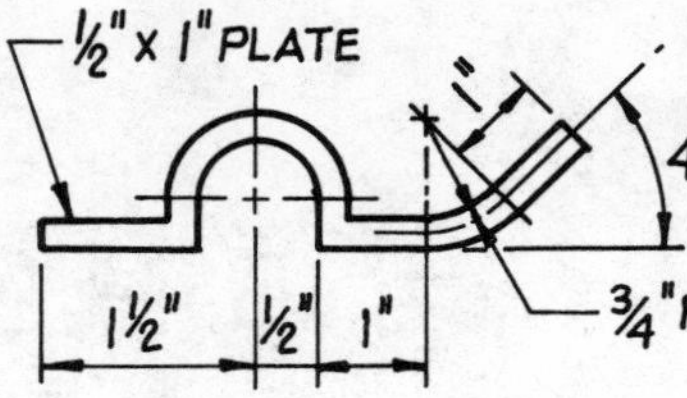

½" x 1" PLATE
1"
45°
¾" R.
1½"
½"
1"

H.

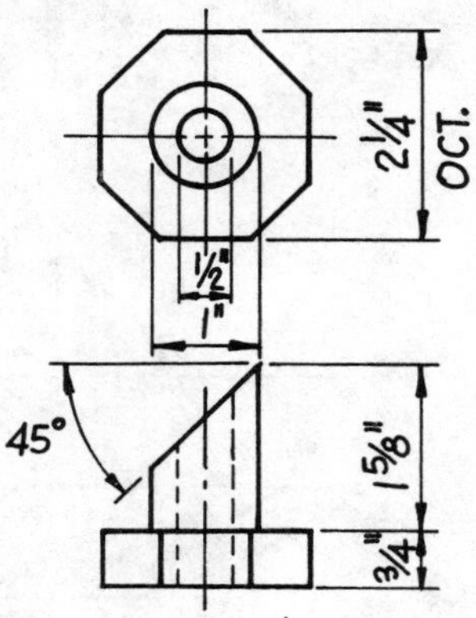

2¼"
OCT.
½"
1"
45°
1⅝"
¾"

I.

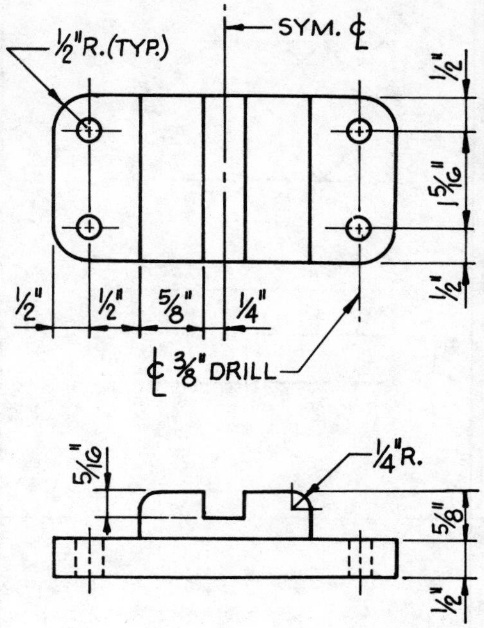

SYM. ℄
½" R. (TYP.)
½"
1 5/16"
½"
½"
½"
⅝"
¼"
℄ ⅜" DRILL
5/16"
¼" R.
⅝"
½"

J.

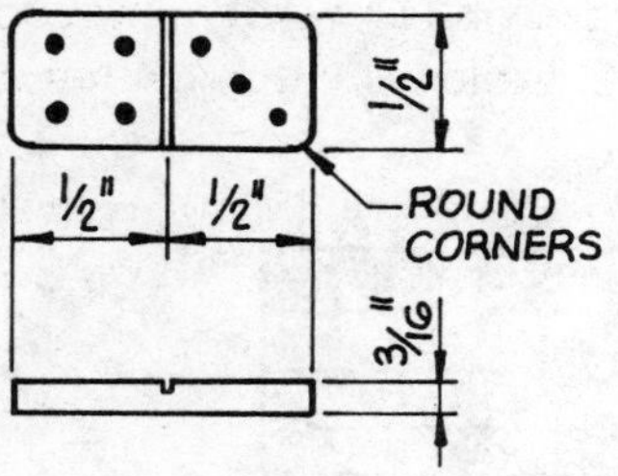

½"
½"
½"
ROUND CORNERS
3/16"
DOMINO

K.

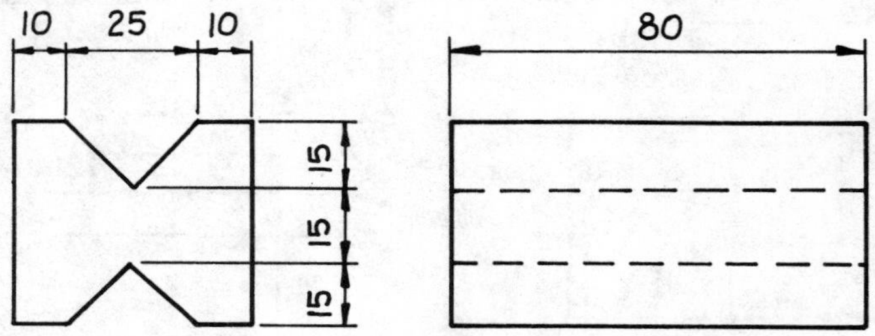

10
25
10
15
15
15
80
DIMENSIONS ARE IN MM.

L.

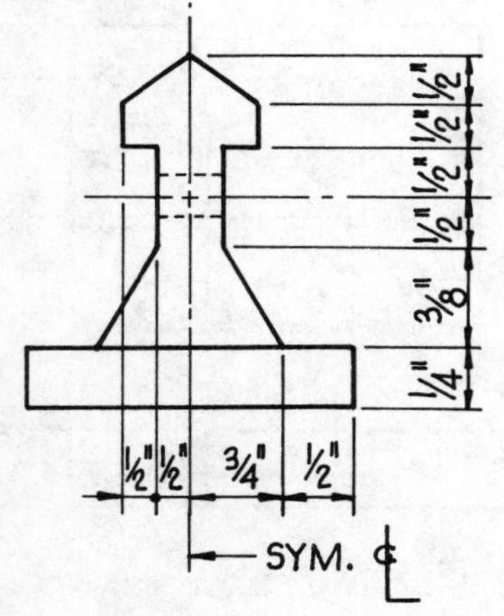

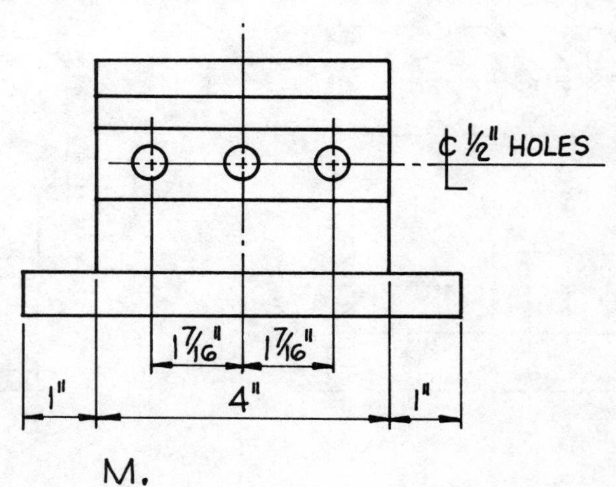

M.

figure continued on next page

CHAPTER 8

Multiview Drawings and Sections

Drawings are prepared to illustrate and communicate the views and dimensions of three-dimensional objects. The communication must have sufficient detail to ensure that the object can be manufactured correctly without any verbal communication. To accomplish this, drawings are normally made with more than one view, each view showing one side of the object. This is called orthographic projection, another way of describing multiview drawings.

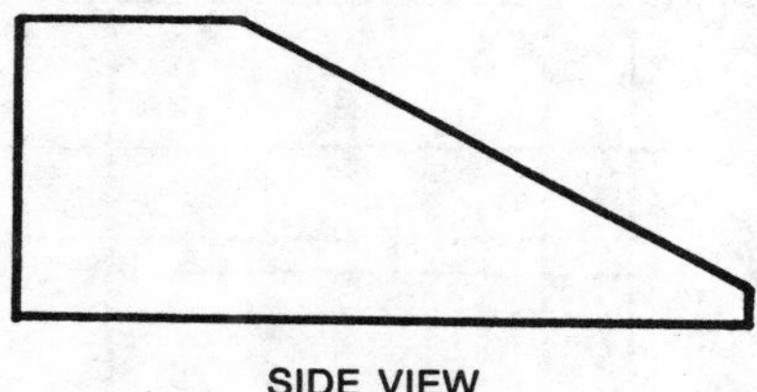

Figure 8-1. Orthographic view of wedge.

Orthographic Projections

Orthographic projection means extending the various views of an object into their respective vision planes. Unlike the isometric, which shows a three-dimensional object, the orthographic view shows only two dimensions. Figure 8-1 is an orthographic view of a wedge. The height and width can be seen in this side view, but the depth or any shape change of the depth is unknown.

To present the full picture, other orthographic views must be made. Figure 8-2 shows how views are projected to provide the necessary picture. While four views are shown, normally only the right end view would be shown because it provides all the necessary information. The other three views are given to enable the reader to see how views are projected.

To understand orthographic projection, the reader must visualize the object as being within a box. Then, views of the side of the box are visualized.

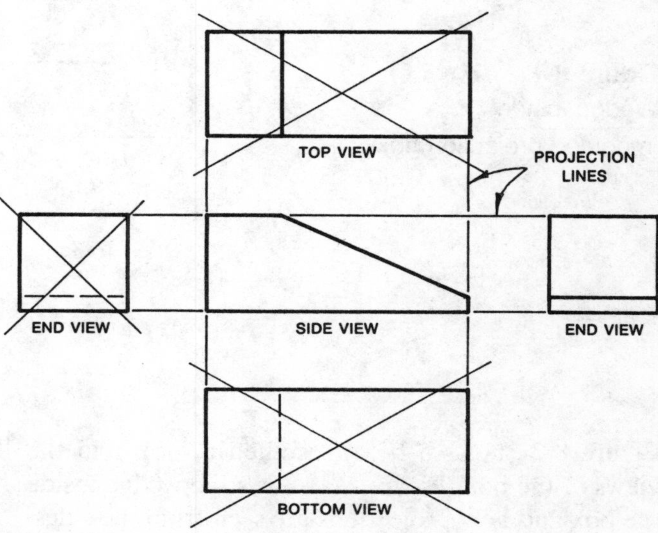

Figure 8-2. Orthographic views of wedge.

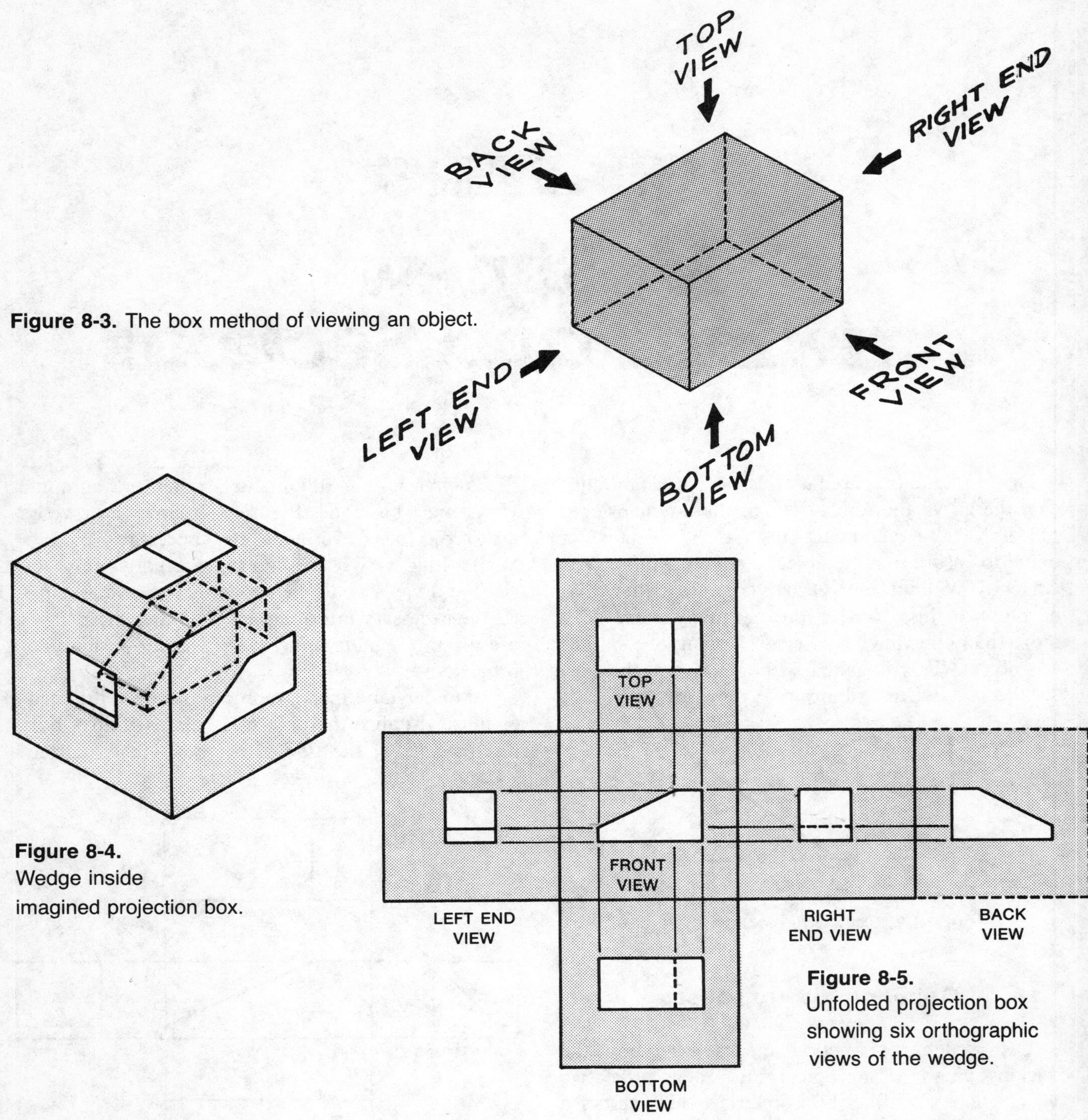

Figure 8-3. The box method of viewing an object.

Figure 8-4. Wedge inside imagined projection box.

Figure 8-5. Unfolded projection box showing six orthographic views of the wedge.

Figure 8-3 shows a box in isometric form and the views of the box. Figure 8-4 shows the wedge inside the box and how projections are seen from the sides. Figure 8-5 shows the box, unfolded, and the orthographic projection views of the wedge. So, the unfolded box principle helps one to understand orthographic projections.

To help understand how projections are made by the drafter, Figure 8-6 presents several views of isometric objects and shows how views would appear on the print being read. Note that curved lines in the object appear as straight lines on the drawing. The reader must have other orthographic views to know that lines are curved. Also note that hidden

(text continued on page 67)

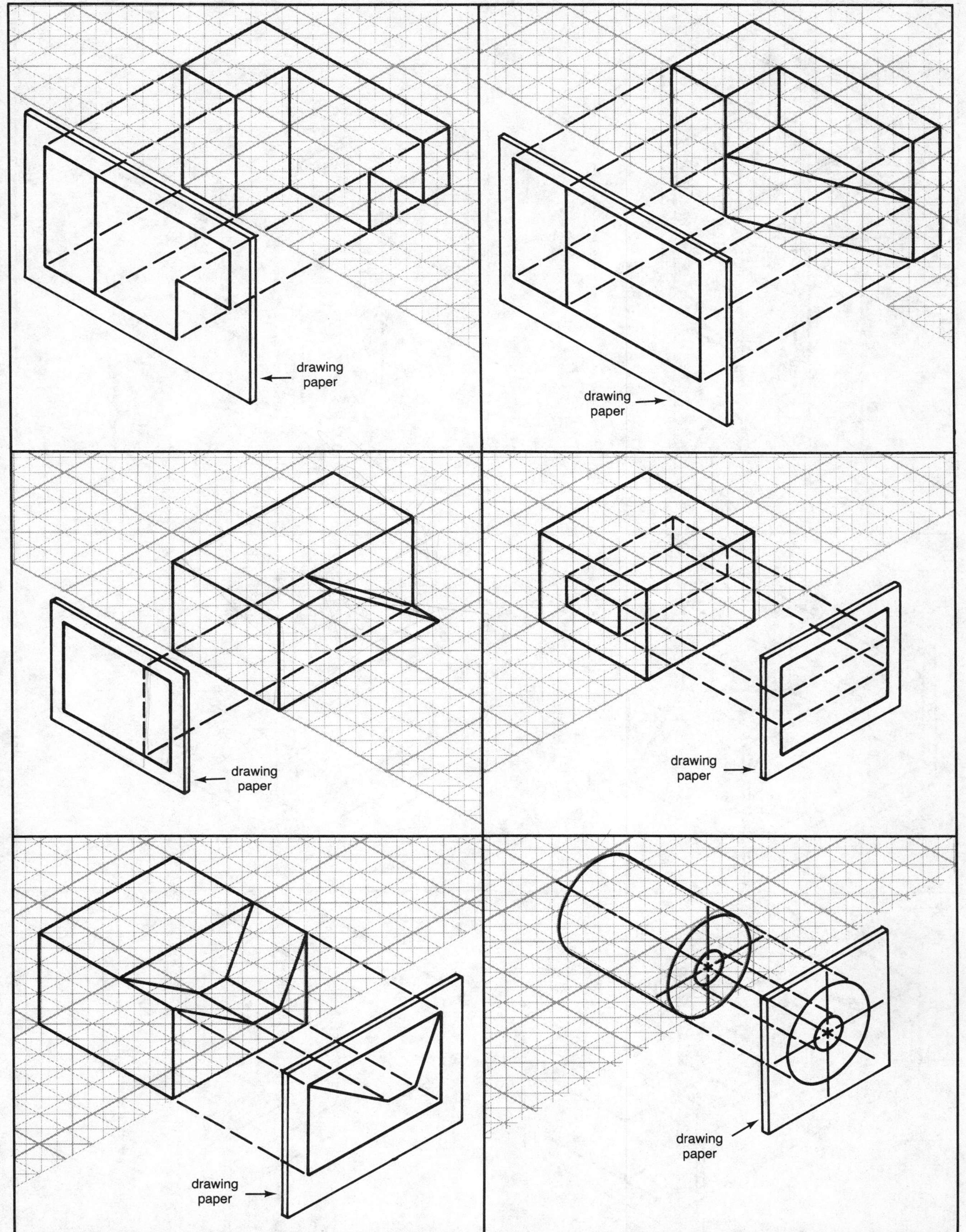

Figure 8-6. Projected views from isometric objects.

figure continued on next page

Figure 8-6 continued

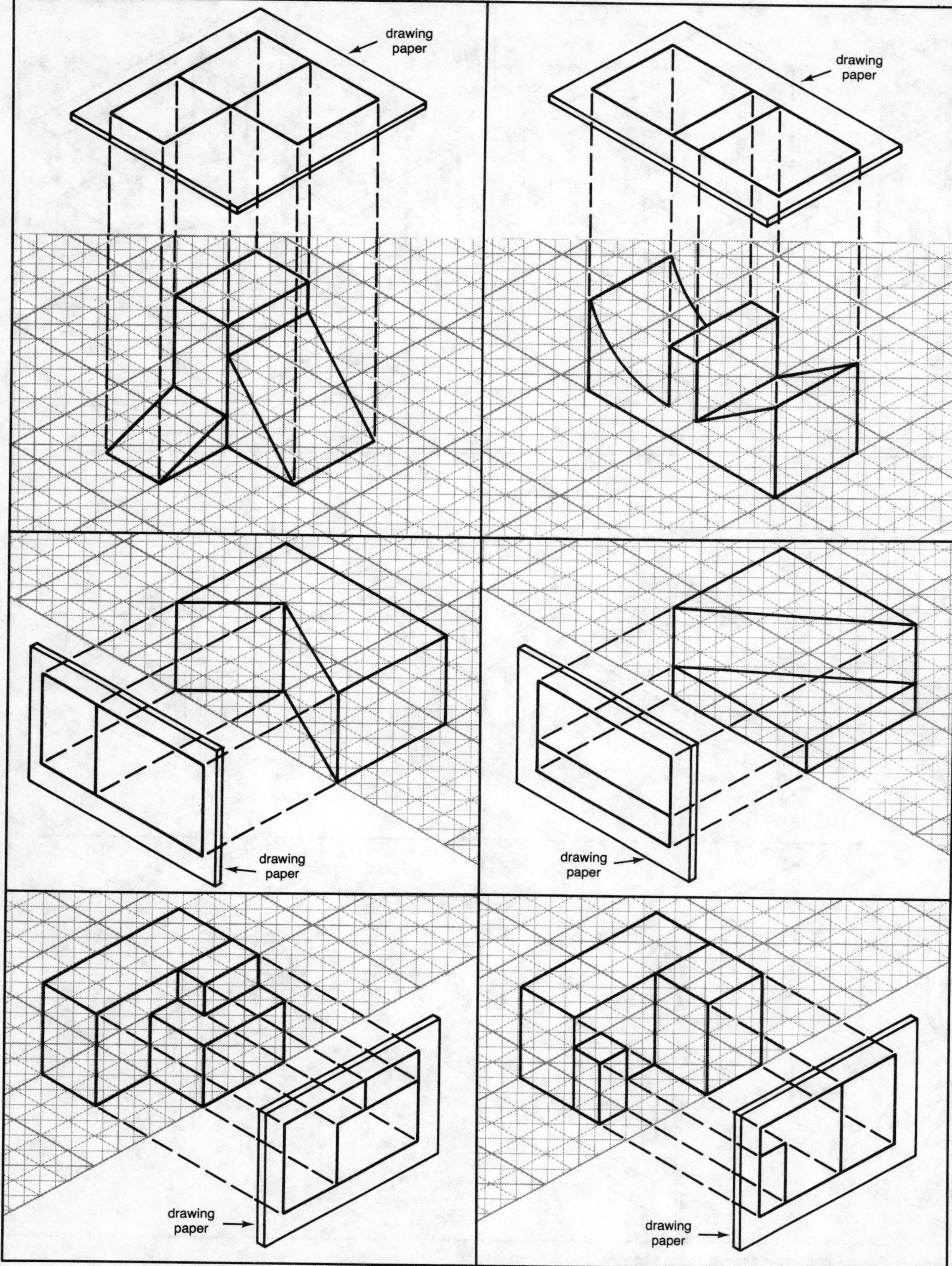

continued from page 64

lines appear as dashed lines on drawings. Without seeing other projections, the reader doesn't know the meaning of these dashed lines.

Figure 8-7 has several isometric objects of which the reader may practice sketching orthographic projections. These need not be to scale but should be to proportion, showing all hidden lines as dashed lines. Draw the view shown in plan and the necessary side or end views to complete the drawing picture. Do not sketch more views than necessary. The instructor will assign the objects to be sketched.

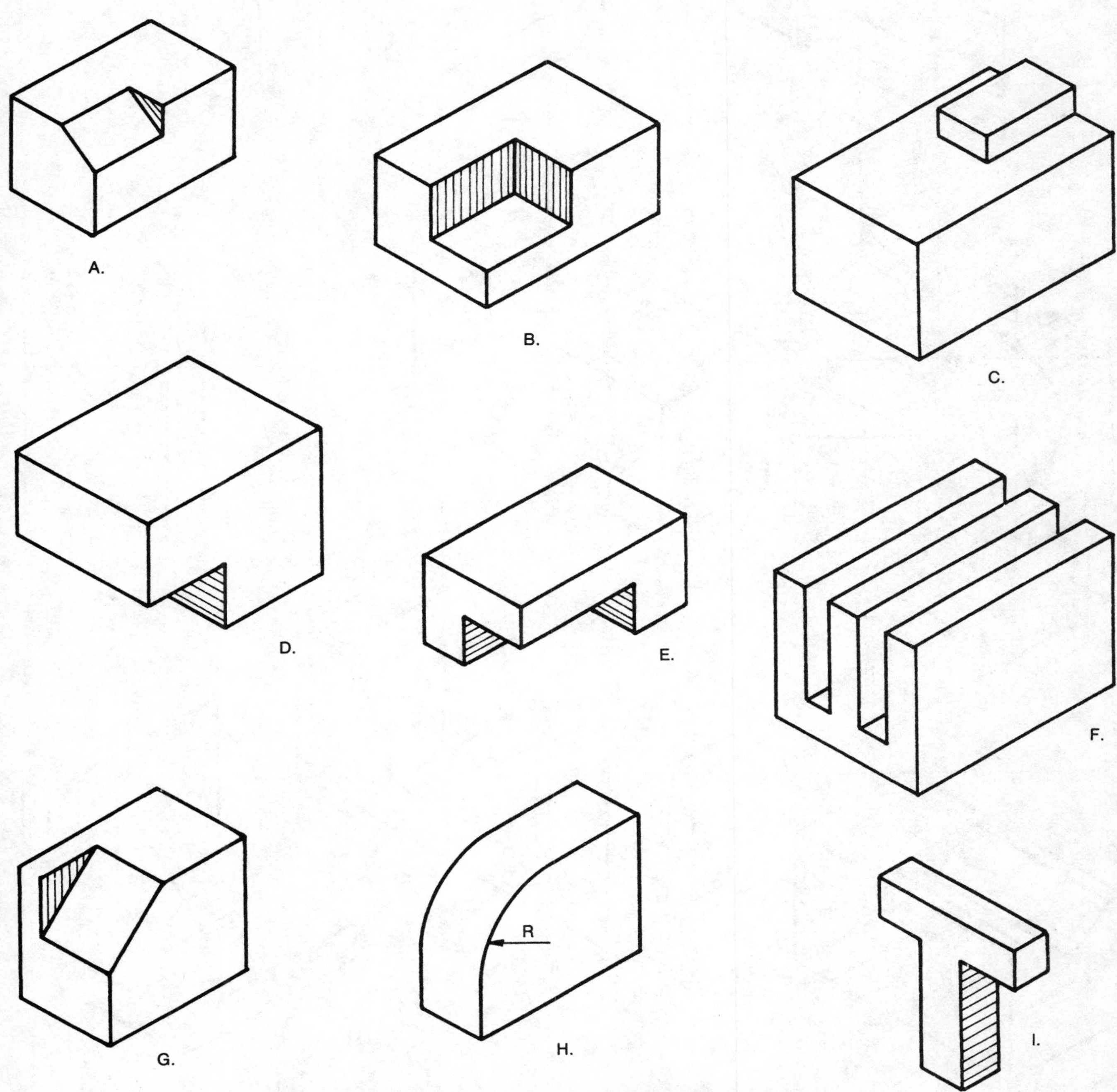

Figure 8-7. Study problems for orthographic projections.

Figure 8-7 continued

Figure 8-7 continued

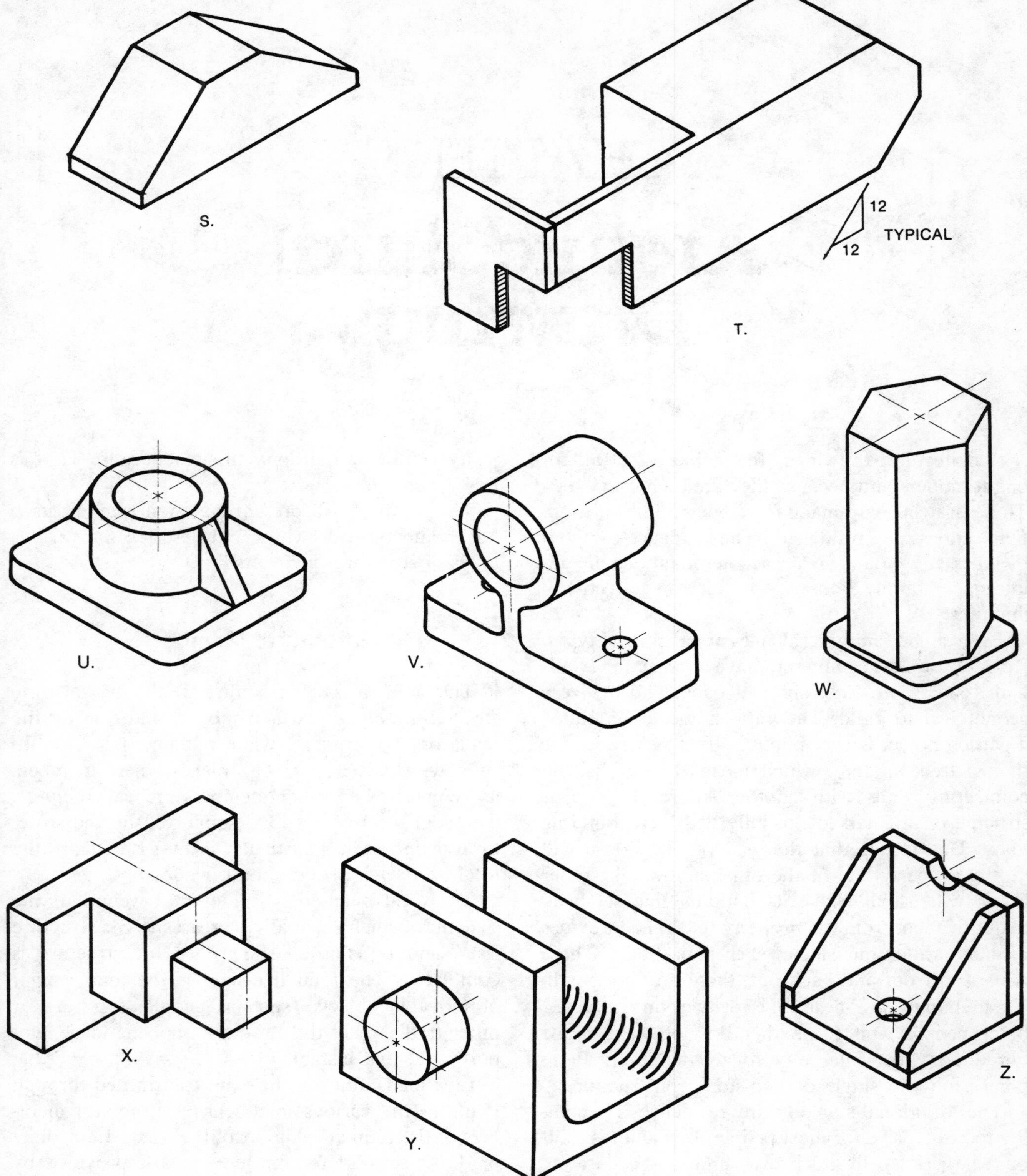

CHAPTER 9

Architectural Drawings

Architectural drawings are the easiest of all for one to understand because they are seen every day. These drawings are made for office buildings, shopping centers, and residences. The residence category includes any type of living accommodations including single family houses, apartments, and condominiums.

Even single family residences are of several types. Row houses are common, have connecting walls and a small yard in front and back, and are very economical to build. The patio home, often called a garden home, is very popular. This is a residence that is free-standing, which means that it has no connecting walls with another house. It is constructed on a small lot, usually with zero building lines. This means that the wall of one house will be the property line of the other house. The other house will usually have its building line set back about five feet from the property line. This provides a small yard on one side of each patio home. There is yard in front and back but, these are also small. These homes are desirable for working couples, older people, and those that do not want to be burdened with a large yard to maintain. Patio homes may be single or two-story construction.

The trauitional single family residence is becoming too expensive for most people. It is a home built on a large lot with a driveway and land separating the homes. This lot is normally 65 to 100 feet wide. As cities and population grow, land becomes more scarce, which prices these homes out of the range of most people. Smaller towns still construct a majority of these traditional houses because land is less expensive.

This chapter will present common symbols used on architectural drawings and describe the various types of structural members.

Structural Members

The main parts of a structure are the *load-bearing structural members*, which support and transfer the loads on the structure while remaining in equilibrium with each other. The places where members are connected to other members are called *joints*. The sum total of the load supported by the structural members at a particular instant is equal to the total *dead load* plus the total *live load*.

The total dead load is the total weight of the structure, which gradually increases as the structure rises, and remains constant once the structure is completed. The total live load is the total weight of movable objects (such as people, furniture or other traffic) that the structure happens to be supporting at that instant.

Live loads in a structure are transmitted through joints to the various load-bearing structural members to the ultimate support of the earth. Immediate or direct support for the live loads is provided by horizontal members. These loads are transmitted to vertical members, which are supported by foundations and/or footings. The foundation is supported by earth.

Vertical Members

Columns, also called *pillars*, are the vertical members in buildings. Columns in walls usually rest directly on footings. The chief vertical structural members in houses and other light construction are called *studs*. They are supported on horizontal members called *sills* or *sole plates*, and are topped by horizontal members called *top plates* or *stud caps*.

Corner posts are enlarged studs, located at the building corners. In the past, corner posts were a solid piece of large timber. Today *built-up* corner posts are used, consisting of various numbers of studs nailed together to form a strong corner post.

Horizontal Members

In technical terminology a horizontal load-bearing structural member that spans a space, and which is supported at both ends, is called a *beam*. A member that is *fixed* at one end only is called a *cantilever beam*. Steel members consisting of solid pieces of the regular structural steel shapes are called beams, but a type of steel member that is actually a light truss (discussed later) is called an *open-web steel joist* or a *bar steel joist*.

Horizontal structural members that support the ends of floor beams or joists in wood frame construction are called *sills*, *girts*, or *girders*, depending on the type of framing being done and the location of the member in the structure. Horizontal members that support studs are called *sill* or *sole plates*. Horizontal members that support the wall-ends of rafters are called *rafter plates* or *top plates*, depending on the type of framing. And horizontal members that assume the weight of concrete or masonry walls above door and window openings are called *lintels*.

Trusses

A beam of given strength, without intermediate supports below, can support a given load only a certain maximum span. If the span desired is wider than this maximum, intermediate supports, such as columns, must be provided for the beam. When this is not feasible or possible, a *truss* is used in place of a beam.

A beam consists of a single horizontal member. A truss is a framework, consisting of members joined together by a number of vertical and/or inclined members forming a series of triangles, the loads being applied at the joints. Figure 9-1 is a steel truss that might be used in a shopping center. Only one-half is drawn for brevity. Trusses may also be constructed of wood.

Roof Members

The horizontal or inclined members that provide support to a roof are called *rafters*. A *ridge* is defined as the intersection of two surfaces forming an outward projecting angle, as at the top of a roof where the two slopes meet. The lengthwise (horizontal) member at the top of a roof, to which the upper ends of the rafters of the two slopes are fastened, is called the *ridge board*, *ridge pole*, or *ridge piece*. In some light construction, the ridge board is omitted and the upper ends of the rafters are nailed together.

Symbols, Abbreviations, and Terms

The ability to read architectural drawings and have a good knowledge of architectural terms and symbology is necessary for people in any construction field, such as carpenters, plumbers, electricians, stone masons, roofers, etc. It is a definite aid to thousands of other people that service the construction industry, such as lumber yard, hardware store, and sales people. Real estate sales people must also know how to read these drawings.

Architectural symbols, terms, and abbreviations are used in this chapter. For immediate student reference, architectural symbols are shown in Figure 9-2, abbreviations in Table 9-1 and definitions are in the glossary at the end of this chapter.

Roof Types

Roof types are considered to be functional and economical. The *hip roof* rises up from all building walls to a center ridge so there are no gables to paint. Figure 9-3 shows the hip roof. This type of roof also provides an overhang all around the house, which shades windows and reduces heating and cooling costs.

The *gable roof* has large gables at each end and, although less expensive for initial cost, it requires

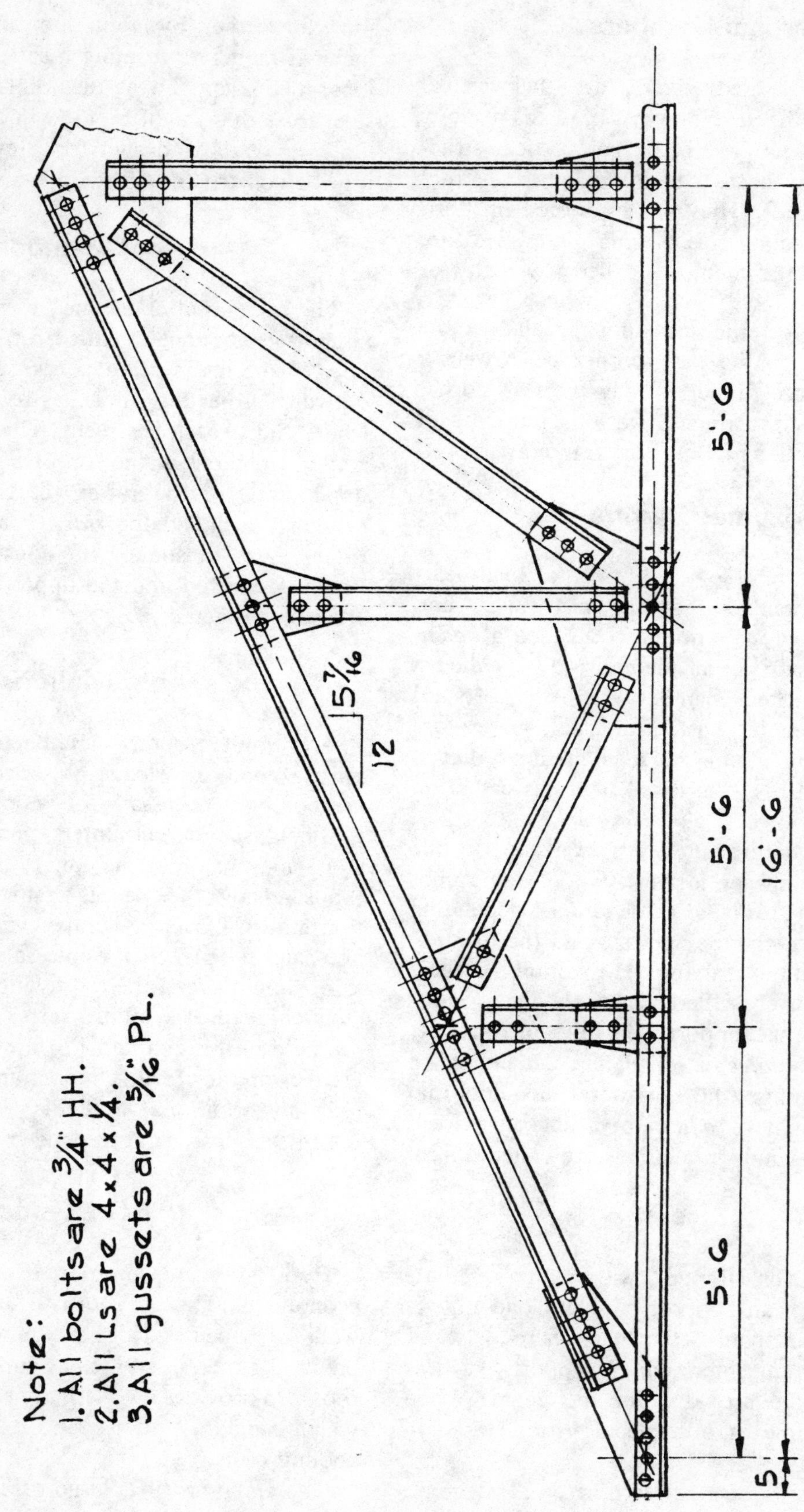

Figure 9-1. Typical truss detail.

Table 9-1. Architectural Abbreviations

Abbreviation	Meaning
AB	anchor bolt
AC	alternating current
ACST	acoustic
AGGR	aggregate
A.I.A.	American Institute of Architects
AIEE	American Institute of Electrical Engineers
AISC	American Institute of Steel Construction
ANT	antenna
ASB	asbestos
A.S.H.V.E.	American Society of Heating and Ventilating Engineers
ASTM	American Society for Testing Materials
AT	asphalt tile
B	bedroom
BD	board
BL	building line
BLDG	building
B.M.	board measure
BM	bench mark
BOT	bottom
BR	bedroom
BSMT	basement
BTU	British thermal unit
C, CL or CLO	closet
CEL	ceiling
CEM	cement
CI	cast iron
CL	centerline
CLG	ceiling
CLKG	caulking
CO	cleanout
COL	column
CONC	concrete
CONC B	concrete block
CONTR	contractor
CU FT	cubic foot
CU IN	cubic inch
CU YD	cubic yard
D or DR	drain
DB	drain board or door bell
DH or DHW	double hung window
DMPR	damper
D R	dining room
DS	downspout
DW	dishwasher
DWG	drawing
ENT	entrance
EXC	excavate
EXP JT	expansion joint
FAI	fresh air intake (or inlet)
FC	furred ceiling
FD	floor drain
FDN	foundation
FL	flashing or floor
FPM	feet per minute
FPRF	fireproof
FT	floor tile
FTG	footing
GT	grease trap
GYP	gypsum
HT	height
I	I beam
J	joist
K	kitchen
KS	kitchen sink (sometimes S)
L	leader
LAU	laundry
L CL	linen closet
LDG	landing
Lino	linoleum
LR	living room
MATL	material
MR	marble
NBFU	National Board of Fire Underwriters
OD	Outside diameter or outside dimension
PL or PLAS	plaster
PLAT	platform
PL GL	plate glass
PLMG	plumbing
PRCST	precast
PTN	partition
R	range
RD	roof drain or rod
REF	refrigerator or reference
REG	register
RF	roof
RFG	roofing
RUB	rubber
SDG	siding
SHTHG	sheathing

table continued on next page

Table 9-1 continued

ST	stairs	TER	terrazzo
STA	station	TV	television
STD	standard		
STN	stone	UR	urinal
STL	steel		
STR	structural	V	vent
SYM	symbol or symmetrical		
SYS	system	WC	water closet
		WDW	window
T or TR	tread	WH	weep hole
TEL	telephone	WP	waterproof
TEMP	temperature	WS	weather stripping

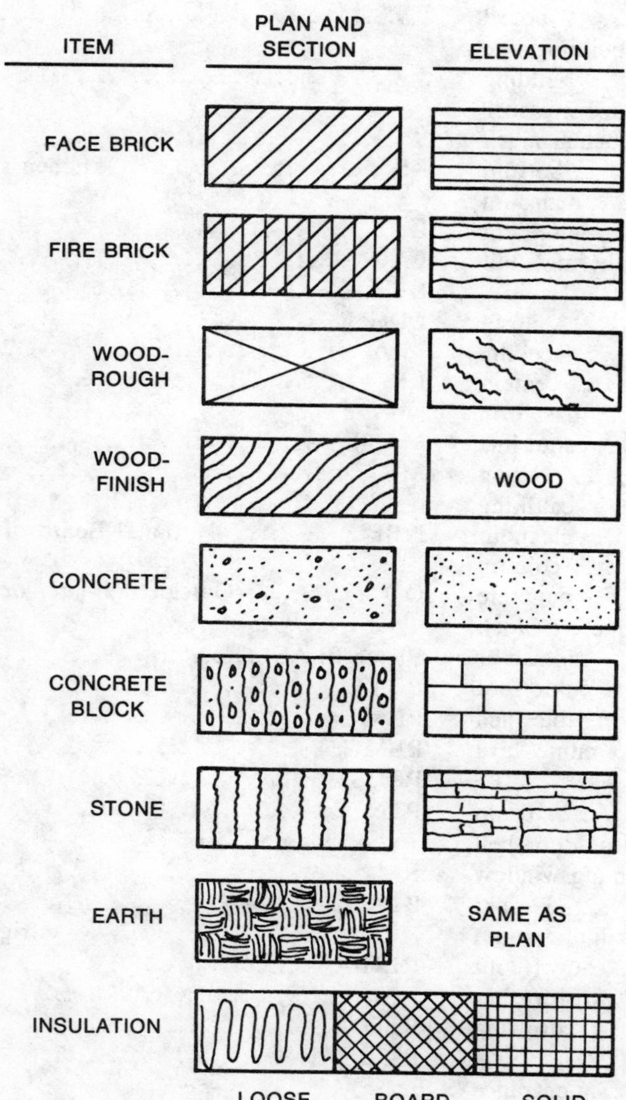

Figure 9-2.
Architectural symbols.

figure continued on next page

Figure 9-2 continued

WEATHERPROOF OUTLET

S_{CB}

CIRCUIT BREAKER

S_3

THREE-WAY SWITCH

S_{WP}

WEATHERPROOF SWITCH

TELEPHONE OUTLET

TELEPHONE JACK

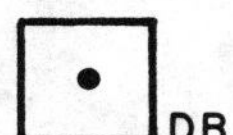

DOOR BELL

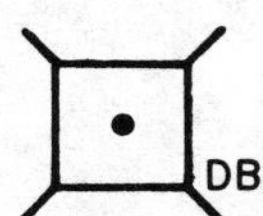

LIGHTED DOOR BELL

DOOR CHIME

RECESSED CEILING LIGHT

CEILING LIGHT W/ PULL SWITCH

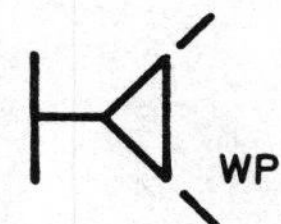

WEATHERPROOF OUTSIDE LIGHT

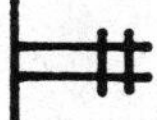

HOSE BIB

GAS OUTLET

SHOWER HEAD

figure continued on next page

Figure 9-2 continued

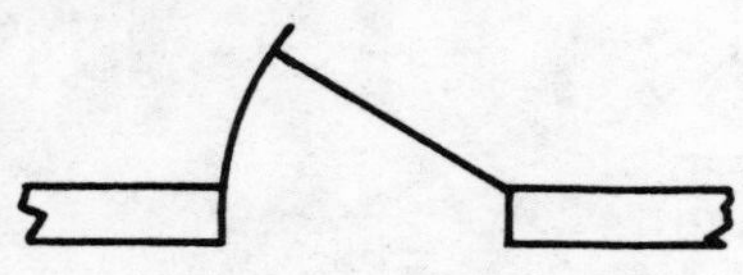
INTERIOR DOOR

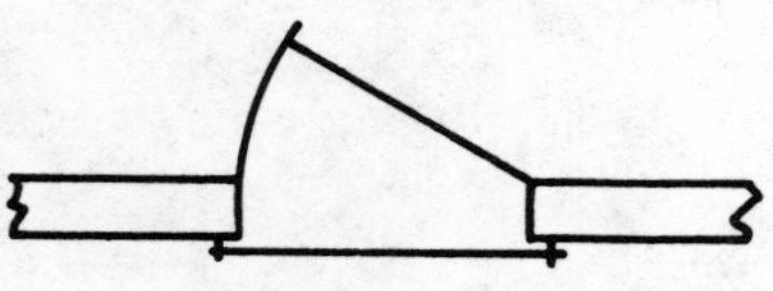
SINGLE EXTERIOR DOOR

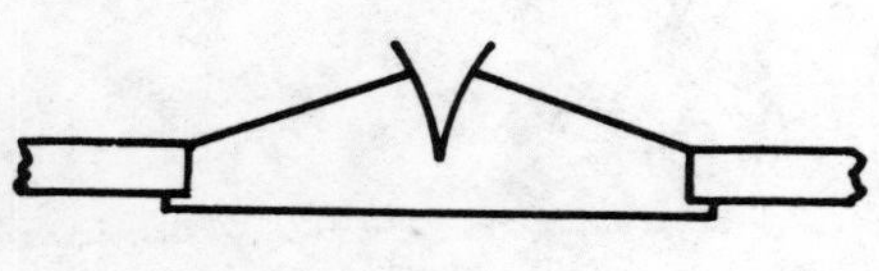
DOUBLE EXTERIOR DOOR

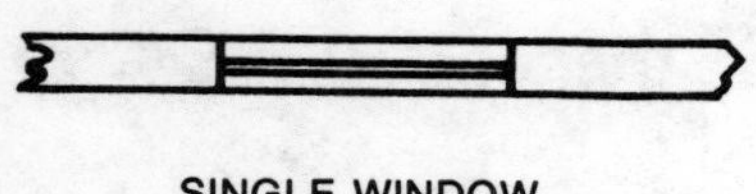
SINGLE WINDOW

DOUBLE WINDOW

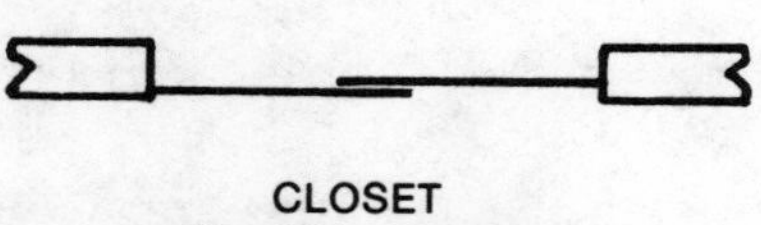
CLOSET SLIDING DOORS

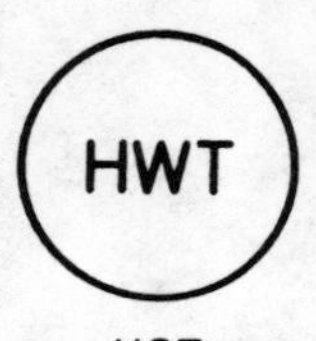

HOT WATER TANK

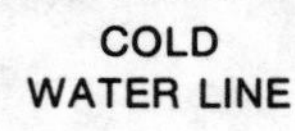
COLD WATER LINE

HOT WATER LINE

GAS LINE

FLOOR DRAIN

SHOWER DRAIN

BUILT-IN BATH

WATER CLOSET

RADIATOR

REGISTER

CEILING LIGHT

WALL LIGHT

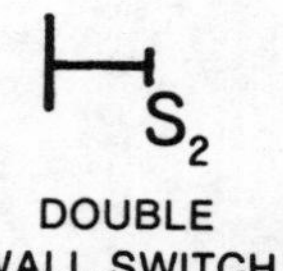

SINGLE WALL SWITCH

DOUBLE WALL SWITCH

DUPLEX ELECT. OUTLET

220 V. ELECT. OUTLET

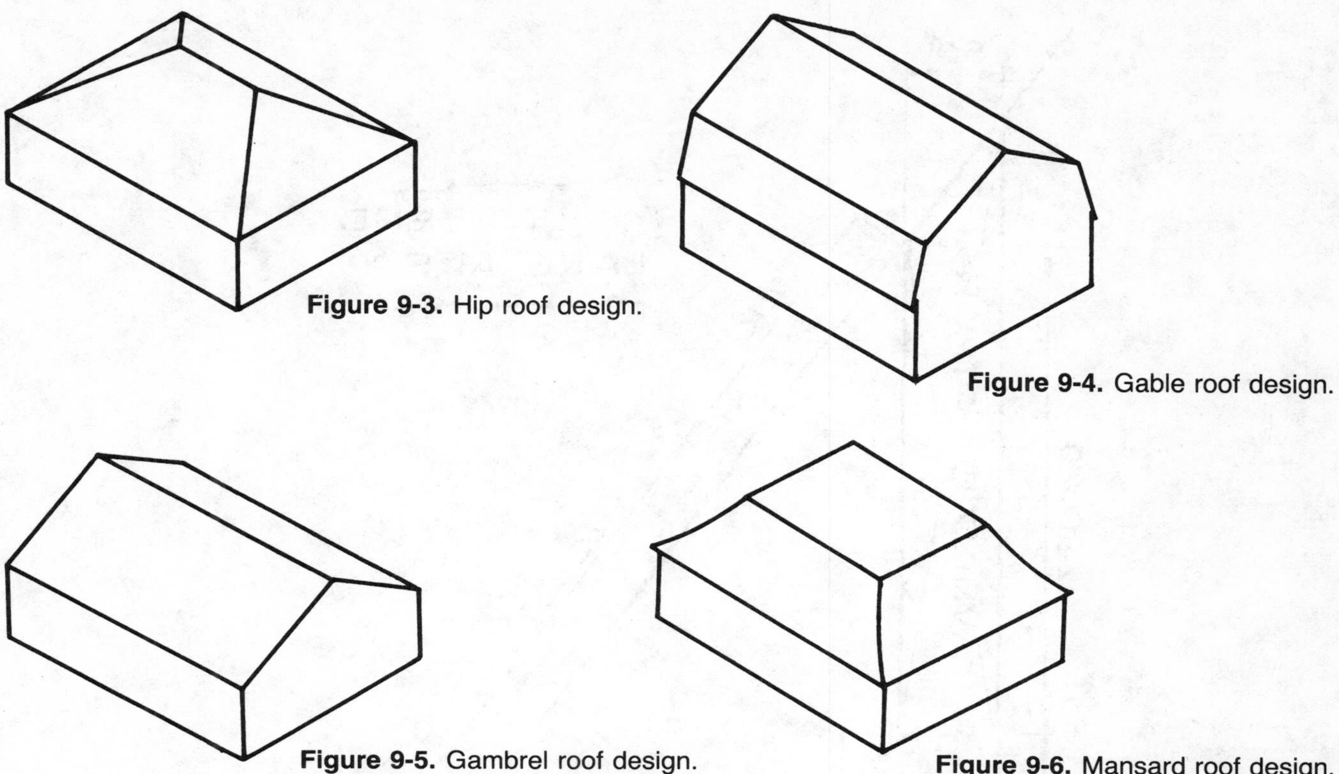

Figure 9-3. Hip roof design.

Figure 9-4. Gable roof design.

Figure 9-5. Gambrel roof design.

Figure 9-6. Mansard roof design.

more maintenance. Figure 9-4 shows the gable roof design.

Roof slope will vary per climatic area. Gable and hip roofs are in the *pitched roof* category, which means any roof steeper than 4:12. The numbers 4 and 12 refer to the rise and base dimensions of the triangle formed by the roof. The 4:12 sloped roof is at an angle of 18°-26′ from the horizontal. Most roofs use a slope of 5:12 or greater. A much steeper incline is specified in climates that have heavy snows to keep snow from building up on the roof.

Other roof types are the *flat roof*, which has no slope; the *sloped flat roof*, which is flat and slightly sloped to one end of the house; the *Gambrel roof* shown in Figure 9-5; and the *Mansard roof* shown in Figure 9-6.

Construction Drawings

Construction drawings are drawings that show as much construction details as possible. These are presented graphically, or by means of drawn pictures. Most construction drawings consist of orthographic views and are presented as plans and elevations, usually drawn to ¼″ = 1′-0″ scale. Site plans may be drawn to a much smaller scale, while details of the plan and elevations are usually drawn to a much larger scale.

Plans

A *plan view* is a view of an object or area as it would appear if projected onto a horizontal plane passed through or held above the object or area. The most common architectural plans are plot plans (also called site plans), foundation plans, floor plans, and roof plans.

A *site plan* may or may not show the contours of grade. Figure 9-7 is a site plan of 2.75 acres of waterfront land and shows the planned home site location. Because no grading is necessary, contour lines are omitted. The site plan does give the property limits by dimensions.

A *foundation plan* is a plan view of a structure projected on a horizontal plane passed through at the level of the foundation top. The foundation shown in Figure 9-8 shows that the main foundation of this structure will consist of a rectangular 8″ concrete block wall, 22′ × 28′, centered on a con-

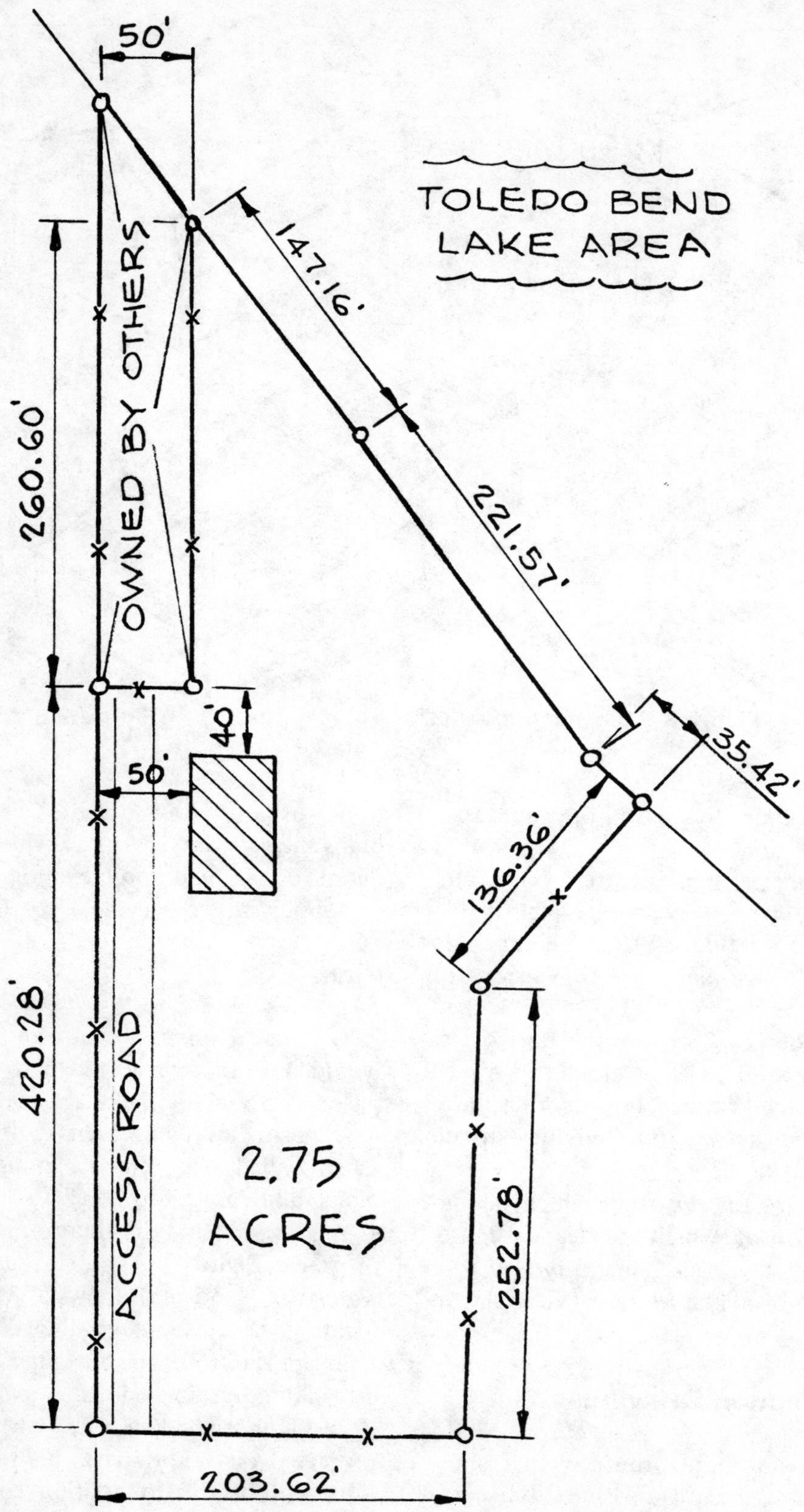

Figure 9-7. Site plan.

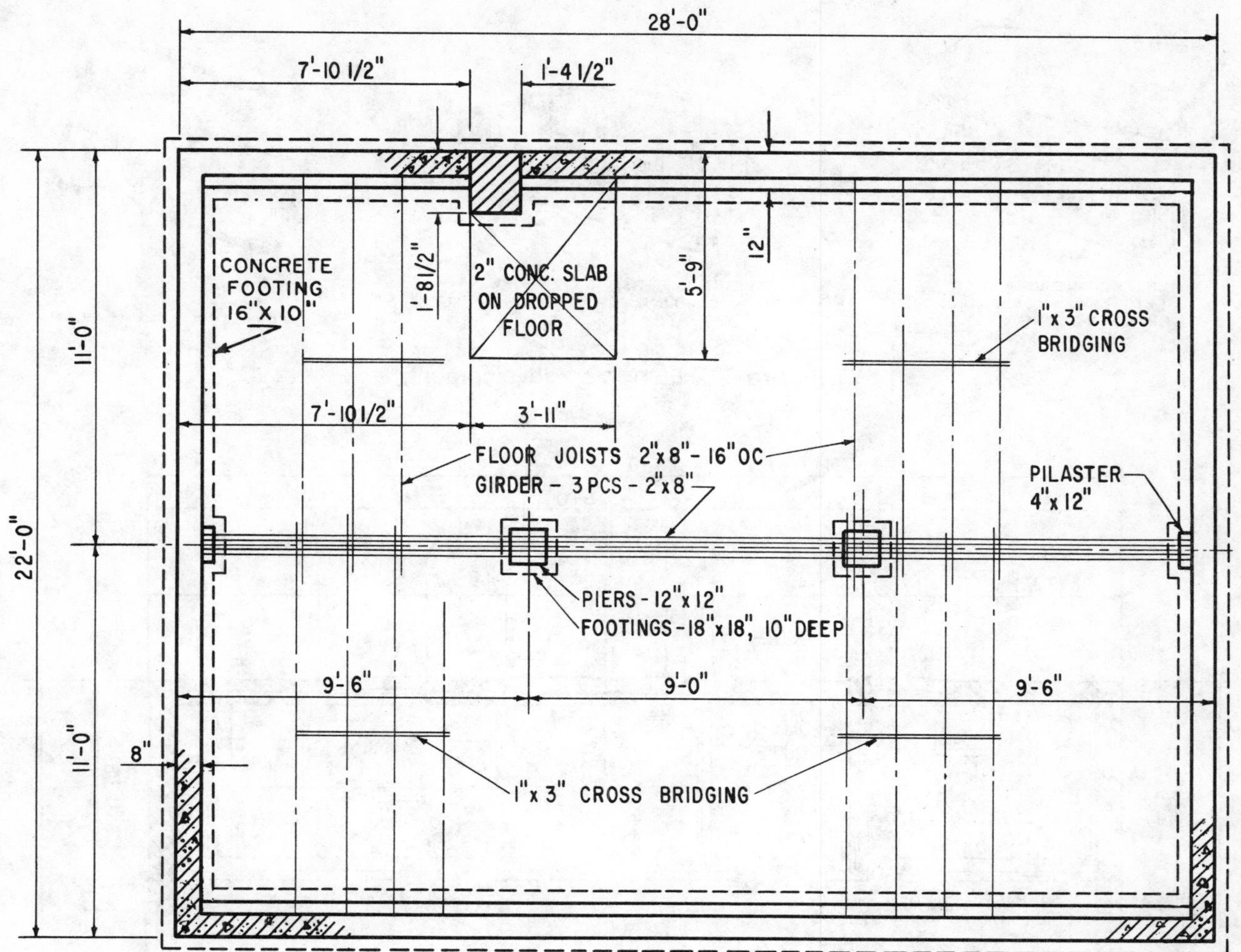

Figure 9-8. Foundation plan.

crete footing 12″ wide. Besides the outside wall and footing, there will be two 12″ square piers, centered on 18″ square footings, and located on center 9′-6″ from the end wall building lines. These piers will support a ground floor center-line girder.

A *floor plan* is developed as shown in Figure 9-9. Figure 9-10 shows the floor plan that rests on the foundation plan of Figure 9-8.

Framing plans show the dimensions, numbers, and arrangement of structural members in wood frame construction. A simple floor framing plan is superimposed on the foundation plan shown in Figure 9-8. From this foundation plan it can be seen that the ground floor joists will consist of 2 × 8's, lapped at the girder, and spaced at 16″ O.C. (on center). The plan also shows that each row of joist is to be braced by a row of 1 × 3 cross bridging. For a more complicated floor framing problem, a separate framing plan like the one shown in Figure 9-11 would be prepared. This type of plan will show how joists are boxed around stairwell and other floor openings. When a floor framing plan is not prepared, carpenters will design one and install it.

Wall framing plans are sometimes prepared. They locate studs, corner posts, bracing, sills, plates, and boxing for doors and windows. Because it is a vertical plane view, a wall framing plan is not a plan drawing in the true technical sense. It is a proposed plan of construction and the carpenter does have some leeway during construction.

Roof framing plans give similar information for rafters, ridges, purlins (a horizontal member that supports the rafters), and other structural members

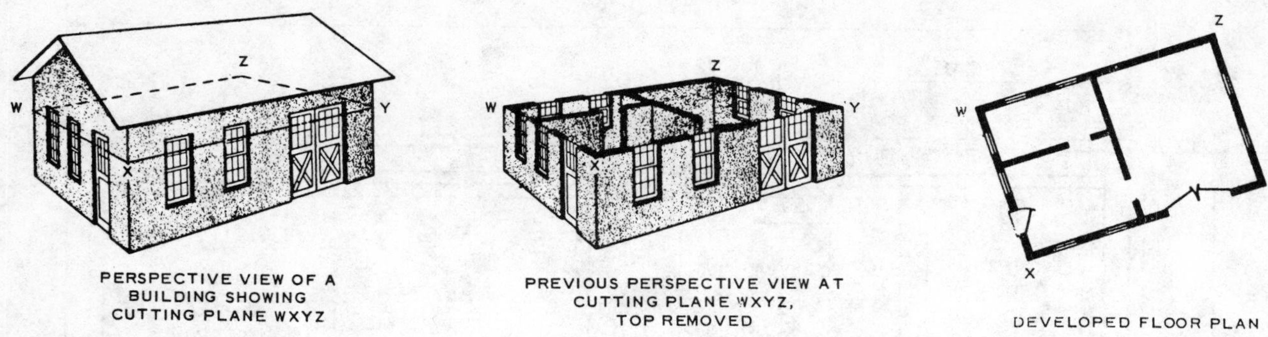

Figure 9-9. Floor plan development.

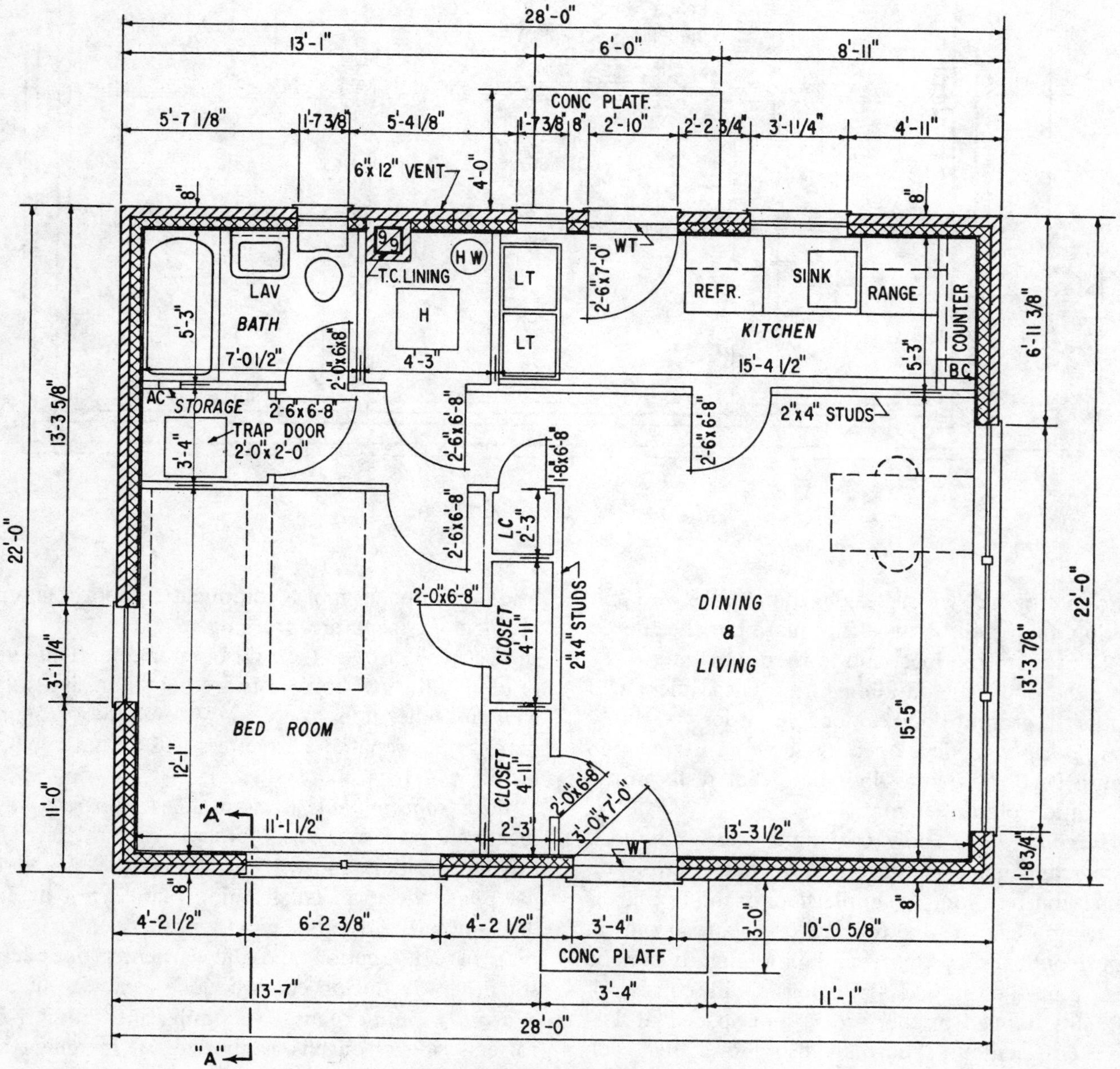

Figure 9-10. Floor plan.

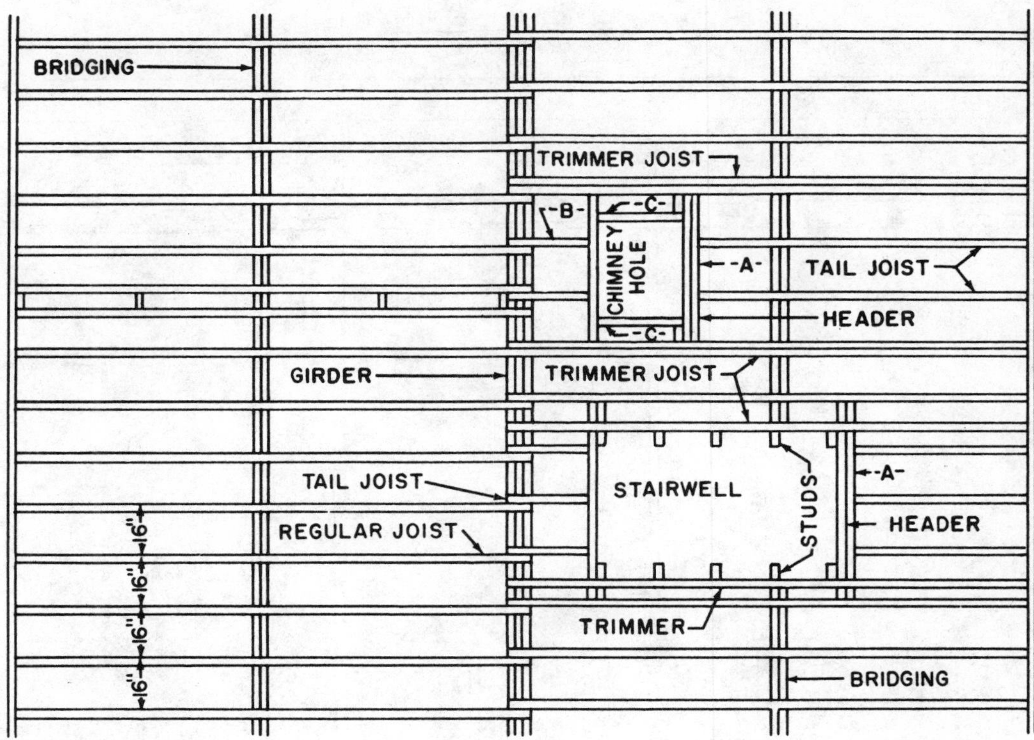

Figure 9-11. Floor framing plan.

in the roof. Except for special cases, these plans are usually not drawn and the carpenter builds the roof framing by calling on experience.

Utility plans are also made, usually using a reproducible of the floor plan. These plans will show wiring, plumbing, heating, and other necessary details. If the house is air-conditioned, the unit and all ducts will be shown.

rear, and end elevations and the direction one is looking will be given.

Figure 9-12 is a proposed house plan for the lake front property shown in Figure 9-7. Figure 9-13 provides the elevations for this plan. The specifications will provide written construction details, such as it is to be on a floating slab, type of shingles, type of brick, etc.

Elevations

Elevations show the front, rear, and sides of the structure projected on vertical planes parallel to the planes of the sides. Sometimes these are referred to as front, rear, and side elevations. When the direction is known, they are referred to as front,

Detail Drawings

Detail drawings are drawn to a much larger scale than plan and elevation drawings so precise details can be shown and noted, if necessary. Figures 9-14 through 9-17 are typical details that may be prepared for house plans.

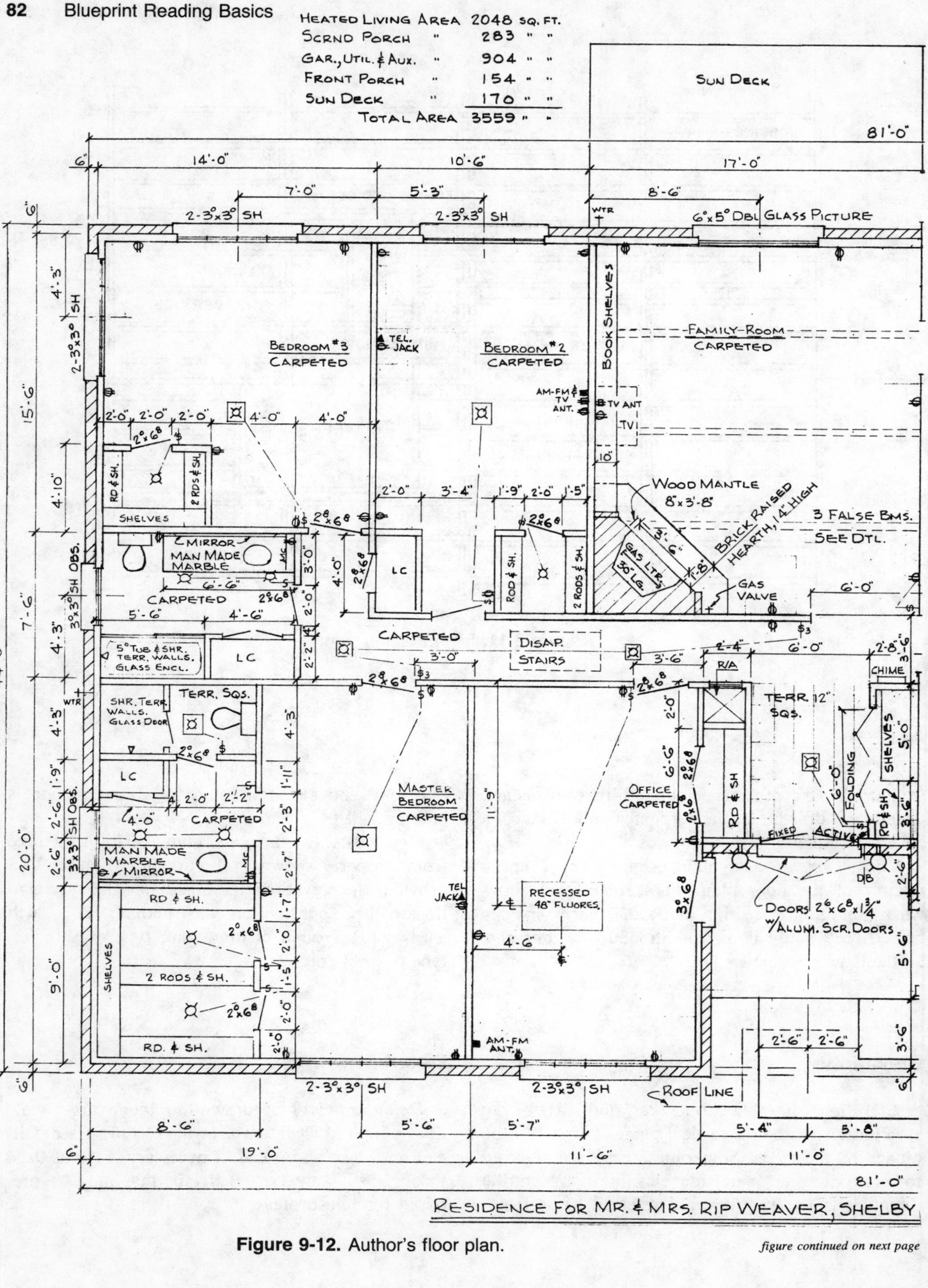

HEATED LIVING AREA	2048	SQ. FT.
SCRND PORCH "	283	" "
GAR., UTIL. & AUX. "	904	" "
FRONT PORCH "	154	" "
SUN DECK "	170	" "
TOTAL AREA	3559	" "

RESIDENCE FOR MR. & MRS. RIP WEAVER, SHELBY

Figure 9-12. Author's floor plan.

figure continued on next page

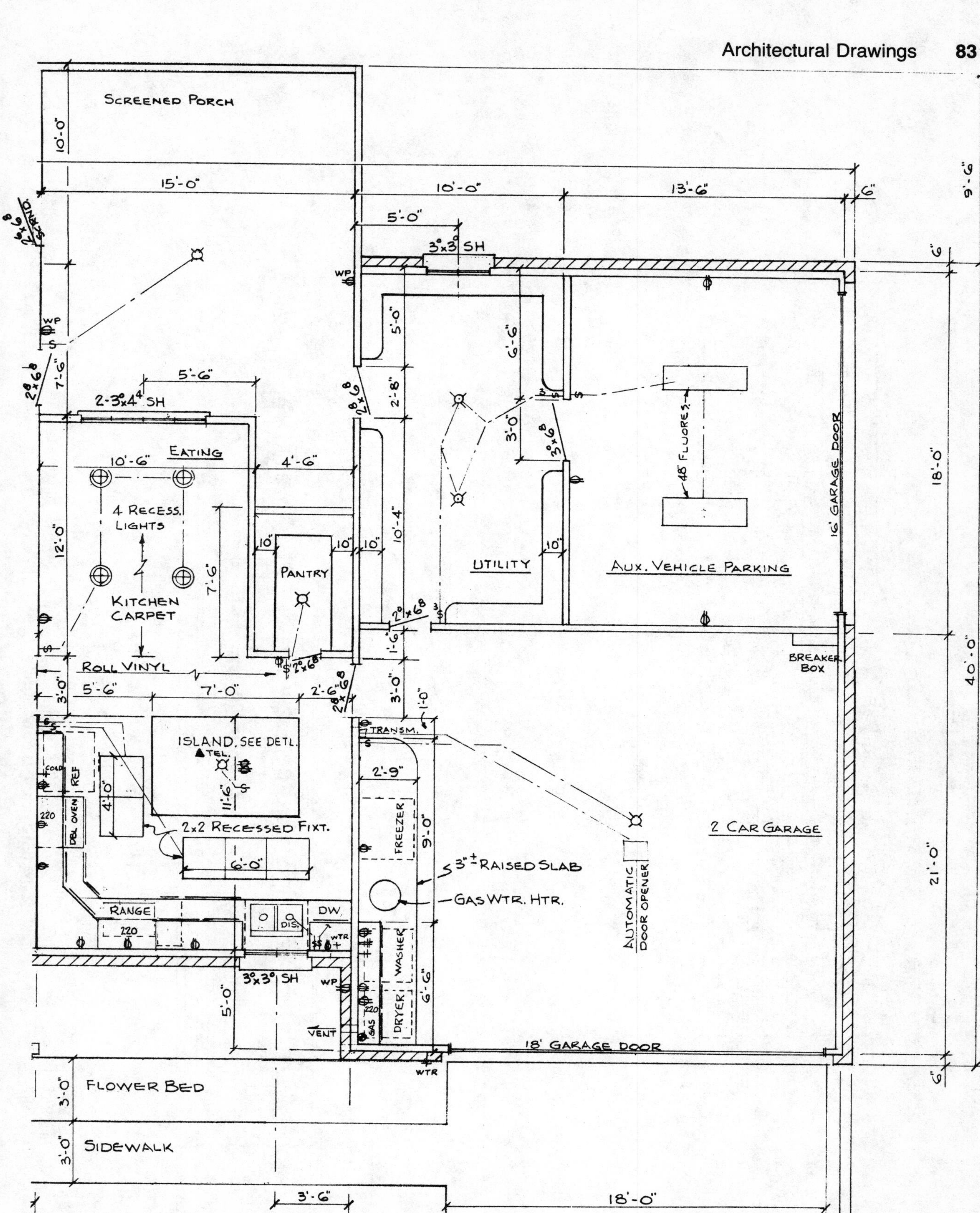
SCREENED PORCH
EATING
4 RECESS. LIGHTS
KITCHEN CARPET
PANTRY
UTILITY
AUX. VEHICLE PARKING
48" FLUORES.
16' GARAGE DOOR
ROLL VINYL
ISLAND, SEE DETL.
2x2 RECESSED FIXT.
REF
DBL OVEN
RANGE
DW
FREEZER
3" ± RAISED SLAB
GAS WTR. HTR.
WASHER
DRYER
2 CAR GARAGE
AUTOMATIC DOOR OPENER
BREAKER BOX
18' GARAGE DOOR
FLOWER BED
SIDEWALK
CO., TEXAS

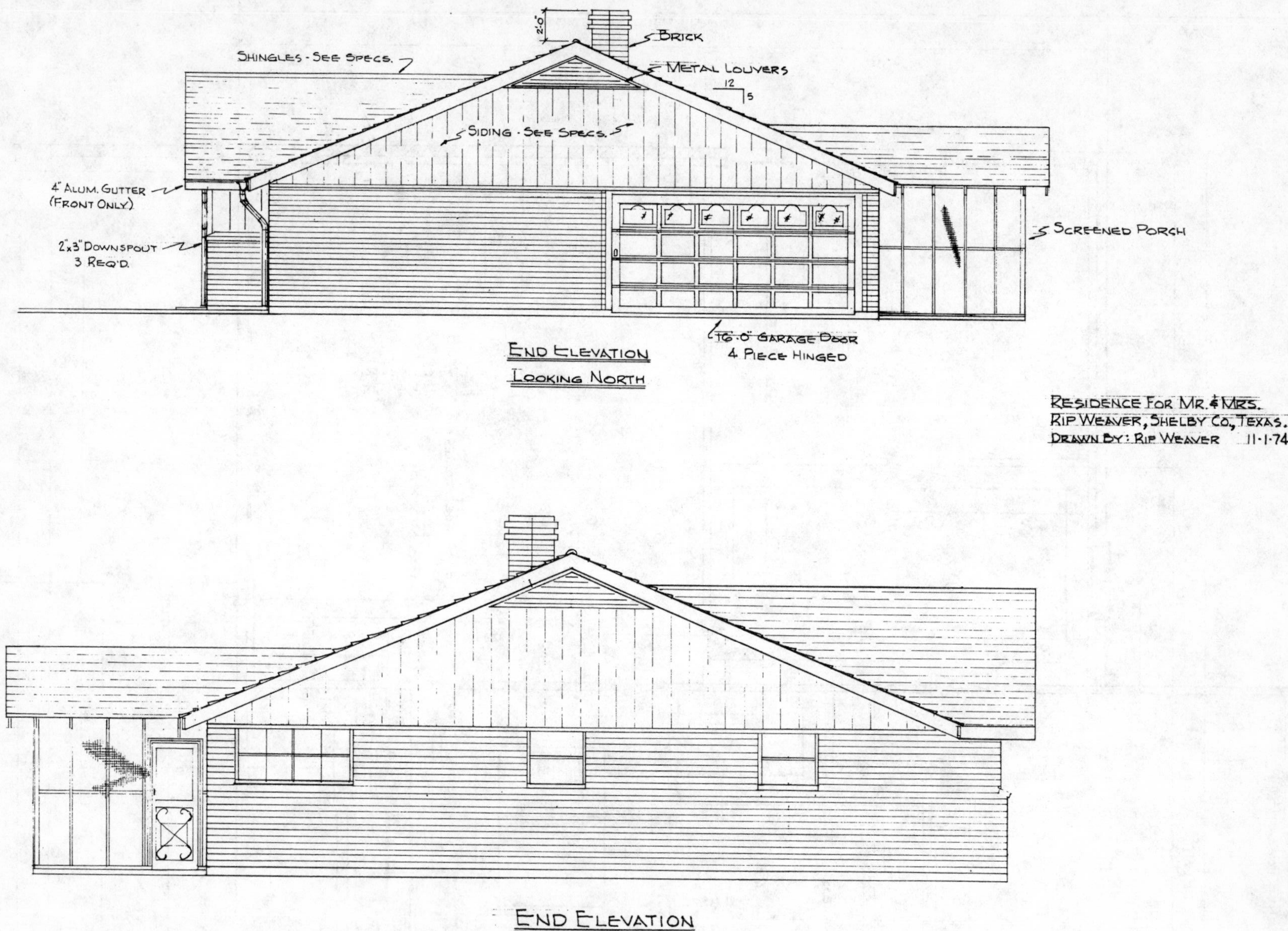

Figure 9-13. House elevation.

figure continued on next page

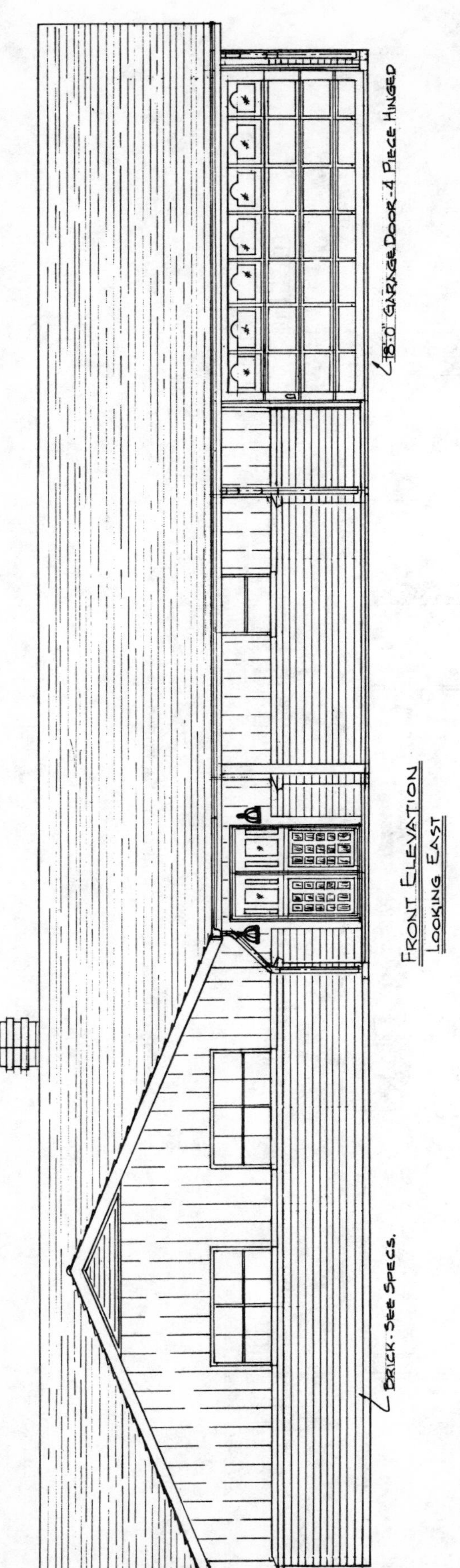

REAR ELEVATION
LOOKING WEST

Figure 9-13 continued

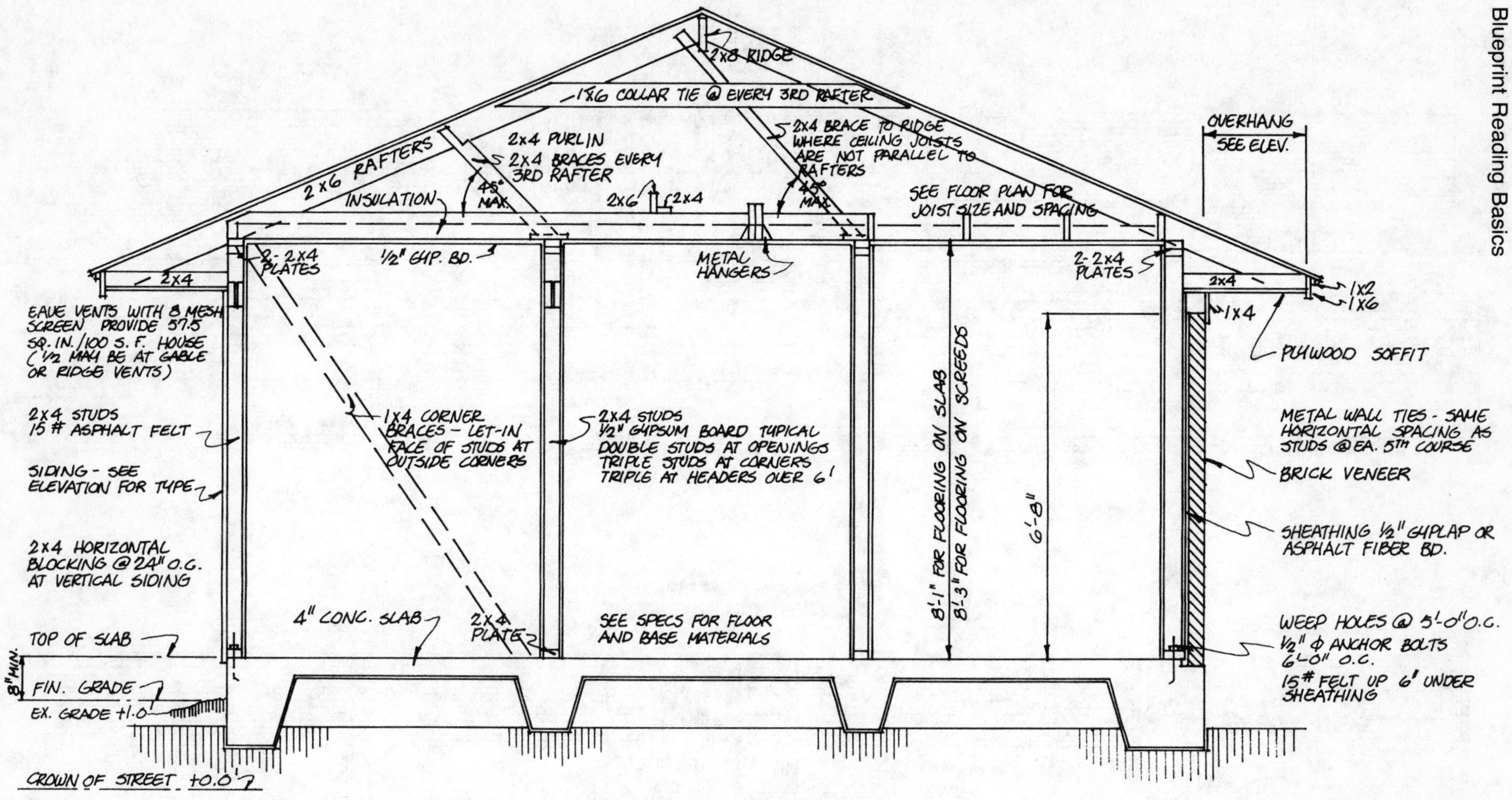

Figure 9-14. Typical construction details.

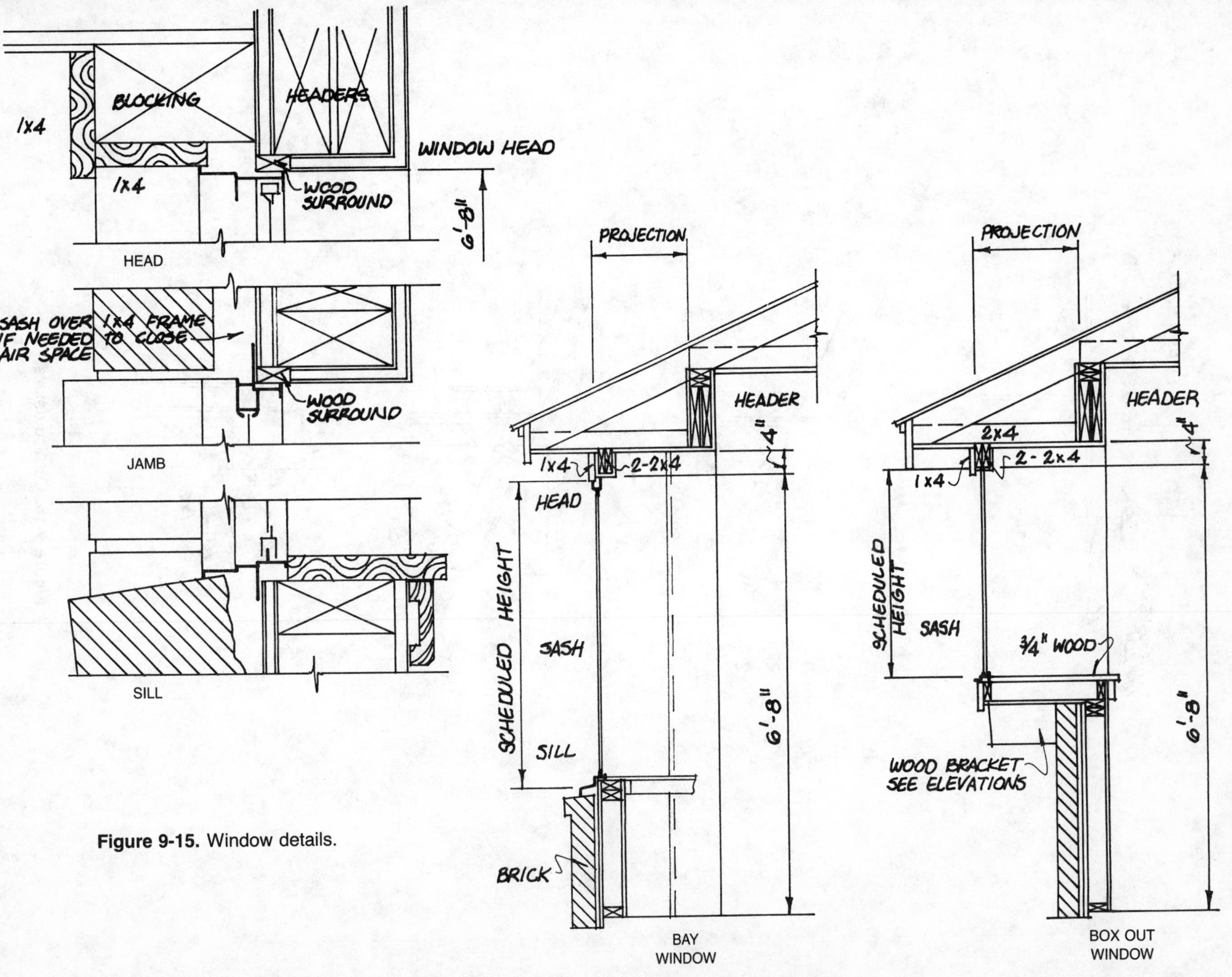

Figure 9-15. Window details.

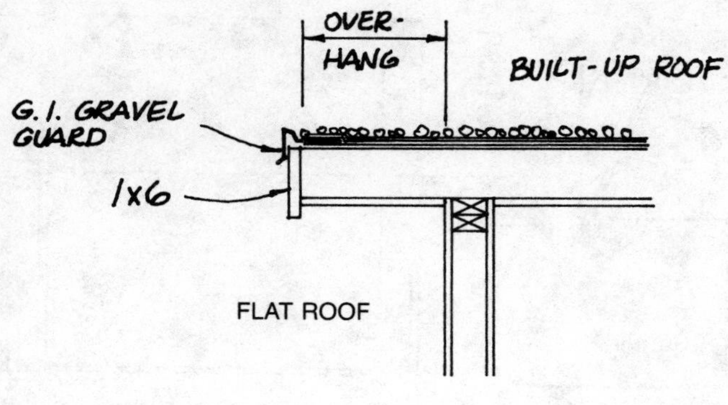

FLAT ROOF

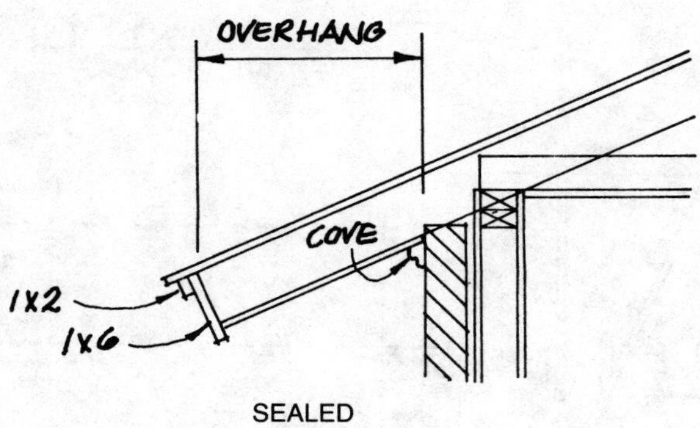

SEALED

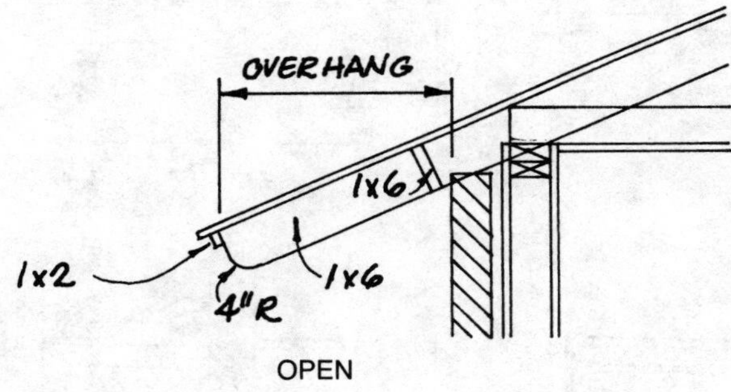

OPEN

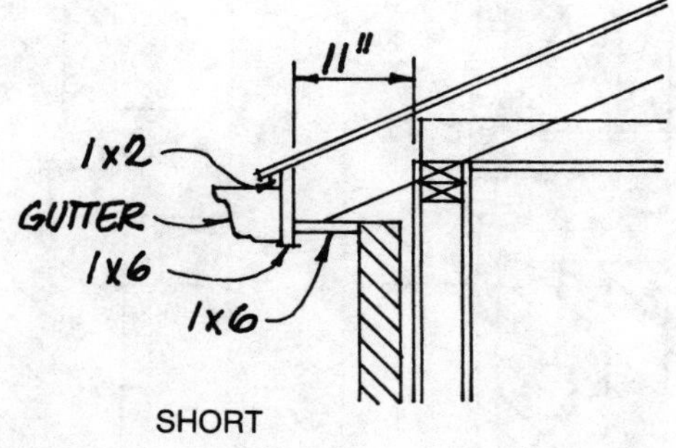

SHORT

Figure 9-16. Cornice details.

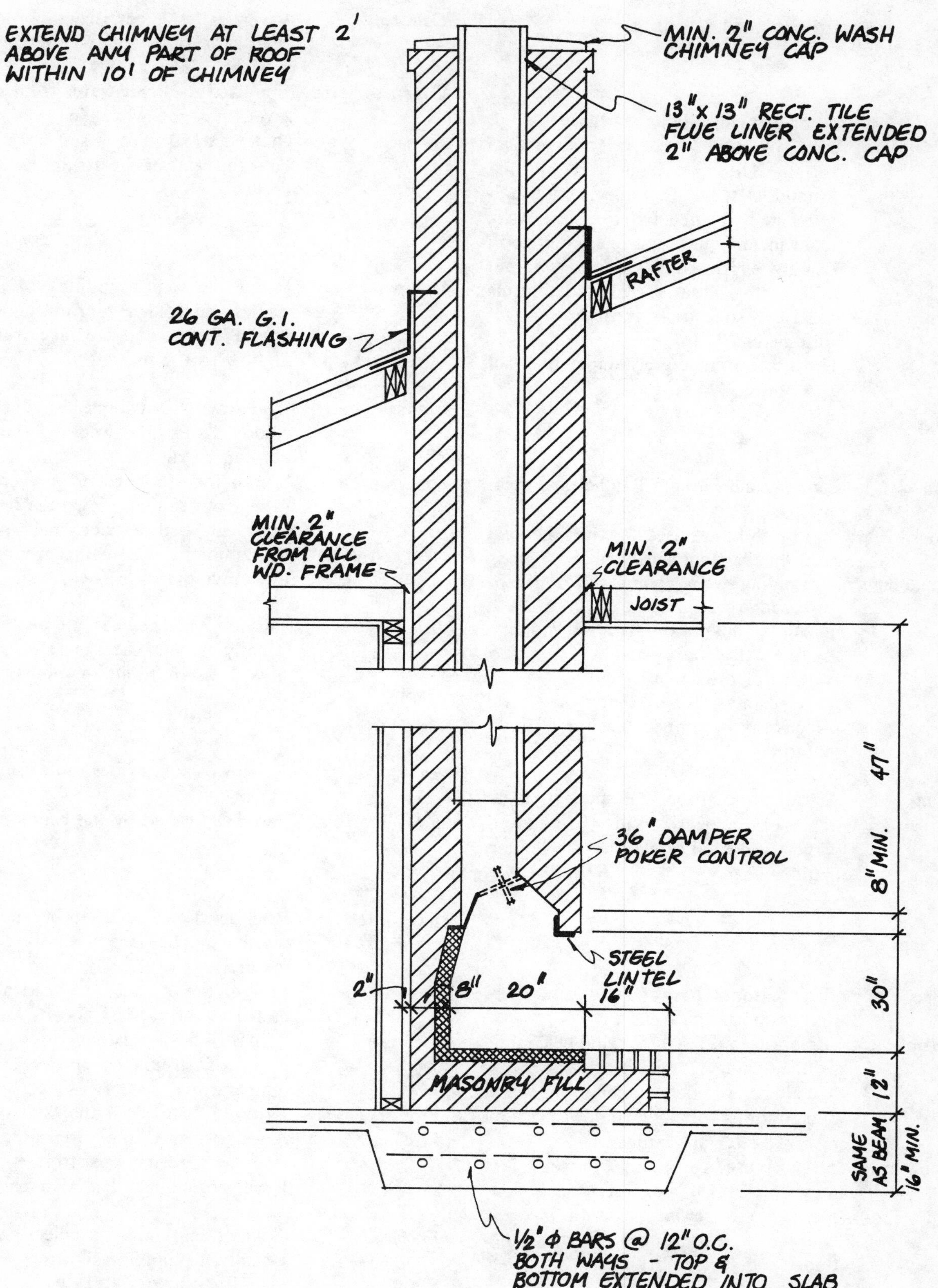

EXTEND CHIMNEY AT LEAST 2' ABOVE ANY PART OF ROOF WITHIN 10' OF CHIMNEY
MIN. 2" CONC. WASH CHIMNEY CAP
13"x 13" RECT. TILE FLUE LINER EXTENDED 2" ABOVE CONC. CAP
RAFTER
26 GA. G.I. CONT. FLASHING
MIN. 2" CLEARANCE FROM ALL WD. FRAME
MIN. 2" CLEARANCE
JOIST
36" DAMPER POKER CONTROL
47"
8" MIN.
STEEL LINTEL
2"
8"
20"
16"
30"
12"
MASONRY FILL
SAME AS BEAM
16" MIN.
1/2" ϕ BARS @ 12" O.C. BOTH WAYS - TOP & BOTTOM EXTENDED INTO SLAB

Figure 9-17. Section through fireplace.

Architectural Glossary

A

Apron Wood trim used with a window stool.

Arch Usually a curved structural area over an opening. Often carries a load.

Area Amount of space. In homes, area usually refers to heated living space. Also, could mean square footage under roof. Usually expressed as square feet, but cubic feet of area is used to determine heating and air-conditioning requirements.

Attic The space immediately under the roof of a house.

B

Backfill Earth or sand used to fill in holes or excavations.

Baseboard The finish board used to cover the joint between the floor and wall.

Base molding A molding which trims the upper or lower edge of baseboards. May be a quarter-round piece. The lower molding is often called *base shoe.*

Bat A piece of broken brick.

Beam A horizontal, load-bearing structural piece. May be timber, steel or of other material.

Bearing partition A wall or partition that supports load A non-bearing partition may be removed without structural damage but the bearing partition cannot.

C

Carriage The horizontal part of a stair's stringers that supports the treads.

Casement A glass frame, used for windows, made to open using vertical hinges.

Casing Metal or wooden members, around openings of doors and windows, applied to give better appearance.

Catch basin A device used to catch water run-off. It may be metal or concrete and is connected to a sewer or drain pipe. It collects water from roof guttering, floors and surface drainage.

Circuit An electrical current path.

Circuit breaker A device used for opening or closing an electrical circuit.

Column A vertical member, usually round or square. May or may not be load-bearing.

Conductor pipe Pipe used to direct water from roof to sewer, at or below grade.

Cornice That part of the roof that projects out from the wall of the structure.

D

Damper A movable item, usually metal, used to regulate air flow. Normally used to regulate heater and fireplace drafts and air flow to rooms.

Double-hung window A window with two sashes. The upper window half will lower and the lower half can be raised.

Drip cap A molding designed to prevent rain water from running down the face of a wall. Also used on doors and windows.

Dry well A man-made hole in the ground used for liquids disposal.

Duct Square or rectangular air passageway. Usually made of sheet metal. In heating units ducts route air to rooms.

E

Eave The lower part of the roof, usually the edge.

Eave line The line formed by the roof's eave.

F

Face brick Brick used on outside to provide an attractive appearance with its color or texture.

Fascia Part of the cornice; the vertical board nailed to the rafter's ends.

Fire brick A hard, heat-resistant brick used in parts of fireplaces subject to high temperatures.

Fire door A heavy metal door that resists fire.

Fire wall A wall designed to be fireproof. It is installed to prevent fires from spreading. Used around vaults. Also called *fire stop.*

Fixture A water containing or collecting device. Usually a plumbing fixture.

Flashing Usually sheet-metal work around openings, such as chimneys, vents, doors and windows, to prevent leaking.

Floor plan A drawing or sketch showing the layout of floors, walls, partitions, windows, doors, etc. in a house or building, as if the ceiling were removed. May be drawn as isometric or orthographic.

Flue The air duct in a chimney. Provides draft and discharges smoke to the atmosphere.

Footing The part of the foundation that bears directly on the soil. May be wider than foundation wall. Drilled, bell-bottomed footings are used in poor soil areas, including filled areas.

Framing A building's skeletal parts.

Furring The building-out or leveling-up of a ceiling or wall with lumber framing.

Fuse A piece of soft metal that is designed to melt should the current exceed the designed value. Circuit breakers have replaced fuses in new homes and buildings.

G

Gable The part of a roof design that forms an inverted *V*. Part of a gable-roof design.

Girder The large horizontal member used to support ends of beams and joists. Supports loading over openings.

Gutter A trough used to collect and direct roof drainage. May be made of galvanized sheet-metal or aluminum.

H

Header One or more pieces of lumber that support ends of joists. Used for framing openings, such as chimneys and stairwells.

Hearth The part of the fireplace flooring on which the fire is built. Also the flooring directly in front of the fireplace, usually made of fireproof material.

Hip rafter The rafter at the junction of two sloping roofs and part of the hip roof design.

Hip roof A roof sloping up from all exterior walls, especially the four main walls. Normally a hip roof has no gables.

I

Insulating ceiling A ceiling made of porous material designed to dampen sound.

Insulation A preparation placed between walls, under floors, above ceilings and around air ducts to contain heat or cold.

J

Jamb The exposed surface of a door or window opening.

Joist Pieces of timber that support the floor. Usually called floor joist.

L

Landing A platform at the top of a flight of stairs. *Intermediate landings* are located to break up long flights of stairs or ladders.

Lath Wood and metal construction. Nailed to studs to serve as plaster base support.

Lavatory A basin or fixture used for washing.

Linear foot A measurement of one foot in a straight line.

Load-bearing Any item that supports part of a structure's load or weight.

Lot lines A lot's limits. Boundary lines.

Louvers Shutters used for air flow.

M

Mantel The shelf over a fireplace.

Masonry Material such as brick, concrete blocks, stone, stucco, etc. used in construction.

Mastic An asphaltic material used around flashing to seal off leakage. Synthetic mastic is also available.

Metal wall ties Strips of metal used to tie a brick veneer wall to the house framing.

Miter A beveled cut on the ends of wood.

Mortar A mixture of sand and cement, sometimes with lime added, serving as a masonry bonding agent.

Mullion The vertical bar separating two adjoining windows.

Muntin The small window bars separating glass panes.

N

Nailers 1″ x 4″ lumber placed on rafters. Shingles are nailed to these.

Nosing The rounded edge of a stair tread.

O

Outlet Where something exits. Air outlets are called registers. Electrical outlets are where appliance plugs are connected to obtain electricity.

Overhang The roof portion that extends past the exterior walls. Usually referred to with a dimension, such as "″24″ overhang."

P

Panel A flat thin piece. May be wood or other material.

Paneling Wood wall finish, usually stained to enrich the natural texture and grain.

Partition A wall dividing two rooms or areas. Usually not load-bearing partition (see *bearing partition*).

Pitch Slope, such as roof pitch or sewer line pitch.

Plank Term applied to lumber 1½″ or more in thickness.

Plate The horizontal 2″ x 4″ member located on top of a row of studs.

Priming Preparing a surface by applying a coat of sealer to surface pores.

Purlin The structural member spanning from truss to truss supporting roof rafters.

R

Rafter A member used to support the roof. Runs from roof ridge to cornice plate.

Ridge The top edge of the roof where two slopes meet.

Rise The vertical distance from the center of a span to the roof ridge.

Riser The vertical piece of a stair.

S

Sash A frame for holding glass in a door or window.

Septic tank A tank buried in the earth which collects sewage, discharging to a septic field.

Sheetrock An inexpensive board material made of gypsum, used for interior walls.

Shingles A roof covering made of small pieces. Shingles may be wood, asphalt, aluminum or other material.

Siding The final boards on a building's exterior walls.

Sill Wood, stone or metal member across the bottom of a door or window opening.

Soffit The underside of an arch.

Sole The horizontal wooden member on which the frame's studs rest.

Stool The wood shelf across the inside bottom of a window.

Stringer The inclined main stairway support.

Studs The vertical members, usually 2″ x 4″, that form the house skeleton in a frame building.

Subflooring Flooring nailed directly to floor joists, which supports the finished flooring.

T

Terra cotta A mixture of baked clay and sand.

Terrazzo A mixture of marble chips and cement ground and highly polished. Used as floors, shower floors and walls and sometimes as room walls. Usually mixed and formed in place, but manufactured 12″ terrazzo squares are available to be grouted together in place.

Tie beam A framing member used to strengthen rafter spans by connecting two or more rafters.

Trap A U-shaped device located beneath plumbing fixtures. The trap stays full of liquid and forms a liquid seal which prevents sewer odors from passing.

Tread The horizontal stair member that is stepped on.

Truss A braced framework used for long spans between supports.

U

Unfinished lumber Lumber that is rough finished, cut and ready for finishing.

Used brick Brick that has been salvaged from another building. It is very attractive but costly. New brick is now made to resemble used brick.

V

Veneer In wood, a thin layer of an expensive wood applied over a less expensive wood, such as plywood. In masonry, a thin layer of masonry over wood or other masonry. Many brick homes are brick veneer, having only one layer (or course) of brick outside of a wood frame.

W

Wainscot A wood or brick lower wall with upper wall of a different material.

Weather-stripping A metal or fabric strip inserted around door and window edges to keep out rain and outside hot or cold air.

CHAPTER 10

Concrete Foundation Drawings

The normal sequence of events of any construction project is to excavate, set forms, locate reinforcing bars within the forms, fix the anchor bolts in a jig, and then pour the concrete foundation. Until the foundation is poured, no above-grade work can begin. Naturally, the field forces want concrete foundation drawings early in the schedule so drawing room effort is applied early in the job to issue these drawings.

Concrete Defined

The ACI (American Concrete Institute) defines concrete as a *mixture of portland cement, fine aggregate, coarse aggregate and water.* The fine aggregate is usually sand and the coarse aggregate is normally gravel. This solution dries to form a solid that weighs about 4,000 pounds per cubic yard. Many people mistakenly refer to concrete as cement. Actually, cement is a dry powder used in making concrete. The ratio of cement to water, per cubic yard of concrete, determines the concrete's strength. The more sacks of cement used, the stronger the concrete becomes.

Cement weighs 94 pounds per sack. A home foundation will normally have four and one-half sacks of cement per cubic yard of concrete; so by weight this concrete is about 10% cement. It will carry a specification rating of 2,500 pounds at 28 days. The 2,500 pounds refers to the concrete's compressive strength and is specified in pounds per square inch. The 28 days refers to the time span from the date the concrete was poured. ACI Standard 318, *Building Code Requirements for Reinforced Concrete,* supplies complete specifications and testing procedures for concrete construction.

For a construction project, the concrete specifications will call for a certain strength rating at a set number of days, usually 28 days. This determines the number of cement sacks required per cubic yard of concrete, water volume, and aggregate size. As the concrete is being poured, the concrete truck driver will fill several test cylinders from the mix. These are marked and sent to a testing laboratory. In seven days the first cylinder is tested. To pass, it must withstand 60% of the specified 28 day compression strength. If it was to be 3,000 pound test concrete at 28 days, the seven day test would have to withstand 1,800 pounds per square inch of compression. When the test cylinders are 28 days old, two more cylinders are tested and their average compression test must be 3,000 pounds per square inch minimum.

Reinforced Concrete

Concrete is classified as plain or reinforced. Plain concrete has no mesh (see below) or reinforcing bars. Plain concrete is used for cases where only compression stresses occur, such as a pump foundation. However, many pump foundations have stresses other than compression and some reinforcement is required. Most foundations are subject to various types of forces transmitting stresses into the concrete. Rebars are placed at strategic points in the foundation to withstand stresses, primarily tension and shear forces, but reinforcing steel also helps resist compression. Reinforcement also helps the concrete to resist cracking due to temperature differences, which causes expansion and contraction, and vibration.

Reinforcing Bars

Round rolled deformed steel bars are used to reinforce concrete. As prevously mentioned, rebars are identified by a number which indicates their size in eighths of an inch. A #7 bar is a deformed bar with a nominal diameter of ⅞″. Rebars range in size from #2 to #11 with #14S and #18S as special sizes sometimes available.

All bars are deformed except #2, which are plain round bars. The deformations are regular spaced knots or rises on the bar, which increase the bond between the bar and the concrete. Rebars are made by a steel mill using a rolling process. Normal mill rolling length is sixty feet. The ASTM (American Society for Testing and Materials) publishes specifications which govern bar material, strength, and deformations. Most construction jobs require rebars to have a minimum yield strength of 40,000 psi (pounds per square inch). Table 10-1 provides rebar data.

Wire Mesh

Welded wire fabric, often called wire mesh, is used as reinforcement for concrete slabs, such as home foundations and paving. This mesh is cold drawn steel wires electrically welded together at right angles forming a mesh of wire squares or rectangles. Wire fabric made of deformed wire is also available but most wire mesh is of plain wire.

Welded wire fabric is designated by WWF, the wire spacing, and the wire gage. For instance, WWF 6 x 6-10/10 is welded wire fabric, wires spaced six inches each way (6″ square) and which are #10 gage thick in both directions, about ⅛″ in diameter. This is the size of wire mesh most commonly used in road and slab construction.

Anchor Bolts

Anchor bolts are embedded in the concrete foundation and project outside and above the top of grout. They firmly anchor equipment such as ves-

Table 10-1. Rebar Data

Bar Size number	Weight # Per Foot	Diameter, Inches	Cross Sectional Area, Sq. Inches	Perimeter, Inches
Standard ASTM A-305 Reinforcing Bars				
2	0.167	0.250	0.05	0.786
3	0.376	0.375	0.11	1.178
4	0.668	0.500	0.20	1.571
5	1.043	0.625	0.31	1.963
6	1.502	0.750	0.44	2.356
7	2.044	0.875	0.60	2.749
8	2.670	1.000	0.79	3.142
9	3.400	1.128	1.00	3.544
10	4.303	1.270	1.27	3.990
11	5.313	1.410	1.56	4.430
Special ASTM A-408 Reinforcing Bars				
14 S	7.65	1.693	2.25	5.32
18 S	13.60	2.257	4.00	7.09

sels, pumps, compressors, or steel structures to their foundations. Anchor bolts are made from round steel rod. The J-bolt (see Figure 5-1) is bent to form a J shape, or sometimes an L shape, and the bent end is embedded in the concrete to resist pulling out. The end projecting out of the concrete is threaded for one or usually two hex head nuts.

Square head machine bolts are also used as anchor bolts for smaller forces. The square head, with a washer, is embedded in the concrete. Figure 10-1 shows the square head machine bolt.

Anchor bolt diameters are designated by a number which indicates the bolt's diameter in eighths of an inch. A #8 anchor bolt is 1″ in diameter and a #12 bolt is 1½″ in diameter. Except for very special cases, #12 anchor bolts are the largest size used.

Foundation Parts

Figure 10-2 is an octagon-shaped foundation for a vertical vessel. Note that the **pedestal** rests on the **spread footing**, often referred to as the **footer.** If this foundation had no spread footing, the pedestal would become the bearing member and would rest on the earth. Pedestals for vessels and pumps stop one foot above grade, usually, but this isn't a universal dimension. Some companies use six or nine inches above grade as the top of grout. For exchangers and some other equipment, pedestals may extend several feet above grade.

While Figure 10-2 shows the octagon-shaped pedestal and spread footing, they may be square, rectangular, or any other shape that is easy to form and fits the supported equipment.

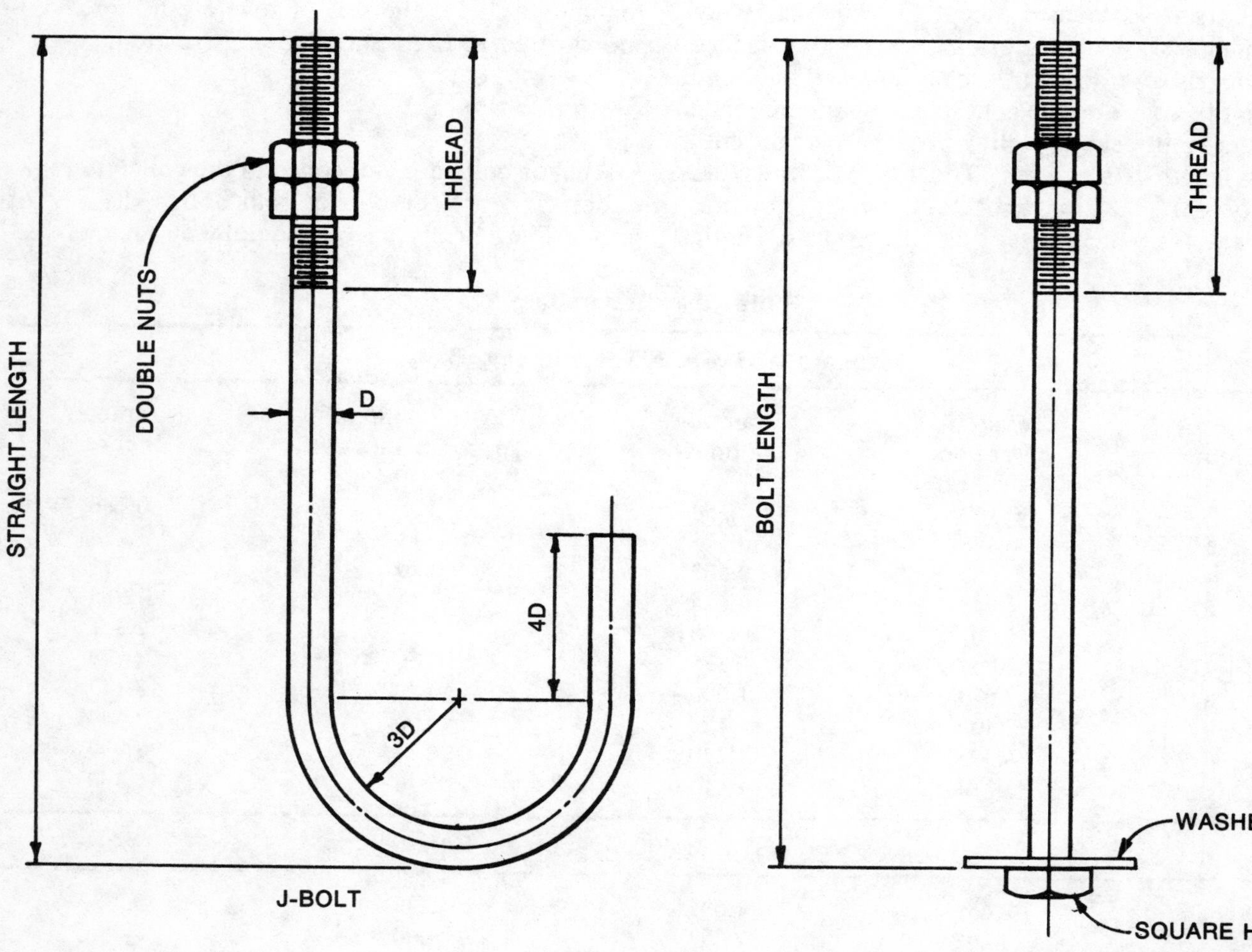

Figure 10-1. Anchor bolts.

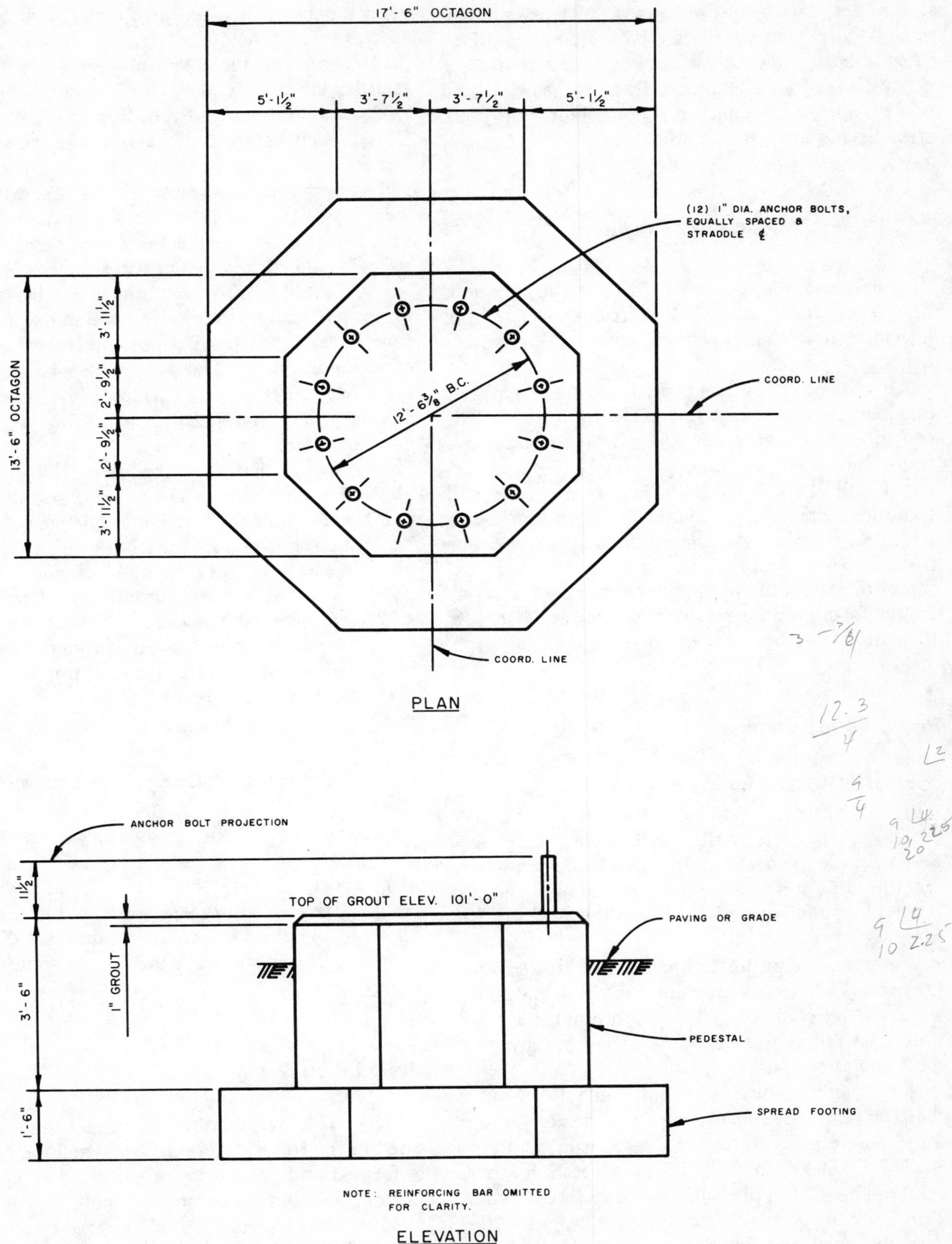

Figure 10-2. Octagon-shaped foundation for a vertical vessel.

For vertical cylindrical items, such as vessels, anchor bolts are located on a **bolt circle,** called BC. Anchor bolts are evenly spaced and are increased in quantities of four so each quadrant of the bolt circle will have the same number of bolts. Note that bolts straddle the normal centerlines.

Foundation Drawings

Concrete foundations serve two main purposes. They keep the supported item from slowly sinking into the ground and prevent overturning during high winds.

All soils have a certain resistance to weight, but this resistance varies with soil composition. This resistance is called **allowable soil bearing pressure** and is expressed in pounds per square foot at a certain depth. This depth will vary by area and soil conditions but it is always at or below the area's **frost line,** the maximum depth that soil freezes. The frost line varies from ten feet in the northern United States to one foot in the southern regions. However, depths for allowable soil bearing are rarely set less than three feet below finished surface elevation.

Pump Foundations

Figure 10-3 is a horizontal pump foundation. It projects one foot above the high point of area paving, so the bottom of the foundation is three feet below the paving's high point. This is the depth that the allowable soil bearing is set for this foundation. As a rule, good soil will have an allowable of 3,000 psf.

The pump's coordinate lines (crosshair lines) are the centerline of discharge and the main centerline, which is the shaft centerline. Coordinates are not shown on foundation drawings. They are given on the foundation location plan.

The foundation in Figure 10-3 contains 3.7 cubic yards of concrete. Calculating the concrete yardage for all items is the draftsman's responsibility. Pump P-1 is 7.5′ x 3.33′ x 4′ = 99.99 cubic feet. Since there are 27 cubic feet in one cubic yard, 99.99 divided by 27 is 3.7 cubic yards. Since concrete weighs 4,000# per cubic yard, this foundation weighs 14,800#. The pump and motor it supports weighs 2,600# so the total weight bearing on the 3,000# allowable soil is 17,400#. The concrete exterior dimensions are 7.5′ x 3.33′ which provide a bearing area of 24.97 square feet. Dividing 17,400# by 24.97 gives 696.84 pounds per square foot actually bearing against the soil. This is well below the 3,000# per square foot allowable.

The pump rebar is shown as MK 402 or 403. The 4 indicates the bar is 4/8″ or ½″ in diameter. The other two digits are the mark numbers. This is a straight, unbent, rebar but if a bar is to be bent, it will be detailed on a reinforcing bar bending diagram. Normally wire mesh is used in pump foundations at the top to prevent chipping, especially at the edges and corners. The pump's weight, or load, is relatively small and provides only compression stresses, which are applied straight toward the foundation.

The six 1″ diameter anchor bolts project 3¼″ above the top of grout. The pump base will have six bolt holes matching these bolts. The pump base will be about 1¾″ thick and the bolts will project 1½″ above the base so that a nut and lock nut may be applied. Note that each anchor bolt is enclosed in a 3″ x 12″ long sheet metal sleeve. This sleeve is fixed in the form and concrete is poured around it but not in it. This allows some tolerance in setting bolts and in matching the pump base holes.

Concrete Paving at Equipment Elevations

Concrete paving, or finished grade, is shown at equipment elevations but no elevation number is given because paving or grade is sloped towards the catch basin and the exact elevation at the equipment isn't calculated because it isn't important. The only elevation of importance is the top of grout, which is always given.

Foundation Grout

Grout is a rich mixture of cement and sand applied to the top of a foundation. The base concrete is formed and poured to an elevation equal to the bottom of grout. Shim stock (small, thin pieces of steel) is added to the top of the base concrete and the equipment is placed on the foundation. Necessary shims are added to ensure the equipment is

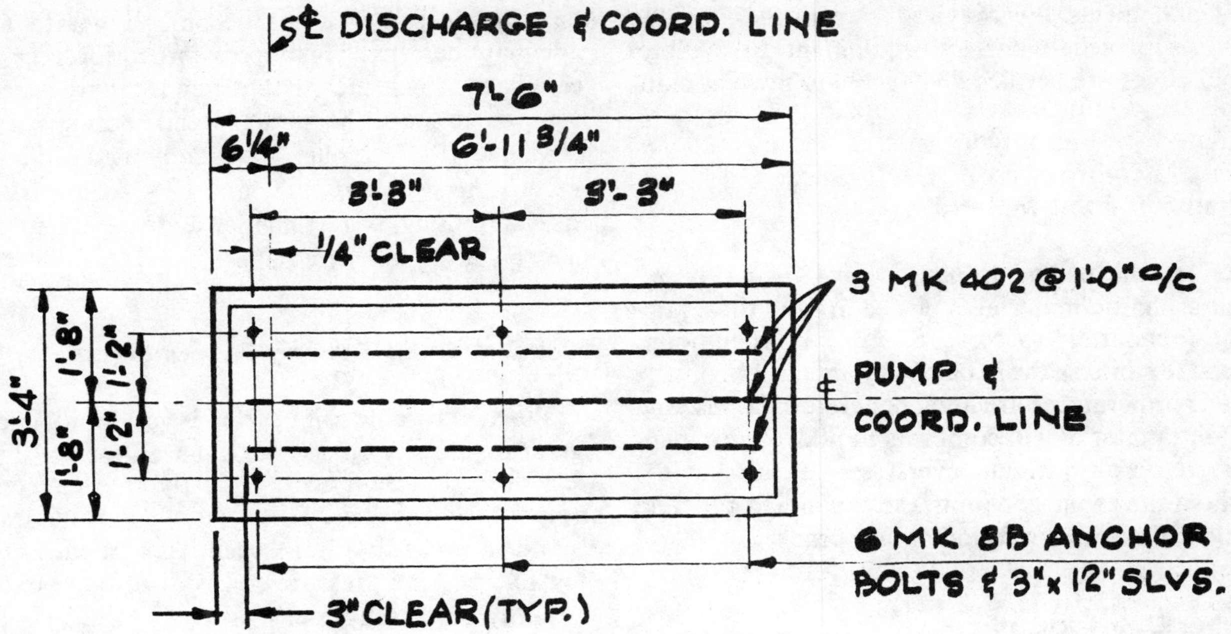

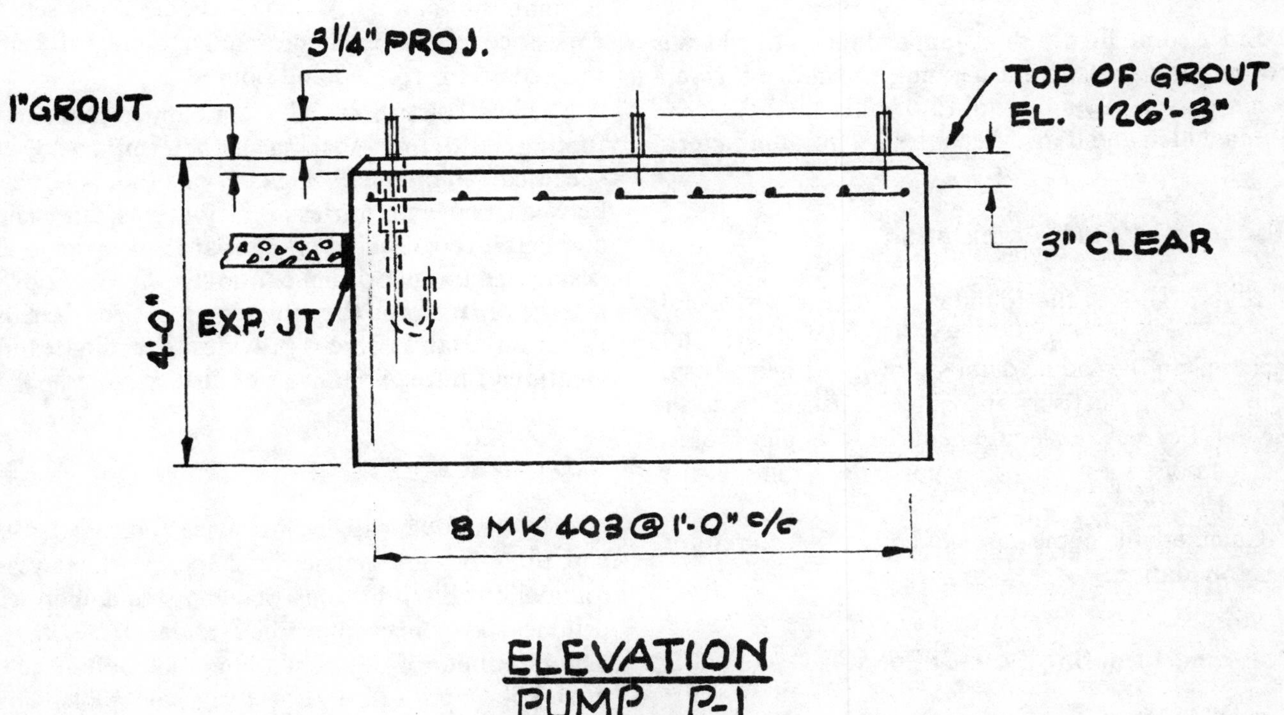

Figure 10-3. Horizontal pump foundation.

level and meets the required elevation. Then the grout is placed under the equipment base plate. Grout edges are beveled 45 degrees to prevent chipping.

Expansion Joint Material

Before the paving is poured, a 4″ wide by ½″ thick asphaltic material is placed around the equipment foundation to serve as an expansion joint, much the same as the expansion joints in a highway. When temperatures are high, concrete expands, and the joint material will compress to allow this expansion to occur without overstressing the paving. Without the expansion joint, the expanding concrete would be overstressed and would crack.

Anchor Bolt Projection

Anchor bolts should project above the top of grout a dimension equal to the sum of

1. The equipment base plate thickness.
2. Two anchor bolt diameters for two nuts.
3. One anchor bolt diameter.

Since some firms use a standard lock nut which is not as wide as a regular hex nut, the above formula is not universally adherred to. In this case, item 3 is not included and item 2 becomes 1¾ bolt diameters.

Horizontal Vessel Foundation

Figure 10-4 is the foundation for the horizontal vessel shown in Figure 10-5. The bolt holes and piers match the saddle details shown in Figure 10-6. The top of grout is 14′-6″ above finished grade or paving. For elevated equipment the term **pier** is used (instead of pedestal) for the concrete resting on the spread footings.

Calculate the cubic yards of concrete required for this foundation.

Combined Footing Foundations

When equipment is very congested or allowable soil bearing is very low, spread footings often interfere with each other. With low allowable soil bearing, spread footings become larger to attain adequate bearing. This forces spread footings to be **combined** which means that two or more pieces of equipment share the same spread footing.

Figure 10-7 is a plan of the combined footing for vertical vessel V-10 and pumps P-2A and B. This spread footing is rectangular but spread footings may be any geometric shape.

Bell-bottomed Footing Foundations

When allowable soil bearing is very small and at a great depth, say 1,000# per square foot at 10 to 20 feet deep, it would be very costly to excavate and pour spread footings. This type of soil is commonly found in areas that have been back filled. Low allowable bearing pressure soil is also found in some swampy or marshy areas.

For this type of soil, bell-bottomed footings are used. Figure 10-8 illustrates this type of foundation. The area is excavated to an elevation equal to the shaft's top. Then holes are drilled and belled out at the bottom to provide more bearing area. The drill is removed and necessary rebars are placed in the drilled hole. Then, the hole is filled with concrete, forming the footing. After this concrete has set, the **cap**, or cover concrete, is formed and poured. Later, the pedestal is formed and poured.

The cap distributes the load uniformly to the footings. Footing spacing is determined by the specified bell diameter. Usually one foot is left clear between bell extremities. Bell footings dimensions and details are located on a separate drawing which is supplied to the drilling contractor. Another drawing is prepared for the contractor called, "Drilling Location Plan" which provides coordinates and locations for each shaft to be drilled.

Cap Shapes

A rectangular cap is shown in Figure 10-8 but caps may be any geometric shape. Vertical vessel foundations often have an octagon-shaped cap with drillings on concentric circles under the octagon. The proximity of adjacent foundations often determine the cap's shape. For congested plots, combined footings are common with many shafts and several pedestals or piers located on the same cap.

(text continued on page 108)

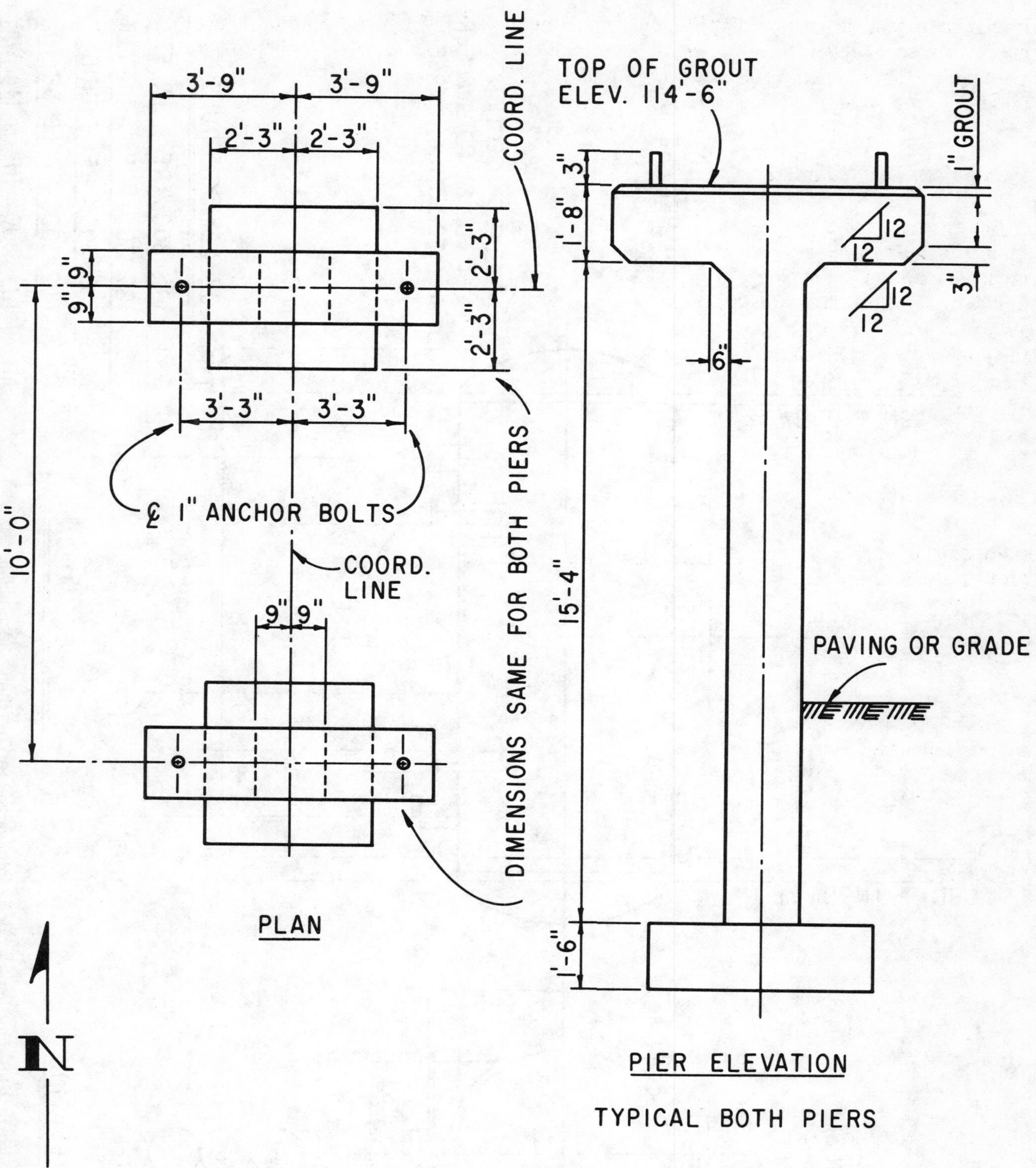

(NOTE: REINFORCING BAR OMITTED FOR CLARITY)

Figure 10-4. Foundation for a horizontal vessel shown in Figure 10-5.

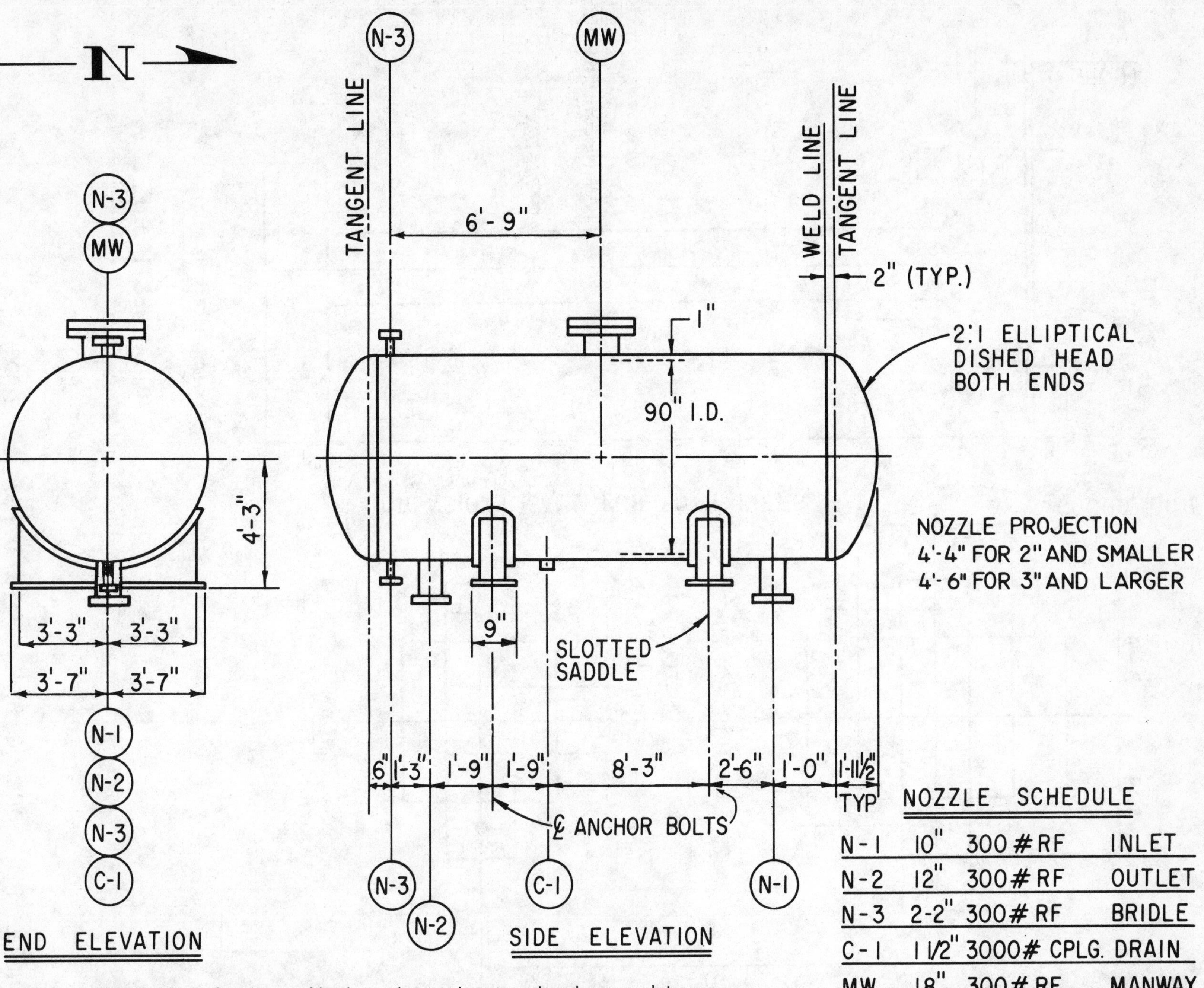

Figure 10-5. One type of horizontal vessel, an overhead accumulator.

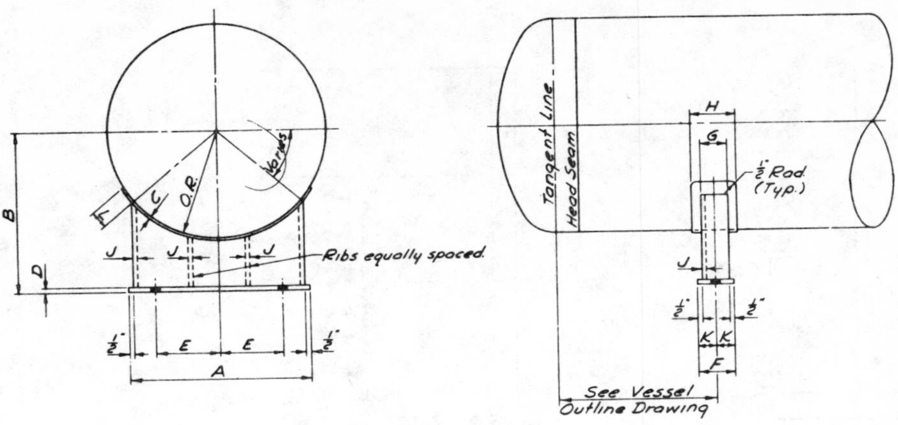

TYPICAL SADDLE DETAILS—EXCLUSIVE OF MARK 24a, 24b & 30a

VESSEL DIAM. IN INCHES	SADDLE MARK Nº	MAX. VESSEL OPER. WEIGHT LBS.	A	B	C	D	E	F	G	H	J	K	L	BOLT DIAM.	Nº OF RIBS	WEIGHT PER SET LBS.
24	24a	1,070	1'-10"	1'-3"	1/4"	1/2"	7"	4"	—	3"	3/8"	2"	1 1/4"	1"		70
	24b	3,770	"	"	"	"	"	"	—	"	1/2"	"	"	"		80
	24c	20,000	"	"	"	"	9"	"	3"	5"	1/4"	"	"	"		80
	24d	31,000	"	"	"	"	"	"	"	"	3/8"	"	"	"		90
	24e	42,000	"	"	"	"	"	"	"	"	1/2"	"	"	"		100
30	30a	2,840	2'-2"	1'-6"	1/4"	1/2"	9"	4"	—	3"	1/2"	2"	1 1/2"	1"		120
	30b	20,000	"	"	"	"	11"	"	3"	5"	1/4"	"	"	"		100
	30c	31,000	"	"	"	"	"	"	"	"	3/8"	"	"	"		130
	30d	42,000	"	"	"	"	"	"	"	"	1/2"	"	"	"		150
36	36a	20,500	2'-8"	2'-0"	1/4"	1/2"	1'-0"	6"	5"	7"	1/4"	3"	2"	1"		170
	36b	32,500	"	"	"	"	"	"	"	"	3/8"	"	"	"		200
	36c	42,500	"	"	"	"	"	"	"	"	1/2"	"	"	"		220
	36d	52,500	"	"	"	5/8"	"	"	"	"	3/8"	"	"	"		260
42	42a	25,000	3'-2"	2'-3"	1/4"	1/2"	1'-3"	6"	5"	7"	1/4"	3"	2 1/4"	1"		200
	42b	34,000	"	"	"	"	"	"	"	"	5/16"	"	"	"		220
	42c	48,000	"	"	"	"	"	"	"	"	1/2"	"	"	"		270
	42d	97,500	"	"	"	5/8"	"	"	"	"	3/8"	"	"	"	1	330
48	48a	27,500	3'-8"	2'-6"	1/4"	1/2"	1'-6"	6"	5"	7"	1/4"	3"	2 1/2"	1"		230
	48b	39,000	"	"	"	"	"	"	"	"	5/16"	"	"	"		250
	48c	62,500	"	"	"	"	"	"	"	"	1/2"	"	"	"		310
	48d	116,250	"	"	"	5/8"	"	"	"	"	5/8"	"	"	"	1	380
54	54a	25,000	4'-2"	2'-9"	1/4"	1/2"	1'-9"	6"	5"	7"	1/4"	3"	2 3/4"	1"		270
	54b	47,500	"	"	"	"	"	"	"	"	3/8"	"	"	"		310
	54c	75,000	"	"	"	"	"	"	"	"	1/2"	"	"	"		360
	54d	135,000	"	"	"	5/8"	"	"	"	"	5/8"	"	"	"	1	440
60	60a	22,000	4'-8"	3'-0"	1/4"	1/2"	2'-0"	6"	5"	7"	1/4"	3"	3"	1"		310
	60b	51,000	"	"	"	"	"	"	"	"	3/8"	"	"	"		370
	60c	87,500	"	"	"	"	"	"	"	"	1/2"	"	"	"		400
	60d	154,000	"	"	"	5/8"	"	"	"	"	5/8"	"	"	"	1	510
66	66a	21,400	5'-2"	3'-3"	1/4"	1/2"	2'-3"	6"	5"	7"	1/4"	3"	3 1/2"	1"		350
	66b	36,000	"	"	"	"	"	"	"	"	5/16"	"	"	"		380
	66c	55,000	"	"	"	"	"	"	"	"	3/8"	"	"	"		410
	66d	90,000	"	"	"	"	"	"	"	"	1/2"	"	"	"		470
	66e	124,000	"	"	"	5/8"	"	"	"	"	5/8"	"	"	"		570
	66f	185,000	"	"	"	"	"	"	"	"	"	"	"	"	1	580
	66g	248,000	"	"	"	"	"	"	"	"	"	"	"	"	2	590
72	72a	33,500	5'-8"	3'-6"	1/4"	1/2"	2'-6"	6"	5"	7"	5/16"	3"	3 3/4"	1"		420
	72b	49,250	"	"	"	"	"	"	"	"	3/8"	"	"	"		460
	72c	99,000	"	"	"	"	"	"	"	"	1/2"	"	"	"		540
	72d	121,500	"	"	"	5/8"	"	"	"	"	5/8"	"	"	"		640
	72e	192,500	"	"	"	"	"	"	"	"	"	"	"	"	1	650
	72f	305,000	"	"	"	"	"	"	"	"	"	"	"	"	3	680
78	78a	53,000	6'-2"	3'-9"	5/16"	3/4"	2'-9"	8"	7"	9"	3/8"	4"	4"	1"		710
	78b	110,000	"	"	"	"	"	"	"	"	1/2"	"	"	"		790
	78c	141,000	"	"	"	"	"	"	"	"	5/8"	"	"	"		870
	78d	210,000	"	"	"	"	"	"	"	"	"	"	"	"	1	880
	78e	352,000	"	"	"	"	"	"	"	"	"	"	"	"	3	910
84	84a	50,000	6'-8"	4'-0"	5/16"	3/4"	3'-0"	8"	7"	9"	3/8"	4"	4 1/4"	1"		810
	84b	116,000	"	"	"	"	"	"	"	"	1/2"	"	"	"		910
	84c	150,000	"	"	"	"	"	"	"	"	5/8"	"	"	"		1020
	84d	225,000	"	"	"	"	"	"	"	"	"	"	"	"	1	1030
	84e	375,000	"	"	"	"	"	"	"	"	"	"	"	"	3	1050
90	90a	47,500	7'-2"	4'-3"	5/16"	3/4"	3'-3"	8"	7"	9"	3/8"	4"	4 1/2"	1"		880
	90b	115,000	"	"	"	"	"	"	"	"	1/2"	"	"	"		990
	90c	162,500	"	"	"	"	"	"	"	"	5/8"	"	"	"		1100
	90d	243,000	"	"	"	"	"	"	"	"	"	"	"	"	1	1110
	90e	325,000	"	"	"	"	"	"	"	"	"	"	"	"	2	1130
	90f	487,500	"	"	"	"		"	"	"	"	"	"	"	4	1160
96	96a	44,250	7'-8"	4'-6"	5/16"	3/4"	3'-6"	8"	7"	9"	3/8"	4"	5"	1"		940
	96b	100,000	"	"	"	"	"	"	"	"	1/2"	"	"	"		1050
	96c	175,000	"	"	"	"	"	"	"	"	5/8"	"	"	"		1170
	96d	262,500	"	"	"	"	"	"	"	"	"	"	"	"	1	1180
	96e	350,000	"	"	"	"	"	"	"	"	"	"	"	"	2	1200
	96f	532,000	"	"	"	"	"	"	"	"	"	"	"	"	4	1230
102	102a	94,500	8'-2"	4'-9"	5/16"	3/4"	3'-9"	10"	9"	11"	1/2"	5"	5 1/4"	1 1/4"		1350
	102b	199,000	"	"	"	"	"	"	"	"	5/8"	"	"	"		1500
	102c	298,000	"	"	"	"	"	"	"	"	"	"	"	"	1	1510
	102d	398,000	"	"	"	"	"	"	"	"	"	"	"	"	2	1530
	102e	476,000	"	"	"	"	"	"	"	"	3/4"	"	"	"	"	1690
	102f	714,000	"	"	"	"	"	"	"	"	"	"	"	"	4	1730
108	108a	85,000	8'-8"	5'-0"	5/16"	3/4"	4'-0"	10"	9"	11"	1/2"	5"	5 1/2"	1 1/4"		1430
	108b	187,000	"	"	"	"	"	"	"	"	5/8"	"	"	"		1590
	108c	290,000	"	"	"	"	"	"	"	"	"	"	"	"	1	1610
	108d	386,000	"	"	"	"	"	"	"	"	"	"	"	"	2	1630
	108e	466,000	"	"	"	"	"	"	"	"	3/4"	"	"	"	"	1810
	108f	699,000	"	"	"	"	"	"	"	"	"	"	"	"	4	1870
114	114a	164,000	9'-2"	5'-3"	5/16"	3/4"	4'-3"	10"	9"	11"	5/8"	5"	5 3/4"	1 1/4"		1760
	114b	300,000	"	"	"	"	"	"	"	"	"	"	"	"	1	1780
	114c	400,000	"	"	"	"	"	"	"	"	"	"	"	"	2	1800
	114d	514,000	"	"	"	"	"	"	"	"	3/4"	"	"	"	"	1980
	114e	600,000	"	"	"	7/8"	"	"	"	"	7/8"	"	"	"	"	2260
	114f	900,000	"	"	"	"	"	"	"	"	"	"	"	"	4	2330

Figure 10-6. Typical saddle details for horizontal vessels. Courtesy of Fluor Engineers and Constructers, Inc.

Figure 10-6 continued

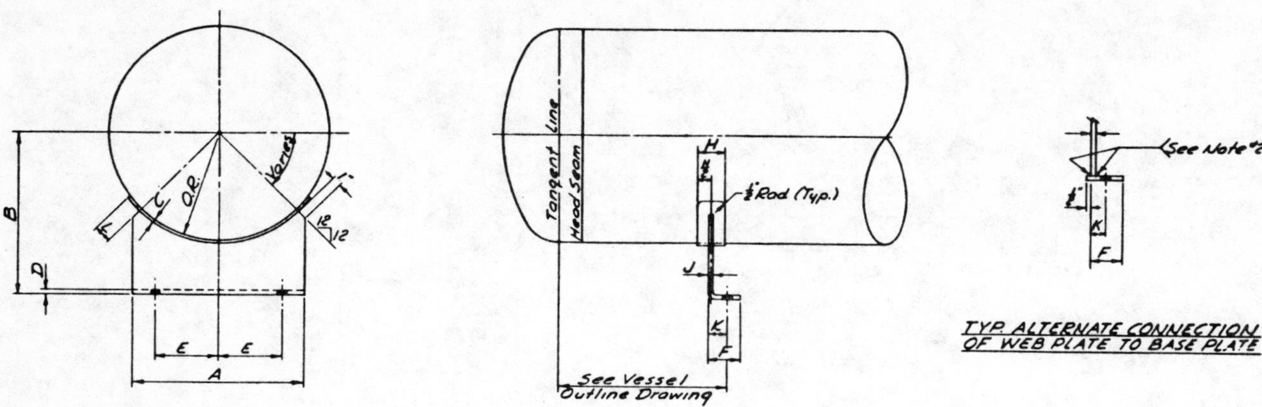

SADDLE DETAILS FOR MARK 24a, 24b & 30a ONLY

VESSEL DIAM. IN INCHES	SADDLE MARK NO.	MAX. VESSEL OPER. WEIGHT LBS	A	B	C	D	E	F	G	H	J	K	L	BOLT DIAM.	NO. OF RIBS	WEIGHT PER SET LBS.
120	120a	150,000	9'-8"	5'-6"	5/16"	3/4"	4'-6"	10"	9"	11"	5/8"	5"	6"	1 1/4"		1860
	120b	300,000	"	"	"	"	"	"	"	"	"	"	"	"	1	1880
	120c	410,000	"	"	"	"	"	"	"	"	"	"	"	"	2	1900
	120d	520,000	"	"	"	"	"	"	"	"	3/4"	"	"	"	"	2100
	120e	666,000	"	"	"	7/8"	"	"	"	"	7/8"	"	"	"	"	2380
	120f	999,000	"	"	"	"	"	"	"	"	"	"	"	"	4	2440
126	126a	146,000	10'-2"	5'-9"	3/8"	3/4"	4'-9"	10"	9"	11"	5/8"	5"	6 1/2"	1 1/4"		2080
	126b	294,000	"	"	"	"	"	"	"	"	"	"	"	"	1	2100
	126c	380,000	"	"	"	"	"	"	"	"	"	"	"	"	2	2120
	126d	536,000	"	"	"	"	"	"	"	"	3/4"	"	"	"	"	2340
	126e	646,000	"	"	"	7/8"	"	"	"	"	7/8"	"	"	"	"	2640
	126f	969,000	"	"	"	"	"	"	"	"	"	"	"	"	4	2700
132	132a	127,500	10'-8"	6'-0"	3/8"	3/4"	5'-0"	10"	9"	11"	5/8"	5"	6 3/4"	1 1/4"		2180
	132b	292,000	"	"	"	"	"	"	"	"	"	"	"	"	1	2220
	132c	400,000	"	"	"	"	"	"	"	"	"	"	"	"	2	2240
	132d	550,000	"	"	"	"	"	"	"	"	3/4"	"	"	"	"	2480
	132e	708,000	"	"	"	7/8"	"	"	"	"	7/8"	"	"	"	"	2820
	132f	1,058,000	"	"	"	"	"	"	"	"	"	"	"	"	4	2880
138	138a	124,500	11'-2"	6'-3"	3/8"	3/4"	5'-3"	10"	9"	11"	5/8"	5"	7"	1 1/4"		2340
	138b	289,000	"	"	"	"	"	"	"	"	"	"	"	"	1	2360
	138c	386,000	"	"	"	"	"	"	"	"	"	"	"	"	2	2400
	138d	544,000	"	"	"	"	"	"	"	"	3/4"	"	"	"	"	2640
	138e	768,000	"	"	"	7/8"	"	"	"	"	7/8"	"	"	"	"	3000
	138f	1,182,000	"	"	"	"	"	"	"	"	"	"	"	"	4	3060
144	144a	280,000	11'-8"	6'-6"	3/8"	3/4"	5'-6"	10"	9"	11"	5/8"	5	7 1/4"	1 1/4	1	2500
	144b	365,000	"	"	"	"	"	"	"	"	"	"	"	"	2	2540
	144c	540,000	"	"	"	"	"	"	"	"	3/4"	"	"	"	"	2800
	144d	728,000	"	"	"	7/8"	"	"	"	"	7/8"	"	"	"	"	3160
	144e	1,092,000	"	"	"	"	"	"	"	"	"	"	"	"	4	3200
	144f	1,290,000	"	"	"	1"	"	"	"	"	1"	"	"	"	"	3400
150	150a	275,000	12'-2"	6'-9"	3/8"	3/4"	5'-9"	10"	9"	11"	5/8"	5	7 1/2"	1 1/4	1	2580
	150b	385,000	"	"	"	"	"	"	"	"	"	"	"	"	2	2620
	150c	560,000	"	"	"	"	"	"	"	"	3/4"	"	"	"	"	2900
	150d	764,000	"	"	"	7/8"	"	"	"	"	7/8"	"	"	"	"	3260
	150e	1,156,000	"	"	"	"	"	"	"	"	"	"	"	"	4	3300
	150f	1,282,000	"	"	"	1"	"	"	"	"	1"	"	"	"	"	3700
156	156a	266,000	12'-8"	7'-0"	3/8"	3/4"	6'-0"	10"	9"	11	5/8"	5	8"	1 1/4"	1	2730
	156b	398,000	"	"	"	"	"	"	"	"	"	"	"	"	2	2760
	156c	576,000	"	"	"	"	"	"	"	"	3/4"	"	"	"	"	3120
	156d	864,000	"	"	"	"	"	"	"	"	"	"	"	"	4	3160
	156e	1,150,000	"	"	"	7/8"	"	"	"	"	7/8"	"	"	"	"	3560
	156f	1,276,000	"	"	"	1"	"	"	"	"	1"	"	"	"	"	3980

GENERAL NOTES

1. Vessel diameters listed are inside diameters. The listed standard dimensions may change in the case of very thick vessel shells, or if very thick insulation is employed. Dimensions shown on vessel drawing shall take precedent.
2. Saddle to vessel attachment shall be a continuous full fillet weld all around. Other welds shall be 1/4" continuous fillet all around, up to & including 72" diameter vessel & 3/8" continuous fillet all around for 78" diameter vessel & larger.
3. Provide 1/8" pipe tap in wear plate if vessel is stress relieved or where plate over laps longitudinal seams.
4. Anchor bolt holes
   - 4a. Provide holes in anchor end as follows:
     1 1/4" ⌀ for 1" ⌀ bolts
     1 1/2" ⌀ for 1 1/4" ⌀ bolts.
   - 4b. Provide slotted holes in expansion end as follows:
     1 1/4" ⌀ x 2" Long for 1" ⌀ bolts
     1 1/2" ⌀ x 2 1/2" Long for 1 1/4" ⌀ bolts
   - 4c. Slots not required for cold vessels less than 10'-0" long.

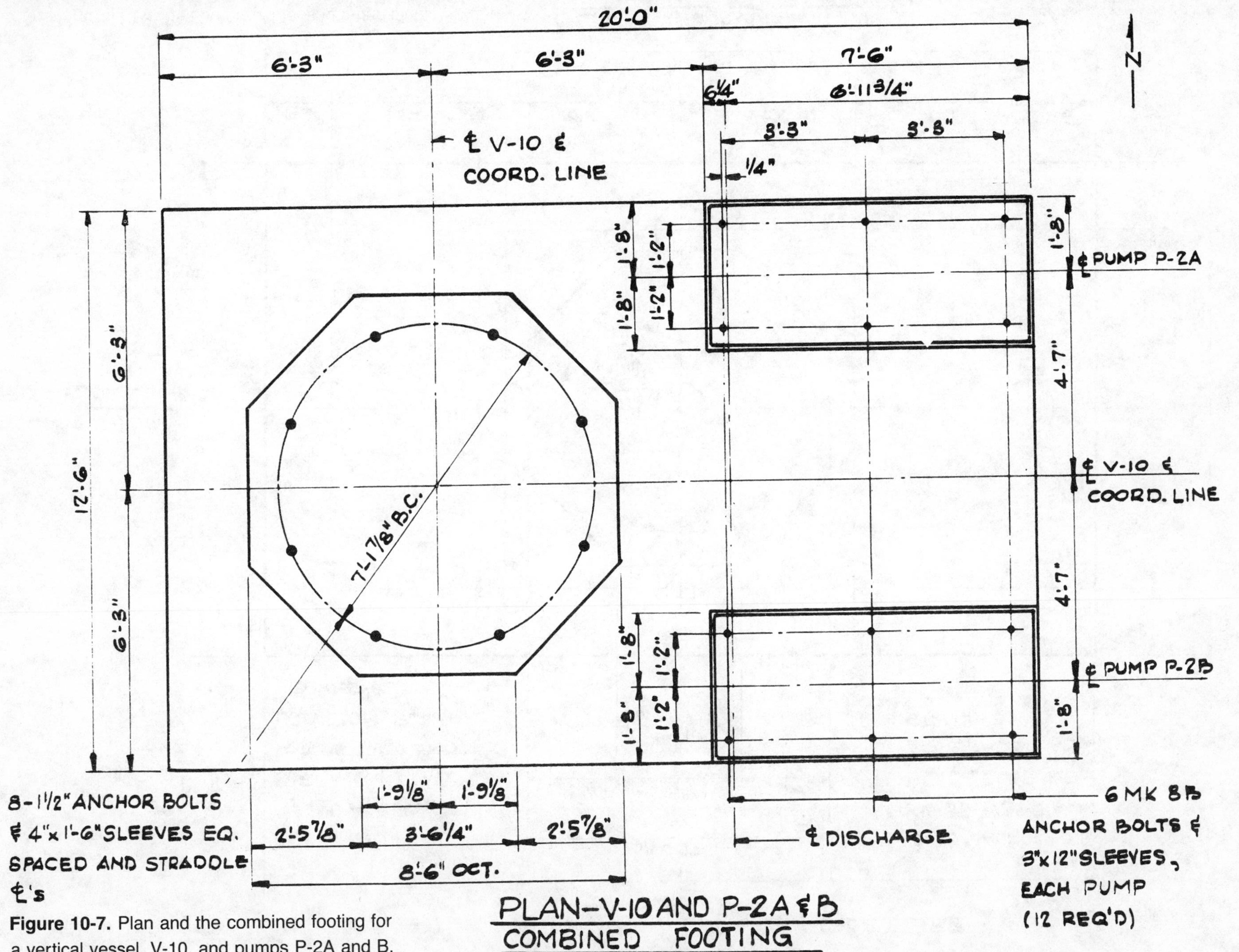

Figure 10-7. Plan and the combined footing for a vertical vessel, V-10, and pumps P-2A and B.

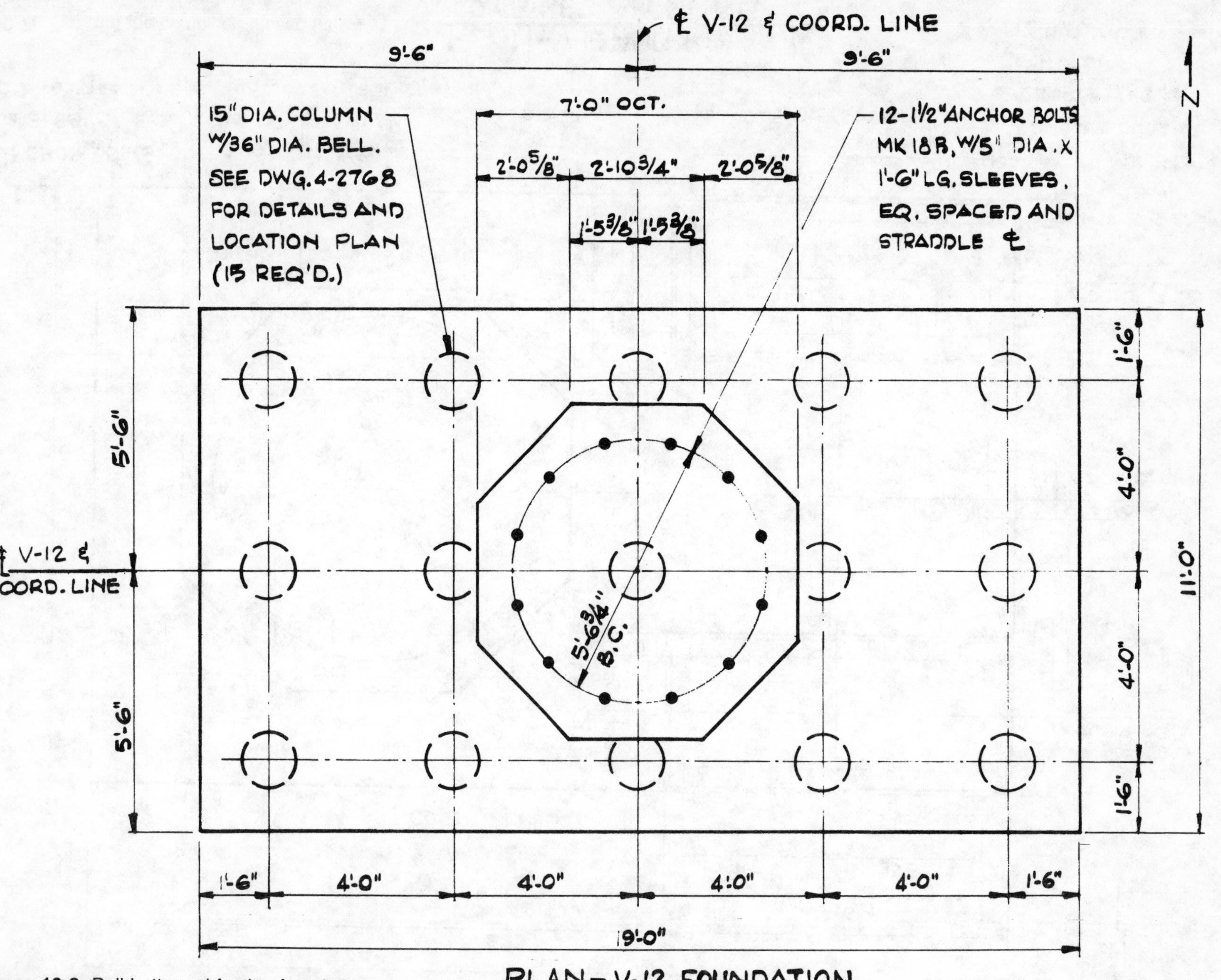

Figure 10-8. Bell-bottomed footing foundation.

figure continued on next page

Figure 10-8 continued

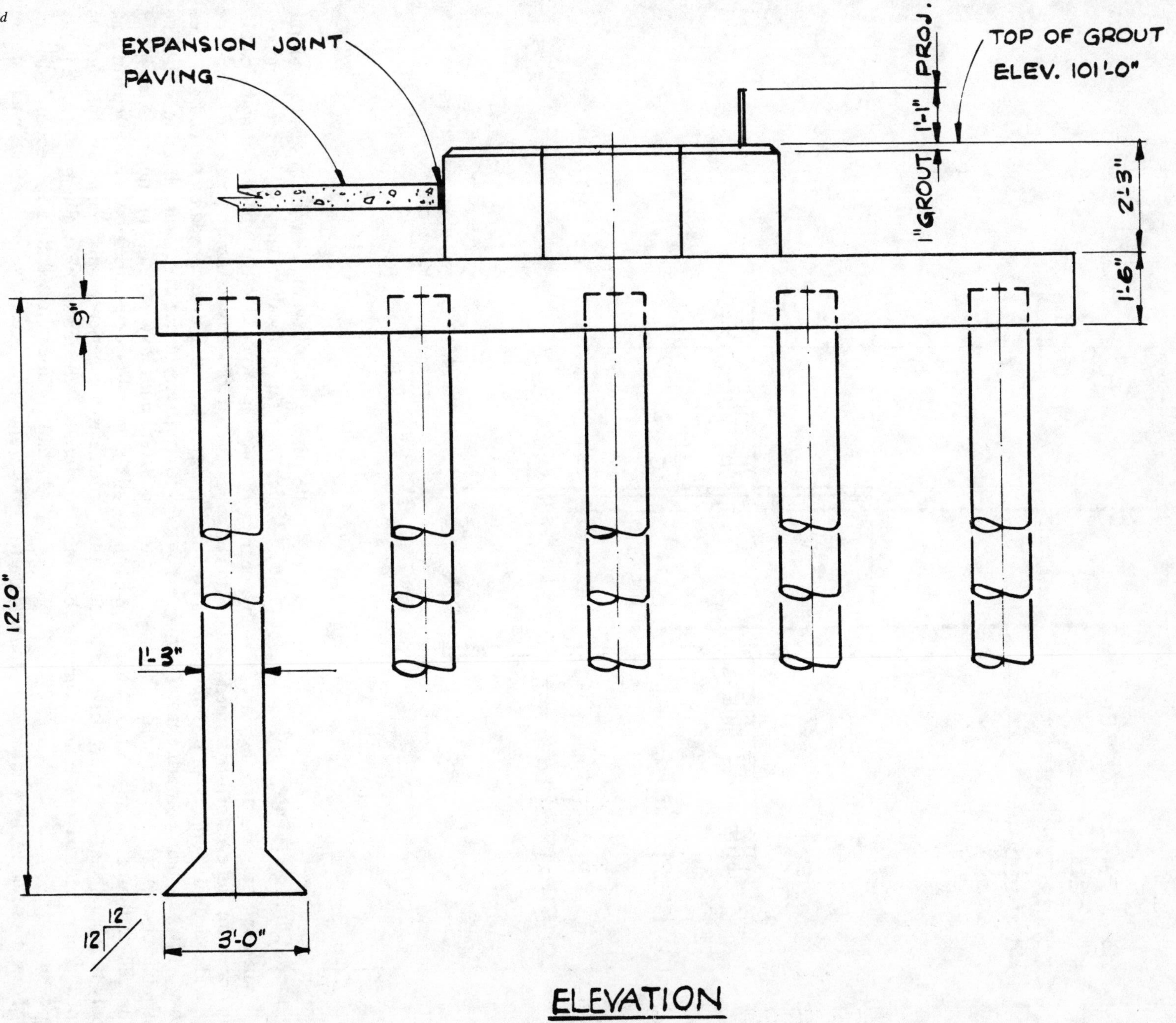

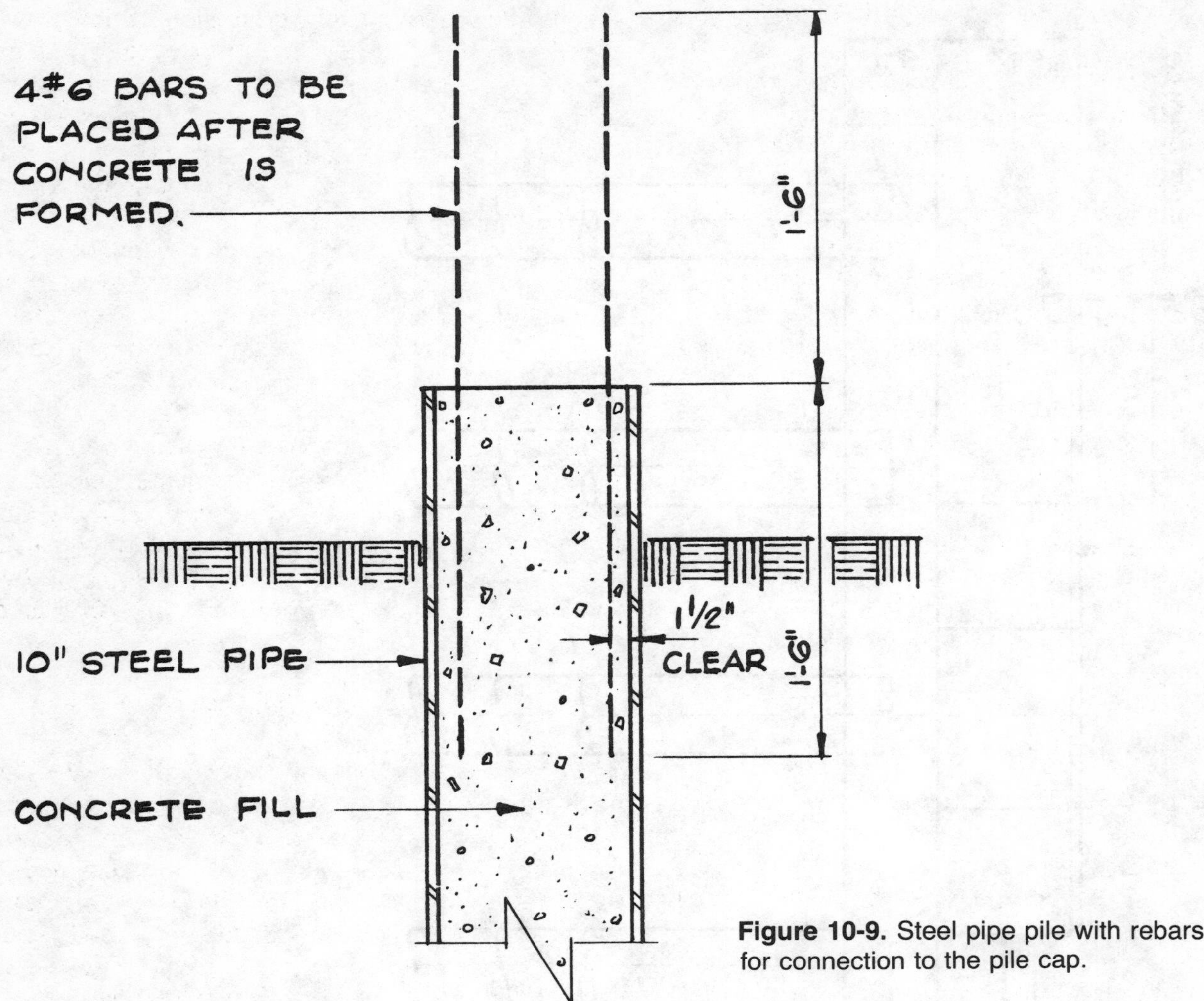

Figure 10-9. Steel pipe pile with rebars for connection to the pile cap.

(text continued from page 100)

Foundations on Piling

When soil has an extremely low allowable bearing pressure and even bell-bottomed footings do not provide adequate support, or perhaps there is no soil to work with, such as marine installations, expensive pile construction is required. This is similar to the bell-bottomed footings except there is no bell, only the round shaft. Piles are driven into the earth to a depth required to attain acceptable bearing pressure.

How does one know what kind of foundation to install? When it is decided to construct something on a plot, a soil analysis is made by an organization that specializes in it. They drill test holes at various locations on the plot and perform soil tests that determine its allowable bearing pressure and the depth of this allowable. They make recommendations to the owner as to the type of footing to install.

Pile material may be lumber, concrete (precast or prestressed), steel pipe capped on the end and filled with concrete, or one of several other materials. The concrete piles may be round or square. Selection of pile material is determined by the soil conditions, contractor experience, and costs of the various materials at the job site. Extra long piles have joints which are field-connected after one section is driven. Both piles and bells have rebars that extend from the top to connect with the cap. Figure 10-9 shows a steel pipe pile with rebars for connection to the pile cap. When the cap is formed, the rebars will be bent horizontally to fit in the cap.

Drawing Types and Scales

Many firms use **preprinted drawings** for foundations. These drawings will have the foundation shown and all required dimension lines located at the proper place. The draftsman will only add the dimension above the given dimension line, fill in the bolt quantity and size, and assign rebar mark numbers and sizes. Preprinted drawings for typical foundations save time and allow quick issue of information. Since one drawing fits all sizes, the foundations are not to any scale.

Some firms use a computer-plotter to make some foundation drawings. Again, these are usually standard types of foundations and are not to scale. However, computer foundation drawings usually are produced fully dimensioned.

Most foundation drawings are drawn by hand by draftsmen and are to scale. Most firms desire each major foundation to be drawn on a separate sheet of drawing paper. Several small foundations may be placed on the same drawing. These would be pours of ten yards or less.

Scale Selection

The draftsman selects the scale the foundation will be drawn to. As a minimum, the scale should be ⅜″ = 1′-0″. If the drawing paper size permits, a larger scale should be used. Almost all drawings are microfilmed and often reproduced to a larger size from the microfilm. This larger size will be less than the original size so sharp lines and good detail are a must. Prints get dirty in the field, but they still must be readable. And the smaller the scale is, the harder the drawing is to read.

Engineering vs Placing Drawings

The ACI Standard Practice (ACI 315) calls for engineering and placing drawings to be separate. **Engineering drawings** are similar to those shown in this chapter where concrete dimensions are shown but rebar details are omitted or shown in only small foundations. **Placing drawings** show the foundation's outline without outline dimensions but all rebars are shown and located by dimension. Placing drawings are issued to the rebar fabricator and to the field. Engineering drawings are issued to the field.

While separate engineering and placing drawings relieve congestion, they are more expensive to make and reproduce. Consequently, unless the concrete drawing is extremely congested, almost all firms make the engineering and placing drawings as one drawing. Figure 10-10 is a combination engineering-placing drawing.

When possible, all rebar placement is shown in the section or elevation, leaving the plan to show most of the engineering dimensions. Note that section A-A shows only two piers although the true section would show four piers. Two piers are not drawn because it would make the drawing more cluttered and would serve no useful purpose. The plan is very clear that six piers are required.

A small section through a pier is drawn to aid the field craftsman to properly interpret the rebar location shown in elevation. The #2 ties are set by the field but usually there is only one pair per each pier.

The Mud Mat

When foundations are excavated and ready to be formed, many times the foundation's base line will be below the area's water table elevation. When this occurs, water will seep into the excavated hole resulting in water and mud that has to be pumped out. The hole will be too muddy for workers to enter and work. Too, heavy rains often cause the same problem. When the excavation becomes too cumbersome for workmen to enter, about two inches of unreinforced concrete, called **raw concrete,** is poured to cover the entire excavated area. This raw concrete is called a **mud mat.** In two or three days the mud mat has hardened enough that the workmen can stand on it and erect the foundation's forms.

Rebar Schedules or Lists

Reinforcing bars are manufactured in lengths to sixty feet long. This bar is shipped to job sites, for field installation, and to rebar fabricators, who cut, bend, and mark the pieces for field installation. Many firms have all rebar shop cut and bent. Some firms, which have bending machines in the field, choose to field fabricate all bent bar up to 1″ in size and field cut all straight bar regardless of size.

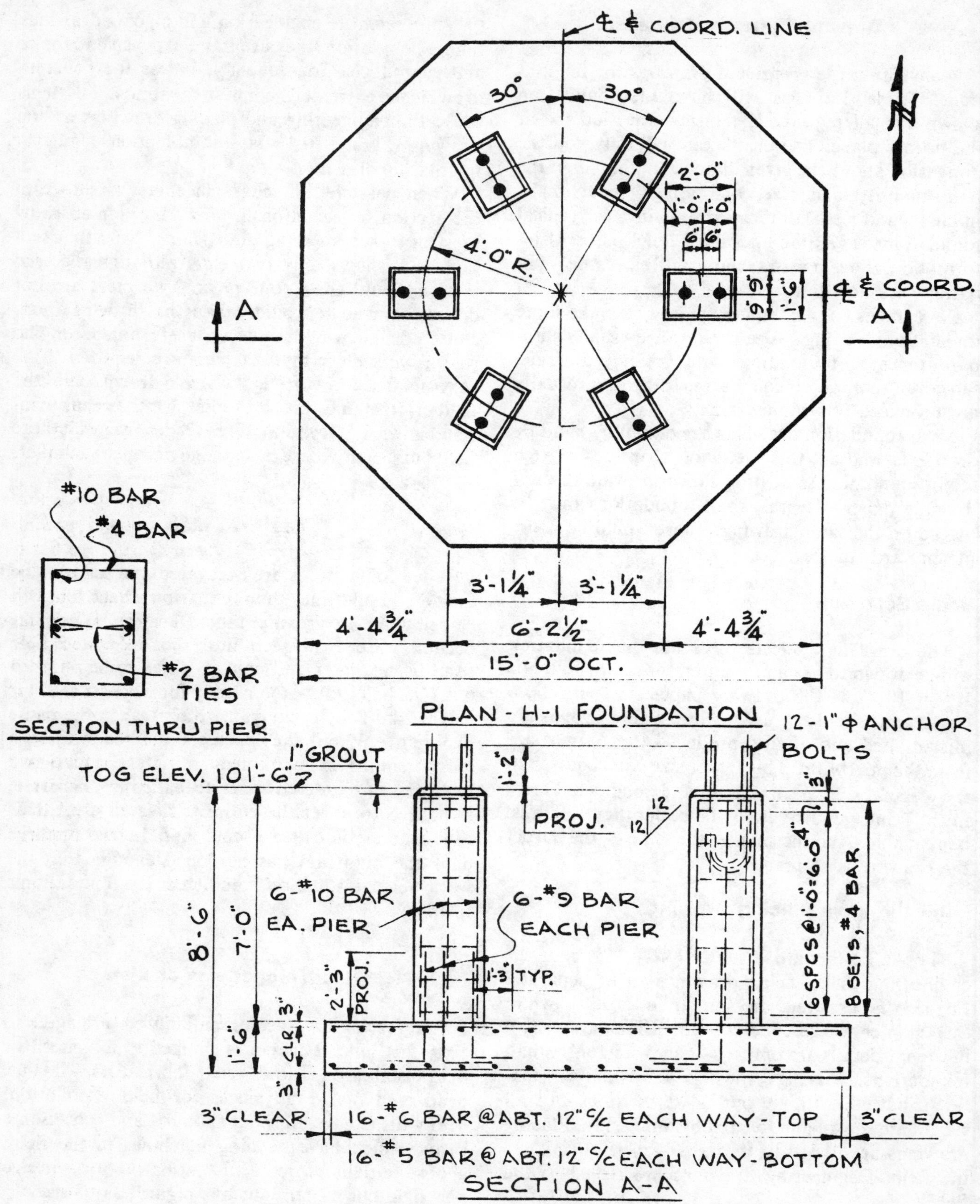

Figure 10-10. Combination engineering-placing drawing.

When field personnel cut bar, they usually use a welding torch. Shops may also choose to cut with a torch but many of them have a shearing machine that will shear the rebars to required lengths. When rebars are cut, dropoff occurs and long lengths of dropoff are welded together to prevent waste. Figure 10-11 shows how rebars are spliced by welding. At times, rebars are spliced in place as they are being erected. Number 5 and smaller rebars are often spliced by lapping the bars and welding them together. When this is not possible, splice bars of the same size are located on each side of the bar and welded to the bars to be joined.

The ACI has established practices for bending of reinforcing bars. These practices are followed by rebar fabricators and contractors and establish methods and procedures that govern all bending. Figure 10-12 shows standard rebar bending details.

Rebar Dimensioning

It is standard industry practice to show all rebar dimensions as out-to-out and consider the bar lengths as the sum of all detailed dimensions, including hooks. Note that Figure 10-12 dimensions to the outside of the bar.

Rebar Slants

Rebars are bent at an angle, especially when forming trusses. This angle is normally 45 degrees. Figure 10-13 shows the standard bent bar. The 45 degree portion is called the bar **slant**. Note that dimensions are to the outside, not centerlines, in every case.

Rebar schedules always supply the total developed length of bar required. This is determined by adding the detailed dimensions but to do this, the slant must be known. Table 10-2 provides dimensions of slants and increments for 45 degree bar bends. Increments are slants minus the formed angle's leg. The table provides the dimension for two increments to assist in determining the developed length for bars that have two slants.

To determine the rebar's total developed length, the straight bar length required to form the bent bar, add the overall length O to two increments, if two are needed, plus the bar needed for hooks A and G per Table 10-2.

Rebar Schedule Form

Figure 10-14 is a typical reinforcing steel schedule form. Standard rebar shapes are shown with letter designations for dimensions. The lower part of the form is completed by the structural draftsman and all dimensions, developed length and piece weights are shown. Should rebars be bent to some other shape than shown, the blank area is provided for these bars to be drawn.

Each bar is marked with an identifying number. There are many ways of marking bars but a common one is to divide the mark number into two sets of numbers, such as 104-601. The first three digits refer to the concrete drawing number, as 4-104. The 4 is the drawing size. The digits 601 refer to the bar size (6) and the actual bar number which begins with 01 and runs to 99. When the concrete has more than 99 pieces, four digits are used in the mark number but this rarely happens.

When shop fabricated, the bar bender bends the steel in accordance with ACI recommendations and adds a tag to the fabricated bar indicating the bar's mark number. At the job site, the bar is sorted into piles which have the same first three digits. For instance, when drawing number 104 is ready to have the rebars placed, the installer locates the stockpile with 104 prefix mark numbers.

When bars are field bent, the field fabrication of bent bars will be tagged and stored in the same manner. Straight bars are often cut in place from stock lengths and would have no tag. When the draftsman prepares the reinforcing steel schedule, it isn't usually known if bars are to be shop or field furnished so all bars are listed and given a mark number.

At the end of drafting for all concrete drawings, a summation of all rebars required for the job is made. This is easily done from the totals, by size, length and weight, from all of the rebar schedules.

Horizontal Heater Foundation

Figure 10-15, 3 pages, is the foundation plan, sections and details for a horizontal fired heater. The plan is basically an engineering drawing which shows construction dimensions, bolt sizes and locations, and the block outs for heater leg base plates. Note that all bolts are assigned a mark number be-

(text continued on page 119)

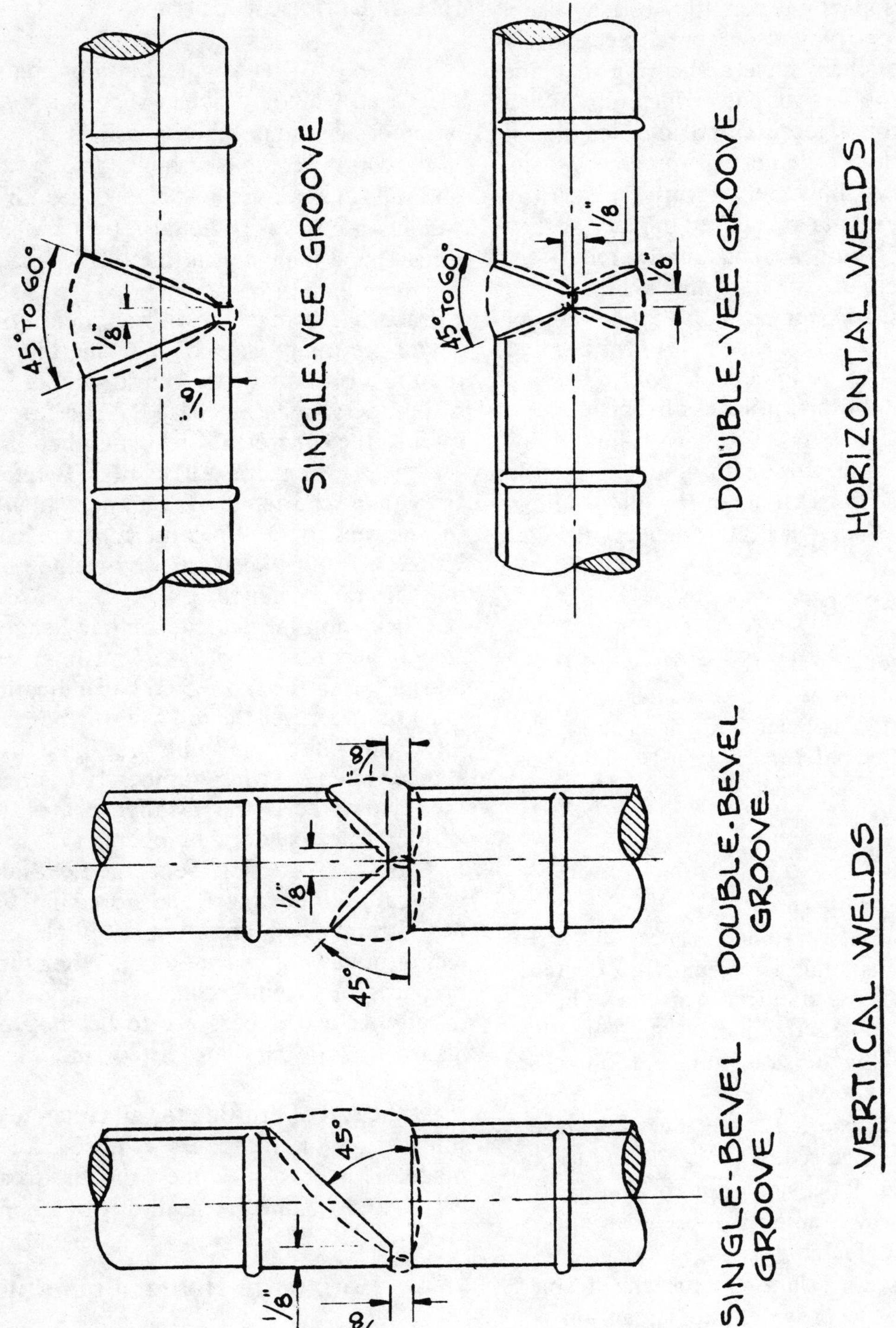

Figure 10-11. Rebars can be spliced by welding.

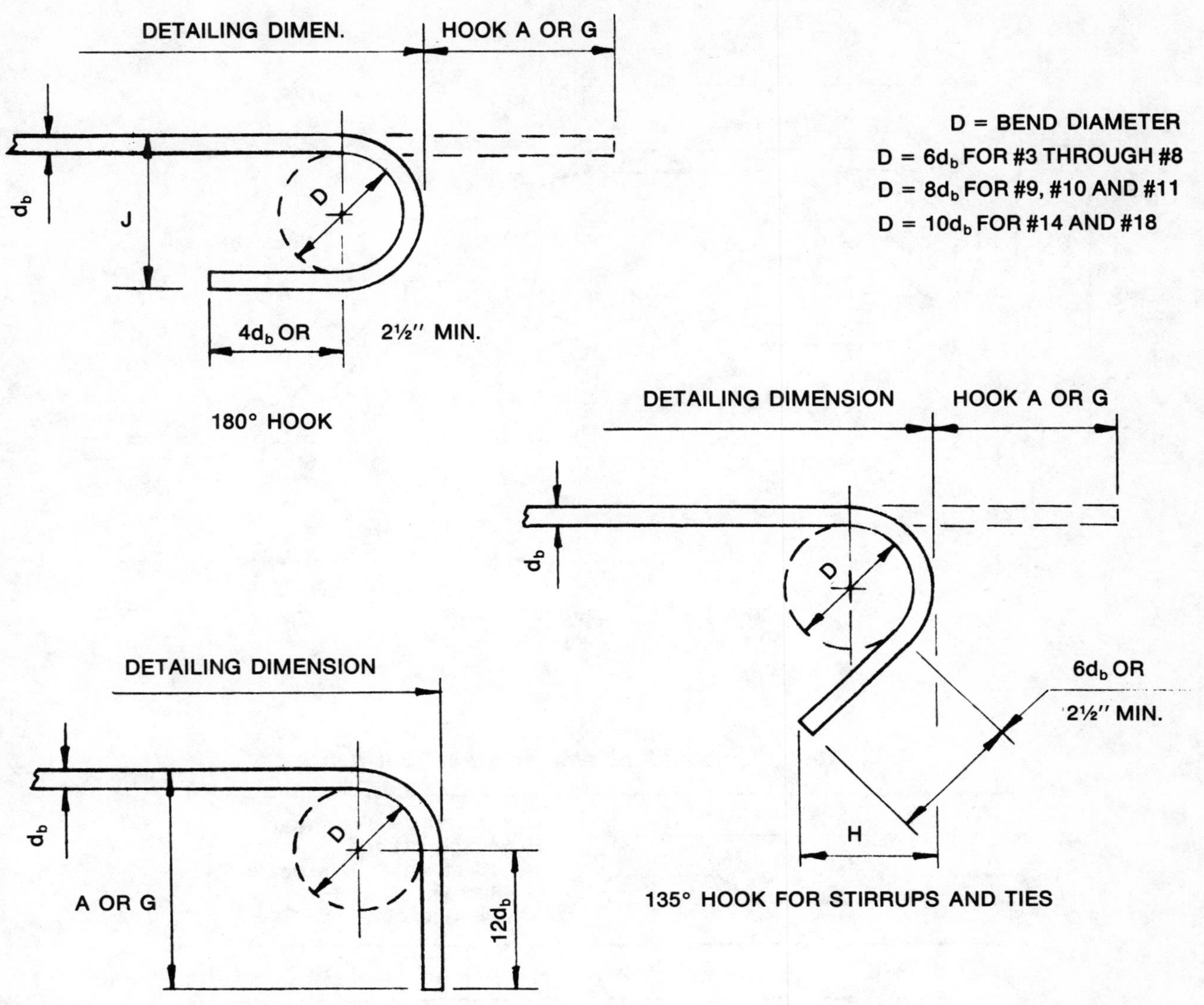

Figure 10-12. Standard rebar bending details.

db or Bar size	180° Hooks			90° Hooks		135° Tie Hooks
	A or G	J	D	A or G	D	H
# 3	5″	3″	2¼″	6″	2¼″	2½″
# 4	6″	4″	3″	8″	3″	3″
# 5	7″	5″	3¾″	10″	3¾″	3¾″
# 6	8″	6″	4½″	1′- 0″	4½″	—
# 7	10″	7″	5¼″	1′- 2″	5¼″	—
# 8	11″	8″	6″	1′- 4″	6″	—
# 9	1′- 3″	11¼″	9″	1′- 7″	9″	—
#10	1′- 5″	1′- 0¾″	10¼″	1′-10″	10¼″	—
#11	1′- 7″	1′- 2¼″	11¼″	2′- 0″	11¼″	—
#14	2′- 2″	1′- 8½″	1′- 5″	2′- 7″	1′- 5″	—
#18	2′-11″	2′- 3″	1′-10¾″	3′- 5″	1′-10¾″	—

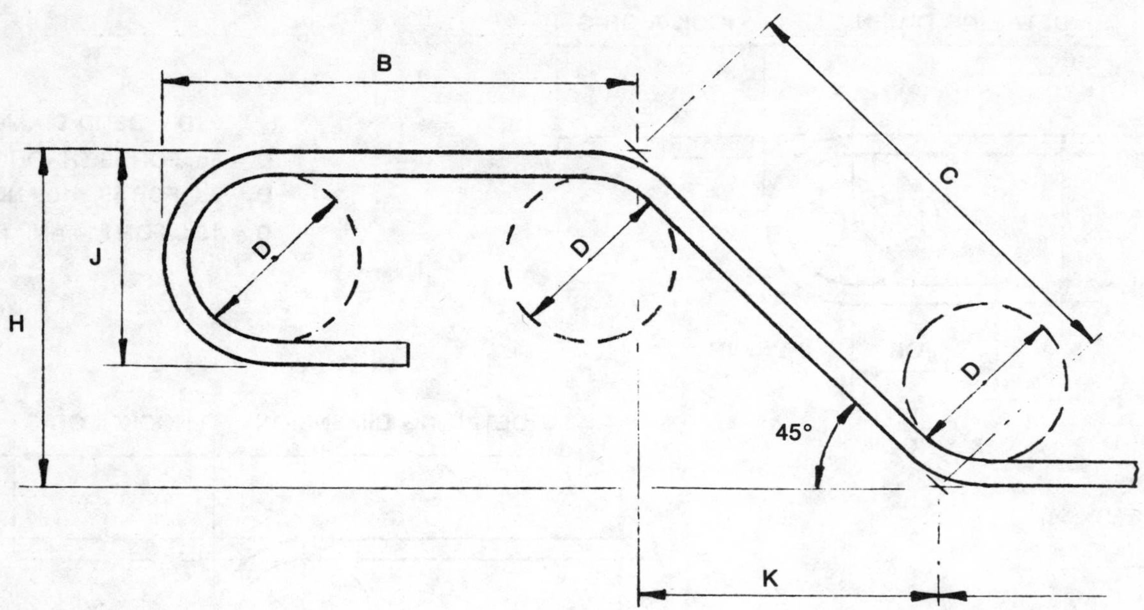

Figure 10-13. Standard bent bar.

Table 10-2. Slants and Increments for 45° Bar Bends

O = Overall bar dimension
H = Height of bend
S = Slant = 1.414H (to nearest ½").
I = Increment = S − H
Developed length = O + 2I + A + G

H	S	2I	H	S	2I	H	S	2I
2"	3"	2"	11½"	1'- 4"	9"	2'- 6"	3'- 6½"	2'- 1"
2½"	3½"	2"	1'- 0"	1'- 5"	10"	2'- 7"	3'- 8"	2'- 2"
3"	4"	2"	1'- 1"	1'- 6½"	11"	2'- 8"	3'- 9"	2'- 2"
3½"	5"	3"	1'- 2"	1'- 8"	1'- 0"	2'- 9"	3'-10½"	2'- 3"
4"	5½"	3"	1'- 3"	1'- 9"	1'- 0"	2'-10"	4'- 0"	2'- 4"
4½"	6½"	4"	1'- 4"	1'-10½"	1'- 1"	2'-11"	4'- 1½"	2'- 5"
5"	7"	4"	1'- 5"	2'- 0"	1'- 2"	3'- 0"	4'- 3"	2'- 6"
5½"	7½"	4"	1'- 6"	2'- 1½"	1'- 3"	3'- 1"	4'- 4½"	2'- 7"
6"	8½"	5"	1'- 7"	2'- 3"	1'- 4"	3'- 2"	4'- 5½"	2'- 7"
6½"	9"	5"	1'- 8"	2'- 4"	1'- 4"	3'- 3"	4'- 7"	2'- 8"
7"	10"	6"	1'- 9"	2'- 5½"	1'- 5"	3'- 4"	4'- 8½"	2'- 9"
7½"	10½"	6"	1'-10"	2'- 7"	1'- 6"	3'- 5"	4'-10"	2'-10"
8"	11½"	7"	1'-11"	2'- 8½"	1'- 7"	3'- 6"	4'-11½"	2'-11"
8½"	1'- 0"	7"	2'- 0"	2'-10"	1'- 8"	3'- 7"	5'- 1"	3'- 0"
9"	1'- 0½"	7"	2'- 1"	2'-11½"	1'- 9"	3'- 8"	5'- 2"	3'- 0"
9½"	1'- 1½"	8"	2'- 2"	3'- 1"	1'-10"	3'- 9"	5'- 3½"	3'- 1"
10"	1'- 2"	8"	2'- 3"	3'- 2"	1'-10"	3'-10"	5'- 5"	3'- 2"
10½"	1'- 3"	9"	2'- 4"	3'- 3½"	1'-11"	3'-11"	5'- 6½"	3'- 3"
11"	1'- 3½"	9"	2'- 5"	3'- 5"	2'- 0"	4'- 0"	5'- 8"	3'- 4"

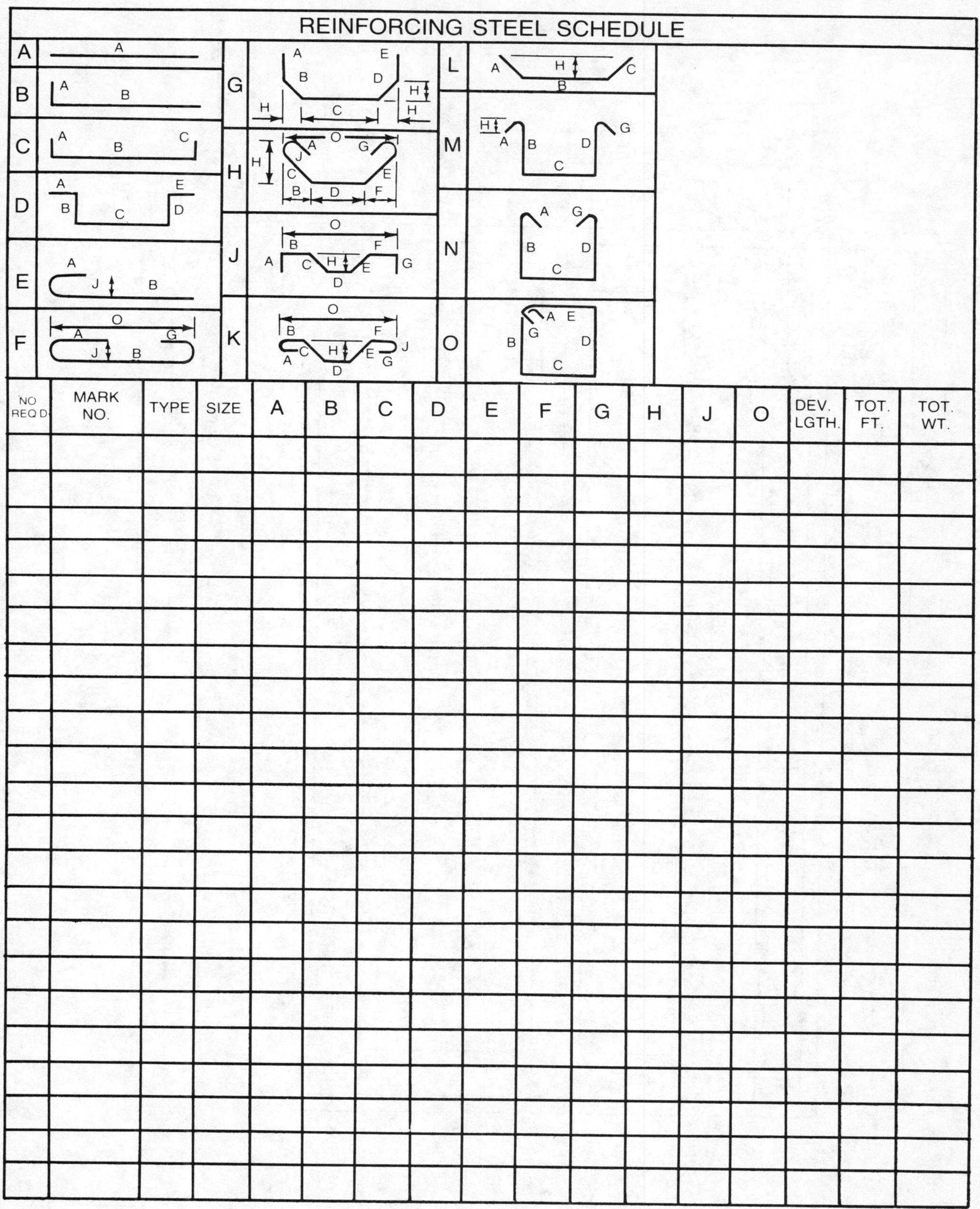

REINFORCING STEEL SCHEDULE

NO REQ D	MARK NO.	TYPE	SIZE	A	B	C	D	E	F	G	H	J	O	DEV. LGTH.	TOT. FT.	TOT. WT.

Figure 10-14. Typical reinforcing steel schedule form.

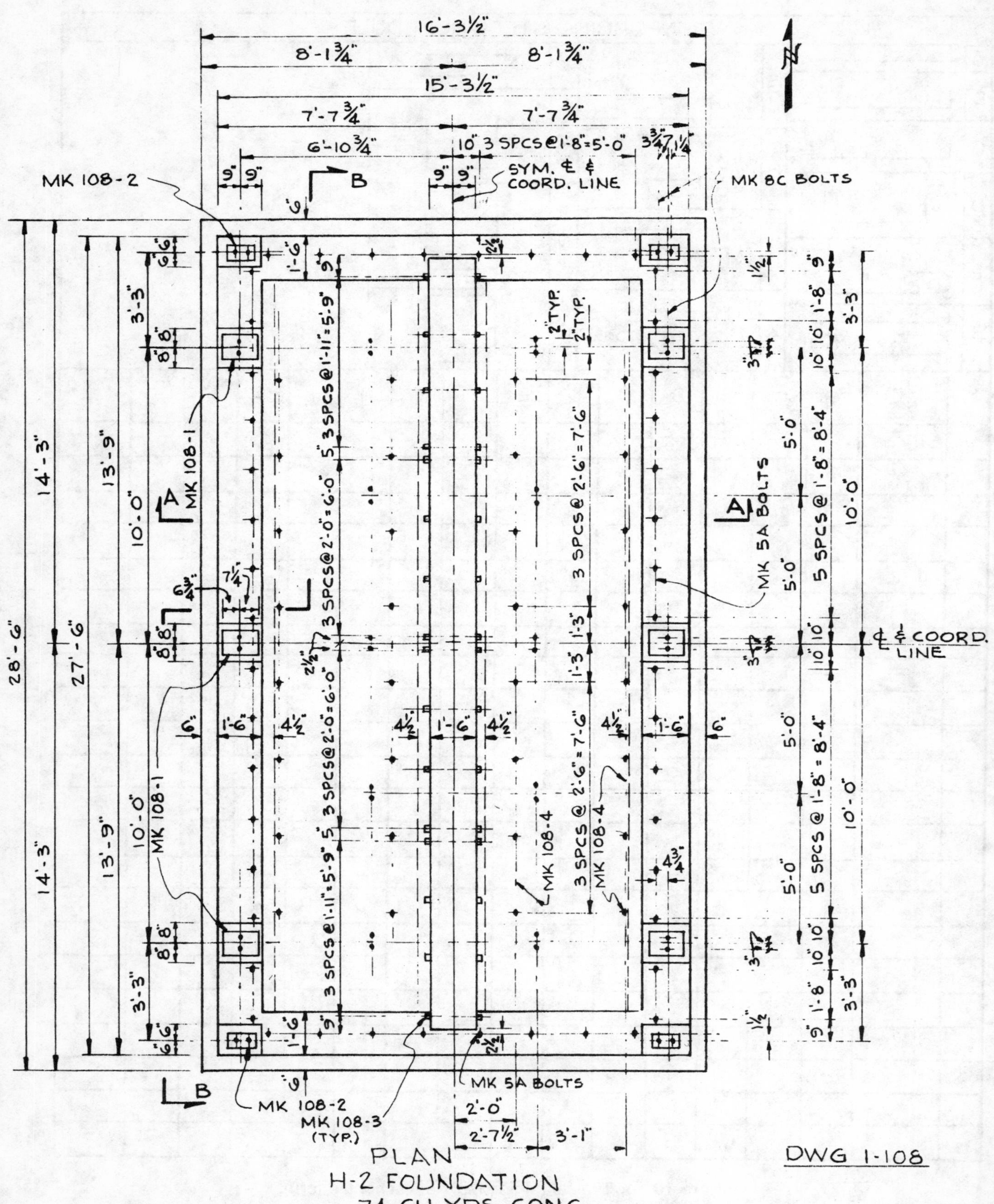

Figure 10-15. Foundation plan, sections, and details for a horizontal fired heater.

figure continued on next page

Figure 10-15 continued

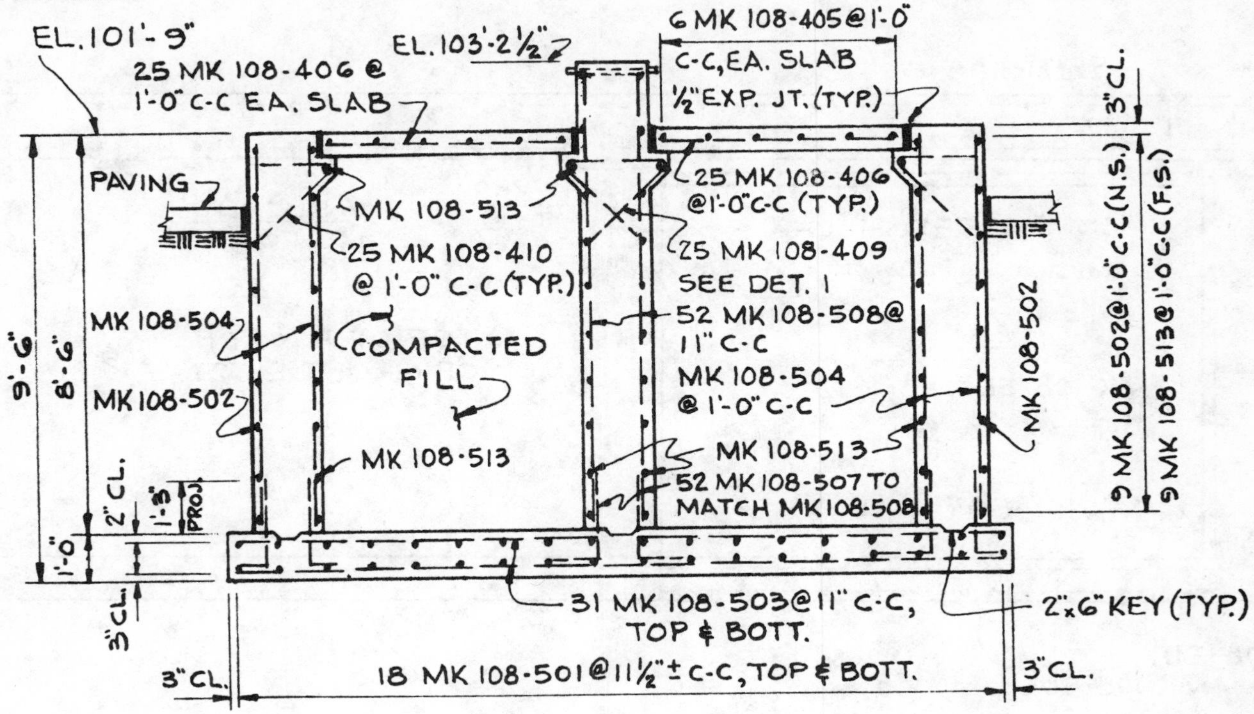

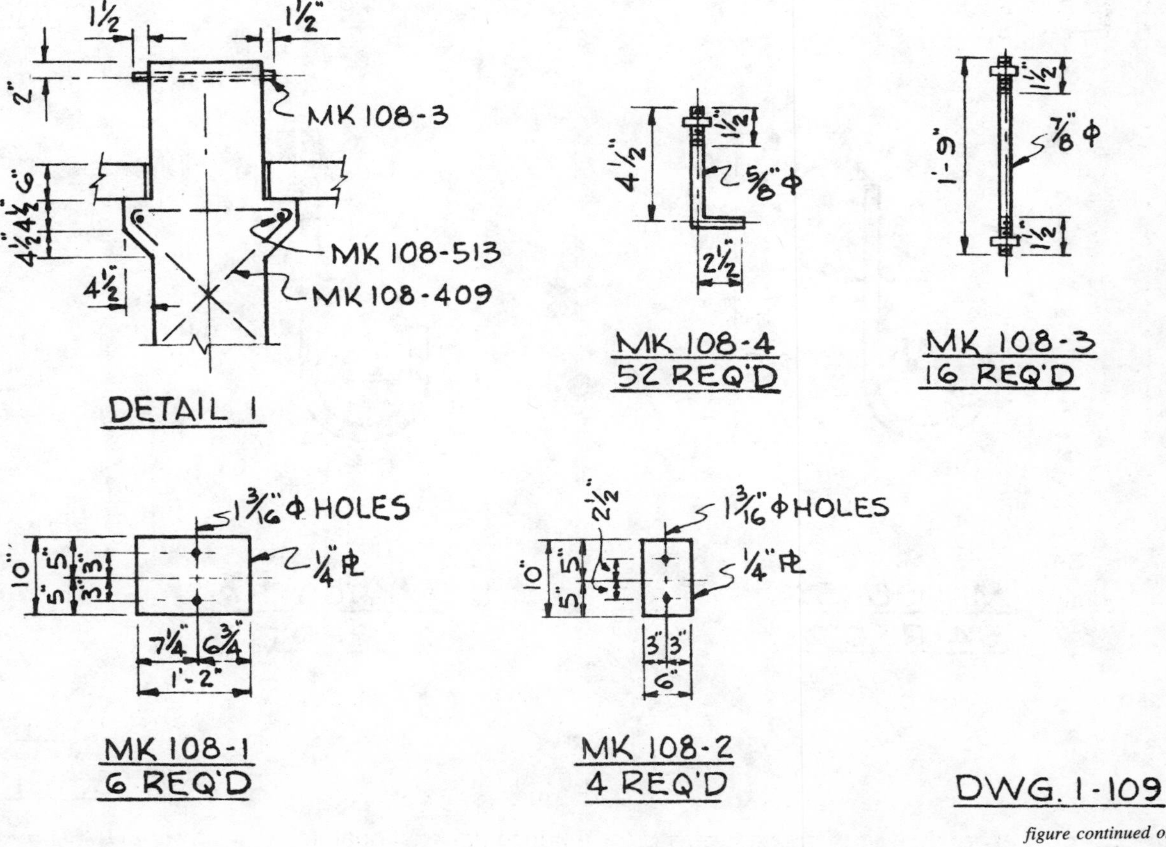

figure continued on next page

Figure 10-15 continued

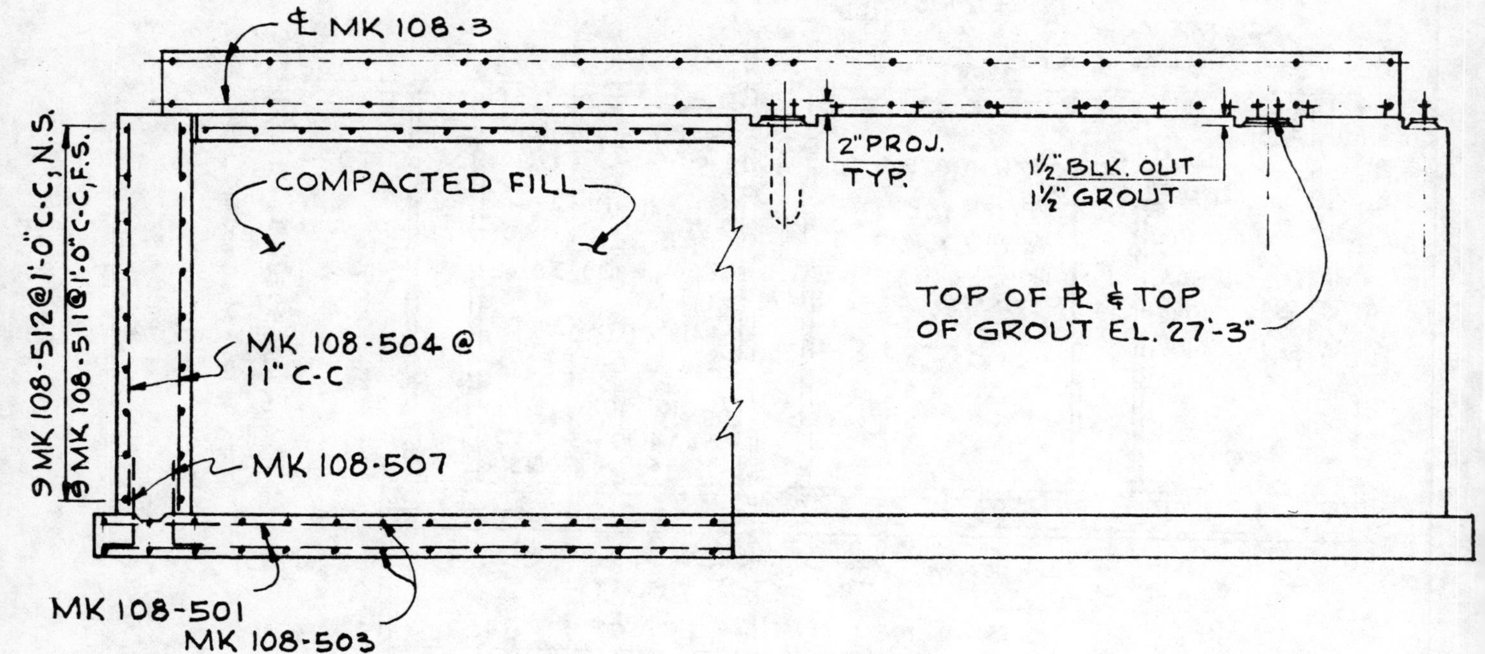

SECTION B-B

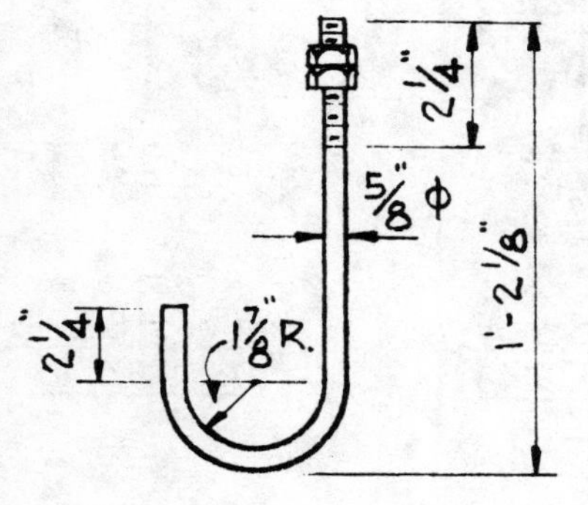

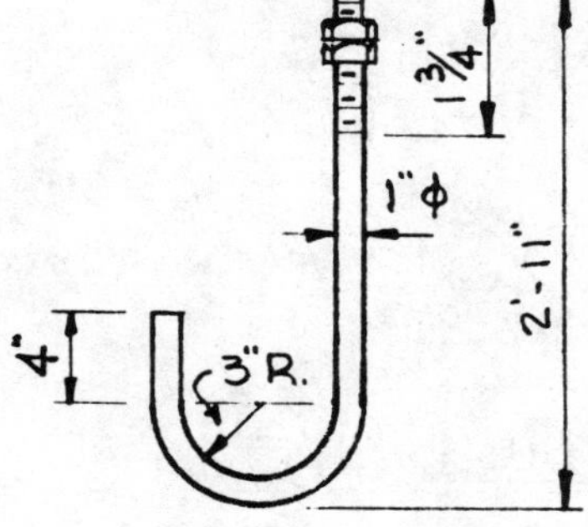

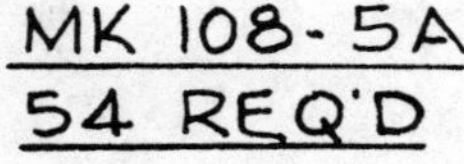

MK 108-8C
20 REQ'D

DWG. 1-110

(text continued from page 111)

ginning with the drawing number and a single number, which is their sequential number. When anchor bolts are identified in this manner, their details are shown on the foundation drawings, not on an anchor bolt schedule.

The plan shows the north arrow and the centerlines that are designated the coordinate lines. Also, at the top of the drawing, the vertical centerline is called out as SYM., which is the abbreviation for symmetrical. By calling this centerline a symmetrical centerline, dimensions may be shown on one half and the other half will have identical dimensions.

Sections A-A and B-B are combination engineering and placing drawings. Construction dimensions and rebar locations are shown. Rebar mark numbers, bolt details, and other necessary details complete the foundation drawing. Note that details are enlarged.

Figure 10-16 is the reinforcing steel schedule for the heater foundation. Since the form had no standard details for two rebars needed in the foundation, details P and R were added. This foundation requires a total of 7,553# of reinforcing steel.

REINFORCING STEEL SCHEDULE

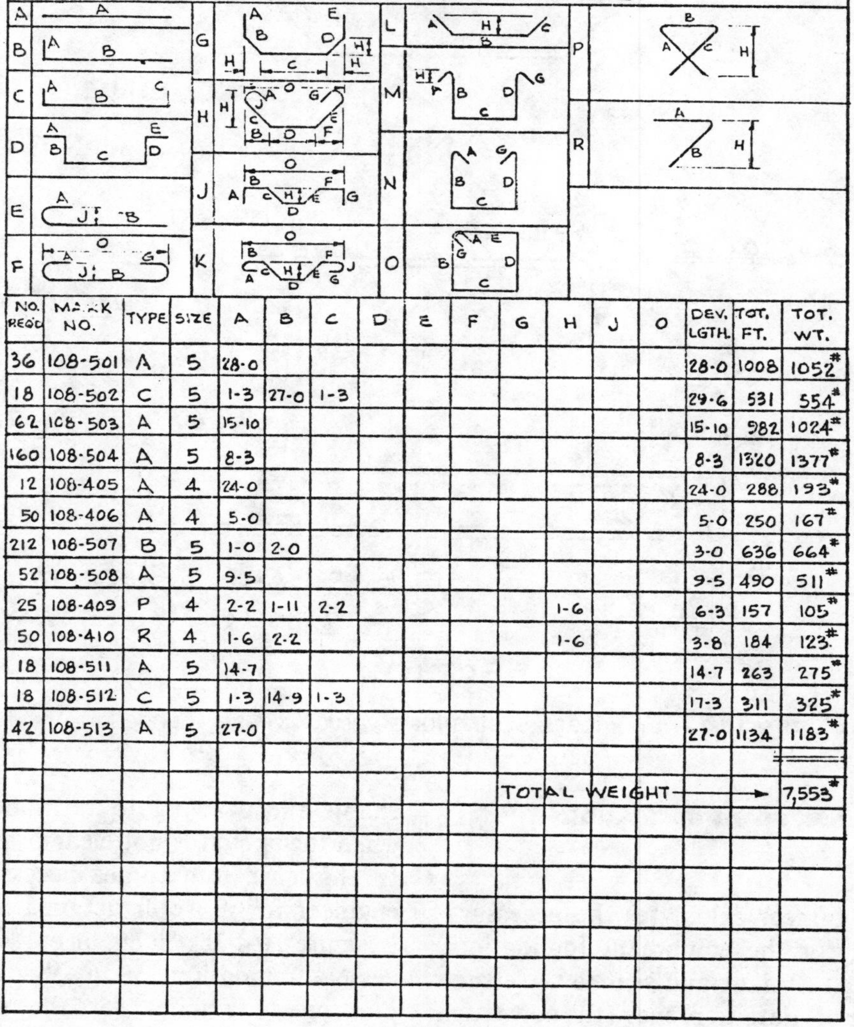

NO. REQ'D	MARK NO.	TYPE	SIZE	A	B	C	D	E	F	G	H	J	O	DEV. LGTH.	TOT. FT.	TOT. WT.
36	108-501	A	5	28-0										28-0	1008	1052#
18	108-502	C	5	1-3	27-0	1-3								29-6	531	554#
62	108-503	A	5	15-10										15-10	982	1024#
160	108-504	A	5	8-3										8-3	1320	1377#
12	108-405	A	4	24-0										24-0	288	193#
50	108-406	A	4	5-0										5-0	250	167#
212	108-507	B	5	1-0	2-0									3-0	636	664#
52	108-508	A	5	9-5										9-5	490	511#
25	108-409	P	4	2-2	1-11	2-2					1-6			6-3	157	105#
50	108-410	R	4	1-6	2-2						1-6			3-8	184	123#
18	108-511	A	5	14-7										14-7	263	275#
18	108-512	C	5	1-3	14-9	1-3								17-3	311	325#
42	108-513	A	5	27-0										27-0	1134	1183#
									TOTAL WEIGHT →							7,553#

WORK WITH DWG. NO. 1-108

Figure 10-16. Reinforcing steel schedule for the heater foundation.

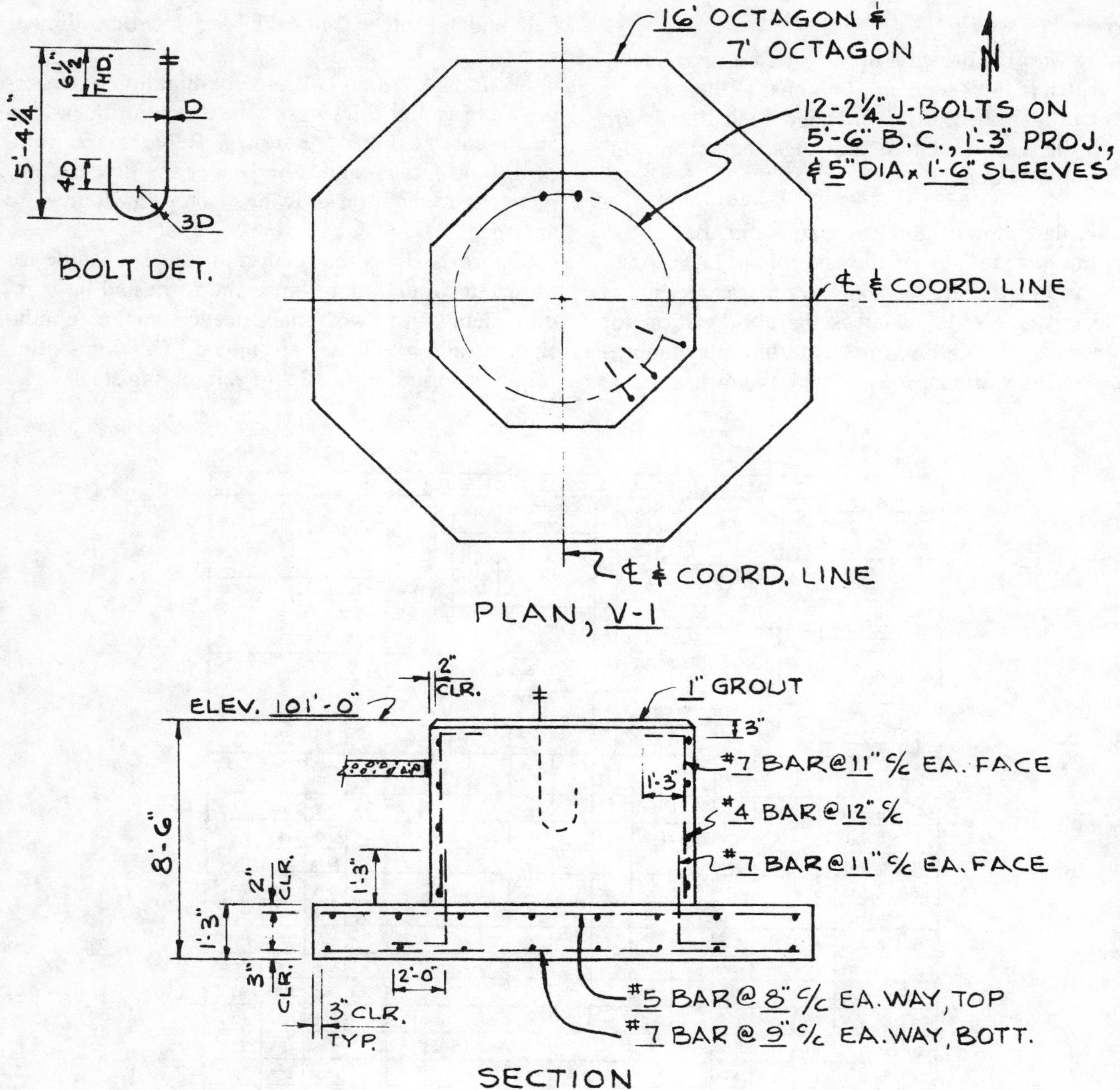

Figure 10-17. Engineer's sketch for a vertical vessel foundation, V-1.

The Draftsman's Information

The structural engineer calculates the necessary structural strength for the equipment foundation. An engineer's sketch is made and sent to the structural draftsman for his data to prepare the construction drawing. This sketch may be in tabular form or it may contain a sketch. The engineer will prepare enough information, in the easiest form for him, to enable the drawing to be done. There are times when the sketch is not clear or perhaps incomplete. In this case it becomes necessary to contact the engineer for more or detailed information.

Figure 10-17 is an engineer's sketch for a vertical vessel foundation, item V-1. Most firms have preprinted sketches for the standard geometric figures and the engineer merely fills in some dimensions and data as indicated by the underlined numbers.

Review Test

This is a test on Chapter 10. All questions should be answered without referring to the text. If the student cannot answer all of them without problem, the entire chapter should be reread. It is imperative that the student know the information in this chapter.

1. Define the purpose of a foundation.
2. Define concrete.
3. How is concrete's compressive strength specified?
4. What is ACI?
5. Concrete test cylinders are tested when they are ______ and ______ days old.
6. Define rebar and its use.
7. Define ASTM.
8. Define footer.
9. What is the difference between a pedestal and a pier?
10. Allowable soil bearing is expressed as ______ per square foot.
11. A foundation having 56.5 cubic yards of concrete would weigh ______#.
12. Define grout.
13. Define combined footing.
14. What is a concrete cap?
15. How deep are piles driven?
16. What is a placing drawing?
17. What is a mud mat?
18. What is a rebar slant?
19. A rebar is marked 109-518. What does this mean?
20. What is the engineer's sketch?

CHAPTER 11
Structural Steel Drawings

The Two Drawing Types

Structural steel draftsmen fall into two main categories. The structural design draftsman receives the engineer's design sketch and prepares steel arrangement design drawings, often sizing some of the members and determining joint types. Arrangement design drawings are prepared by engineering-construction companies, architects and general engineering contractors. These drawings supply control dimensions, specify joint types and size all steel members, but do not give detail cut lengths or count and specify all bolts, or in the case of welded structures they do not specify weld details.

The structural detail draftsman usually works for a structural steel fabricator. The fabricator must bid competitively to receive the steel fabrication contract. When awarded the bid, the fabricator will receive copies of all structural design arrangement drawings, typical details (not specific details) and full narrative structural steel specifications. With this information fabrication draftsmen prepare structural steel detail drawings for their shop to cut, weld, locate bolt holes and identify with a piece mark number. These hundreds of pieces are then shipped to the job site for field erection. If a steel assembly is small enough the design drawings will specify it to be shop assembled and shipped in one piece. Sometimes large structures are partially assembled, in the shop. These pieces are called *subassemblies.* So, steel fabrication shops will ship a single fabricated piece, a subassembly or, if transportable, a fully assembled item. Each shipping piece must have a plainly marked piece mark number for identification. Piece mark numbers are assigned by the detail draftsman from general guidelines supplied by the design draftsman.

Usually the beginning structural steel draftsman will work for a structural steel fabricator for several years before going to work for an engineering-construction company or an architect. As a steel detailer, experience is gained that enables the draftsman to become a better design draftsman. The detailer has a chance to observe design arrangement drawings from companies all over the world. A good knowledge of fabricator office and shop procedures is an important asset to becoming a design draftsman, since the fabricator usually must submit all of their drawings, detail specifications and bills-of-material to the design draftsman for approval before fabrication begins. Many structural steel detail draftsmen enjoy their work so much they never change.

A Structure Is Born

Before going into the details of just how the two types of structural draftmen function, it is important to understand the complete cycle a set of structural drawings must go through, what makes the complete package and the team work required to complete the item being built. The item may be the steel framework of a huge office building, a pipe rack for a refinery or chemical plant, a huge bridge spanning a wide river or a small T-support which supports piping and electrical conduits. Whether of several thousand tons or a few hundred pounds, each structural steel drawing showing steel to be fabricated must follow the same sequence.

Figure 11-1 shows the steps necessary to build a structural steel item. The numbers shown are explained below.

1. The contractor's engineer calculates steel member sizes, establishes overall dimensions and prepares a design sketch which is sent to a design draftsman.
2. The design draftsman prepares arrangement outline drawings and typical detail sheets. At times the design draftsman will prepare some narrative specifications for the engineer's approval, but these are usually prepared by the engineer.
3. The engineer checks a print of the drawing for accuracy, completeness and compliance with customer specifications. Narrative specifications are written. Any corrections are sent back to the draftsman to be incorporated on the original tracing. At times a heavily experienced Senior Designer Draftsman may do the detail checking instead of the engineer. In this case the engineer will review a print after corrections have been made. This engineer has final approval authority.

   After all specifications and drawings are corrected and complete (in some cases when in preliminary stages) the engineer prepares a *request for quote,* asking several steel fabricators to quote their best price and delivery for items covered by the drawings and specifications.
4. A fabricator estimating engineer receives the request for quote, prints of all drawings and specifications and prepares quotation. To do this a material take-off is done, each member's weight is calculated and priced, and fabrication time and cost is estimated. Steel mills are contacted to determine current price and delivery for sizes and shapes involved. Engineering, drafting and other office time is estimated, and the cost established. Overhead and profit is added. The cost of shipping the fabricated steel to the job site must be estimated. Depending on the job site location, transportation may be by truck, rail or water.

   The quotation time will be between two and ten weeks, depending upon the size of the order being estimated and the number of other quotes in the office at the same time.

   After the quotation is complete it is reviewed by the manager and officially transmitted to the contractor's engineer, usually through their purchasing department.
5. The contractor engineer tabulates price and delivery quotations submitted by all steel fabricators. Prime criteria is past performance of each vendor, delivery to meet job needs and quoted price. The order of importance is often performance, delivery and price.

Structural steel bids are submitted on three bases. The least popular type is *cost-plus,* where the client reimburses the steel fabricator for all material and labor costs, direct and indirect, plus a fee for profit. Sometimes the engineering costs are in this fee, and only shop labor and material costs are reimbursable.

The second most popular bid type is *pound-price.* Here the vendor quotes a price per pound for fabricated steel, usually in two categories—heavy shapes and light shapes. Heavy shapes might be 12″ and larger, while 10″ and smaller would be considered light shapes. Included in the fabricator's pound-price would be all material and labor costs plus their profit. Each fabricated piece has its weight estimated by the structural draftsman using standard weight charts and the client is billed by weight only. Of course, the client will check the bill by having their structural steel draftsmen also estimate the weight of the fabricated items.

The most popular bid type is *lump-sum.* With this bid type the fabricator uses the contractor's design drawing and quotes a single price for the engineering, material purchasing and handling, fabrication and delivery to the job site. This contract type has the advantage of allowing the contractor and their client to know exactly how much money to budget for fabricated steel. It also has several disadvantages. Contractor drawings must be complete in every detail prior to bidding. Changes which may occur later in the job will upset the fabricator's bid and costly change-orders must be negotiated to allow for these changes. The pound-price bid can be made early in the design from preliminary and incomplete drawings. This allows requests-for-quotation to be issued earlier, often meaning the complete time-consuming cycle will be finished earlier; thus, job site delivery would be earlier, often by two or three months. But if an early bid and early steel delivery is not critical to the job completion, the lump-sum bid is preferred.

After the successful bidder is selected the contractor's engineer will write a requisition for the structural steel fabrication and transmit it to their purchasing department, where the purchase order is

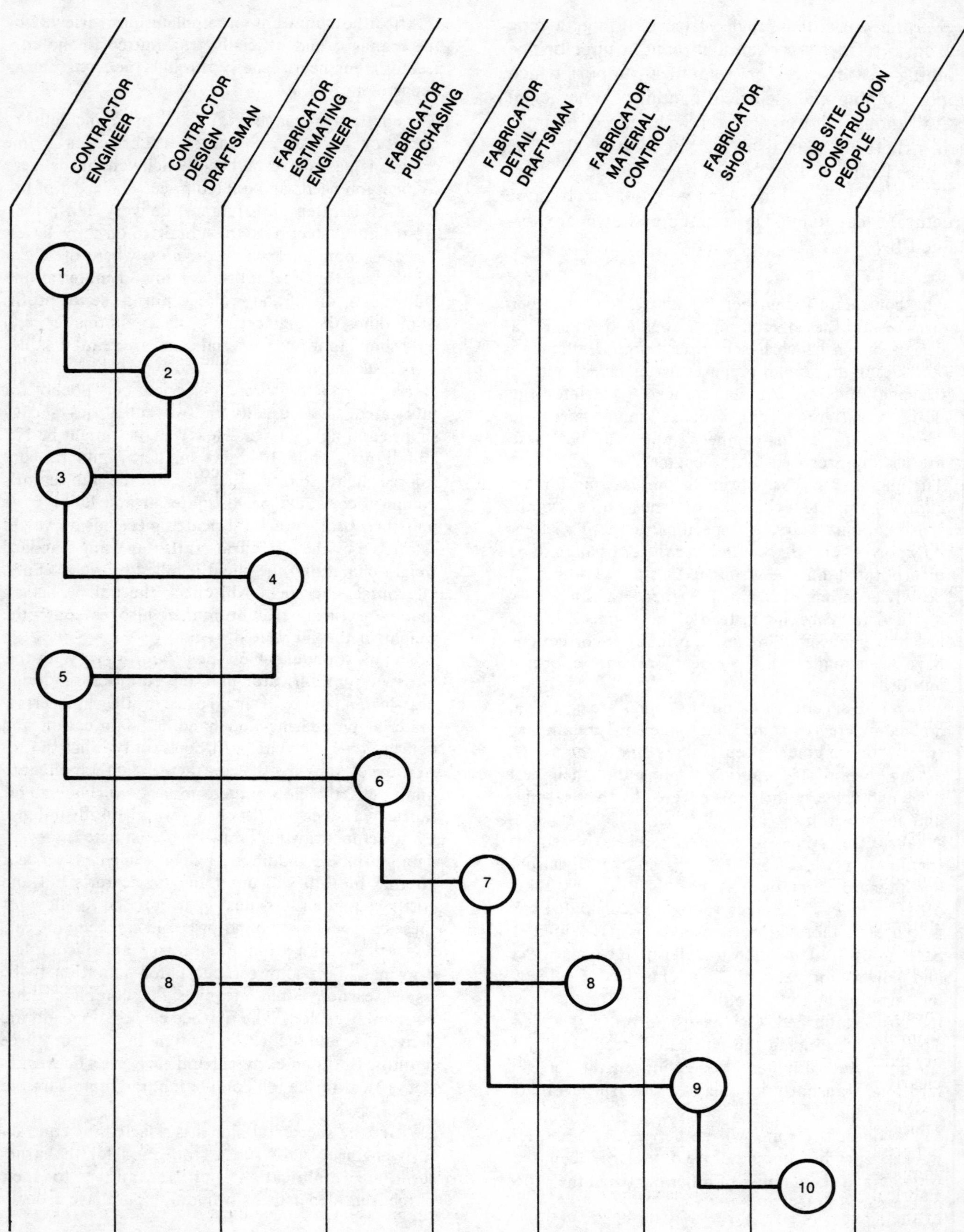

Figure 11-1. Sequence of a structural item or items. (See text for number explanation.)

typed, terms listed and legal paragraphs included. The PO (purchase order) is then mailed to the selected fabricator. Unsuccessful bidders are notified that the contract was given to another bidder.

6. When the fabricator receives the order, a review is made of the original estimating material take-off, and, if it is still good for sizes, shapes and weights, material procurement begins. First the in-house material stock is checked to see what is on hand. Other material is ordered from the steel mill. Actual fabrication cannot begin until a good percentage of all material is in the material stockpile.
7. While the mill is making the steel shapes, detail drafting starts. Hopefully all drafting can be completed, drawings approved by the client and prints issued to the shop before the steel arrives. The fabricator's draftsmen prepare drawings of the entire erected assembly, calculate shippable pieces and assign piece mark numbers. Then a drawing is made for each piece mark number. This *steel detail drawing* is used by the shop to fabricate the piece.
8. Once the drawings are complete they may have to be submitted to the client's drafting department for approval. If not, prints are issued to the fabricator's material control department where each foot of every size and shape on the drawing is recorded as consumed, and a constant check is kept on how much of each item is used compared to how much is either in stock or on order. Anytime the consumed quantities come near the total ordered and in-stock quantity, a new order is placed to supplement the original order.
9. Prints are often issued to the shop at the same time they are issued to the material control department. This issue is to allow the shop foremen and supervisors to begin planning. Normally the print used for fabrication will be issued after material control certifies and stamps the drawing showing that all required material is on hand. If the material shown on the detail drawing is in-house, the drawing is stamped and issued to the shop. If the material was part of the mill order, drawings will be stamped and issued for fabrication as the material is received.

   When a good quantity of these stamped detail sheets are issued, fabrication begins. Steel is selected from the stockpile and cut per the detail sheet, holes are punched as required and piece mark numbers are clearly marked on each section

   When a full shipping load has been fabricated, the fabricator's traffic department, often called shipping department, makes arrangements to have the pieces shipped to the client's job site.
10. Fabricated structural steel is received and stockpiled at the jobsite. A large structure will have hundreds, possibly thousands, of fabricated steel pieces. By careful coordination and full cooperation between fabricator and client, the first steel shipment will be the pieces required at the bottom part of the structure. It does very little good to receive the top floor steel first.

    The field will receive a copy of all steel detail drawings, the arrangement drawing showing all piece mark numbers and the contractor's design drawing. The arrangement drawing with piece marks serves as the road map, and the structure is erected piece by piece as one might build a structure with a toy erector set. But, in the field giant cranes lift each fabricated piece and hold it in place while welds are made or bolts are installed. Should the fabricated piece be misfabricated or should the detail draftsman make a drawing error, the piece will not fit. Then workmen wait while the crane lowers the piece to the ground where jobsite steel fabricators attempt to modify the piece, hoping to make it fit. Often this cannot be done, and the fabricator's shop must make a new piece and ship it to the job site by special truck. Any drafting error can cost hundreds or even thousands of dollars in lost worker and equipment time.

The ten steps noted have been simplified for ease of comprehension. The steel fabrication business is highly competitive and each company has procedures that they believe make them better than their competition. The students entering this field will soon learn their company's procedures.

Structural Steel Shapes

Structural steel shapes are manufactured in a wide variety of shapes, sizes and weights per linear foot. Steel mills roll these sections in six basic steel materials. The most common material used is ASTM A36 specification, an all purpose steel with high strength and good welding characteristics.

The American Institute of Steel Construction (AISC) *Manual of Steel Construction* lists dimensions, weights and properties for all structural steel shapes. All professional structural steel draftsmen own a copy of this manual. This chapter cannot cover all structural steel shapes and will be limited to shapes commonly used.

Common shapes are pictured in Figure 11-2 and explained below.

American Standard Beams (S) are generally called I-beams because of their resemblance to that capital letter. Used as columns and struts.

American Standard Channels (C) are used as struts and in trusses when light loadings are required. They are often used for steel platforming load-bearing members.

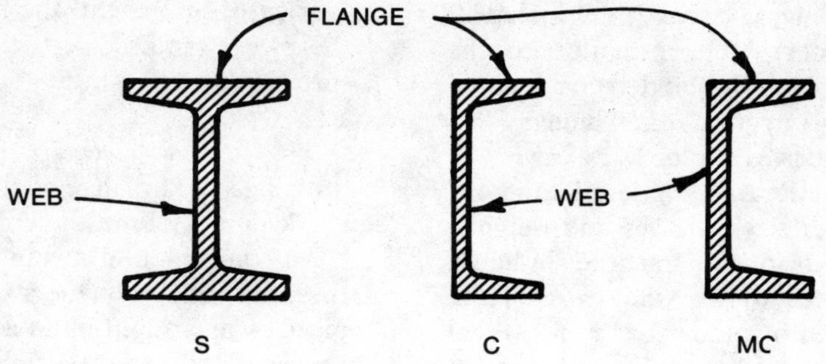

AMERICAN STANDARD BEAMS (S), CHANNELS (C) AND MISCELLANEOUS CHANNELS (MC)

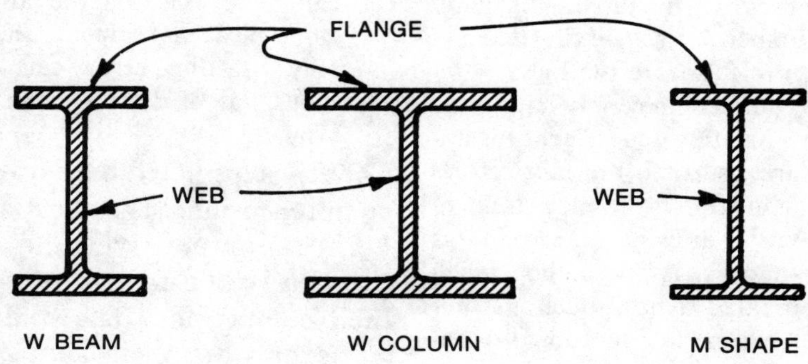

WIDE FLANGE SHAPES (W)
MISCELLANEOUS SHAPES (M)

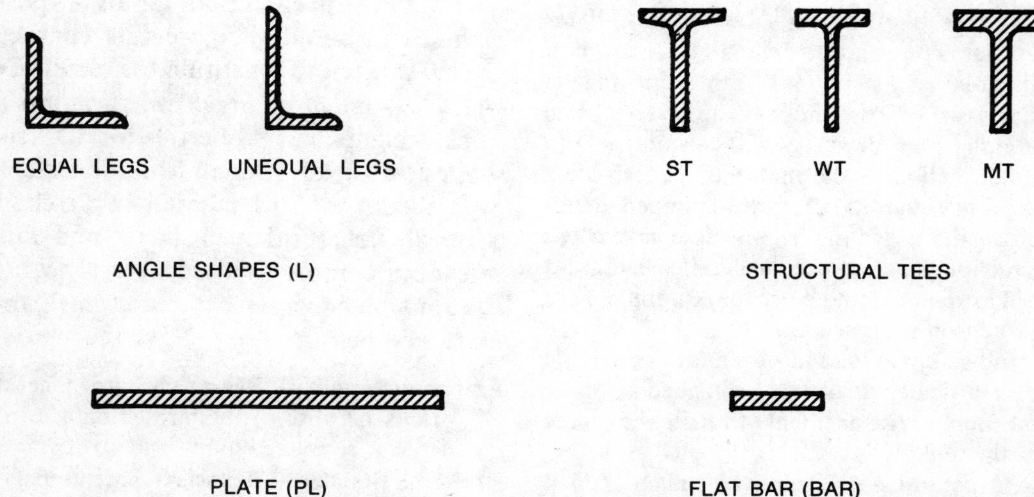

Figure 11-2. Plain material shapes shown in section.

Wide-Flange Shapes (W) are used as both beams and columns and are furnished with constant thickness flanges.

Miscellaneous Shapes (M) are similar in shape to W shapes.

Structural Tees (WT, MT and ST) are made by splitting S, W and M shapes, usually at mid-distance of their webs. Most structural steel fabricators order S, W and M shapes and cut the webs themselves to form Tees.

Angles (L) are used for struts, platforms, to add framing strength and for many other items. They have two legs set at right angles to each other. These legs may be equal or unequal widths.

Flat Bars (Bar) have a rectangular cross-section, and are rolled in many widths and thicknesses, but widths are normally limited to 6″ or 8″ depending on the thickness. If wider bars are needed a sheet of plate is cut to form it.

Plate (PL or ℄) is also rectangular in cross-section and comes in varied widths and thicknesses, but in larger pieces than bars. Plate widths start at 10″ and are rolled up to 200″ wide depending on thickness. Lengths are as long as shipping will allow.

Floor Plate (Floor PL or Floor ℄) is a skid resistant raised pattern on one side piece of plate used as a walking surface. Floor plate is made in a variety of designs and, although thicknesses up to ⅞″ are available, the most common thickness specified is ¼″. The specified thickness does not include the raised pattern thickness.

All structural steel material shipped from the mill is referred to as *plain material*—meaning not fabricated material.

In Figure 11-2 note that the S, C and MC shapes have tapered flanges on either side of the web. W shapes usually have parallel, equal thickness flanges, although a few mills put a 3° taper on their flanges inner surfaces. M shapes may have either parallel or tapered inner surfaces, depending on the section.

Structural steel shapes are available in lengths to 100′. Purchasers usually specify cut lengths to suit shipping methods or warehouse storage and handling facilities.

Structural Nomenclature

The structural steel business has terms peculiar to its business. Many of these terms are also used by concrete designers and draftsmen. Listed below are terms which the structural draftsman must know.

Column is a vertical steel or concrete member that supports structures, pipe racks, buildings, et cetera.

Strut is a structural member which carries load, axially only, in tension or compression.

Truss is a fabricated load-bearing structure used mainly for long spans, such as a roof truss in a building or a truss made for an elevated road crossing.

Slide plate is plate placed on concrete to allow a vessel or exchanger saddle to slide on to take care of expansion and contraction.

Floor plate is sometimes called checkered plate because it has a raised pattern in a checked pattern to form a non-slip surface. It is used for platforms and floors in compressor buildings.

Gage line is the line through the center of holes punched in a structural member for bolting.

Kip is 1,000 pounds.

Beam is a horizontal or inclined member that carries a load. A *simple beam* is supported at two points, usually the ends, and tends to deflect in the middle. A *fixed beam* is one with both ends firmly anchored, usually bolted or welded, to another member or to concrete which prevents the ends from rotating as is possible with the simple beam which is not fixed, only supported. A *cantilever beam* has one end fixed. The other end is unsupported. The *continuous beam* is supported intermittently for its entire length preventing deflection. Figure 11-3 pictures beam types.

Loads are forces imposed on a structure or member. Design must resist these loads. *Dead load* is the combined weight of all items being supported. *Live load* is the force applied by a moving object, such as a person's weight, a truck, traveling cranes and elevators or stored material and equipment and other vehicles. *Wind load* is the force caused by wind blowing against surfaces, tending to overturn them. In hurricane areas wind loading is a critical item. *Earthquake load* is designed for areas subject to earthquakes, usually as a factor added to wind load totals. *Impact load* is a heavy short-lived load that must be considered such as the force a truck would cause when running over a manhole.

Uniform loads are loads evenly distributed over the length of a beam.

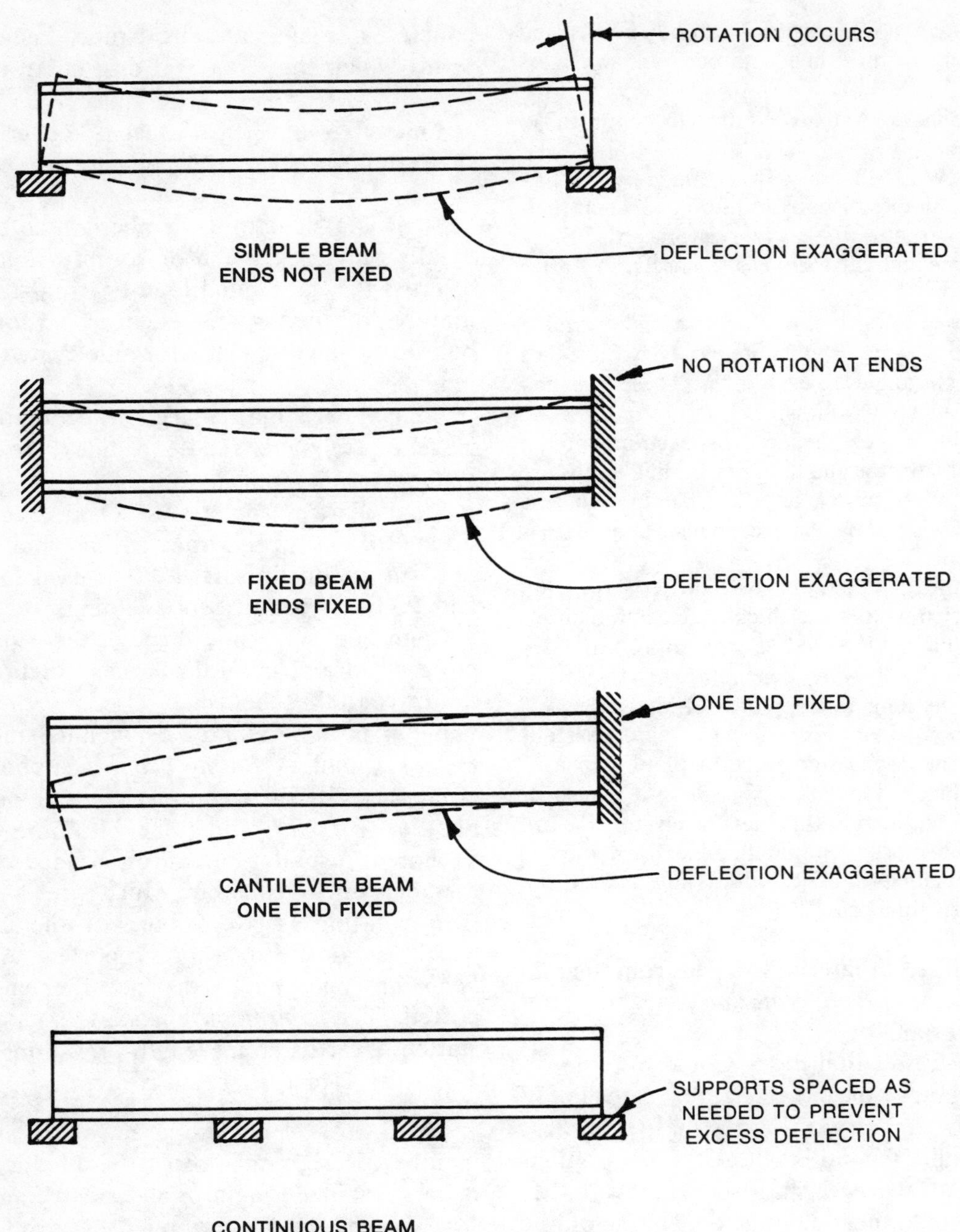

Figure 11-3. Beam types.

Concentrated loads are loads considered to react at one point of the beam or column.

Moment is the product of a force stated in units of weight (kips or pounds) times a distance of length (feet or inches). Hence, moments are expressed in units of force and distance such as kip-feet, kip-inches, pound-feet and pound-inches.

Reactions are total forces in effect at a steel joint or support point. Reactions are expressed in kips.

Stress is the internal force acting on a unit of cross-sectional area (usually expressed in square inches). The selected member must have a total *allowable stress* equal to or preferably greater than the calculated stress.

Shape Designation

Structural shapes have short designations which are used on all drawings, in most drafting room conversations and in letters. Table 11-1 supplies many of the most common shape types and their designations as established by AISC. The AISC *Manual of Steel Construction* supplies a full listing of shapes, weights and engineering properties.

Shape Sizes and Dimensions

All shapes are manufactured to size and thickness as specified by AISC. The size and thickness determines their weight per foot of length. On special order many steel mills will roll sizes and thicknesses not normally commercially rolled.

Table 11-2 lists commercially available wide-flange shape sizes and detail dimensions.

Table 11-3 lists commercially available M shape sizes and detail dimensions.

Table 11-4 lists commercially available S shape (I-beam) sizes and detail dimensions. Table 11-5 lists commercially available American standard channel sizes and detail dimensions. Table 11-6 lists commercially available miscellaneous channel sizes and ensions.

Table 11-7 supplies data for angles with equal legs. Table 11-8 supplies data for angles with unequal legs. Table 11-9 supplies data for square and round bars. Table 11-10 supplies weights for steel plate. Table 11-11 supplies area of rectangular sections.

(text continued on page 161)

Table 11-1. Shape Designations

Shape Type	Designation	Definition
W	W14 x 87	Wide-flange, 14″ nominal depth, weighing 87 pounds per linear foot.
W	W36 x 300	Wide-flange, 36″ nominal depth, weighing 300 pounds per linear foot.
S	S12 x 35	I-beam, 12″ depth, weighing 35 pounds per linear foot.
M	M14 x 17.2	Miscellaneous shape, 14″ nominal depth, weighing 17.2 pounds per linear foot.
C	C10 x 30	Channel, 10″ wide, weighing 30 pounds per linear foot.
MC	MC12 x 50	Miscellaneous channel, 12″ wide, weighing 50 pounds per linear foot.
L	L6 x 6 x ¾	Angle, equal legs 6″ long, ¾″ thickness.
L	L8 x 6 x ½	Angle, unequal legs 8″ and 6″ long, ½″ thickness.
WT	WT18 x 80	Tee cut from wide-flange, 18″ nominal depth, weighing 80 pounds per linear foot.
ST	ST12 x 50	Tee cut from I-beam, 12″ depth, weighing 50 pounds per linear foot.
MT	MT6 x 5.9	Tee cut from miscellaneous shape, 6″ nominal depth, weighing 5.9 pounds per linear foot.
PL	PL½ x 18	Plate, ½″ thick, 18″ wide.
Bar	Bar 1 □	Bar, 1″ square.
Bar	Bar 1 ϕ	Bar, 1″ round.
Bar	Bar 4 x ½	Flat bar, 4″ wide, ½″ thick.

Table 11-2. W Shapes

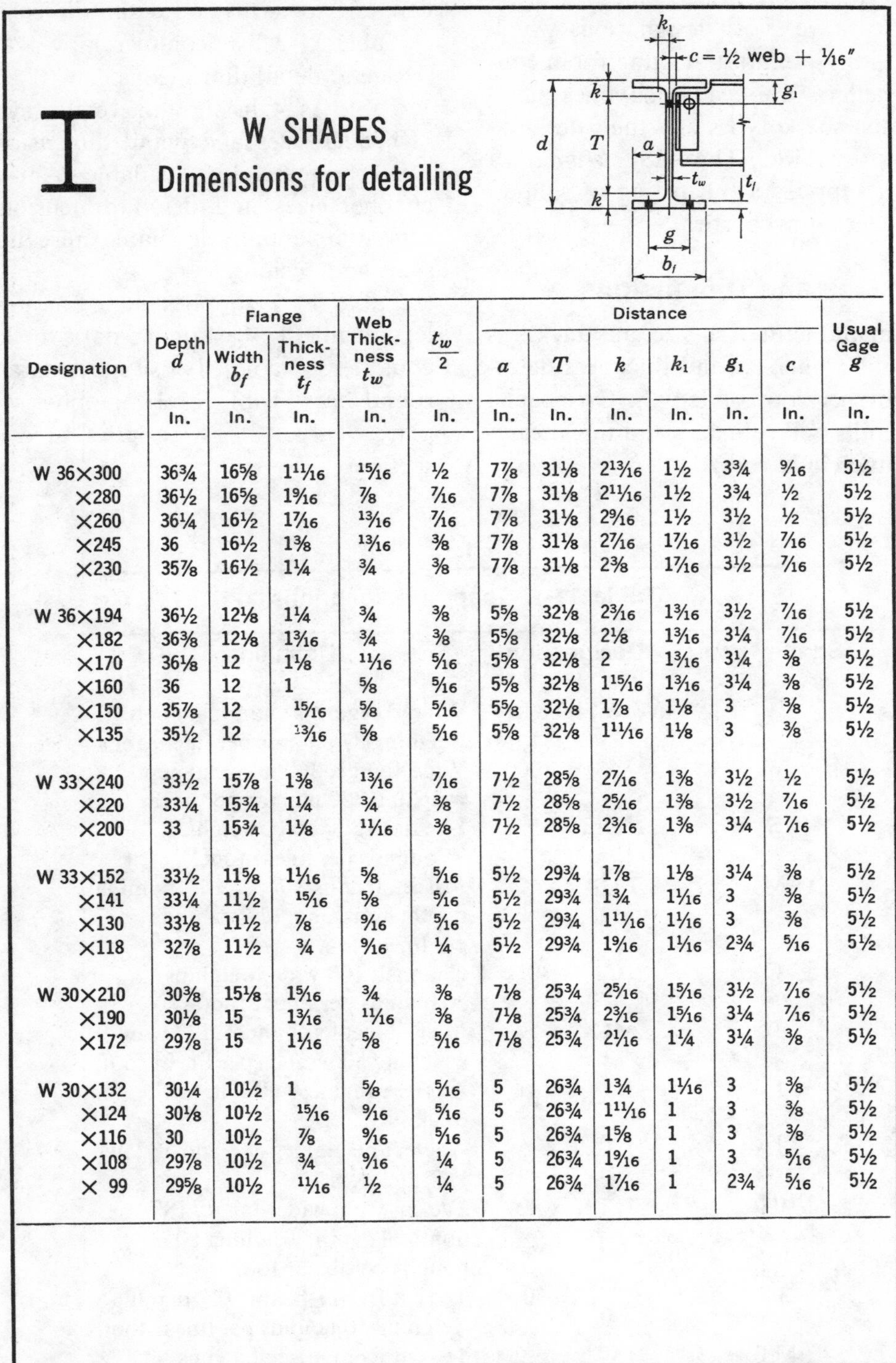

Designation	Depth d	Flange Width b_f	Flange Thickness t_f	Web Thickness t_w	$\frac{t_w}{2}$	Distance a	Distance T	Distance k	Distance k_1	Distance g_1	Distance c	Usual Gage g
	In.	In.	In.	In.	In.	In.	In.	In.	In.	In.	In.	In.
W 36×300	36 3/4	16 5/8	1 11/16	15/16	1/2	7 7/8	31 1/8	2 13/16	1 1/2	3 3/4	9/16	5 1/2
×280	36 1/2	16 5/8	1 9/16	7/8	7/16	7 7/8	31 1/8	2 11/16	1 1/2	3 3/4	1/2	5 1/2
×260	36 1/4	16 1/2	1 7/16	13/16	7/16	7 7/8	31 1/8	2 9/16	1 1/2	3 1/2	1/2	5 1/2
×245	36	16 1/2	1 3/8	13/16	3/8	7 7/8	31 1/8	2 7/16	1 7/16	3 1/2	7/16	5 1/2
×230	35 7/8	16 1/2	1 1/4	3/4	3/8	7 7/8	31 1/8	2 3/8	1 7/16	3 1/2	7/16	5 1/2
W 36×194	36 1/2	12 1/8	1 1/4	3/4	3/8	5 5/8	32 1/8	2 3/16	1 3/16	3 1/2	7/16	5 1/2
×182	36 3/8	12 1/8	1 3/16	3/4	3/8	5 5/8	32 1/8	2 1/8	1 3/16	3 1/4	7/16	5 1/2
×170	36 1/8	12	1 1/8	11/16	5/16	5 5/8	32 1/8	2	1 3/16	3 1/4	3/8	5 1/2
×160	36	12	1	5/8	5/16	5 5/8	32 1/8	1 15/16	1 3/16	3 1/4	3/8	5 1/2
×150	35 7/8	12	15/16	5/8	5/16	5 5/8	32 1/8	1 7/8	1 1/8	3	3/8	5 1/2
×135	35 1/2	12	13/16	5/8	5/16	5 5/8	32 1/8	1 11/16	1 1/8	3	3/8	5 1/2
W 33×240	33 1/2	15 7/8	1 3/8	13/16	7/16	7 1/2	28 5/8	2 7/16	1 3/8	3 1/2	1/2	5 1/2
×220	33 1/4	15 3/4	1 1/4	3/4	3/8	7 1/2	28 5/8	2 5/16	1 3/8	3 1/2	7/16	5 1/2
×200	33	15 3/4	1 1/8	11/16	3/8	7 1/2	28 5/8	2 3/16	1 3/8	3 1/4	7/16	5 1/2
W 33×152	33 1/2	11 5/8	1 1/16	5/8	5/16	5 1/2	29 3/4	1 7/8	1 1/8	3 1/4	3/8	5 1/2
×141	33 1/4	11 1/2	15/16	5/8	5/16	5 1/2	29 3/4	1 3/4	1 1/16	3	3/8	5 1/2
×130	33 1/8	11 1/2	7/8	9/16	5/16	5 1/2	29 3/4	1 11/16	1 1/16	3	3/8	5 1/2
×118	32 7/8	11 1/2	3/4	9/16	1/4	5 1/2	29 3/4	1 9/16	1 1/16	2 3/4	5/16	5 1/2
W 30×210	30 3/8	15 1/8	1 5/16	3/4	3/8	7 1/8	25 3/4	2 5/16	1 5/16	3 1/2	7/16	5 1/2
×190	30 1/8	15	1 3/16	11/16	3/8	7 1/8	25 3/4	2 3/16	1 5/16	3 1/4	7/16	5 1/2
×172	29 7/8	15	1 1/16	5/8	5/16	7 1/8	25 3/4	2 1/16	1 1/4	3 1/4	3/8	5 1/2
W 30×132	30 1/4	10 1/2	1	5/8	5/16	5	26 3/4	1 3/4	1 1/16	3	3/8	5 1/2
×124	30 1/8	10 1/2	15/16	9/16	5/16	5	26 3/4	1 11/16	1	3	3/8	5 1/2
×116	30	10 1/2	7/8	9/16	5/16	5	26 3/4	1 5/8	1	3	3/8	5 1/2
×108	29 7/8	10 1/2	3/4	9/16	1/4	5	26 3/4	1 9/16	1	3	5/16	5 1/2
× 99	29 5/8	10 1/2	11/16	1/2	1/4	5	26 3/4	1 7/16	1	2 3/4	5/16	5 1/2

Courtesy of American Institute of Steel Construction

Table 11-2 continued

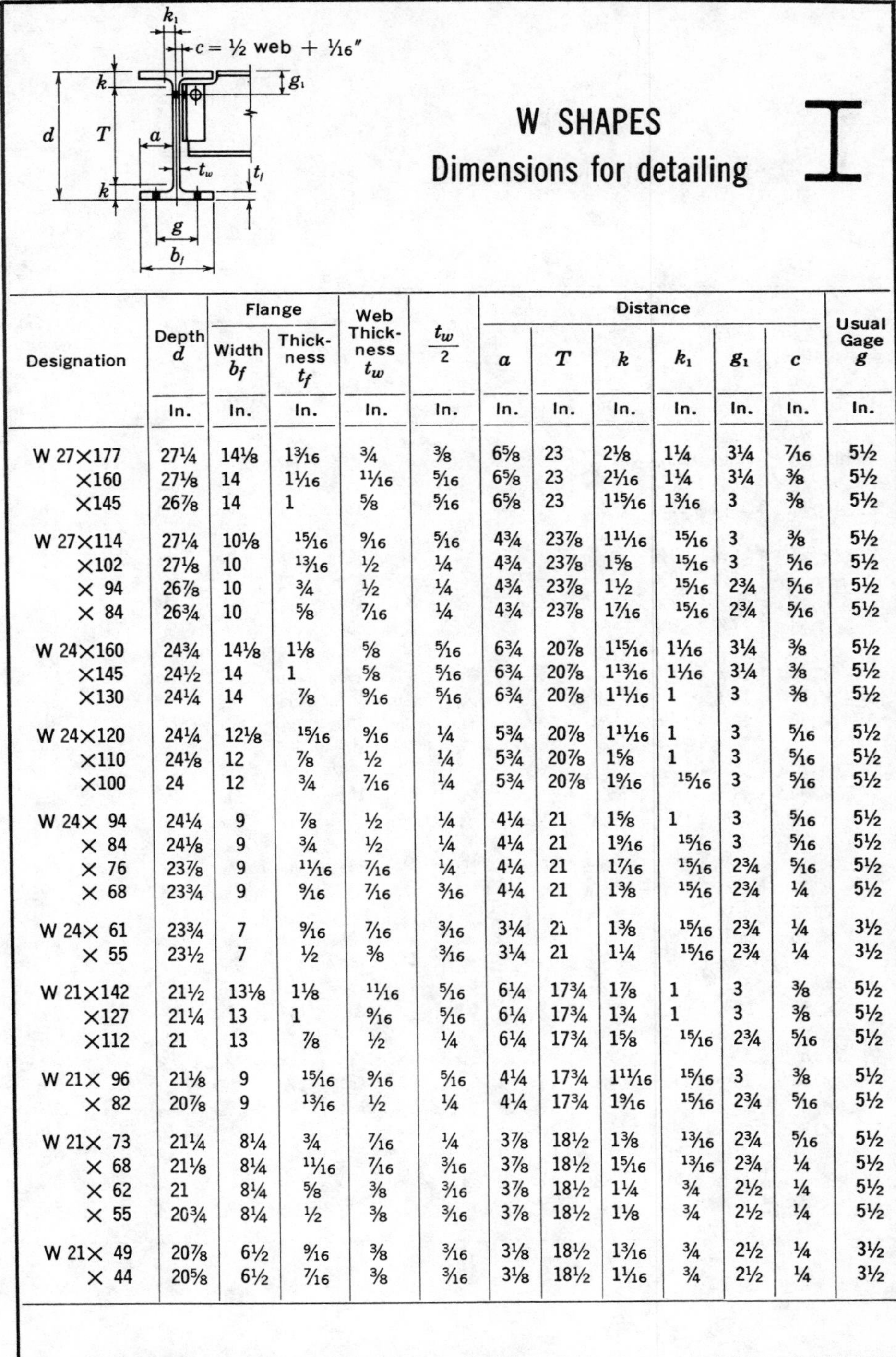

W SHAPES
Dimensions for detailing

Designation	Depth d	Flange Width b_f	Flange Thickness t_f	Web Thickness t_w	$\frac{t_w}{2}$	Distance a	Distance T	Distance k	Distance k_1	Distance g_1	Distance c	Usual Gage g
	In.	In.	In.	In.	In.	In.	In.	In.	In.	In.	In.	In.
W 27×177	27 1/4	14 1/8	1 3/16	3/4	3/8	6 5/8	23	2 1/8	1 1/4	3 1/4	7/16	5 1/2
×160	27 1/8	14	1 1/16	11/16	5/16	6 5/8	23	2 1/16	1 1/4	3 1/4	3/8	5 1/2
×145	26 7/8	14	1	5/8	5/16	6 5/8	23	1 15/16	1 3/16	3	3/8	5 1/2
W 27×114	27 1/4	10 1/8	15/16	9/16	5/16	4 3/4	23 7/8	1 11/16	15/16	3	3/8	5 1/2
×102	27 1/8	10	13/16	1/2	1/4	4 3/4	23 7/8	1 5/8	15/16	3	5/16	5 1/2
× 94	26 7/8	10	3/4	1/2	1/4	4 3/4	23 7/8	1 1/2	15/16	2 3/4	5/16	5 1/2
× 84	26 3/4	10	5/8	7/16	1/4	4 3/4	23 7/8	1 7/16	15/16	2 3/4	5/16	5 1/2
W 24×160	24 3/4	14 1/8	1 1/8	5/8	5/16	6 3/4	20 7/8	1 15/16	1 1/16	3 1/4	3/8	5 1/2
×145	24 1/2	14	1	5/8	5/16	6 3/4	20 7/8	1 13/16	1 1/16	3 1/4	3/8	5 1/2
×130	24 1/4	14	7/8	9/16	5/16	6 3/4	20 7/8	1 11/16	1	3	3/8	5 1/2
W 24×120	24 1/4	12 1/8	15/16	9/16	1/4	5 3/4	20 7/8	1 11/16	1	3	5/16	5 1/2
×110	24 1/8	12	7/8	1/2	1/4	5 3/4	20 7/8	1 5/8	1	3	5/16	5 1/2
×100	24	12	3/4	7/16	1/4	5 3/4	20 7/8	1 9/16	15/16	3	5/16	5 1/2
W 24× 94	24 1/4	9	7/8	1/2	1/4	4 1/4	21	1 5/8	1	3	5/16	5 1/2
× 84	24 1/8	9	3/4	1/2	1/4	4 1/4	21	1 9/16	15/16	3	5/16	5 1/2
× 76	23 7/8	9	11/16	7/16	1/4	4 1/4	21	1 7/16	15/16	2 3/4	5/16	5 1/2
× 68	23 3/4	9	9/16	7/16	3/16	4 1/4	21	1 3/8	15/16	2 3/4	1/4	5 1/2
W 24× 61	23 3/4	7	9/16	7/16	3/16	3 1/4	21	1 3/8	15/16	2 3/4	1/4	3 1/2
× 55	23 1/2	7	1/2	3/8	3/16	3 1/4	21	1 1/4	15/16	2 3/4	1/4	3 1/2
W 21×142	21 1/2	13 1/8	1 1/8	11/16	5/16	6 1/4	17 3/4	1 7/8	1	3	3/8	5 1/2
×127	21 1/4	13	1	9/16	5/16	6 1/4	17 3/4	1 3/4	1	3	3/8	5 1/2
×112	21	13	7/8	1/2	1/4	6 1/4	17 3/4	1 5/8	15/16	2 3/4	5/16	5 1/2
W 21× 96	21 1/8	9	15/16	9/16	5/16	4 1/4	17 3/4	1 11/16	15/16	3	3/8	5 1/2
× 82	20 7/8	9	13/16	1/2	1/4	4 1/4	17 3/4	1 9/16	15/16	2 3/4	5/16	5 1/2
W 21× 73	21 1/4	8 1/4	3/4	7/16	1/4	3 7/8	18 1/2	1 3/8	13/16	2 3/4	5/16	5 1/2
× 68	21 1/8	8 1/4	11/16	7/16	3/16	3 7/8	18 1/2	1 5/16	13/16	2 3/4	1/4	5 1/2
× 62	21	8 1/4	5/8	3/8	3/16	3 7/8	18 1/2	1 1/4	3/4	2 1/2	1/4	5 1/2
× 55	20 3/4	8 1/4	1/2	3/8	3/16	3 7/8	18 1/2	1 1/8	3/4	2 1/2	1/4	5 1/2
W 21× 49	20 7/8	6 1/2	9/16	3/8	3/16	3 1/8	18 1/2	1 3/16	3/4	2 1/2	1/4	3 1/2
× 44	20 5/8	6 1/2	7/16	3/8	3/16	3 1/8	18 1/2	1 1/16	3/4	2 1/2	1/4	3 1/2

Courtesy of American Institute of Steel Construction

Table 11-2 continued

W SHAPES
Dimensions for detailing

Designation	Depth d	Flange Width b_f	Flange Thickness t_f	Web Thickness t_w	$\frac{t_w}{2}$	Distance a	Distance T	Distance k	Distance k_1	Distance g_1	Distance c	Usual Gage g
	In.	In.	In.	In.	In.	In.	In.	In.	In.	In.	In.	In.
W 18×114	18½	11⅞	1	⅝	5/16	5⅝	15⅛	1 11/16	15/16	3	⅜	5½
×105	18⅜	11¾	15/16	9/16	¼	5⅝	15⅛	1⅝	15/16	3	5/16	5½
× 96	18⅛	11¾	13/16	½	¼	5⅝	15⅛	1½	⅞	2¾	5/16	5½
W 18× 85	18⅜	8⅞	15/16	½	¼	4⅛	15⅛	1⅝	⅞	3	5/16	5½
× 77	18⅛	8¾	13/16	½	¼	4⅛	15⅛	1½	⅞	2¾	5/16	5½
× 70	18	8¾	¾	7/16	¼	4⅛	15⅛	1 7/16	⅞	2¾	5/16	5½
× 64	17⅞	8¾	11/16	⅜	3/16	4⅛	15⅛	1⅜	13/16	2¾	¼	5½
W 18× 60	18¼	7½	11/16	7/16	3/16	3⅝	15⅞	1 3/16	11/16	2¾	¼	3½
× 55	18⅛	7½	⅝	⅜	3/16	3⅝	15⅞	1⅛	⅝	2¾	¼	3½
× 50	18	7½	9/16	⅜	3/16	3⅝	15⅞	1 1/16	⅝	2½	¼	3½
× 45	17⅞	7½	½	5/16	3/16	3⅝	15⅞	1	⅝	2½	¼	3½
W 18× 40	17⅞	6	½	5/16	3/16	2⅞	15¾	1 1/16	⅝	2½	¼	3½
× 35	17¾	6	7/16	5/16	⅛	2⅞	15¾	1	⅝	2½	3/16	3½
W 16× 96	16⅜	11½	⅞	9/16	¼	5½	13⅛	1⅝	⅞	2¾	5/16	5½
× 88	16⅛	11½	13/16	½	¼	5½	13⅛	1½	⅞	2¾	5/16	5½
W 16× 78	16⅜	8⅝	⅞	½	¼	4	13⅛	1⅝	⅞	2¾	5/16	5½
× 71	16⅛	8½	13/16	½	¼	4	13⅛	1½	⅞	2¾	5/16	5½
× 64	16	8½	11/16	7/16	¼	4	13⅛	1 7/16	⅞	2¾	5/16	5½
× 58	15⅞	8½	⅝	7/16	3/16	4	13⅛	1⅜	13/16	2¾	¼	5½
W 16× 50	16¼	7⅛	⅝	⅜	3/16	3⅜	13¾	1¼	¾	2¾	¼	3½
× 45	16⅛	7	9/16	⅜	3/16	3⅜	13¾	1 3/16	11/16	2½	¼	3½
× 40	16	7	½	5/16	⅛	3⅜	13¾	1⅛	11/16	2½	3/16	3½
× 36	15⅞	7	7/16	5/16	⅛	3⅜	13¾	1 1/16	11/16	2½	3/16	3½
W 16× 31	15⅞	5½	7/16	¼	⅛	2⅝	13¾	1 1/16	11/16	2½	3/16	2¾
× 26	15⅝	5½	⅜	¼	⅛	2⅝	13¾	15/16	⅝	2¼	3/16	2¾

Courtesy of American Institute of Steel Construction

Table 11-2 continued

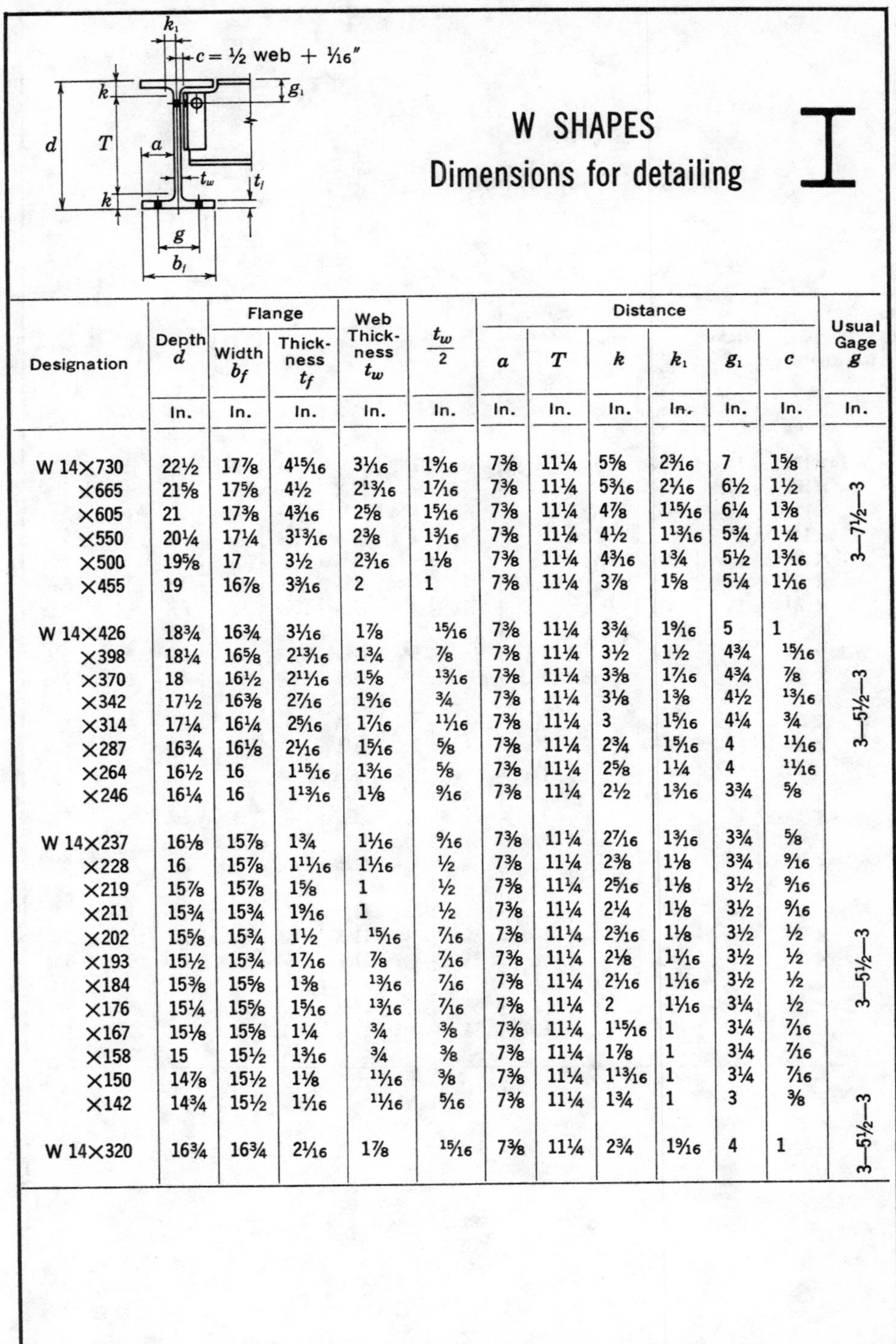

W SHAPES
Dimensions for detailing

Designation	Depth d	Flange Width b_f	Flange Thickness t_f	Web Thickness t_w	$\frac{t_w}{2}$	Distance a	Distance T	Distance k	Distance k_1	Distance g_1	Distance c	Usual Gage g
	In.	In.	In.	In.	In.	In.	In.	In.	In.	In.	In.	In.
W 14×730	22½	17⅞	4 15/16	3 1/16	1 9/16	7⅜	11¼	5⅝	2 3/16	7	1⅝	
×665	21⅝	17⅝	4½	2 13/16	1 7/16	7⅜	11¼	5 3/16	2 1/16	6½	1½	
×605	21	17⅜	4 3/16	2⅝	1 5/16	7⅜	11¼	4⅞	1 15/16	6¼	1⅜	3—7½—3
×550	20¼	17¼	3 13/16	2⅜	1 3/16	7⅜	11¼	4½	1 13/16	5¾	1¼	
×500	19⅝	17	3½	2 3/16	1⅛	7⅜	11¼	4 3/16	1¾	5½	1 3/16	
×455	19	16⅞	3 3/16	2	1	7⅜	11¼	3⅞	1⅝	5¼	1 1/16	
W 14×426	18¾	16¾	3 1/16	1⅞	15/16	7⅜	11¼	3¾	1 9/16	5	1	
×398	18¼	16⅝	2 13/16	1¾	⅞	7⅜	11¼	3½	1½	4¾	15/16	
×370	18	16½	2 11/16	1⅝	13/16	7⅜	11¼	3⅜	1 7/16	4¾	⅞	
×342	17½	16⅜	2 7/16	1 9/16	¾	7⅜	11¼	3⅛	1⅜	4½	13/16	3—5½—3
×314	17¼	16¼	2 5/16	1 7/16	11/16	7⅜	11¼	3	1 5/16	4¼	¾	
×287	16¾	16⅛	2 1/16	1 5/16	⅝	7⅜	11¼	2¾	1 5/16	4	11/16	
×264	16½	16	1 15/16	1 3/16	⅝	7⅜	11¼	2⅝	1¼	4	11/16	
×246	16¼	16	1 13/16	1⅛	9/16	7⅜	11¼	2½	1 3/16	3¾	⅝	
W 14×237	16⅛	15⅞	1¾	1 1/16	9/16	7⅜	11¼	2 7/16	1 3/16	3¾	⅝	
×228	16	15⅞	1 11/16	1 1/16	½	7⅜	11¼	2⅜	1⅛	3¾	9/16	
×219	15⅞	15⅞	1⅝	1	½	7⅜	11¼	2 5/16	1⅛	3½	9/16	
×211	15¾	15¾	1 9/16	1	½	7⅜	11¼	2¼	1⅛	3½	9/16	
×202	15⅝	15¾	1½	15/16	7/16	7⅜	11¼	2 3/16	1⅛	3½	½	
×193	15½	15¾	1 7/16	⅞	7/16	7⅜	11¼	2⅛	1 1/16	3½	½	3—5½—3
×184	15⅜	15⅝	1⅜	13/16	7/16	7⅜	11¼	2 1/16	1 1/16	3½	½	
×176	15¼	15⅝	1 5/16	13/16	7/16	7⅜	11¼	2	1 1/16	3¼	½	
×167	15⅛	15⅝	1¼	¾	⅜	7⅜	11¼	1 15/16	1	3¼	7/16	
×158	15	15½	1 3/16	¾	⅜	7⅜	11¼	1⅞	1	3¼	7/16	
×150	14⅞	15½	1⅛	11/16	⅜	7⅜	11¼	1 13/16	1	3¼	7/16	
×142	14¾	15½	1 1/16	11/16	5/16	7⅜	11¼	1¾	1	3	⅜	
W 14×320	16¾	16¾	2 1/16	1⅞	15/16	7⅜	11¼	2¾	1 9/16	4	1	3—5½—3

Courtesy of American Institute of Steel Construction

Table 11-2 continued

W SHAPES
Dimensions for detailing

Designation	Depth d	Flange Width b_f	Flange Thickness t_f	Web Thickness t_w	$\frac{t_w}{2}$	Distance a	T	k	k_1	g_1	c	Usual Gage g
	In.	In.	In.	In.	In.	In.	In.	In.	In.	In.	In.	In.
W 14×136	14¾	14¾	1¹⁄₁₆	¹¹⁄₁₆	⁵⁄₁₆	7	11¼	1¾	¹⁵⁄₁₆	3	⅜	5½
×127	14⅝	14¾	1	⅝	⁵⁄₁₆	7	11¼	1¹¹⁄₁₆	¹⁵⁄₁₆	3	⅜	5½
×119	14½	14⅝	¹⁵⁄₁₆	⁹⁄₁₆	⁵⁄₁₆	7	11¼	1⅝	¹⁵⁄₁₆	3	⅜	5½
×111	14⅜	14⅝	⅞	⁹⁄₁₆	¼	7	11¼	1⁹⁄₁₆	⅞	2¾	⁵⁄₁₆	5½
×103	14¼	14⅝	¹³⁄₁₆	½	¼	7	11¼	1½	⅞	2¾	⁵⁄₁₆	5½
× 95	14⅛	14½	¾	⁷⁄₁₆	¼	7	11¼	1⁷⁄₁₆	⅞	2¾	⁵⁄₁₆	5½
× 87	14	14½	¹¹⁄₁₆	⁷⁄₁₆	³⁄₁₆	7	11¼	1⅜	¹³⁄₁₆	2¾	¼	5½
W 14× 84	14⅛	12	¾	⁷⁄₁₆	¼	5¾	11¼	1⁷⁄₁₆	⅞	2¾	⁵⁄₁₆	5½
× 78	14	12	¹¹⁄₁₆	⁷⁄₁₆	³⁄₁₆	5¾	11¼	1⅜	⅞	2¾	¼	5½
W 14× 74	14¼	10⅛	¹³⁄₁₆	⁷⁄₁₆	¼	4¾	11¼	1½	⅞	2¾	⁵⁄₁₆	5½
× 68	14	10	¹¹⁄₁₆	⁷⁄₁₆	³⁄₁₆	4¾	11¼	1⅜	¹³⁄₁₆	2¾	¼	5½
× 61	13⅞	10	⅝	⅜	³⁄₁₆	4¾	11¼	1⁵⁄₁₆	¹³⁄₁₆	2¾	¼	5½
W 14× 53	14	8	¹¹⁄₁₆	⅜	³⁄₁₆	3⅞	11¼	1⅜	¹³⁄₁₆	2¾	¼	5½
× 48	13¾	8	⁹⁄₁₆	⁵⁄₁₆	³⁄₁₆	3⅞	11¼	1¼	¹³⁄₁₆	2½	¼	5½
× 43	13⅝	8	½	⁵⁄₁₆	⅛	3⅞	11¼	1³⁄₁₆	¹³⁄₁₆	2½	³⁄₁₆	5½
W 14× 38	14⅛	6¾	½	⁵⁄₁₆	³⁄₁₆	3¼	11⅞	1⅛	¹¹⁄₁₆	2½	¼	3½
× 34	14	6¾	⁷⁄₁₆	⁵⁄₁₆	⅛	3¼	11⅞	1¹⁄₁₆	¹¹⁄₁₆	2½	³⁄₁₆	3½
× 30	13⅞	6¾	⅜	¼	⅛	3¼	11⅞	1	¹¹⁄₁₆	2½	³⁄₁₆	3½
W 14× 26	13⅞	5	⁷⁄₁₆	¼	⅛	2⅜	11⅞	1	¹¹⁄₁₆	2½	³⁄₁₆	2¾
× 22	13¾	5	⁵⁄₁₆	¼	⅛	2⅜	11⅞	¹⁵⁄₁₆	⅝	2¼	³⁄₁₆	2¾

Courtesy of American Institute of Steel Construction

Table 11-2 continued

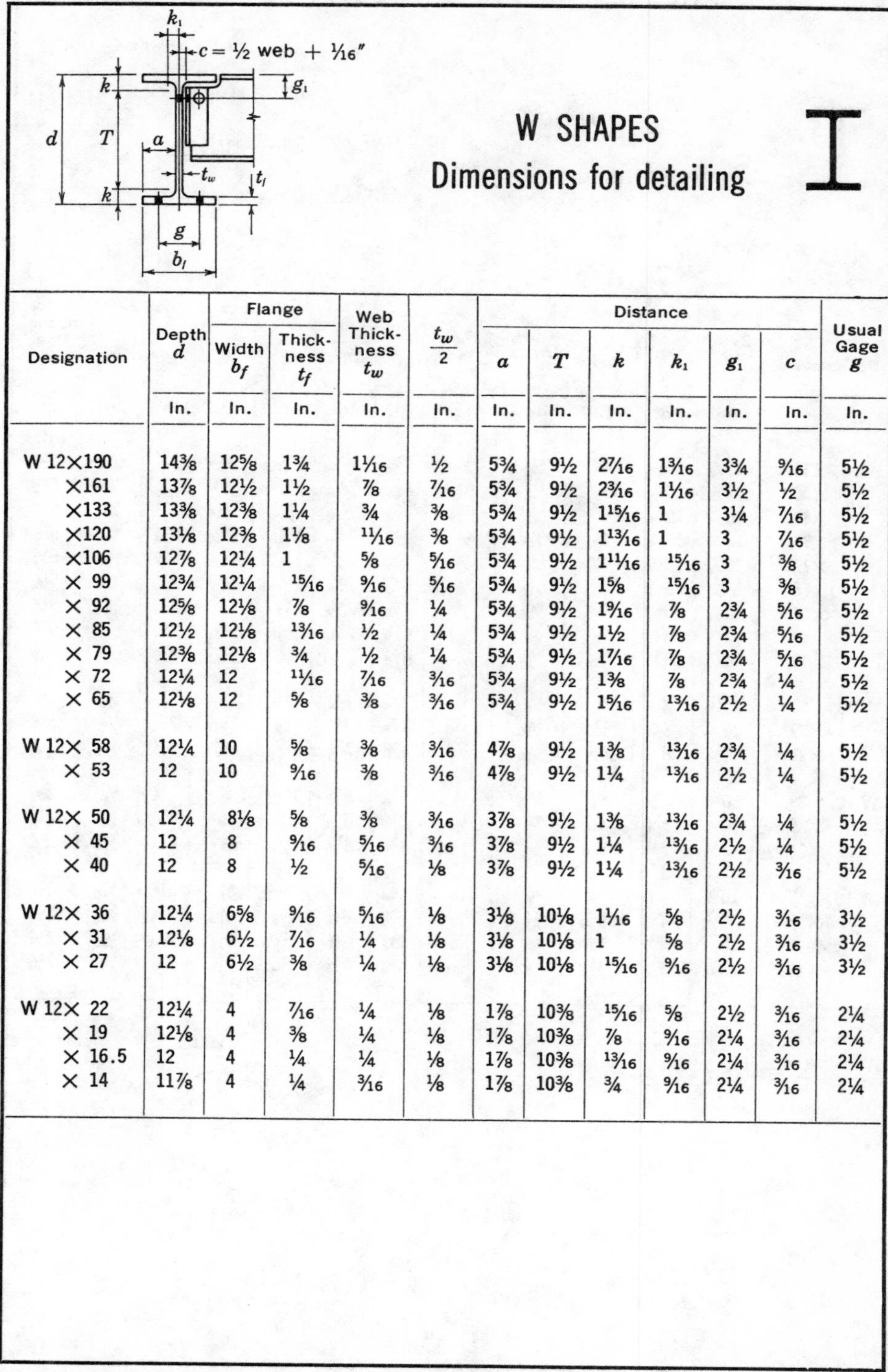

W SHAPES
Dimensions for detailing

Designation	Depth d	Flange Width b_f	Flange Thickness t_f	Web Thickness t_w	$\frac{t_w}{2}$	Distance a	Distance T	Distance k	Distance k_1	Distance g_1	Distance c	Usual Gage g
	In.	In.	In.	In.	In.	In.	In.	In.	In.	In.	In.	In.
W 12×190	14 3/8	12 5/8	1 3/4	1 1/16	1/2	5 3/4	9 1/2	2 7/16	1 3/16	3 3/4	9/16	5 1/2
×161	13 7/8	12 1/2	1 1/2	7/8	7/16	5 3/4	9 1/2	2 3/16	1 1/16	3 1/2	1/2	5 1/2
×133	13 3/8	12 3/8	1 1/4	3/4	3/8	5 3/4	9 1/2	1 15/16	1	3 1/4	7/16	5 1/2
×120	13 1/8	12 3/8	1 1/8	11/16	3/8	5 3/4	9 1/2	1 13/16	1	3	7/16	5 1/2
×106	12 7/8	12 1/4	1	5/8	5/16	5 3/4	9 1/2	1 11/16	15/16	3	3/8	5 1/2
× 99	12 3/4	12 1/4	15/16	9/16	5/16	5 3/4	9 1/2	1 5/8	15/16	3	3/8	5 1/2
× 92	12 5/8	12 1/8	7/8	9/16	1/4	5 3/4	9 1/2	1 9/16	7/8	2 3/4	5/16	5 1/2
× 85	12 1/2	12 1/8	13/16	1/2	1/4	5 3/4	9 1/2	1 1/2	7/8	2 3/4	5/16	5 1/2
× 79	12 3/8	12 1/8	3/4	1/2	1/4	5 3/4	9 1/2	1 7/16	7/8	2 3/4	5/16	5 1/2
× 72	12 1/4	12	11/16	7/16	3/16	5 3/4	9 1/2	1 3/8	7/8	2 3/4	1/4	5 1/2
× 65	12 1/8	12	5/8	3/8	3/16	5 3/4	9 1/2	1 5/16	13/16	2 1/2	1/4	5 1/2
W 12× 58	12 1/4	10	5/8	3/8	3/16	4 7/8	9 1/2	1 3/8	13/16	2 3/4	1/4	5 1/2
× 53	12	10	9/16	3/8	3/16	4 7/8	9 1/2	1 1/4	13/16	2 1/2	1/4	5 1/2
W 12× 50	12 1/4	8 1/8	5/8	3/8	3/16	3 7/8	9 1/2	1 3/8	13/16	2 3/4	1/4	5 1/2
× 45	12	8	9/16	5/16	3/16	3 7/8	9 1/2	1 1/4	13/16	2 1/2	1/4	5 1/2
× 40	12	8	1/2	5/16	1/8	3 7/8	9 1/2	1 1/4	13/16	2 1/2	3/16	5 1/2
W 12× 36	12 1/4	6 5/8	9/16	5/16	1/8	3 1/8	10 1/8	1 1/16	5/8	2 1/2	3/16	3 1/2
× 31	12 1/8	6 1/2	7/16	1/4	1/8	3 1/8	10 1/8	1	5/8	2 1/2	3/16	3 1/2
× 27	12	6 1/2	3/8	1/4	1/8	3 1/8	10 1/8	15/16	9/16	2 1/2	3/16	3 1/2
W 12× 22	12 1/4	4	7/16	1/4	1/8	1 7/8	10 3/8	15/16	5/8	2 1/2	3/16	2 1/4
× 19	12 1/8	4	3/8	1/4	1/8	1 7/8	10 3/8	7/8	9/16	2 1/4	3/16	2 1/4
× 16.5	12	4	1/4	1/4	1/8	1 7/8	10 3/8	13/16	9/16	2 1/4	3/16	2 1/4
× 14	11 7/8	4	1/4	3/16	1/8	1 7/8	10 3/8	3/4	9/16	2 1/4	3/16	2 1/4

Courtesy of American Institute of Steel Construction

Table 11-2 continued

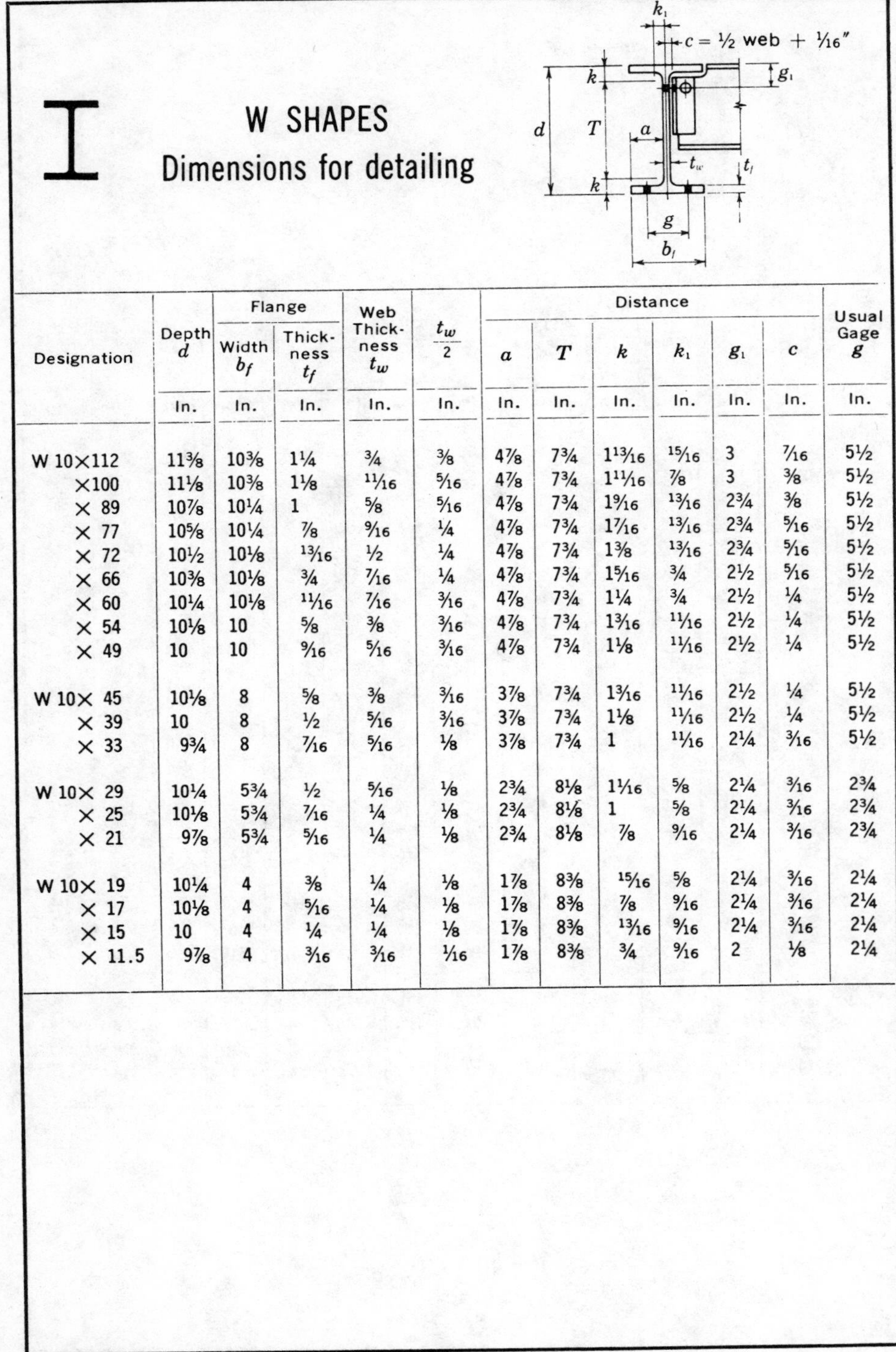

W SHAPES
Dimensions for detailing

Designation	Depth d	Flange Width b_f	Flange Thickness t_f	Web Thickness t_w	$\frac{t_w}{2}$	Distance a	Distance T	Distance k	Distance k_1	Distance g_1	Distance c	Usual Gage g
	In.	In.	In.	In.	In.	In.	In.	In.	In.	In.	In.	In.
W 10×112	11 3/8	10 3/8	1 1/4	3/4	3/8	4 7/8	7 3/4	1 13/16	15/16	3	7/16	5 1/2
×100	11 1/8	10 3/8	1 1/8	11/16	5/16	4 7/8	7 3/4	1 11/16	7/8	3	3/8	5 1/2
× 89	10 7/8	10 1/4	1	5/8	5/16	4 7/8	7 3/4	1 9/16	13/16	2 3/4	3/8	5 1/2
× 77	10 5/8	10 1/4	7/8	9/16	1/4	4 7/8	7 3/4	1 7/16	13/16	2 3/4	5/16	5 1/2
× 72	10 1/2	10 1/8	13/16	1/2	1/4	4 7/8	7 3/4	1 3/8	13/16	2 3/4	5/16	5 1/2
× 66	10 3/8	10 1/8	3/4	7/16	1/4	4 7/8	7 3/4	1 5/16	3/4	2 1/2	5/16	5 1/2
× 60	10 1/4	10 1/8	11/16	7/16	3/16	4 7/8	7 3/4	1 1/4	3/4	2 1/2	1/4	5 1/2
× 54	10 1/8	10	5/8	3/8	3/16	4 7/8	7 3/4	1 3/16	11/16	2 1/2	1/4	5 1/2
× 49	10	10	9/16	5/16	3/16	4 7/8	7 3/4	1 1/8	11/16	2 1/2	1/4	5 1/2
W 10× 45	10 1/8	8	5/8	3/8	3/16	3 7/8	7 3/4	1 3/16	11/16	2 1/2	1/4	5 1/2
× 39	10	8	1/2	5/16	3/16	3 7/8	7 3/4	1 1/8	11/16	2 1/2	1/4	5 1/2
× 33	9 3/4	8	7/16	5/16	1/8	3 7/8	7 3/4	1	11/16	2 1/4	3/16	5 1/2
W 10× 29	10 1/4	5 3/4	1/2	5/16	1/8	2 3/4	8 1/8	1 1/16	5/8	2 1/4	3/16	2 3/4
× 25	10 1/8	5 3/4	7/16	1/4	1/8	2 3/4	8 1/8	1	5/8	2 1/4	3/16	2 3/4
× 21	9 7/8	5 3/4	5/16	1/4	1/8	2 3/4	8 1/8	7/8	9/16	2 1/4	3/16	2 3/4
W 10× 19	10 1/4	4	3/8	1/4	1/8	1 7/8	8 3/8	15/16	5/8	2 1/4	3/16	2 1/4
× 17	10 1/8	4	5/16	1/4	1/8	1 7/8	8 3/8	7/8	9/16	2 1/4	3/16	2 1/4
× 15	10	4	1/4	1/4	1/8	1 7/8	8 3/8	13/16	9/16	2 1/4	3/16	2 1/4
× 11.5	9 7/8	4	3/16	3/16	1/16	1 7/8	8 3/8	3/4	9/16	2	1/8	2 1/4

Courtesy of American Institute of Steel Construction

Table 11-2 continued

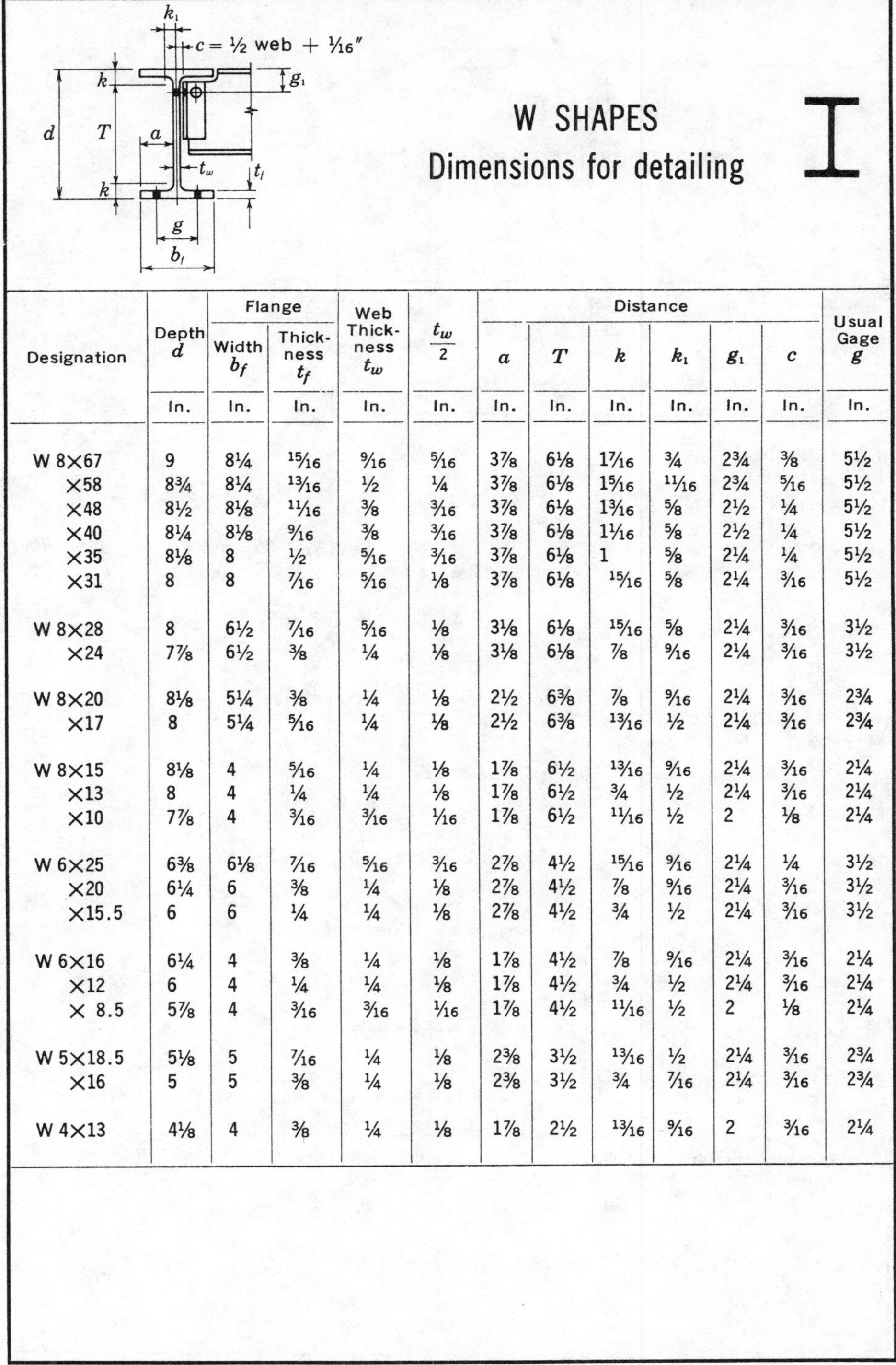

W SHAPES
Dimensions for detailing

Designation	Depth d	Flange		Web Thickness t_w	$\frac{t_w}{2}$	Distance						Usual Gage g
		Width b_f	Thickness t_f			a	T	k	k_1	g_1	c	
	In.	In.	In.	In.	In.	In.	In.	In.	In.	In.	In.	In.
W 8×67	9	8 1/4	15/16	9/16	5/16	3 7/8	6 1/8	1 7/16	3/4	2 3/4	3/8	5 1/2
×58	8 3/4	8 1/4	13/16	1/2	1/4	3 7/8	6 1/8	1 5/16	11/16	2 3/4	5/16	5 1/2
×48	8 1/2	8 1/8	11/16	3/8	3/16	3 7/8	6 1/8	1 3/16	5/8	2 1/2	1/4	5 1/2
×40	8 1/4	8 1/8	9/16	3/8	3/16	3 7/8	6 1/8	1 1/16	5/8	2 1/2	1/4	5 1/2
×35	8 1/8	8	1/2	5/16	3/16	3 7/8	6 1/8	1	5/8	2 1/4	1/4	5 1/2
×31	8	8	7/16	5/16	1/8	3 7/8	6 1/8	15/16	5/8	2 1/4	3/16	5 1/2
W 8×28	8	6 1/2	7/16	5/16	1/8	3 1/8	6 1/8	15/16	5/8	2 1/4	3/16	3 1/2
×24	7 7/8	6 1/2	3/8	1/4	1/8	3 1/8	6 1/8	7/8	9/16	2 1/4	3/16	3 1/2
W 8×20	8 1/8	5 1/4	3/8	1/4	1/8	2 1/2	6 3/8	7/8	9/16	2 1/4	3/16	2 3/4
×17	8	5 1/4	5/16	1/4	1/8	2 1/2	6 3/8	13/16	1/2	2 1/4	3/16	2 3/4
W 8×15	8 1/8	4	5/16	1/4	1/8	1 7/8	6 1/2	13/16	9/16	2 1/4	3/16	2 1/4
×13	8	4	1/4	1/4	1/8	1 7/8	6 1/2	3/4	1/2	2 1/4	3/16	2 1/4
×10	7 7/8	4	3/16	3/16	1/16	1 7/8	6 1/2	11/16	1/2	2	1/8	2 1/4
W 6×25	6 3/8	6 1/8	7/16	5/16	3/16	2 7/8	4 1/2	15/16	9/16	2 1/4	1/4	3 1/2
×20	6 1/4	6	3/8	1/4	1/8	2 7/8	4 1/2	7/8	9/16	2 1/4	3/16	3 1/2
×15.5	6	6	1/4	1/4	1/8	2 7/8	4 1/2	3/4	1/2	2 1/4	3/16	3 1/2
W 6×16	6 1/4	4	3/8	1/4	1/8	1 7/8	4 1/2	7/8	9/16	2 1/4	3/16	2 1/4
×12	6	4	1/4	1/4	1/8	1 7/8	4 1/2	3/4	1/2	2 1/4	3/16	2 1/4
× 8.5	5 7/8	4	3/16	3/16	1/16	1 7/8	4 1/2	11/16	1/2	2	1/8	2 1/4
W 5×18.5	5 1/8	5	7/16	1/4	1/8	2 3/8	3 1/2	13/16	1/2	2 1/4	3/16	2 3/4
×16	5	5	3/8	1/4	1/8	2 3/8	3 1/2	3/4	7/16	2 1/4	3/16	2 3/4
W 4×13	4 1/8	4	3/8	1/4	1/8	1 7/8	2 1/2	13/16	9/16	2	3/16	2 1/4

Courtesy of American Institute of Steel Construction

Table 11-3. M Shapes

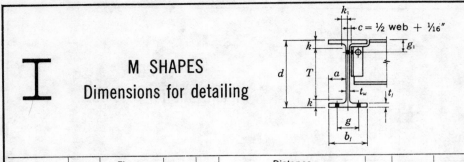

M SHAPES
Dimensions for detailing

Designation	Depth d	Flange: Width b_f	Flange: Thickness t_f	Web Thickness t_w	$\frac{t_w}{2}$	Distance: a	Distance: T	Distance: k	Distance: k_1	Distance: g_1	Distance: c	Grip	Max. Flange Fastener	Usual Flange Gage g
	In.	In.	In.	In.	In.	In.	In.	In.	In.	In.	In.	In.	In.	In.
M 14×17.2	14	4	¼	3/16	⅛	1⅞	12¾	⅝	⅜	2¼	3/16	¼	¾	2¼
M 12×11.8	12	3⅛	¼	3/16	1/16	1½	10⅞	9/16	⅜	2¼	⅛	¼	—	—
M 10×29.1	9⅞	5⅞	⅜	7/16	3/16	2¾	8⅛	⅞	½	2½	¼	7/16	⅞	2¾
×22.9	9⅞	5¾	⅜	¼	⅛	2¾	8⅛	⅞	7/16	2½	3/16	⅜	⅞	2¾
M 10× 9	10	2¾	3/16	3/16	1/16	1¼	9	½	5/16	2	⅛	3/16	—	—
M 8×34.3	8	8	7/16	⅜	3/16	3¾	5⅞	1 1/16	⅝	2½	¼	7/16	⅞	5½
×32.6	8	8	7/16	5/16	3/16	3¾	5⅞	1 1/16	⅝	2½	¼	7/16	⅞	5½
M 8×22.5	8	5⅜	⅜	⅜	3/16	2½	6¼	⅞	½	2¼	¼	⅜	⅞	2¾
×18.5	8	5¼	⅜	¼	⅛	2½	6¼	⅞	7/16	2¼	3/16	⅜	⅞	2¾
M 8× 6.5	8	2¼	3/16	⅛	1/16	1⅛	7	½	¼	2	⅛	3/16	—	—
M 7× 5.5	7	2⅛	3/16	⅛	1/16	1	6⅛	7/16	¼	2	⅛	3/16	—	—
M 6×22.5	6	6	⅜	⅜	3/16	2⅞	4⅜	13/16	½	2¼	¼	⅜	⅞	3½
×20	6	6	⅜	¼	⅛	2⅞	4⅜	13/16	7/16	2¼	3/16	⅜	⅞	3½
M 6× 4.4	6	1⅞	3/16	⅛	1/16	⅞	5¼	⅜	¼	2	⅛	3/16	—	—
M 5×18.9	5	5	7/16	5/16	3/16	2⅜	3¼	⅞	½	2½	¼	7/16	⅞	2¾
M 4×13.8	4	4	⅜	5/16	3/16	1⅞	2⅜	13/16	½	2	¼	⅜	¾	2¼
×13	4	4	⅜	¼	⅛	1⅞	2⅜	13/16	7/16	2	3/16	⅜	¾	2¼

Gage g permissible near beam ends; elsewhere Specification Sect. 1.16.5 may require reduction in fastener size.

Courtesy of American Institute of Steel Construction

Table 11-4. S Shapes

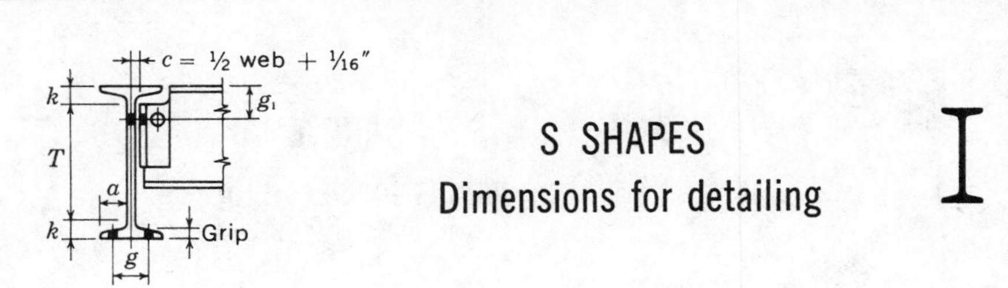

Designation	Depth d	Flange Width b_f	Flange Thickness t_f	Web Thickness t_w	$\frac{t_w}{2}$	Distance a	Distance T	Distance k	Distance g_1	Distance c	Grip	Max. Flange Fastener	Usual Flange Gage g
	In.	In.	In.	In.	In.	In.	In.	In.	In.	In.	In.	In.	In.
S 24×120	24	8	1⅛	13/16	⅜	3⅝	20	2	3¼	7/16	1⅛	1	4
×105.9	24	7⅞	1⅛	⅝	5/16	3⅝	20	2	3¼	⅜	1⅛	1	4
S 24×100	24	7¼	⅞	¾	⅜	3¼	20½	1¾	3	7/16	⅞	1	4
× 90	24	7⅛	⅞	⅝	5/16	3¼	20½	1¾	3	⅜	⅞	1	4
× 79.9	24	7	⅞	½	¼	3¼	20½	1¾	3	5/16	⅞	1	4
S 20× 95	20	7¼	15/16	13/16	⅜	3¼	16¼	1⅞	3	7/16	15/16	1	4
× 85	20	7	15/16	⅝	5/16	3¼	16¼	1⅞	3	⅜	⅞	1	4
S 20× 75	20	6⅜	13/16	⅝	5/16	2⅞	16¾	1⅝	3	⅜	13/16	⅞	3½
× 65.4	20	6¼	13/16	½	¼	2⅞	16¾	1⅝	3	5/16	¾	⅞	3½
S 18× 70	18	6¼	11/16	11/16	⅜	2¾	15	1½	2¾	7/16	11/16	⅞	3½
× 54.7	18	6	11/16	7/16	¼	2¾	15	1½	2¾	5/16	11/16	⅞	3½
S 15× 50	15	5⅝	⅝	9/16	¼	2½	12¼	1⅜	2¾	5/16	9/16	¾	3½
× 42.9	15	5½	⅝	7/16	3/16	2½	12¼	1⅜	2¾	¼	9/16	¾	3½
S 12× 50	12	5½	11/16	11/16	5/16	2⅜	9⅛	1 7/16	2¾	⅜	11/16	¾	3
× 40.8	12	5¼	11/16	7/16	¼	2⅜	9⅛	1 7/16	2¾	5/16	⅝	¾	3
S 12× 35	12	5⅛	9/16	7/16	3/16	2⅜	9⅝	1 3/16	2½	¼	½	¾	3
× 31.8	12	5	9/16	⅜	3/16	2⅜	9⅝	1 3/16	2½	¼	½	¾	3
S 10× 35	10	5	½	⅝	5/16	2⅛	7¾	1⅛	2½	⅜	½	¾	2¾
× 25.4	10	4⅝	½	5/16	⅛	2⅛	7¾	1⅛	2½	3/16	½	¾	2¾
S 8× 23	8	4⅛	7/16	7/16	¼	1⅞	6	1	2½	5/16	7/16	¾	2¼
× 18.4	8	4	7/16	¼	⅛	1⅞	6	1	2½	3/16	7/16	¾	2¼
S 7× 20	7	3⅞	⅜	7/16	¼	1¾	5¼	⅞	2½	5/16	⅜	⅝	2¼
× 15.3	7	3⅝	⅜	¼	⅛	1¾	5¼	⅞	2½	3/16	⅜	⅝	2¼
S 6× 17.25	6	3⅝	⅜	7/16	¼	1½	4⅜	13/16	2¼	5/16	⅜	⅝	2
× 12.5	6	3⅜	⅜	¼	⅛	1½	4⅜	13/16	2¼	3/16	5/16	—	—
S 5× 14.75	5	3¼	5/16	½	¼	1⅜	3½	¾	2¼	5/16	5/16	—	—
× 10	5	3	5/16	3/16	⅛	1⅜	3½	¾	2¼	3/16	5/16	—	—
S 4× 9.5	4	2¾	5/16	5/16	3/16	1¼	2⅝	11/16	2	¼	5/16	—	—
× 7.7	4	2⅝	5/16	3/16	⅛	1¼	2⅝	11/16	2	3/16	5/16	—	—
S 3× 7.5	3	2½	¼	⅜	3/16	1⅛	1¾	⅝	—	¼	¼	—	—
× 5.7	3	2⅜	¼	3/16	1/16	1⅛	1¾	⅝	—	⅛	¼	—	—

Gage g permissible near beam ends; elsewhere Specification Sect. 1.16.5 may require reduction in fastener size.

Courtesy of American Institute of Steel Construction

Table 11-5. American Standard Channels

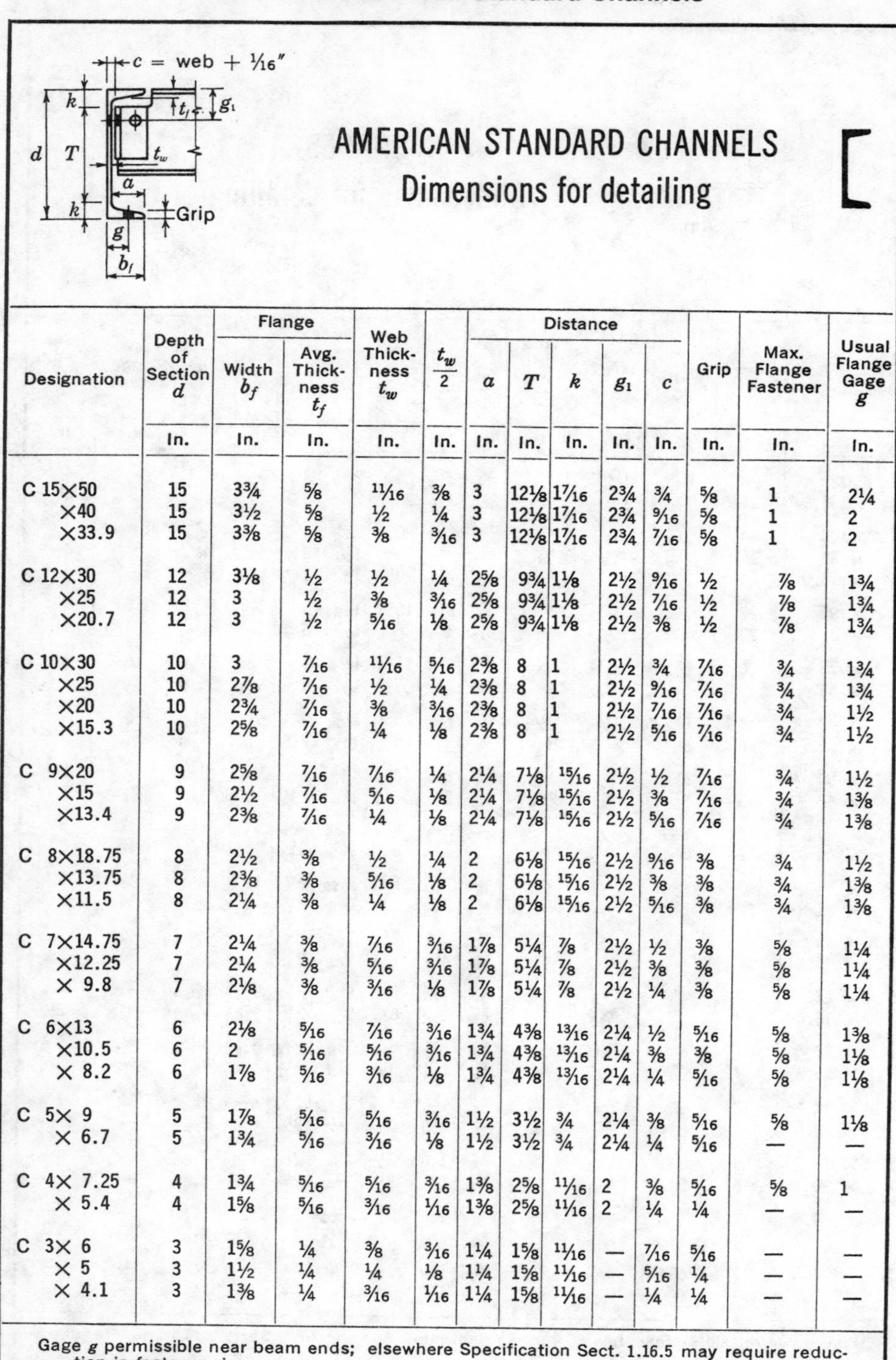

AMERICAN STANDARD CHANNELS
Dimensions for detailing

Designation	Depth of Section d	Flange: Width b_f	Flange: Avg. Thickness t_f	Web Thickness t_w	$\frac{t_w}{2}$	Distance: a	Distance: T	Distance: k	Distance: g_1	Distance: c	Grip	Max. Flange Fastener	Usual Flange Gage g
	In.	In.	In.	In.	In.	In.	In.	In.	In.	In.	In.	In.	In.
C 15×50	15	3¾	⅝	11/16	⅜	3	12⅛	1 7/16	2¾	¾	⅝	1	2¼
×40	15	3½	⅝	½	¼	3	12⅛	1 7/16	2¾	9/16	⅝	1	2
×33.9	15	3⅜	⅝	⅜	3/16	3	12⅛	1 7/16	2¾	7/16	⅝	1	2
C 12×30	12	3⅛	½	½	¼	2⅝	9¾	1⅛	2½	9/16	½	⅞	1¾
×25	12	3	½	⅜	3/16	2⅝	9¾	1⅛	2½	7/16	½	⅞	1¾
×20.7	12	3	½	5/16	⅛	2⅝	9¾	1⅛	2½	⅜	½	⅞	1¾
C 10×30	10	3	7/16	11/16	5/16	2⅜	8	1	2½	¾	7/16	¾	1¾
×25	10	2⅞	7/16	½	¼	2⅜	8	1	2½	9/16	7/16	¾	1¾
×20	10	2¾	7/16	⅜	3/16	2⅜	8	1	2½	7/16	7/16	¾	1½
×15.3	10	2⅝	7/16	¼	⅛	2⅜	8	1	2½	5/16	7/16	¾	1½
C 9×20	9	2⅝	7/16	7/16	¼	2¼	7⅛	15/16	2½	½	7/16	¾	1½
×15	9	2½	7/16	5/16	⅛	2¼	7⅛	15/16	2½	⅜	7/16	¾	1⅜
×13.4	9	2⅜	7/16	¼	⅛	2¼	7⅛	15/16	2½	5/16	7/16	¾	1⅜
C 8×18.75	8	2½	⅜	½	¼	2	6⅛	15/16	2½	9/16	⅜	¾	1½
×13.75	8	2⅜	⅜	5/16	⅛	2	6⅛	15/16	2½	⅜	⅜	¾	1⅜
×11.5	8	2¼	⅜	¼	⅛	2	6⅛	15/16	2½	5/16	⅜	¾	1⅜
C 7×14.75	7	2¼	⅜	7/16	3/16	1⅞	5¼	⅞	2½	½	⅜	⅝	1¼
×12.25	7	2¼	⅜	5/16	3/16	1⅞	5¼	⅞	2½	⅜	⅜	⅝	1¼
× 9.8	7	2⅛	⅜	3/16	⅛	1⅞	5¼	⅞	2½	¼	⅜	⅝	1¼
C 6×13	6	2⅛	5/16	7/16	3/16	1¾	4⅜	13/16	2¼	½	5/16	⅝	1⅜
×10.5	6	2	5/16	5/16	3/16	1¾	4⅜	13/16	2¼	⅜	⅜	⅝	1⅛
× 8.2	6	1⅞	5/16	3/16	⅛	1¾	4⅜	13/16	2¼	¼	5/16	⅝	1⅛
C 5× 9	5	1⅞	5/16	5/16	3/16	1½	3½	¾	2¼	⅜	5/16	⅝	1⅛
× 6.7	5	1¾	5/16	3/16	⅛	1½	3½	¾	2¼	¼	5/16	—	—
C 4× 7.25	4	1¾	5/16	5/16	3/16	1⅜	2⅝	11/16	2	⅜	5/16	⅝	1
× 5.4	4	1⅝	5/16	3/16	1/16	1⅜	2⅝	11/16	2	¼	¼	—	—
C 3× 6	3	1⅝	¼	⅜	3/16	1¼	1⅝	11/16	—	7/16	5/16	—	—
× 5	3	1½	¼	¼	⅛	1¼	1⅝	11/16	—	5/16	¼	—	—
× 4.1	3	1⅜	¼	3/16	1/16	1¼	1⅝	11/16	—	¼	¼	—	—

Gage g permissible near beam ends; elsewhere Specification Sect. 1.16.5 may require reduction in fastener size

Courtesy of American Institute of Steel Construction

Table 11-6. Miscellaneous Channels

MISCELLANEOUS CHANNELS
Dimensions for detailing

c = web + 1/16″

Designation	Depth of Section d	Flange: Width b_f	Flange: Avg. Thickness t_f	Web Thickness t_w	$\frac{t_w}{2}$	Distance: a	Distance: T	Distance: k	Distance: g_1	Distance: c	Grip	Max. Flange Fastener	Usual Flange Gage g
	In.	In.	In.	In.	In.	In.	In.	In.	In.	In.	In.	In.	In.
MC 18×58	18	4¼	⅝	11/16	⅜	3½	15¼	1⅜	2½	¾	⅝	1	2½
×51.9	18	4⅛	⅝	⅝	5/16	3½	15¼	1⅜	2½	11/16	⅝	1	2½
×45.8	18	4	⅝	½	¼	3½	15¼	1⅜	2½	9/16	⅝	1	2½
×42.7	18	4	⅝	7/16	¼	3½	15¼	1⅜	2½	½	⅝	1	2½
MC 13×50	13	4⅜	⅝	13/16	⅜	3⅝	10¼	1⅜	2¾	⅞	⅝	1	2½
×40	13	4⅛	⅝	9/16	¼	3⅝	10¼	1⅜	2¾	⅝	9/16	1	2½
×35	13	4⅛	⅝	7/16	¼	3⅝	10¼	1⅜	2¾	½	9/16	1	2½
×31.8	13	4	⅝	⅜	3/16	3⅝	10¼	1⅜	2¾	7/16	9/16	1	2½
MC 12×50	12	4⅛	11/16	13/16	7/16	3¼	9⅜	1 5/16	2½	⅞	11/16	1	2½
×45	12	4	11/16	11/16	⅜	3¼	9⅜	1 5/16	2½	¾	11/16	1	2½
×40	12	3⅞	11/16	9/16	5/16	3¼	9⅜	1 5/16	2½	⅝	11/16	1	2½
×35	12	3¾	11/16	7/16	¼	3¼	9⅜	1 5/16	2½	½	11/16	1	2½
MC 12×37	12	3⅝	⅝	⅝	5/16	3	9⅜	1 5/16	2½	11/16	⅝	⅞	2¼
×32.9	12	3½	⅝	½	¼	3	9⅜	1 5/16	2½	9/16	9/16	⅞	2¼
×30.9	12	3½	⅝	7/16	¼	3	9⅜	1 5/16	2½	½	9/16	⅞	2¼
MC 12×10.6	12	1½	5/16	3/16	⅛	1¼	10⅝	11/16	2¼	¼	¼	—	—
MC 10×41.1	10	4⅜	9/16	13/16	⅜	3½	7½	1¼	2½	⅞	9/16	⅞	2½
×33.6	10	4⅛	9/16	9/16	5/16	3½	7½	1¼	2½	⅝	9/16	⅞	2½
×28.5	10	4	9/16	7/16	3/16	3½	7½	1¼	2½	½	9/16	⅞	2½
MC 10×28.3	10	3½	9/16	½	¼	3	7½	1¼	2½	9/16	9/16	⅞	2
×25.3	10	3½	½	7/16	3/16	3⅛	7¾	1⅛	2½	½	½	⅞	2
×24.9	10	3⅜	9/16	⅜	3/16	3	7½	1¼	2½	7/16	9/16	⅞	2
×21.9	10	3½	½	5/16	3/16	3⅛	7¾	1⅛	2½	⅜	½	⅞	2
MC 10× 8.4	10	1½	¼	3/16	1/16	1⅜	8⅝	11/16	2¼	¼	¼	—	—
MC 10× 6.5	10	1⅛	3/16	⅛	1/16	1	9⅛	7/16	2¼	3/16	3/16	—	—

Gage g permissible near beam ends; elsewhere Specification Sect. 1.16.5 may require reduction in fastener size.

Courtesy of American Institute of Steel Construction

Table 11-6 continued

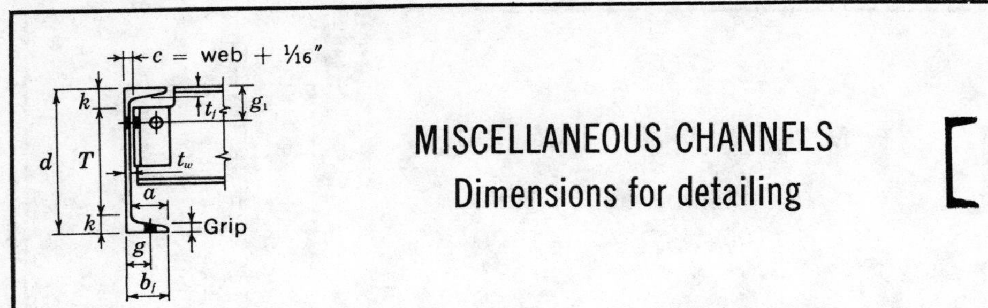

MISCELLANEOUS CHANNELS
Dimensions for detailing

Designation	Depth of Section d	Flange Width b_f	Flange Avg. Thickness t_f	Web Thickness t_w	$\frac{t_w}{2}$	Distance a	Distance T	Distance k	Distance g_1	Distance c	Grip	Max. Flange Fastener	Usual Flange Gage g
	In.	In.	In.	In.	In.	In.	In.	In.	In.	In.	In.	In.	In.
MC 9×25.4	9	3½	9/16	7/16	¼	3	6⅝	1 3/16	2½	½	9/16	⅞	2
9×23.9	9	3½	9/16	⅜	3/16	3	6⅝	1 3/16	2½	7/16	9/16	⅞	2
MC 8×22.8	8	3½	½	7/16	3/16	3⅛	5⅝	1 3/16	2½	½	½	⅞	2
×21.4	8	3½	½	⅜	3/16	3⅛	5⅝	1 3/16	2½	7/16	½	⅞	2
MC 8×20	8	3	½	⅜	3/16	2⅝	5¾	1⅛	2½	7/16	½	⅞	2
×18.7	8	3	½	⅜	3/16	2⅝	5¾	1⅛	2½	7/16	½	⅞	2
MC 8× 8.5	8	1⅞	5/16	3/16	1/16	1¾	6½	¾	2¼	¼	5/16	⅝	1⅛
MC 7×22.7	7	3⅝	½	½	¼	3⅛	4¾	1⅛	2½	9/16	½	⅞	2
×19.1	7	3½	½	⅜	3/16	3⅛	4¾	1⅛	2½	7/16	½	⅞	2
MC 7×17.6	7	3	½	⅜	3/16	2⅝	4⅞	1 1/16	2½	7/16	½	¾	1¾
MC 6×18	6	3½	½	⅜	3/16	3⅛	3⅞	1 1/16	2½	7/16	½	⅞	2
×15.3	6	3½	⅜	5/16	3/16	3⅛	4¼	⅞	2¼	⅜	⅜	⅞	2
MC 6×16.3	6	3	½	⅜	3/16	2⅝	3⅞	1 1/16	2½	7/16	½	¾	1¾
×15.1	6	3	½	5/16	3/16	2⅝	3⅞	1 1/16	2½	⅜	½	¾	1¾
MC 6×12	6	2½	⅜	5/16	⅛	2⅛	4⅜	13/16	2¼	⅜	⅜	⅝	1½
MC 3× 9	3	2⅛	⅜	½	¼	1⅝	1¾	⅝	—	9/16	—	—	—
× 7.1	3	2	⅜	5/16	⅛	1⅝	1¾	⅝	—	⅜	—	—	—

Gage g permissible near beam ends; elsewhere Specification Sect. 1.16.5 may require reduction in fastener size.

Courtesy of American Institute of Steel Construction

Table 11-7. Angles with Equal Legs

ANGLES
Equal legs
Properties for designing

Size and Thickness	k	Weight per Foot	Area	AXIS X-X AND AXIS Y-Y				AXIS Z-Z
				I	S	r	x or y	r
In.	In.	Lb.	In.^{2}	In.^{4}	In.^{3}	In.	In.	In.
L 8 × 8 × 1 1/8	1 3/4	56.9	16.7	98.0	17.5	2.42	2.41	1.56
1	1 5/8	51.0	15.0	89.0	15.8	2.44	2.37	1.56
7/8	1 1/2	45.0	13.2	79.6	14.0	2.45	2.32	1.57
3/4	1 3/8	38.9	11.4	69.7	12.2	2.47	2.28	1.58
5/8	1 1/4	32.7	9.61	59.4	10.3	2.49	2.23	1.58
9/16	1 3/16	29.6	8.68	54.1	9.34	2.50	2.21	1.59
1/2	1 1/8	26.4	7.75	48.6	8.36	2.50	2.19	1.59
L 6 × 6 × 1	1 1/2	37.4	11.0	35.5	8.57	1.80	1.86	1.17
7/8	1 3/8	33.1	9.73	31.9	7.63	1.81	1.82	1.17
3/4	1 1/4	28.7	8.44	28.2	6.66	1.83	1.78	1.17
5/8	1 1/8	24.2	7.11	24.2	5.66	1.84	1.73	1.18
9/16	1 1/16	21.9	6.43	22.1	5.14	1.85	1.71	1.18
1/2	1	19.6	5.75	19.9	4.61	1.86	1.68	1.18
7/16	15/16	17.2	5.06	17.7	4.08	1.87	1.66	1.19
3/8	7/8	14.9	4.36	15.4	3.53	1.88	1.64	1.19
5/16	13/16	12.4	3.65	13.0	2.97	1.89	1.62	1.20
L 5 × 5 × 7/8	1 3/8	27.2	7.98	17.8	5.17	1.49	1.57	.973
3/4	1 1/4	23.6	6.94	15.7	4.53	1.51	1.52	.975
5/8	1 1/8	20.0	5.86	13.6	3.86	1.52	1.48	.978
1/2	1	16.2	4.75	11.3	3.16	1.54	1.43	.983
7/16	15/16	14.3	4.18	10.0	2.79	1.55	1.41	.986
3/8	7/8	12.3	3.61	8.74	2.42	1.56	1.39	.990
5/16	13/16	10.3	3.03	7.42	2.04	1.57	1.37	.994
L 4 × 4 × 3/4	1 1/8	18.5	5.44	7.67	2.81	1.19	1.27	.778
5/8	1	15.7	4.61	6.66	2.40	1.20	1.23	.779
1/2	7/8	12.8	3.75	5.56	1.97	1.22	1.18	.782
7/16	13/16	11.3	3.31	4.97	1.75	1.23	1.16	.785
3/8	3/4	9.8	2.86	4.36	1.52	1.23	1.14	.788
5/16	11/16	8.2	2.40	3.71	1.29	1.24	1.12	.791
1/4	5/8	6.6	1.94	3.04	1.05	1.25	1.09	.795

Courtesy of American Institute of Steel Construction

Table 11-7 continued

ANGLES
Equal legs
Properties for designing

Size and Thickness	k	Weight per Foot	Area	AXIS X-X AND AXIS Y-Y				AXIS Z-Z
				I	S	r	x or y	r
In.	In.	Lb.	In.^{2}	In.^{4}	In.^{3}	In.	In.	In.
L 3½ × 3½ × ½	⅞	11.1	3.25	3.64	1.49	1.06	1.06	.683
7/16	13/16	9.8	2.87	3.26	1.32	1.07	1.04	.684
⅜	¾	8.5	2.48	2.87	1.15	1.07	1.01	.687
5/16	11/16	7.2	2.09	2.45	.976	1.08	.990	.690
¼	⅝	5.8	1.69	2.01	.794	1.09	.968	.694
L 3 × 3 × ½	13/16	9.4	2.75	2.22	1.07	.898	.932	.584
7/16	¾	8.3	2.43	1.99	.954	.905	.910	.585
⅜	11/16	7.2	2.11	1.76	.833	.913	.888	.587
5/16	⅝	6.1	1.78	1.51	.707	.922	.865	.589
¼	9/16	4.9	1.44	1.24	.577	.930	.842	.592
3/16	½	3.71	1.09	.962	.441	.939	.820	.596
L 2½ × 2½ × ½	13/16	7.7	2.25	1.23	.724	.739	.806	.487
⅜	11/16	5.9	1.73	.984	.566	.753	.762	.487
5/16	⅝	5.0	1.46	.849	.482	.761	.740	.489
¼	9/16	4.1	1.19	.703	.394	.769	.717	.491
3/16	½	3.07	.902	.547	.303	.778	.694	.495
L 2 × 2 × ⅜	11/16	4.7	1.36	.479	.351	.594	.636	.389
5/16	⅝	3.92	1.15	.416	.300	.601	.614	.390
¼	9/16	3.19	.938	.348	.247	.609	.592	.391
3/16	½	2.44	.715	.272	.190	.617	.569	.394
⅛	7/16	1.65	.484	.190	.131	.626	.546	.398
L 1¾ × 1¾ × ¼	½	2.77	.813	.227	.186	.529	.529	.341
3/16	7/16	2.12	.621	.179	.144	.537	.506	.343
⅛	⅜	1.44	.422	.126	.099	.546	.484	.347
L 1½ × 1½ × ¼	7/16	2.34	.688	.139	.134	.449	.466	.292
3/16	⅜	1.80	.527	.110	.104	.457	.444	.293
5/32	⅜	1.52	.444	.094	.088	.461	.433	.295
⅛	5/16	1.23	.359	.078	.072	.465	.421	.296
L 1¼ × 1¼ × ¼	7/16	1.92	.563	.077	.091	.369	.403	.243
3/16	⅜	1.48	.434	.061	.071	.377	.381	.244
⅛	5/16	1.01	.297	.044	.049	.385	.359	.246
L 1 × 1 × ¼	⅜	1.49	.438	.037	.056	.290	.339	.196
3/16	5/16	1.16	.340	.030	.044	.297	.318	.195
⅛	¼	.80	.234	.022	.031	.304	.296	.196

Courtesy of American Institute of Steel Construction

Table 11-8. Angles with Unequal Legs

ANGLES
Unequal legs
Properties for designing

Size and Thickness	k	Weight per Foot	Area	AXIS X-X				AXIS Y-Y				AXIS Z-Z	
				I	S	r	y	I	S	r	x	r	Tan α
In.	In.	Lb.	In.²	In.⁴	In.³	In.	In.	In.⁴	In.³	In.	In.	In.	
L 9 × 4 × 1	1½	40.8	12.0	97.0	17.6	2.84	3.50	12.0	4.00	1.00	1.00	.834	.203
⅞	1⅜	36.1	10.6	86.8	15.7	2.86	3.45	10.8	3.56	1.01	.953	.836	.208
¾	1¼	31.3	9.19	76.1	13.6	2.88	3.41	9.63	3.11	1.02	.906	.841	.212
⅝	1⅛	26.3	7.73	64.9	11.5	2.90	3.36	8.32	2.65	1.04	.858	.847	.216
9/16	1 1/16	23.8	7.00	59.1	10.4	2.91	3.33	7.63	2.41	1.04	.834	.850	.218
½	1	21.3	6.25	53.2	9.34	2.92	3.31	6.92	2.17	1.05	.810	.854	.220
L 8 × 6 × 1	1½	44.2	13.0	80.8	15.1	2.49	2.65	38.8	8.92	1.73	1.65	1.28	.543
⅞	1⅜	39.1	11.5	72.3	13.4	2.51	2.61	34.9	7.94	1.74	1.61	1.28	.547
¾	1¼	33.8	9.94	63.4	11.7	2.53	2.56	30.7	6.92	1.76	1.56	1.29	.551
⅝	1⅛	28.5	8.36	54.1	9.87	2.54	2.52	26.3	5.88	1.77	1.52	1.29	.554
9/16	1 1/16	25.7	7.56	49.3	8.95	2.55	2.50	24.0	5.34	1.78	1.50	1.30	.556
½	1	23.0	6.75	44.3	8.02	2.56	2.47	21.7	4.79	1.79	1.47	1.30	.558
7/16	15/16	20.2	5.93	39.2	7.07	2.57	2.45	19.3	4.23	1.80	1.45	1.31	.560
L 8 × 4 × 1	1½	37.4	11.0	69.6	14.1	2.52	3.05	11.6	3.94	1.03	1.05	.846	.247
⅞	1⅜	33.1	9.73	62.5	12.5	2.53	3.00	10.5	3.51	1.04	.999	.848	.253
¾	1¼	28.7	8.44	54.9	10.9	2.55	2.95	9.36	3.07	1.05	.953	.852	.258
⅝	1⅛	24.2	7.11	46.9	9.21	2.57	2.91	8.10	2.62	1.07	.906	.857	.262
9/16	1 1/16	21.9	6.43	42.8	8.35	2.58	2.88	7.43	2.38	1.07	.882	.861	.265
½	1	19.6	5.75	38.5	7.49	2.59	2.86	6.74	2.15	1.08	.859	.865	.267
7/16	15/16	17.2	5.06	34.1	6.60	2.60	2.83	6.02	1.90	1.09	.835	.869	.269
L 7 × 4 × ⅞	1⅜	30.2	8.86	42.9	9.65	2.20	2.55	10.2	3.46	1.07	1.05	.856	.318
¾	1¼	26.2	7.69	37.8	8.42	2.22	2.51	9.05	3.03	1.09	1.01	.860	.324
⅝	1⅛	22.1	6.48	32.4	7.14	2.24	2.46	7.84	2.58	1.10	.963	.865	.329
9/16	1 1/16	20.0	5.87	29.6	6.48	2.24	2.44	7.19	2.35	1.11	.940	.868	.332
½	1	17.9	5.25	26.7	5.81	2.25	2.42	6.53	2.12	1.11	.917	.872	.335
7/16	15/16	15.8	4.62	23.7	5.13	2.26	2.39	5.83	1.88	1.12	.893	.876	.337
⅜	⅞	13.6	3.98	20.6	4.44	2.27	2.37	5.10	1.63	1.13	.870	.880	.340

Courtesy of American Institute of Steel Construction

Table 11-8 continued

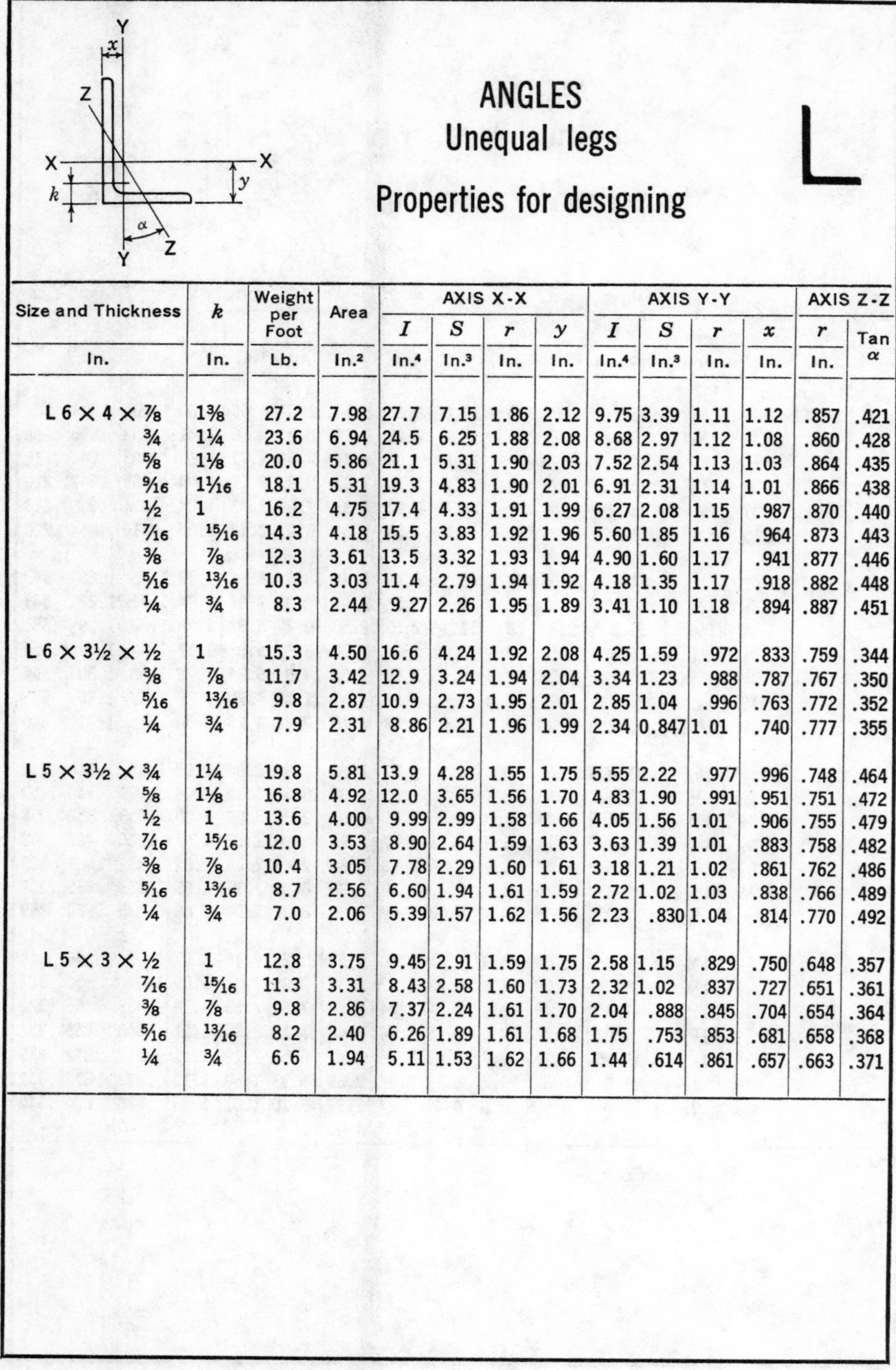

ANGLES
Unequal legs
Properties for designing

Size and Thickness	k	Weight per Foot	Area	AXIS X-X				AXIS Y-Y				AXIS Z-Z	
				I	*S*	*r*	*y*	*I*	*S*	*r*	*x*	*r*	Tan α
In.	In.	Lb.	In.²	In.⁴	In.³	In.	In.	In.⁴	In.³	In.	In.	In.	
L 6 × 4 × ⅞	1⅜	27.2	7.98	27.7	7.15	1.86	2.12	9.75	3.39	1.11	1.12	.857	.421
¾	1¼	23.6	6.94	24.5	6.25	1.88	2.08	8.68	2.97	1.12	1.08	.860	.428
⅝	1⅛	20.0	5.86	21.1	5.31	1.90	2.03	7.52	2.54	1.13	1.03	.864	.435
9/16	1 1/16	18.1	5.31	19.3	4.83	1.90	2.01	6.91	2.31	1.14	1.01	.866	.438
½	1	16.2	4.75	17.4	4.33	1.91	1.99	6.27	2.08	1.15	.987	.870	.440
7/16	15/16	14.3	4.18	15.5	3.83	1.92	1.96	5.60	1.85	1.16	.964	.873	.443
⅜	⅞	12.3	3.61	13.5	3.32	1.93	1.94	4.90	1.60	1.17	.941	.877	.446
5/16	13/16	10.3	3.03	11.4	2.79	1.94	1.92	4.18	1.35	1.17	.918	.882	.448
¼	¾	8.3	2.44	9.27	2.26	1.95	1.89	3.41	1.10	1.18	.894	.887	.451
L 6 × 3½ × ½	1	15.3	4.50	16.6	4.24	1.92	2.08	4.25	1.59	.972	.833	.759	.344
⅜	⅞	11.7	3.42	12.9	3.24	1.94	2.04	3.34	1.23	.988	.787	.767	.350
5/16	13/16	9.8	2.87	10.9	2.73	1.95	2.01	2.85	1.04	.996	.763	.772	.352
¼	¾	7.9	2.31	8.86	2.21	1.96	1.99	2.34	0.847	1.01	.740	.777	.355
L 5 × 3½ × ¾	1¼	19.8	5.81	13.9	4.28	1.55	1.75	5.55	2.22	.977	.996	.748	.464
⅝	1⅛	16.8	4.92	12.0	3.65	1.56	1.70	4.83	1.90	.991	.951	.751	.472
½	1	13.6	4.00	9.99	2.99	1.58	1.66	4.05	1.56	1.01	.906	.755	.479
7/16	15/16	12.0	3.53	8.90	2.64	1.59	1.63	3.63	1.39	1.01	.883	.758	.482
⅜	⅞	10.4	3.05	7.78	2.29	1.60	1.61	3.18	1.21	1.02	.861	.762	.486
5/16	13/16	8.7	2.56	6.60	1.94	1.61	1.59	2.72	1.02	1.03	.838	.766	.489
¼	¾	7.0	2.06	5.39	1.57	1.62	1.56	2.23	.830	1.04	.814	.770	.492
L 5 × 3 × ½	1	12.8	3.75	9.45	2.91	1.59	1.75	2.58	1.15	.829	.750	.648	.357
7/16	15/16	11.3	3.31	8.43	2.58	1.60	1.73	2.32	1.02	.837	.727	.651	.361
⅜	⅞	9.8	2.86	7.37	2.24	1.61	1.70	2.04	.888	.845	.704	.654	.364
5/16	13/16	8.2	2.40	6.26	1.89	1.61	1.68	1.75	.753	.853	.681	.658	.368
¼	¾	6.6	1.94	5.11	1.53	1.62	1.66	1.44	.614	.861	.657	.663	.371

Courtesy of American Institute of Steel Construction

Table 11-8 continued

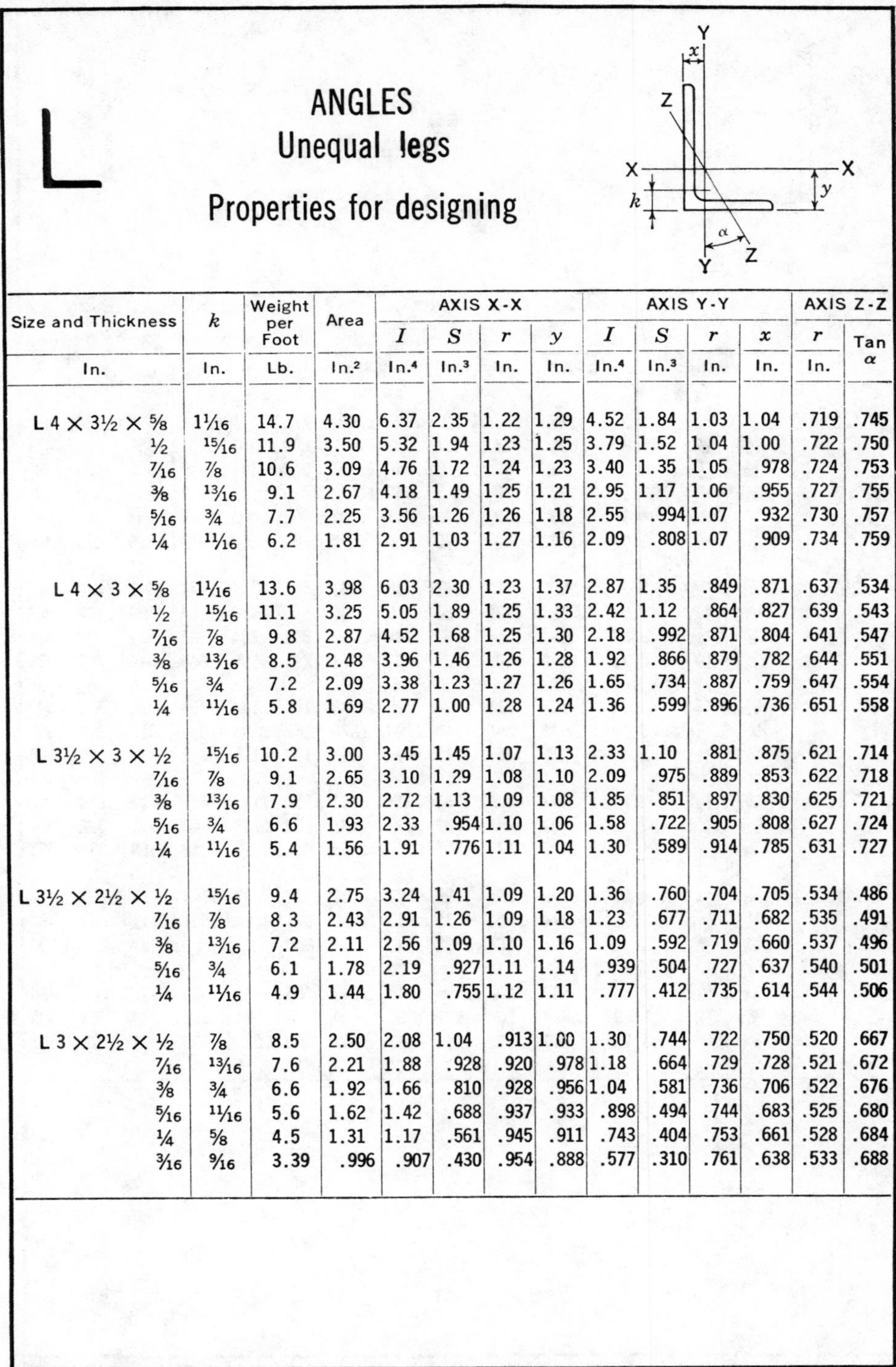

L

ANGLES
Unequal legs
Properties for designing

Size and Thickness	k	Weight per Foot	Area	AXIS X-X				AXIS Y-Y				AXIS Z-Z	
				I	S	r	y	I	S	r	x	r	Tan α
In.	In.	Lb.	In.²	In.⁴	In.³	In.	In.	In.⁴	In.³	In.	In.	In.	
L 4 × 3½ × ⅝	1¹⁄₁₆	14.7	4.30	6.37	2.35	1.22	1.29	4.52	1.84	1.03	1.04	.719	.745
½	¹⁵⁄₁₆	11.9	3.50	5.32	1.94	1.23	1.25	3.79	1.52	1.04	1.00	.722	.750
⁷⁄₁₆	⅞	10.6	3.09	4.76	1.72	1.24	1.23	3.40	1.35	1.05	.978	.724	.753
⅜	¹³⁄₁₆	9.1	2.67	4.18	1.49	1.25	1.21	2.95	1.17	1.06	.955	.727	.755
⁵⁄₁₆	¾	7.7	2.25	3.56	1.26	1.26	1.18	2.55	.994	1.07	.932	.730	.757
¼	¹¹⁄₁₆	6.2	1.81	2.91	1.03	1.27	1.16	2.09	.808	1.07	.909	.734	.759
L 4 × 3 × ⅝	1¹⁄₁₆	13.6	3.98	6.03	2.30	1.23	1.37	2.87	1.35	.849	.871	.637	.534
½	¹⁵⁄₁₆	11.1	3.25	5.05	1.89	1.25	1.33	2.42	1.12	.864	.827	.639	.543
⁷⁄₁₆	⅞	9.8	2.87	4.52	1.68	1.25	1.30	2.18	.992	.871	.804	.641	.547
⅜	¹³⁄₁₆	8.5	2.48	3.96	1.46	1.26	1.28	1.92	.866	.879	.782	.644	.551
⁵⁄₁₆	¾	7.2	2.09	3.38	1.23	1.27	1.26	1.65	.734	.887	.759	.647	.554
¼	¹¹⁄₁₆	5.8	1.69	2.77	1.00	1.28	1.24	1.36	.599	.896	.736	.651	.558
L 3½ × 3 × ½	¹⁵⁄₁₆	10.2	3.00	3.45	1.45	1.07	1.13	2.33	1.10	.881	.875	.621	.714
⁷⁄₁₆	⅞	9.1	2.65	3.10	1.29	1.08	1.10	2.09	.975	.889	.853	.622	.718
⅜	¹³⁄₁₆	7.9	2.30	2.72	1.13	1.09	1.08	1.85	.851	.897	.830	.625	.721
⁵⁄₁₆	¾	6.6	1.93	2.33	.954	1.10	1.06	1.58	.722	.905	.808	.627	.724
¼	¹¹⁄₁₆	5.4	1.56	1.91	.776	1.11	1.04	1.30	.589	.914	.785	.631	.727
L 3½ × 2½ × ½	¹⁵⁄₁₆	9.4	2.75	3.24	1.41	1.09	1.20	1.36	.760	.704	.705	.534	.486
⁷⁄₁₆	⅞	8.3	2.43	2.91	1.26	1.09	1.18	1.23	.677	.711	.682	.535	.491
⅜	¹³⁄₁₆	7.2	2.11	2.56	1.09	1.10	1.16	1.09	.592	.719	.660	.537	.496
⁵⁄₁₆	¾	6.1	1.78	2.19	.927	1.11	1.14	.939	.504	.727	.637	.540	.501
¼	¹¹⁄₁₆	4.9	1.44	1.80	.755	1.12	1.11	.777	.412	.735	.614	.544	.506
L 3 × 2½ × ½	⅞	8.5	2.50	2.08	1.04	.913	1.00	1.30	.744	.722	.750	.520	.667
⁷⁄₁₆	¹³⁄₁₆	7.6	2.21	1.88	.928	.920	.978	1.18	.664	.729	.728	.521	.672
⅜	¾	6.6	1.92	1.66	.810	.928	.956	1.04	.581	.736	.706	.522	.676
⁵⁄₁₆	¹¹⁄₁₆	5.6	1.62	1.42	.688	.937	.933	.898	.494	.744	.683	.525	.680
¼	⅝	4.5	1.31	1.17	.561	.945	.911	.743	.404	.753	.661	.528	.684
³⁄₁₆	⁹⁄₁₆	3.39	.996	.907	.430	.954	.888	.577	.310	.761	.638	.533	.688

Courtesy of American Institute of Steel Construction

Table 11-8 continued

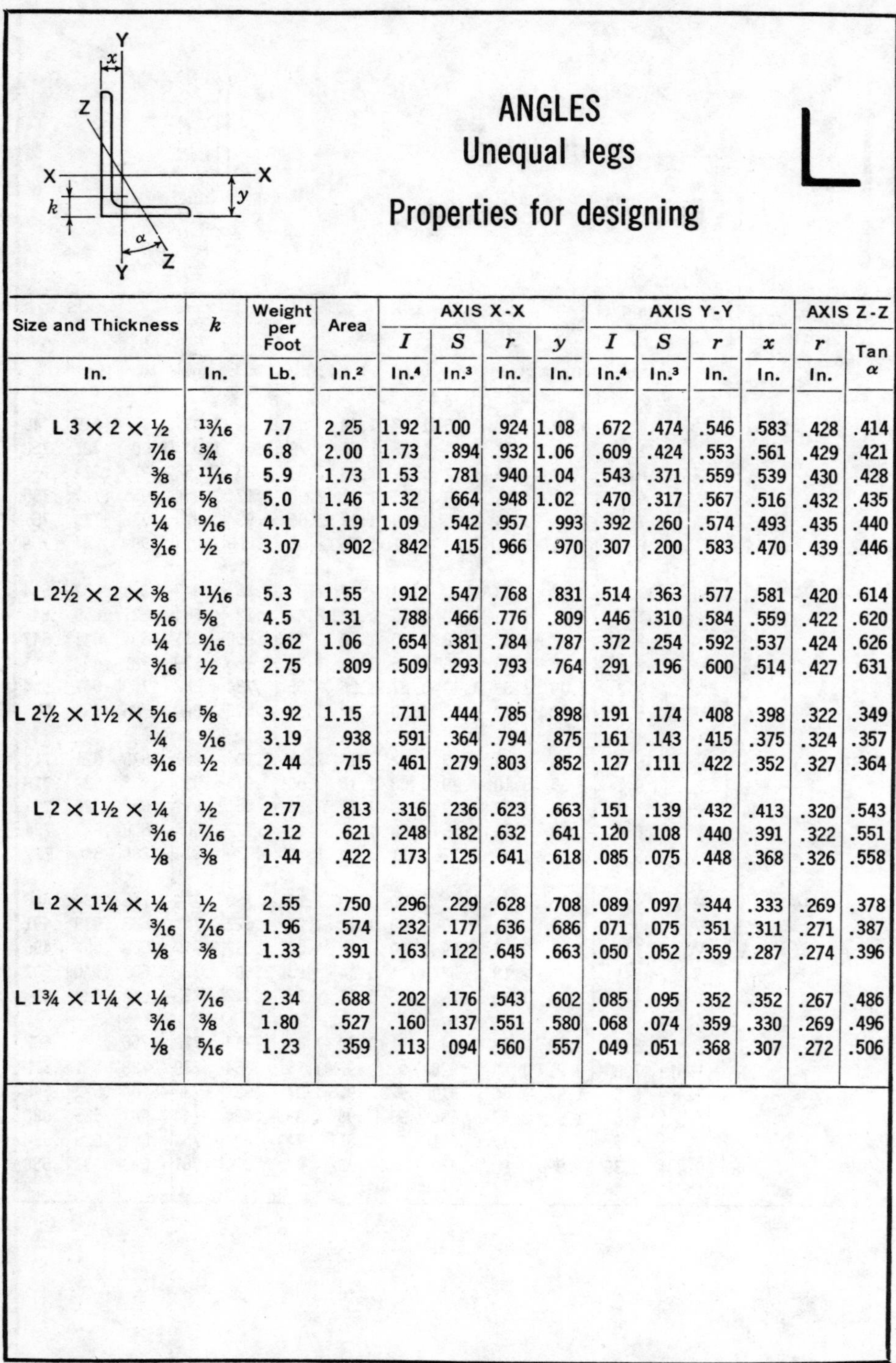

ANGLES
Unequal legs
Properties for designing

Size and Thickness	k	Weight per Foot	Area	AXIS X-X				AXIS Y-Y				AXIS Z-Z	
				I	S	r	y	I	S	r	x	r	Tan α
In.	In.	Lb.	In.²	In.⁴	In.³	In.	In.	In.⁴	In.³	In.	In.	In.	
L 3 × 2 × ½	13/16	7.7	2.25	1.92	1.00	.924	1.08	.672	.474	.546	.583	.428	.414
7/16	¾	6.8	2.00	1.73	.894	.932	1.06	.609	.424	.553	.561	.429	.421
⅜	11/16	5.9	1.73	1.53	.781	.940	1.04	.543	.371	.559	.539	.430	.428
5/16	⅝	5.0	1.46	1.32	.664	.948	1.02	.470	.317	.567	.516	.432	.435
¼	9/16	4.1	1.19	1.09	.542	.957	.993	.392	.260	.574	.493	.435	.440
3/16	½	3.07	.902	.842	.415	.966	.970	.307	.200	.583	.470	.439	.446
L 2½ × 2 × ⅜	11/16	5.3	1.55	.912	.547	.768	.831	.514	.363	.577	.581	.420	.614
5/16	⅝	4.5	1.31	.788	.466	.776	.809	.446	.310	.584	.559	.422	.620
¼	9/16	3.62	1.06	.654	.381	.784	.787	.372	.254	.592	.537	.424	.626
3/16	½	2.75	.809	.509	.293	.793	.764	.291	.196	.600	.514	.427	.631
L 2½ × 1½ × 5/16	⅝	3.92	1.15	.711	.444	.785	.898	.191	.174	.408	.398	.322	.349
¼	9/16	3.19	.938	.591	.364	.794	.875	.161	.143	.415	.375	.324	.357
3/16	½	2.44	.715	.461	.279	.803	.852	.127	.111	.422	.352	.327	.364
L 2 × 1½ × ¼	½	2.77	.813	.316	.236	.623	.663	.151	.139	.432	.413	.320	.543
3/16	7/16	2.12	.621	.248	.182	.632	.641	.120	.108	.440	.391	.322	.551
⅛	⅜	1.44	.422	.173	.125	.641	.618	.085	.075	.448	.368	.326	.558
L 2 × 1¼ × ¼	½	2.55	.750	.296	.229	.628	.708	.089	.097	.344	.333	.269	.378
3/16	7/16	1.96	.574	.232	.177	.636	.686	.071	.075	.351	.311	.271	.387
⅛	⅜	1.33	.391	.163	.122	.645	.663	.050	.052	.359	.287	.274	.396
L 1¾ × 1¼ × ¼	7/16	2.34	.688	.202	.176	.543	.602	.085	.095	.352	.352	.267	.486
3/16	⅜	1.80	.527	.160	.137	.551	.580	.068	.074	.359	.330	.269	.496
⅛	5/16	1.23	.359	.113	.094	.560	.557	.049	.051	.368	.307	.272	.506

Courtesy of American Institute of Steel Construction

Table 11-9. Square and Round Bars

SQUARE AND ROUND BARS
Weight and area

Size Inches	Weight Lb. per Foot ■	Weight Lb. per Foot ●	Area Square Inches ▨	Area Square Inches ◍	Size Inches	Weight Lb. per Foot ■	Weight Lb. per Foot ●	Area Square Inches ▨	Area Square Inches ◍
0					3	30.60	24.03	9.000	7.069
1/16	.013	.010	.0039	.0031	1/16	31.89	25.05	9.379	7.366
1/8	.053	.042	.0156	.0123	1/8	33.20	26.08	9.766	7.670
3/16	.120	.094	.0352	.0276	3/16	34.54	27.13	10.160	7.980
1/4	.213	.167	.0625	.0491	1/4	35.91	28.21	10.563	8.296
5/16	.332	.261	.0977	.0767	5/16	37.31	29.30	10.973	8.618
3/8	.478	.376	.1406	.1105	3/8	38.73	30.42	11.391	8.946
7/16	.651	.511	.1914	.1503	7/16	40.18	31.55	11.816	9.281
1/2	.850	.668	.2500	.1963	1/2	41.65	32.71	12.250	9.621
9/16	1.076	.845	.3164	.2485	9/16	43.15	33.89	12.691	9.968
5/8	1.328	1.043	.3906	.3068	5/8	44.68	35.09	13.141	10.321
11/16	1.607	1.262	.4727	.3712	11/16	46.23	36.31	13.598	10.680
3/4	1.913	1.502	.5625	.4418	3/4	47.81	37.55	14.063	11.045
13/16	2.245	1.763	.6602	.5185	13/16	49.42	38.81	14.535	11.416
7/8	2.603	2.044	.7656	.6013	7/8	51.05	40.10	15.016	11.793
15/16	2.988	2.347	.8789	.6903	15/16	52.71	41.40	15.504	12.177
1	3.400	2.670	1.0000	.7854	4	54.40	42.73	16.000	12.566
1/16	3.838	3.015	1.1289	.8866	1/16	56.11	44.07	16.504	12.962
1/8	4.303	3.380	1.2656	.9940	1/8	57.85	45.44	17.016	13.364
3/16	4.795	3.766	1.4102	1.1075	3/16	59.62	46.83	17.535	13.772
1/4	5.313	4.172	1.5625	1.2272	1/4	61.41	48.23	18.063	14.186
5/16	5.857	4.600	1.7227	1.3530	5/16	63.23	49.66	18.598	14.607
3/8	6.428	5.049	1.8906	1.4849	3/8	65.08	51.11	19.141	15.033
7/16	7.026	5.518	2.0664	1.6230	7/16	66.95	52.58	19.691	15.466
1/2	7.650	6.008	2.2500	1.7671	1/2	68.85	54.07	20.250	15.904
9/16	8.301	6.519	2.4414	1.9175	9/16	70.78	55.59	20.816	16.349
5/8	8.978	7.051	2.6406	2.0739	5/8	72.73	57.12	21.391	16.800
11/16	9.682	7.604	2.8477	2.2365	11/16	74.71	58.67	21.973	17.257
3/4	10.413	8.178	3.0625	2.4053	3/4	76.71	60.25	22.563	17.721
13/16	11.170	8.773	3.2852	2.5802	13/16	78.74	61.85	23.160	18.190
7/8	11.953	9.388	3.5156	2.7612	7/8	80.80	63.46	23.766	18.665
15/16	12.763	10.024	3.7539	2.9483	15/16	82.89	65.10	24.379	19.147
2	13.600	10.681	4.0000	3.1416	5	85.00	66.76	25.000	19.635
1/16	14.463	11.359	4.2539	3.3410	1/16	87.14	68.44	25.629	20.129
1/8	15.353	12.058	4.5156	3.5466	1/8	89.30	70.14	26.266	20.629
3/16	16.270	12.778	4.7852	3.7583	3/16	91.49	71.86	26.910	21.135
1/4	17.213	13.519	5.0625	3.9761	1/4	93.71	73.60	27.563	21.648
5/16	18.182	14.280	5.3477	4.2000	5/16	95.96	75.36	28.223	22.166
3/8	19.178	15.062	5.6406	4.4301	3/8	98.23	77.15	28.891	22.691
7/16	20.201	15.866	5.9414	4.6664	7/16	100.53	78.95	29.566	23.221
1/2	21.250	16.690	6.2500	4.9087	1/2	102.85	80.78	30.250	23.758
9/16	22.326	17.534	6.5664	5.1572	9/16	105.20	82.62	30.941	24.301
5/8	23.428	18.400	6.8906	5.4119	5/8	107.58	84.49	31.641	24.850
11/16	24.557	19.287	7.2227	5.6727	11/16	109.98	86.38	32.348	25.406
3/4	25.713	20.195	7.5625	5.9396	3/4	112.41	88.29	33.063	25.967
13/16	26.895	21.123	7.9102	6.2126	13/16	114.87	90.22	33.785	26.535
7/8	28.103	22.072	8.2656	6.4918	7/8	117.35	92.17	34.516	27.109
15/16	29.338	23.042	8.6289	6.7771	15/16	119.86	94.14	35.254	27.688
3	30.600	24.033	9.0000	7.0686	6	122.40	96.13	36.000	28.274

Courtesy of American Institute of Steel Construction

Table 11-9 continued

SQUARE AND ROUND BARS
Weight and area

Size Inches	Weight Lb. per Foot ■	Weight Lb. per Foot ●	Area Square Inches ▨	Area Square Inches ◍	Size Inches	Weight Lb. per Foot ■	Weight Lb. per Foot ●	Area Square Inches ▨	Area Square Inches ◍
6	122.40	96.13	36.000	28.274	9	275.40	216.30	81.000	63.617
1/16	124.96	98.15	36.754	28.866	1/16	279.24	219.31	82.129	64.504
1/8	127.55	100.18	37.516	29.465	1/8	283.10	222.35	83.266	65.397
3/16	130.17	102.23	38.285	30.069	3/16	286.99	225.41	84.410	66.296
1/4	132.81	104.31	39.063	30.680	1/4	290.91	228.48	85.563	67.201
5/16	135.48	106.41	39.848	31.296	5/16	294.86	231.58	86.723	68.112
3/8	138.18	108.53	40.641	31.919	3/8	298.83	234.70	87.891	69.029
7/16	140.90	110.66	41.441	32.548	7/16	302.83	237.84	89.066	69.953
1/2	143.65	112.82	42.250	33.183	1/2	306.85	241.00	90.250	70.882
9/16	146.43	115.00	43.066	33.824	9/16	310.90	244.18	91.441	71.818
5/8	149.23	117.20	43.891	34.472	5/8	314.98	247.38	92.641	72.760
11/16	152.06	119.43	44.723	35.125	11/16	319.08	250.61	93.848	73.708
3/4	154.91	121.67	45.563	35.785	3/4	323.21	253.85	95.063	74.662
13/16	157.79	123.93	46.410	36.450	13/16	327.37	257.12	96.285	75.622
7/8	160.70	126.22	47.266	37.122	7/8	331.55	260.40	97.516	76.589
15/16	163.64	128.52	48.129	37.800	15/16	335.76	263.71	98.754	77.561
7	166.60	130.85	49.000	38.485	10	340.00	267.04	100.000	78.540
1/16	169.59	133.19	49.879	39.175	1/16	344.26	270.38	101.254	79.525
1/8	172.60	135.56	50.766	39.871	1/8	348.55	273.75	102.516	80.516
3/16	175.64	137.95	51.660	40.574	3/16	352.87	277.14	103.785	81.513
1/4	178.71	140.36	52.563	41.282	1/4	357.21	280.55	105.063	82.516
5/16	181.81	142.79	53.473	41.997	5/16	361.58	283.99	106.348	83.525
3/8	184.93	145.24	54.391	42.718	3/8	365.98	287.44	107.641	84.541
7/16	188.07	147.71	55.316	43.445	7/16	370.40	290.91	108.941	85.563
1/2	191.25	150.21	56.250	44.179	1/2	374.85	294.41	110.250	86.590
9/16	194.45	152.72	57.191	44.918	9/16	379.33	297.92	111.566	87.624
5/8	197.68	155.26	58.141	45.664	5/8	383.83	301.46	112.891	88.664
11/16	200.93	157.81	59.098	46.415	11/16	388.36	305.02	114.223	89.710
3/4	204.21	160.39	60.063	47.173	3/4	392.91	308.59	115.563	90.763
13/16	207.52	162.99	61.035	47.937	13/16	397.49	312.19	116.910	91.821
7/8	210.85	165.60	62.016	48.707	7/8	402.10	315.81	118.266	92.886
15/16	214.21	168.24	63.004	49.483	15/16	406.74	319.45	119.629	93.957
8	217.60	170.90	64.000	50.265	11	411.40	323.11	121.000	95.033
1/16	221.01	173.58	65.004	51.054	1/16	416.09	326.80	122.379	96.116
1/8	224.45	176.29	66.016	51.849	1/8	420.80	330.50	123.766	97.205
3/16	227.92	179.01	67.035	52.649	3/16	425.54	334.22	125.160	98.301
1/4	231.41	181.75	68.063	53.456	1/4	430.31	337.97	126.563	99.402
5/16	234.93	184.52	69.098	54.269	5/16	435.11	341.73	127.973	100.510
3/8	238.48	187.30	70.141	55.088	3/8	439.93	345.52	129.391	101.623
7/16	242.05	190.11	71.191	55.914	7/16	444.78	349.33	130.816	102.743
1/2	245.65	192.93	72.250	56.745	1/2	449.65	353.16	132.250	103.869
9/16	249.28	195.78	73.316	57.583	9/16	454.55	357.00	133.691	105.001
5/8	252.93	198.65	74.391	58.426	5/8	459.48	360.87	135.141	106.139
11/16	256.61	201.54	75.473	59.276	11/16	464.43	364.76	136.598	107.284
3/4	260.31	204.45	76.563	60.132	3/4	469.41	368.68	138.063	108.434
13/16	264.04	207.38	77.660	60.994	13/16	474.42	372.61	139.535	109.591
7/8	267.80	210.33	78.766	61.863	7/8	479.45	376.56	141.016	110.754
15/16	271.59	213.31	79.879	62.737	15/16	484.51	380.54	142.504	111.923
9	275.40	216.30	81.000	63.617	12	489.60	384.53	144.000	113.098

Courtesy of American Institute of Steel Construction

Table 11-10. Weight of Rectangular Sections

WEIGHT OF RECTANGULAR SECTIONS
Pounds per linear foot

Width In.	Thickness, Inches													
	3/16	1/4	5/16	3/8	7/16	1/2	9/16	5/8	11/16	3/4	13/16	7/8	15/16	1
¼	.16	.21	.27	.32	.37	.43	.48	.53	.58	.64	.69	.74	.80	.85
½	.32	.43	.53	.64	.74	.85	.96	1.06	1.17	1.28	1.38	1.49	1.59	1.70
¾	.48	.64	.80	.96	1.12	1.28	1.43	1.59	1.75	1.91	2.07	2.23	2.39	2.55
1	.64	.85	1.06	1.28	1.49	1.70	1.91	2.13	2.34	2.55	2.76	2.98	3.19	3.40
1¼	.80	1.06	1.33	1.59	1.86	2.13	2.39	2.66	2.92	3.19	3.45	3.72	3.98	4.25
1½	.96	1.28	1.59	1.91	2.23	2.55	2.87	3.19	3.51	3.83	4.14	4.46	4.78	5.10
1¾	1.12	1.49	1.86	2.23	2.60	2.98	3.35	3.72	4.09	4.46	4.83	5.21	5.58	5.95
2	1.28	1.70	2.13	2.55	2.98	3.40	3.83	4.25	4.68	5.10	5.53	5.95	6.38	6.80
2¼	1.43	1.91	2.39	2.87	3.35	3.83	4.30	4.78	5.26	5.74	6.22	6.69	7.17	7.65
2½	1.59	2.13	2.66	3.19	3.72	4.25	4.78	5.31	5.84	6.38	6.91	7.44	7.97	8.50
2¾	1.75	2.34	2.92	3.51	4.09	4.68	5.26	5.84	6.43	7.01	7.60	8.18	8.77	9.35
3	1.91	2.55	3.19	3.83	4.46	5.10	5.74	6.38	7.01	7.65	8.29	8.93	9.56	10.2
3¼	2.07	2.76	3.45	4.14	4.83	5.53	6.22	6.91	7.60	8.29	8.98	9.67	10.4	11.1
3½	2.23	2.98	3.72	4.46	5.21	5.95	6.69	7.44	8.18	8.93	9.67	10.4	11.2	11.9
3¾	2.39	3.19	3.98	4.78	5.58	6.38	7.17	7.97	8.77	9.56	10.4	11.2	12.0	12.8
4	2.55	3.40	4.25	5.10	5.95	6.80	7.65	8.50	9.35	10.2	11.1	11.9	12.8	13.6
4¼	2.71	3.61	4.52	5.42	6.32	7.23	8.13	9.03	9.93	10.8	11.7	12.6	13.6	14.5
4½	2.87	3.83	4.78	5.74	6.69	7.65	8.61	9.56	10.5	11.5	12.4	13.4	14.3	15.3
4¾	3.03	4.04	5.05	6.06	7.07	8.08	9.08	10.1	11.1	12.1	13.1	14.1	15.1	16.2
5	3.19	4.25	5.31	6.38	7.44	8.50	9.56	10.6	11.7	12.8	13.8	14.9	15.9	17.0
5¼	3.35	4.46	5.58	6.69	7.81	8.93	10.0	11.2	12.3	13.4	14.5	15.6	16.7	17.9
5½	3.51	4.68	5.84	7.01	8.18	9.35	10.5	11.7	12.9	14.0	15.2	16.4	17.5	18.7
5¾	3.67	4.89	6.11	7.33	8.55	9.78	11.0	12.2	13.4	14.7	15.9	17.1	18.3	19.6
6	3.83	5.10	6.38	7.65	8.93	10.2	11.5	12.8	14.0	15.3	16.6	17.9	19.1	20.4
6¼	3.98	5.31	6.64	7.97	9.30	10.6	12.0	13.3	14.6	15.9	17.3	18.6	19.9	21.3
6½	4.14	5.53	6.91	8.29	9.67	11.1	12.4	13.8	15.2	16.6	18.0	19.3	20.7	22.1
6¾	4.30	5.74	7.17	8.61	10.0	11.5	12.9	14.3	15.8	17.2	18.7	20.1	21.5	23.0
7	4.46	5.95	7.44	8.93	10.4	11.9	13.4	14.9	16.4	17.9	19.3	20.8	22.3	23.8
7¼	4.62	6.16	7.70	9.24	10.8	12.3	13.9	15.4	17.0	18.5	20.0	21.6	23.1	24.7
7½	4.78	6.38	7.97	9.56	11.2	12.8	14.3	15.9	17.5	19.1	20.7	22.3	23.9	25.5
7¾	4.94	6.59	8.23	9.88	11.5	13.2	14.8	16.5	18.1	19.8	21.4	23.1	24.7	26.4
8	5.10	6.80	8.50	10.2	11.9	13.6	15.3	17.0	18.7	20.4	22.1	23.8	25.5	27.2
8¼	5.26	7.01	8.77	10.5	12.3	14.0	15.8	17.5	19.3	21.0	22.8	24.5	26.3	28.1
8½	5.42	7.23	9.03	10.8	12.6	14.5	16.3	18.1	19.9	21.7	23.5	25.3	27.1	28.9
8¾	5.58	7.44	9.30	11.2	13.0	14.9	16.7	18.6	20.5	22.3	24.2	26.0	27.9	29.8
9	5.74	7.65	9.56	11.5	13.4	15.3	17.2	19.1	21.0	23.0	24.9	26.8	28.7	30.6
9¼	5.90	7.86	9.83	11.8	13.8	15.7	17.7	19.7	21.6	23.6	25.6	27.5	29.5	31.5
9½	6.06	8.08	10.1	12.1	14.1	16.2	18.2	20.2	22.2	24.2	26.2	28.3	30.3	32.3
9¾	6.22	8.29	10.4	12.4	14.5	16.6	18.7	20.7	22.8	24.9	26.9	29.0	31.1	33.2
10	6.38	8.50	10.6	12.8	14.9	17.0	19.1	21.3	23.4	25.5	27.6	29.8	31.9	34.0

Courtesy of American Institute of Steel Construction

Table 11-10 continued

WEIGHT OF RECTANGULAR SECTIONS
Pounds per linear foot

Width In.	Thickness, Inches													
	3/16	1/4	5/16	3/8	7/16	1/2	9/16	5/8	11/16	3/4	13/16	7/8	15/16	1
10¼	6.53	8.71	10.9	13.1	15.3	17.4	19.6	21.8	24.0	26.1	28.3	30.5	32.7	34.9
10½	6.69	8.93	11.2	13.4	15.6	17.9	20.1	22.3	24.5	26.8	29.0	31.2	33.5	35.7
10¾	6.85	9.14	11.4	13.7	16.0	18.3	20.6	22.8	25.1	27.4	29.7	32.0	34.3	36.6
11	7.01	9.35	11.7	14.0	16.4	18.7	21.0	23.4	25.7	28.1	30.4	32.7	35.1	37.4
11¼	7.17	9.55	12.0	14.3	16.7	19.1	21.5	23.9	26.3	28.7	31.1	33.5	35.9	38.3
11½	7.33	9.78	12.2	14.7	17.1	19.6	22.0	24.4	26.9	29.3	31.8	34.2	36.7	39.1
11¾	7.49	9.99	12.5	15.0	17.5	20.0	22.5	25.0	27.5	30.0	32.5	35.0	37.5	40.0
12	7.65	10.2	12.8	15.3	17.9	20.4	23.0	25.5	28.1	30.6	33.2	35.7	38.3	40.8
12½	7.97	10.6	13.3	15.9	18.6	21.3	23.9	26.6	29.2	31.9	34.5	37.2	39.8	42.5
13	8.29	11.1	13.8	16.6	19.3	22.1	24.9	27.6	30.4	33.2	35.9	38.7	41.4	44.2
13½	8.61	11.5	14.3	17.2	20.1	23.0	25.8	28.7	31.6	34.4	37.3	40.2	43.0	45.9
14	8.93	11.9	14.9	17.9	20.8	23.8	26.8	29.8	32.7	35.7	38.7	41.7	44.6	47.6
14½	9.24	12.3	15.4	18.5	21.6	24.7	27.7	30.8	33.9	37.0	40.1	43.1	46.2	49.3
15	9.56	12.8	15.9	19.1	22.3	25.5	28.7	31.9	35.1	38.3	41.4	44.6	47.8	51.0
15½	9.88	13.2	16.5	19.8	23.1	26.4	29.6	32.9	36.2	39.5	42.8	46.1	49.4	52.7
16	10.2	13.6	17.0	20.4	23.8	27.2	30.6	34.0	37.4	40.8	44.2	47.6	51.0	54.4
16½	10.5	14.0	17.5	21.0	24.5	28.1	31.6	35.1	38.6	42.1	45.6	49.1	52.6	56.1
17	10.8	14.5	18.1	21.7	25.3	28.9	32.5	36.1	39.7	43.4	47.0	50.6	54.2	57.8
17½	11.2	14.9	18.6	22.3	26.0	29.8	33.5	37.2	40.9	44.6	48.3	52.1	55.8	59.5
18	11.5	15.3	19.1	23.0	26.8	30.6	34.4	38.3	42.1	45.9	49.7	53.6	57.4	61.2
18½	11.8	15.7	19.7	23.6	27.5	31.5	35.4	39.3	43.2	47.2	51.1	55.0	59.0	62.9
19	12.1	16.2	20.2	24.2	28.3	32.3	36.3	40.4	44.4	48.5	52.5	56.5	60.6	64.6
19½	12.4	16.6	20.7	24.9	29.0	33.2	37.3	41.4	45.6	49.7	53.9	58.0	62.2	66.3
20	12.8	17.0	21.3	25.5	29.8	34.0	38.3	42.5	46.8	51.0	55.3	59.5	63.8	68.0
20½	13.1	17.4	21.8	26.1	30.5	34.9	39.2	43.6	47.9	52.3	56.6	61.0	65.3	69.7
21	13.4	17.9	22.3	26.8	31.2	35.7	40.2	44.6	49.1	53.6	58.0	62.5	66.9	71.4
21½	13.7	18.3	22.8	27.4	32.0	36.6	41.1	45.7	50.3	54.8	59.4	64.0	68.5	73.1
22	14.0	18.7	23.4	28.1	32.7	37.4	42.1	46.8	51.4	56.1	60.8	65.5	70.1	74.8
22½	14.3	19.1	23.9	28.7	33.5	38.3	43.0	47.8	52.6	57.4	62.2	66.9	71.7	76.5
23	14.7	19.6	24.4	29.3	34.2	39.1	44.0	48.9	53.8	58.7	63.5	68.4	73.3	78.2
23½	15.0	20.0	25.0	30.0	35.0	40.0	44.9	49.9	54.9	59.9	64.9	69.9	74.9	79.9
24	15.3	20.4	25.5	30.6	35.7	40.8	45.9	51.0	56.1	61.2	66.3	71.4	76.5	81.6
25	15.9	21.3	26.6	31.9	37.2	42.5	47.8	53.1	58.4	63.8	69.1	74.4	79.9	85.0
26	16.6	22.1	27.6	33.2	38.7	44.2	49.7	55.3	60.8	66.3	71.8	77.4	82.9	88.4
27	17.2	23.0	28.7	34.4	40.2	45.9	51.6	57.4	63.1	68.9	74.6	80.3	86.1	91.8
28	17.9	23.8	29.8	35.7	41.7	47.6	53.6	59.5	65.5	71.4	77.4	83.3	89.3	95.2
29	18.5	24.7	30.8	37.0	43.1	49.3	55.5	61.6	67.8	74.0	80.1	86.3	92.4	98.6
30	19.1	25.5	31.9	38.3	44.6	51.0	57.4	63.8	70.1	76.5	82.9	89.3	95.6	102
31	19.8	26.4	32.9	39.5	46.1	52.7	59.3	65.9	72.5	79.1	85.6	92.2	98.8	105
32	20.4	27.2	34.0	40.8	47.6	54.4	61.2	68.0	74.8	81.6	88.4	95.2	102	109

Courtesy of American Institute of Steel Construction

Table 11-10 continued

WEIGHT OF RECTANGULAR SECTIONS
Pounds per linear foot

Width In.	Thickness, Inches													
	3/16	1/4	5/16	3/8	7/16	1/2	9/16	5/8	11/16	3/4	13/16	7/8	15/16	1
33	21.0	28.1	35.1	42.1	49.1	56.1	63.1	70.1	77.1	84.2	91.2	98.2	105	112
34	21.7	28.9	36.1	43.4	50.6	57.8	65.0	72.3	79.5	86.7	93.9	101	108	116
35	22.3	29.8	37.2	44.6	52.1	59.5	66.9	74.4	81.8	89.3	96.1	104	112	119
36	23.0	30.6	38.3	45.9	53.6	61.2	68.9	76.5	84.2	91.8	99.5	107	115	122
37	23.6	31.5	39.3	47.2	55.0	62.9	70.8	78.6	86.5	94.4	102	110	118	126
38	24.2	32.3	40.4	48.5	56.5	64.6	72.7	80.8	88.8	96.9	105	113	121	129
39	24.9	33.2	41.4	49.7	58.0	66.3	74.6	82.9	91.2	99.5	108	116	124	133
40	25.5	34.0	42.5	51.0	59.5	68.0	76.5	85.0	93.5	102	111	119	128	136
41	26.1	34.9	43.6	52.3	61.0	69.7	78.4	87.1	95.8	105	113	122	131	139
42	26.8	35.7	44.6	53.6	62.5	71.4	80.3	89.3	98.2	107	116	125	134	143
43	27.4	36.6	45.7	54.8	64.0	73.1	82.2	91.4	101	110	119	128	137	146
44	28.1	37.4	46.8	56.1	65.5	74.8	84.2	93.5	103	112	122	131	140	150
45	28.7	38.3	47.8	57.4	66.9	76.5	86.1	95.6	105	115	124	134	143	153
46	29.3	39.1	48.9	58.7	68.4	78.2	88.0	97.8	108	117	127	137	147	156
47	30.0	40.0	49.9	59.9	69.9	79.9	89.9	99.9	110	120	130	140	150	160
48	30.6	40.8	51.0	61.2	71.4	81.6	91.8	102	112	122	133	143	153	163
49	31.2	41.7	52.1	62.5	72.9	83.3	93.7	104	115	125	135	146	156	167
50	31.9	42.5	53.1	63.8	74.4	85.0	95.6	106	117	128	138	149	159	170
51	32.5	43.4	54.2	65.0	75.9	86.7	97.5	108	119	130	141	152	163	173
52	33.2	44.2	55.3	66.3	77.4	88.4	99.5	111	122	133	144	155	166	177
53	33.8	45.1	56.3	67.6	78.8	90.1	101	113	124	135	146	158	169	180
54	34.4	45.9	57.4	68.9	80.3	91.8	103	115	126	138	149	161	172	184
55	35.1	46.8	58.4	70.1	81.8	93.5	105	117	129	140	152	164	175	187
56	35.7	47.6	59.5	71.4	83.3	95.2	107	119	131	143	155	167	179	190
57	36.3	48.5	60.6	72.7	84.8	96.9	109	121	133	145	158	170	182	194
58	37.0	49.3	61.6	74.0	86.3	98.6	111	123	136	148	160	173	185	197
59	37.6	50.2	62.7	75.2	87.8	100	113	125	138	151	163	176	188	201
60	38.3	51.0	63.8	76.5	89.3	102	115	128	140	153	166	179	191	204
61	38.9	51.9	64.8	77.8	90.7	104	117	130	143	156	169	182	194	207
62	39.5	52.7	65.9	79.1	92.2	105	119	132	145	158	171	185	198	211
63	40.2	53.6	66.9	80.3	93.7	107	121	134	147	161	174	187	201	214
64	40.8	54.4	68.0	81.6	95.2	109	122	136	150	163	177	190	204	218
65	41.4	55.3	69.1	82.9	96.7	111	124	138	152	166	180	193	207	221
66	42.1	56.1	70.1	84.2	98.2	112	126	140	154	168	182	196	210	224
67	42.7	57.0	71.2	85.4	99.7	114	128	142	157	171	185	199	214	228
68	43.4	57.8	72.3	86.7	101	116	130	145	159	173	188	202	217	231
69	44.0	58.7	73.3	88.0	103	117	132	147	161	176	191	205	220	235
70	44.6	59.5	74.4	89.3	104	119	134	149	164	179	193	208	223	238
71	45.3	60.4	75.4	90.5	106	121	136	151	166	181	196	211	226	241
72	45.9	61.2	76.5	91.8	107	122	138	153	168	184	199	214	230	245

Courtesy of American Institute of Steel Construction

Table 11-10 continued

WEIGHT OF RECTANGULAR SECTIONS
Pounds per linear foot

Width In.	Thickness, Inches													
	3/16	1/4	5/16	3/8	7/16	1/2	9/16	5/8	11/16	3/4	13/16	7/8	15/16	1
73	46.5	62.1	77.6	93.1	109	124	140	155	171	186	202	217	233	248
74	47.2	62.9	78.6	94.4	110	126	142	157	173	189	204	220	236	252
75	47.8	63.8	79.7	95.6	112	128	143	159	175	191	207	223	239	255
76	48.5	64.6	80.8	96.9	113	129	145	162	178	194	210	226	242	258
77	49.1	65.5	81.8	98.2	115	131	147	164	180	196	213	229	245	262
78	49.7	66.3	82.9	99.5	116	133	149	166	182	199	216	232	249	265
79	50.4	67.2	83.9	101	118	134	151	168	185	202	218	235	252	269
80	51.0	68.0	85.0	102	119	136	153	170	187	204	221	238	255	272
81	51.6	68.9	86.1	103	121	138	155	172	189	207	224	241	258	275
82	52.3	69.7	87.1	105	122	139	157	174	192	209	227	244	261	279
83	52.9	70.6	88.2	106	124	141	159	176	194	212	229	247	265	282
84	53.6	71.4	89.3	107	125	143	161	179	196	214	232	250	268	286
85	54.2	72.3	90.3	108	126	145	163	181	199	217	235	253	271	289
86	54.8	73.1	91.4	110	128	146	165	183	201	219	238	256	274	292
87	55.5	74.0	92.4	111	129	148	166	185	203	222	240	259	277	296
88	56.1	74.8	93.5	112	131	150	168	187	206	224	243	262	281	299
89	56.7	75.7	94.6	114	132	151	170	189	208	227	246	265	284	303
90	57.4	76.5	95.6	115	134	153	172	191	210	230	249	268	287	306
91	...	77.4	96.7	116	135	155	174	193	213	232	251	271	290	309
92	...	78.2	97.8	117	137	156	176	196	215	235	254	274	293	313
93	...	79.1	98.8	119	138	158	178	198	217	237	257	277	296	316
94	...	79.9	99.9	120	140	160	180	200	220	240	260	280	300	320
95	...	80.8	101	121	141	162	182	202	222	242	262	283	303	328
96	...	81.6	102	122	143	163	184	204	224	245	265	286	306	326
98	...	83.3	104	125	146	167	187	208	229	250	271	292	312	333
100	...	85.0	106	128	149	170	191	213	234	255	276	298	319	340
102	...	85.7	108	130	152	173	195	217	238	260	282	304	325	347
104	...	88.4	111	133	155	177	199	221	243	265	287	309	332	354
106	...	90.1	113	135	158	180	203	225	248	270	293	315	338	360
108	...	91.8	115	138	161	184	207	230	253	275	298	321	344	367
110	...	93.5	117	140	164	187	210	234	257	281	304	327	351	374
112	...	95.2	119	143	167	190	214	238	262	286	309	333	357	381
114	...	96.9	121	145	170	194	218	242	267	291	315	339	363	388
116	...	98.6	123	148	173	197	222	247	271	296	321	345	370	394
118	...	100	125	151	176	201	226	251	276	301	326	351	376	401
120	...	102	128	153	179	204	230	255	281	306	332	357	383	408
122	...	104	130	156	182	207	233	259	285	311	337	363	389	415
124	...	105	132	158	185	211	237	264	290	316	343	369	395	422
126	...	107	134	161	187	214	241	268	295	321	348	375	402	428
128	...	109	136	163	190	218	245	272	299	326	354	381	408	435

Courtesy of American Institute of Steel Construction

Table 11-10 continued

WEIGHT OF RECTANGULAR SECTIONS
Pounds per linear foot

Width In.	Thickness, Inches												
	1/4	5/16	3/8	7/16	1/2	9/16	5/8	11/16	3/4	13/16	7/8	15/16	1
130	111	138	166	193	221	249	276	304	332	359	387	414	442
132	112	140	168	196	224	252	281	309	337	365	393	421	449
134	114	142	171	199	228	256	285	313	342	370	399	427	456
136	116	145	173	202	231	260	289	318	347	376	405	434	462
138	117	147	176	205	235	264	293	323	352	381	411	440	469
140	119	149	179	208	238	268	298	327	357	387	417	446	476
142	121	151	181	211	241	272	302	332	362	392	422	453	483
144	122	153	184	214	245	275	306	337	367	398	428	459	490
146	124	155	186	217	248	279	310	341	372	403	434	465	496
148	126	157	189	220	252	283	315	346	377	409	440	472	503
150	128	159	191	223	255	287	319	351	383	414	446	478	510
152	129	162	194	226	258	291	323	355	388	420	452	485	517
154	131	164	196	229	262	295	327	360	393	425	458	491	524
156	133	166	199	232	265	298	332	365	398	431	464	497	530
158	134	168	201	235	269	302	336	369	403	436	470	504	537
160	136	170	204	238	272	306	340	374	408	442	476	510	544
162	138	172	207	241	275	310	344	379	413	448	482	516	551
164	139	174	209	244	279	314	349	383	418	453	488	523	558
166	141	176	212	247	282	317	353	388	423	459	494	529	564
168	143	179	214	250	286	321	357	393	428	464	500	536	571
170	145	181	217	253	289	325	361	397	434	470	506	542	578

Courtesy of American Institute of Steel Construction

Table 11-11. Area of Rectangular Sections

AREA OF RECTANGULAR SECTIONS
Square inches

Width In.	Thickness, Inches													
	3/16	¼	5/16	⅜	7/16	½	9/16	⅝	11/16	¾	13/16	⅞	15/16	1
¼	.047	.063	.078	.094	.109	.125	.141	.156	.172	.188	.203	.219	.234	.250
½	.094	.125	.156	.188	.219	.250	.281	.313	.344	.375	.406	.438	.469	.500
¾	.141	.188	.234	.281	.328	.375	.422	.469	.516	.563	.609	.656	.703	.750
1	.188	.250	.313	.375	.438	.500	.563	.625	.688	.750	.813	.875	.938	1.00
1¼	.234	.313	.391	.469	.547	.625	.703	.781	.859	.938	1.02	1.09	1.17	1.25
1½	.281	.375	.469	.563	.656	.750	.844	.938	1.03	1.13	1.22	1.31	1.41	1.50
1¾	.328	.438	.547	.656	.766	.875	.984	1.09	1.20	1.31	1.42	1.53	1.64	1.75
2	.375	.500	.625	.750	.875	1.00	1.13	1.25	1.38	1.50	1.63	1.75	1.88	2.00
2¼	.422	.563	.703	.844	.984	1.13	1.27	1.41	1.55	1.69	1.83	1.97	2.11	2.25
2½	.469	.625	.781	.938	1.09	1.25	1.41	1.56	1.72	1.88	2.03	2.19	2.34	2.50
2¾	.516	.688	.859	1.03	1.20	1.38	1.55	1.72	1.89	2.06	2.23	2.41	2.58	2.75
3	.563	.750	.938	1.13	1.31	1.50	1.69	1.88	2.06	2.25	2.44	2.63	2.81	3.00
3¼	.609	.813	1.02	1.22	1.42	1.63	1.83	2.03	2.23	2.44	2.64	2.84	3.05	3.25
3½	.656	.875	1.09	1.31	1.53	1.75	1.97	2.19	2.41	2.63	2.84	3.06	3.28	3.50
3¾	.703	.938	1.17	1.41	1.64	1.88	2.11	2.34	2.58	2.81	3.05	3.28	3.52	3.75
4	.750	1.00	1.25	1.50	1.75	2.00	2.25	2.50	2.75	3.00	3.25	3.50	3.75	4.00
4¼	.797	1.06	1.33	1.59	1.86	2.13	2.39	2.66	2.92	3.19	3.45	3.72	3.98	4.25
4½	.844	1.13	1.41	1.69	1.97	2.25	2.53	2.81	3.09	3.38	3.66	3.94	4.22	4.50
4¾	.891	1.19	1.48	1.78	2.09	2.38	2.67	2.97	3.27	3.56	3.86	4.16	4.45	4.75
5	.938	1.25	1.56	1.88	2.19	2.50	2.81	3.13	3.44	3.75	4.06	4.38	4.69	5.00
5¼	.984	1.31	1.64	1.97	2.30	2.63	2.95	3.28	3.61	3.94	4.27	4.59	4.92	5.25
5½	1.03	1.38	1.72	2.06	2.41	2.75	3.09	3.44	3.78	4.13	4.47	4.81	5.16	5.50
5¾	1.08	1.44	1.80	2.16	2.52	2.88	3.23	3.59	3.95	4.31	4.67	5.03	5.39	5.75
6	1.13	1.50	1.88	2.25	2.63	3.00	3.38	3.75	4.13	4.50	4.88	5.25	5.63	6.00
6¼	1.17	1.56	1.95	2.34	2.73	3.13	3.52	3.91	4.30	4.69	5.08	5.47	5.86	6.25
6½	1.22	1.63	2.03	2.44	2.84	3.25	3.66	4.06	4.47	4.88	5.28	5.69	6.09	6.50
6¾	1.27	1.69	2.10	2.53	2.95	3.38	3.80	4.22	4.64	5.06	5.48	5.91	6.33	6.75
7	1.31	1.75	2.19	2.63	3.06	3.50	3.94	4.38	4.81	5.25	5.69	6.13	6.56	7.00
7¼	1.36	1.81	2.27	2.72	3.17	3.63	4.08	4.53	4.98	5.44	5.89	6.34	6.80	7.25
7½	1.41	1.88	2.34	2.81	3.28	3.75	4.22	4.69	5.16	5.63	6.09	6.56	7.03	7.50
7¾	1.45	1.94	2.42	2.91	3.39	3.88	4.36	4.84	5.33	5.81	6.30	6.78	7.27	7.75
8	1.50	2.00	2.50	3.00	3.50	4.00	4.50	5.00	5.50	6.00	6.50	7.00	7.50	8.00
8¼	1.55	2.06	2.58	3.09	3.61	4.13	4.64	5.16	5.67	6.19	6.70	7.22	7.73	8.25
8½	1.59	2.13	2.66	3.19	3.72	4.25	4.78	5.31	5.84	6.38	6.91	7.44	7.97	8.50
8¾	1.64	2.19	2.73	3.28	3.83	4.38	4.92	5.47	6.02	6.56	7.11	7.66	8.20	8.75
9	1.69	2.25	2.81	3.38	3.94	4.50	5.06	5.63	6.19	6.75	7.31	7.88	8.44	9.00
9¼	1.73	2.31	2.89	3.47	4.05	4.63	5.20	5.78	6.36	6.94	7.52	8.09	8.67	9.25
9½	1.78	2.38	2.97	3.56	4.16	4.75	5.34	5.94	6.53	7.13	7.72	8.31	8.91	9.50
9¾	1.83	2.44	3.05	3.66	4.27	4.88	5.48	6.09	6.70	7.31	7.92	8.53	9.14	9.75
10	1.88	2.50	3.13	3.75	4.38	5.00	5.63	6.25	6.88	7.50	8.13	8.75	9.38	10.00

Courtesy of American Institute of Steel Construction

Table 11-11 continued

AREA OF RECTANGULAR SECTIONS
Square inches

Width In.	Thickness, Inches													
	3/16	1/4	5/16	3/8	7/16	1/2	9/16	5/8	11/16	3/4	13/16	7/8	15/16	1
10¼	1.92	2.56	3.20	3.84	4.48	5.13	5.77	6.41	7.05	7.69	8.33	8.97	9.61	10.25
10½	1.97	2.63	3.28	3.94	4.59	5.25	5.91	6.56	7.22	7.88	8.53	9.19	9.84	10.50
10¾	2.02	2.69	3.36	4.03	4.70	5.38	6.05	6.72	7.39	8.06	8.73	9.41	10.08	10.75
11	2.06	2.75	3.44	4.13	4.81	5.50	6.19	6.88	7.56	8.25	8.94	9.63	10.31	11.00
11¼	2.11	2.81	3.52	4.22	4.92	5.63	6.33	7.03	7.73	8.44	9.14	9.84	10.55	11.25
11½	2.16	2.88	3.59	4.31	5.03	5.75	6.47	7.19	7.91	8.63	9.34	10.06	10.78	11.50
11¾	2.20	2.94	3.67	4.41	5.14	5.88	6.61	7.34	8.08	8.81	9.55	10.28	11.02	11.75
12	2.25	3.00	3.75	4.50	5.25	6.00	6.75	7.50	8.25	9.00	9.75	10.50	11.25	12.00
12½	2.34	3.13	3.91	4.69	5.47	6.25	7.03	7.81	8.59	9.38	10.16	10.94	11.72	12.50
13	2.44	3.25	4.06	4.88	5.69	6.50	7.31	8.13	8.94	9.75	10.56	11.38	12.19	13.00
13½	2.53	3.38	4.22	5.06	5.91	6.75	7.59	8.44	9.28	10.13	10.97	11.81	12.66	13.50
14	2.63	3.50	4.38	5.25	6.13	7.00	7.88	8.75	9.63	10.50	11.38	12.25	13.13	14.00
14½	2.72	3.63	4.53	5.44	6.34	7.25	8.16	9.06	9.97	10.88	11.78	12.69	13.59	14.50
15	2.81	3.75	4.69	5.63	6.56	7.50	8.44	9.38	10.31	11.25	12.19	13.13	14.06	15.00
15½	2.91	3.88	4.84	5.81	6.78	7.75	8.72	9.69	10.66	11.63	12.59	13.56	14.53	15.50
16	3.00	4.00	5.00	6.00	7.00	8.00	9.00	10.00	11.00	12.00	13.00	14.00	15.00	16.00
16½	3.09	4.13	5.16	6.19	7.22	8.25	9.28	10.31	11.34	12.38	13.41	14.44	15.47	16.50
17	3.19	4.25	5.31	6.38	7.44	8.50	9.56	10.63	11.69	12.75	13.81	14.88	15.94	17.00
17½	3.28	4.38	5.47	6.56	7.66	8.75	9.84	10.94	12.03	13.13	14.22	15.31	16.41	17.50
18	3.38	4.50	5.63	6.75	7.88	9.00	10.13	11.25	12.38	13.50	14.63	15.75	16.88	18.00
18½	3.47	4.63	5.78	6.94	8.09	9.25	10.41	11.56	12.72	13.88	15.03	161.9	17.34	18.50
19	3.56	4.75	5.94	7.13	8.31	9.50	10.69	11.88	13.06	14.25	15.44	16.63	17.81	19.00
19½	3.66	4.88	6.09	7.31	8.53	9.75	10.97	12.19	13.41	14.63	15.84	17.06	18.28	19.50
20	3.75	5.00	6.25	7.50	8.75	10.00	11.25	12.50	13.75	15.00	16.25	17.50	18.75	20.00
20½	3.84	5.13	6.41	7.69	8.97	10.25	11.53	12.81	14.09	15.38	16.66	17.94	19.22	20.50
21	3.94	5.25	6.56	7.88	9.19	10.50	11.81	13.13	14.44	15.75	17.06	18.38	19.69	21.00
21½	4.03	5.38	6.72	8.06	9.41	10.75	12.09	13.44	14.78	16.13	17.47	18.81	20.16	21.50
22	4.13	5.50	6.88	8.25	9.63	11.00	12.38	13.75	15.13	16.50	17.88	19.25	20.63	22.00
22½	4.22	5.63	7.03	8.44	9.84	11.25	12.66	14.06	15.47	16.88	18.28	19.69	21.09	22.50
23	4.31	5.75	7.19	8.63	10.06	11.50	12.94	14.38	15.81	17.25	18.69	20.13	21.56	23.00
23½	4.41	5.88	7.34	8.81	10.28	11.75	13.22	14.69	16.16	17.63	19.09	20.56	22.03	23.50
24	4.50	6.00	7.50	9.00	10.50	12.00	13.50	15.00	16.50	18.00	19.50	21.00	22.50	24.00
25	4.69	6.25	7.81	9.38	10.94	12.50	14.06	15.63	17.19	18.75	20.31	21.88	23.44	25.00
26	4.88	6.50	8.13	9.75	11.38	13.00	14.63	16.25	17.88	19.50	21.13	22.75	24.38	26.00
27	5.06	6.75	8.44	10.13	11.81	13.50	15.19	16.88	18.56	20.25	21.94	23.63	25.31	27.00
28	5.25	7.00	8.75	10.50	12.25	14.00	15.75	17.50	19.25	21.00	22.75	24.50	26.25	28.00
29	5.44	7.25	9.06	10.88	12.69	14.50	16.31	18.13	19.94	21.75	23.56	25.38	27.19	29.00
30	5.63	7.50	9.38	11.25	13.13	15.00	16.88	18.75	20.63	22.50	24.38	26.25	28.13	30.00
31	5.81	7.75	9.69	11.63	13.56	15.50	17.44	19.38	21.31	23.25	25.19	27.13	29.06	31.00
32	6.00	8.00	10.00	12.00	14.00	16.00	18.00	20.00	22.00	24.00	26.00	28.00	30.00	32.00

Courtesy of American Institute of Steel Construction

Table 11-11 continued

AREA OF RECTANGULAR SECTIONS
Square inches

Width In.	Thickness, Inches													
	3/16	1/4	5/16	3/8	7/16	1/2	9/16	5/8	11/16	3/4	13/16	7/8	15/16	1
33	6.19	8.25	10.31	12.38	14.44	16.50	18.56	20.63	22.69	24.75	26.81	28.88	30.94	33.00
34	6.38	8.50	10.63	12.75	14.88	17.00	19.13	21.25	23.38	25.50	27.63	29.75	31.88	34.00
35	6.56	8.75	10.94	13.13	15.31	17.50	19.69	21.88	24.06	26.25	28.44	30.63	32.81	35.00
36	6.75	9.00	11.25	13.50	15.75	18.00	20.25	22.50	24.75	27.00	29.25	31.50	33.75	36.00
37	6.94	9.25	11.56	13.88	16.19	18.50	20.81	23.13	25.44	27.75	30.06	32.38	34.69	37.00
38	7.13	9.50	11.88	14.25	16.63	19.00	21.38	23.75	26.13	28.50	30.88	33.25	35.63	38.00
39	7.31	9.75	12.19	14.63	17.06	19.50	21.94	24.38	26.81	29.25	31.69	34.13	36.56	39.00
40	7.50	10.00	12.50	15.00	17.50	20.00	22.50	25.00	27.50	30.00	32.50	35.00	37.50	40.00
41	7.69	10.25	12.81	15.38	17.94	20.50	23.06	25.63	28.19	30.75	33.31	35.88	38.44	41.00
42	7.88	10.50	13.13	15.75	18.38	21.00	23.63	26.25	28.88	31.50	34.13	36.75	39.38	42.00
43	8.06	10.75	13.44	16.13	18.81	21.50	24.19	26.88	29.56	32.25	34.94	37.63	40.31	43.00
44	8.25	11.00	13.75	16.50	19.25	22.00	24.75	27.50	30.25	33.00	35.75	38.50	41.25	44.00
45	8.44	11.25	14.06	16.88	19.69	22.50	25.31	28.13	30.94	33.75	36.56	39.38	42.19	45.00
46	8.63	11.50	14.38	17.25	20.13	23.00	25.88	28.75	31.63	34.50	37.38	40.25	43.13	46.00
47	8.81	11.75	14.69	17.63	20.56	23.50	26.44	29.38	32.31	35.25	38.19	41.13	44.06	47.00
48	9.00	12.00	15.00	18.00	21.00	24.00	27.00	30.00	33.00	36.00	39.00	42.00	45.00	48.00
49	9.19	12.25	15.31	18.38	21.44	24.50	27.56	30.63	33.69	36.75	39.81	42.88	45.94	49.00
50	9.38	12.50	15.63	18.75	21.88	25.00	28.13	31.25	34.38	37.50	40.63	43.75	46.88	50.00
51	9.56	12.75	15.94	19.13	22.31	25.50	28.69	31.88	35.06	38.25	41.44	44.63	47.81	51.00
52	9.75	13.00	16.25	19.50	22.75	26.00	29.25	32.50	35.75	39.00	42.25	45.50	48.75	52.00
53	9.94	13.25	16.56	19.88	23.19	26.50	29.81	33.13	36.44	39.75	43.06	46.38	49.69	53.00
54	10.13	13.50	16.88	20.25	23.63	27.00	30.38	33.75	37.13	40.50	43.88	47.25	50.63	54.00
55	10.31	13.75	17.19	20.63	24.06	27.50	30.94	34.38	37.81	41.25	44.69	48.13	51.56	55.00
56	10.50	14.00	17.50	21.00	24.50	28.00	31.50	35.00	38.50	42.00	45.50	49.00	52.50	56.00
57	10.69	14.25	17.81	21.38	24.94	28.50	32.06	35.63	39.19	42.75	46.31	49.88	53.44	57.00
58	10.88	14.50	18.13	21.75	25.38	29.00	32.63	36.25	39.88	43.50	47.13	50.75	54.38	58.00
59	11.06	14.75	18.44	22.13	25.81	29.50	33.19	36.88	40.56	44.25	47.94	51.63	55.31	59.00
60	11.25	15.00	18.75	22.50	26.25	30.00	33.75	37.50	41.25	45.00	48.75	52.50	56.25	60.00
61	11.44	15.25	19.06	22.88	26.69	30.50	34.31	38.13	41.94	45.75	49.56	53.38	57.19	61.00
62	11.63	15.50	19.38	23.25	27.13	31.00	34.88	38.75	42.63	46.50	50.38	54.25	58.13	62.00
63	11.81	15.75	19.69	23.63	27.56	31.50	35.44	39.38	43.31	47.25	51.19	55.13	59.06	63.00
64	12.00	16.00	20.00	24.00	28.00	32.00	36.00	40.00	44.00	48.00	52.00	56.00	60.00	64.00
65	12.19	16.25	20.31	24.38	28.44	32.50	36.56	40.63	44.69	48.75	52.81	56.88	60.94	65.00
66	12.38	16.50	20.63	24.75	28.88	33.00	37.13	41.25	45.38	49.50	53.63	57.75	61.88	66.00
67	12.56	16.75	20.94	25.13	29.31	33.50	37.69	41.88	46.06	50.25	54.44	58.63	62.81	67.00
68	12.75	17.00	21.25	25.50	29.75	34.00	38.25	42.50	46.75	51.00	55.25	59.50	63.75	68.00
69	12.94	17.25	21.56	25.88	30.19	34.50	38.81	43.13	47.44	51.75	56.06	60.38	64.69	69.00
70	13.13	17.50	21.88	26.25	30.63	35.00	39.38	43.75	48.13	52.50	56.88	61.25	65.63	70.00
71	13.31	17.75	22.19	26.63	31.06	35.50	39.94	44.38	48.81	53.25	57.69	62.13	66.56	71.00
72	13.50	18.00	22.50	27.00	31.50	36.00	40.50	45.00	49.50	54.00	58.50	63.00	67.50	72.00

Courtesy of American Institute of Steel Construction

Table 11-11 continued

AREA OF RECTANGULAR SECTIONS
Square inches

Width In.	Thickness, Inches													
	3/16	1/4	5/16	3/8	7/16	1/2	9/16	5/8	11/16	3/4	13/16	7/8	15/16	1
73	13.69	18.25	22.81	27.38	31.94	36.50	41.06	45.63	50.19	54.75	59.31	63.88	68.44	73.00
74	13.88	18.50	23.13	27.75	32.38	37.00	41.63	46.25	50.88	55.50	60.13	64.75	69.38	74.00
75	14.06	18.75	23.44	28.13	32.81	37.50	42.19	46.88	51.56	56.25	60.94	65.63	70.31	75.00
76	14.25	19.00	23.75	28.50	33.25	38.00	42.75	47.50	52.25	57.00	61.75	66.50	71.25	76.00
77	14.44	19.25	24.06	28.88	33.69	38.50	43.31	48.13	52.94	57.75	62.56	67.38	72.19	77.00
78	14.63	19.50	24.38	29.25	34.13	39.00	43.88	48.75	53.63	58.50	63.38	68.25	73.13	78.00
79	14.81	19.75	24.69	29.63	34.56	39.50	44.44	49.38	54.31	59.25	64.19	69.13	74.06	79.00
80	15.00	20.00	25.00	30.00	35.00	40.00	45.00	50.00	55.00	60.00	65.00	70.00	75.00	80.00
81	15.19	20.25	25.31	30.38	35.44	40.50	45.56	50.63	55.69	60.75	65.81	70.88	75.94	81.00
82	15.38	20.50	25.63	30.75	35.88	41.00	46.13	51.25	56.38	61.50	66.63	71.75	76.88	82.00
83	15.56	20.75	25.94	31.13	36.31	41.50	46.69	51.88	57.06	62.25	67.44	72.63	77.81	83.00
84	15.75	21.00	26.25	31.50	36.75	42.00	47.25	52.50	57.75	63.00	68.25	73.50	78.75	84.00
85	15.94	21.25	26.56	31.88	37.19	42.50	47.81	53.13	58.44	63.75	69.06	74.38	79.69	85.00
86	16.13	21.50	26.88	32.25	37.63	43.00	48.38	53.75	59.13	64.50	69.88	75.25	80.63	86.00
87	16.31	21.75	27.19	32.63	38.06	43.50	48.94	54.38	59.81	65.25	70.69	76.13	81.56	87.00
88	16.50	22.00	27.50	33.00	38.50	44.00	49.50	55.00	60.50	66.00	71.50	77.00	82.50	88.00
89	16.69	22.25	27.81	33.38	38.94	44.50	50.06	55.63	61.19	66.75	72.31	77.88	83.44	89.00
90	16.88	22.50	28.13	33.75	39.38	45.00	50.63	56.25	61.88	67.50	73.13	78.75	84.38	90.00
91	...	22.75	28.44	34.13	39.81	45.50	51.19	56.88	62.56	68.25	73.94	79.63	85.31	91.00
92	...	23.00	28.75	34.50	40.25	46.00	51.75	57.50	63.25	69.00	74.75	80.50	86.25	92.00
93	...	23.25	29.06	34.88	40.69	46.50	52.31	58.13	63.94	69.75	75.56	81.38	87.19	93.00
94	...	23.50	29.38	35.25	41.13	47.00	52.88	58.75	64.63	70.50	76.38	82.25	88.13	94.00
95	...	23.75	29.69	35.63	41.56	47.50	53.44	59.38	65.31	71.25	77.19	83.13	89.06	95.00
96	...	24.00	30.00	36.00	42.00	48.00	54.00	60.00	66.00	72.00	78.00	84.00	90.00	96.00
98	...	24.50	30.63	36.75	42.88	49.00	55.13	61.25	67.38	73.50	79.63	85.75	91.88	98.00
100	...	25.00	31.25	37.50	43.75	50.00	56.25	62.50	68.75	75.00	81.25	87.50	93.75	100.00
102	...	25.50	31.88	38.25	44.63	51.00	57.38	63.75	70.13	76.50	82.88	89.25	95.63	102.00
104	...	26.00	32.50	39.00	45.50	52.00	58.50	65.00	71.50	78.00	84.50	91.00	97.50	104.00
106	...	26.50	33.13	39.75	46.38	53.00	59.63	66.25	72.88	79.50	86.13	92.75	99.38	106.00
108	...	27.00	33.75	40.50	47.25	54.00	60.75	67.50	74.25	81.00	87.75	94.50	101.25	108.00
110	...	27.50	34.38	41.25	48.13	55.00	61.88	68.75	75.63	82.50	89.38	96.25	103.13	110.00
112	...	28.00	35.00	42.00	49.00	56.00	63.00	70.00	77.00	84.00	91.00	98.00	105.00	112.00
114	...	28.50	35.63	42.75	49.88	57.00	64.13	71.25	78.38	85.50	92.63	99.75	106.88	114.00
116	...	29.00	36.25	43.50	50.75	58.00	65.25	72.50	79.75	87.00	94.25	101.50	108.75	116.00
118	...	29.50	36.88	44.25	51.63	59.00	66.38	73.75	81.13	88.50	95.88	103.25	110.63	118.00
120	...	30.00	37.50	45.00	52.50	60.00	67.50	75.00	82.50	90.00	97.50	105.00	112.50	120.00
122	...	30.50	38.13	45.75	53.38	61.00	68.63	76.25	83.88	91.50	99.13	106.75	114.38	122.00
124	...	31.00	38.75	46.50	54.25	62.00	69.75	77.50	85.25	93.00	100.75	108.50	116.25	124.00
126	...	31.50	39.38	47.25	55.13	63.00	70.88	78.75	86.63	94.50	102.38	110.25	118.13	126.00
128	...	32.00	40.00	48.00	56.00	64.00	72.00	80.00	88.00	96.00	104.00	112.00	120.00	128.00

Courtesy of American Institute of Steel Construction

Table 11-11 continued

AREA OF RECTANGULAR SECTIONS
Square inches

Width In.	Thickness, Inches												
	¼	5/16	⅜	7/16	½	9/16	⅝	11/16	¾	13/16	⅞	15/16	1
130	32.50	40.63	48.75	56.88	65.00	73.13	81.25	89.38	97.50	105.63	113.75	121.88	130.00
132	33.00	41.25	49.50	57.75	66.00	74.25	82.50	90.75	99.00	107.25	115.50	123.75	132.00
134	33.50	41.88	50.25	58.63	67.00	75.38	83.75	92.13	100.50	108.88	117.25	125.63	134.00
136	34.00	42.50	51.00	59.50	68.00	76.50	85.00	93.50	102.00	110.50	119.00	127.50	136.00
138	34.50	43.13	51.75	60.38	69.00	77.63	86.25	94.88	103.50	112.13	120.75	129.38	138.00
140	35.00	43.75	52.50	61.25	70.00	78.75	87.50	96.25	105.00	113.75	122.50	131.25	140.00
142	35.50	44.38	53.25	62.13	71.00	79.88	88.75	97.63	106.50	115.38	124.25	133.13	142.00
144	36.00	45.00	54.00	63.00	72.00	81.00	90.00	99.00	108.00	117.00	126.00	135.00	144.00
146	36.50	45.63	54.75	63.88	73.00	82.13	91.25	100.38	109.50	118.63	127.75	136.88	146.00
148	37.00	46.25	55.50	64.75	74.00	83.25	92.50	101.75	111.00	120.25	129.50	138.75	148.00
150	37.50	46.88	56.25	65.63	75.00	84.38	93.75	103.13	112.50	121.88	131.25	140.63	150.00
152	38.00	47.50	57.00	66.50	76.00	85.50	95.00	104.50	114.00	123.50	133.00	142.50	152.00
154	38.50	48.13	57.75	67.38	77.00	86.63	96.25	105.88	115.50	125.13	134.75	144.38	154.00
156	39.00	48.75	58.50	68.25	78.00	87.75	97.50	107.25	117.00	126.75	136.50	146.25	156.00
158	39.50	49.38	59.25	69.13	79.00	88.88	98.75	108.63	118.50	128.38	138.25	148.13	158.00
160	40.00	50.00	60.00	70.00	80.00	90.00	100.00	110.00	120.00	130.00	140.00	150.00	160.00
162	40.50	50.63	60.75	70.88	81.00	91.13	101.25	111.38	121.50	131.63	141.75	151.88	162.00
164	41.00	51.25	61.50	71.75	82.00	92.25	102.50	112.75	123.00	133.25	143.50	153.75	164.00
166	41.50	51.88	62.25	72.63	83.00	93.38	103.75	114.13	124.50	134.88	145.25	155.63	166.00
168	42.00	52.50	63.00	73.50	84.00	94.50	105.00	115.50	126.00	136.50	147.00	157.50	168.00
170	42.50	53.13	63.75	74.38	85.00	95.63	106.25	116.88	127.50	138.13	148.75	159.38	170.00

Courtesy of American Institute of Steel Construction

continued from page 129

Metric Weights

The structural steel draftsman will have many jobs that are fabricated for overseas shipment by a U.S. contractor or a foreign contractor having the steel supplied and fabricated in the U.S.A. These jobs usually require that all weights be specified in metric units. The AISC tables list weights in pounds per linear foot. In metric units weights are expressed as kilograms per meter, commonly called kilos per meter. The conversion is done by the structural steel draftsman.

Chapter 4 supplies metric conversion data. One foot equals 0.3048 meter, so a steel member 10′-3″ long is first converted to 10.25 feet; this is then multiplied by 0.3048 to equal 3.124 meters long. If this member weighed 105.5 pounds per foot this must be converted to kilos. One pound is 0.4536 kilo, so 105.5 is multiplied by 0.4536 to equal 47.855 kilos per foot. But the result is still not expressed in full metric units, kilos per meter. There are 3.2808 feet in one meter. So, 47.855 must be multiplied by 3.2808, which equals 157 kilograms per meter. In effect to convert pounds per foot to kilos per meter the pounds have been multiplied by 0.4536 and the result has been multiplied by 3.2808.

The short conversion method establishes a factor which can be multiplied by any number of pounds per foot and provide a kilo per meter answer. This factor would be obtained by multiplying 0.4536 by 3.2808 which equals 1.4881708. The last three digits are dropped and the factor is 1.4882. To check this factor the member we had to convert to metric units weighed 105.5 pounds per foot, so 105.5 times 1.4882 equals 157.00. *To convert any pounds per foot to kilos per meter multiply by 1.4882.*

The 105.5 pound member was 10.25 feet long or 3.124 meters long. So 157 times 3.124 equals 490.468 kilos. To check this multiply 105.5 pounds times 10.25 feet which produces 1081.375 pounds. To convert pounds to kilograms multiply by 0.4536, so 1081.4 x 0.4536 equals 490.5 kilos, close enough to 490.468.

Conversion Problems

Convert the following English weight units to metric units. All answers must be in kilograms.

1. What is the metric weight for a piece of steel, W12 x 133, 87′-4″ long?
2. What is the metric weight for an S12 x 40.8 that is 34′-8″ long?
3. What is the metric weight for a C8 x 18.75 that is 126′-6″ long?
4. What is the metric weight for an L2 x 2 x ⅜ that is 17′-2″ long?
5. What is the metric weight for an L5 x 3½ x ½ that is 19′-4″ long?
6. What is the metric weight of a piece of steel plate that is ½″ thick, 40″ wide and 14′-6″ long?
7. What is the total metric weight of a steel structure that has 42′-6″ of W6 x 15.5, 72′-3″ of C3 x 6 and 12½ feet of L1¾ x 1¾ x ¼?
8. What is the total metric weight of a steel structure that has 112′-9″ of W14 x 150, 92.4′ of W6 x 12, 80½ feet of C4 x 7.25 and 20′-6″ of L1¼ x 1¼ x ¼?
9. What is the total metric weight of a steel structure that has 246′-8″ of C6 x 10.5, 88.25 feet of L6 x 4 x ½, and 128.85 sq. ft. of ¼″ plate?
10. What is the total metric weight of a steel structure that has 48¼′ of C8 x 13.75, 90′ of L5 x 3 x ½, 17′-4″ of bar 1 ⌗, 22′ of bar ½ ϕ and 408 sq. ft. of ¼″ plate?

Gage Lines

Any steel shape has standard *gage lines,* the lines where bolt holes are placed. The distance between gage lines or from gage line to back of angle, channel or outside face of flange is called *gage.* The AISC establishes the gage for all steel members. Tables 11-2 through 11-6 show member gage lines. The first bolt hole is placed on the web gage line, and subsequent bolt holes spaced 3″ from it and 3″ apart. Flange bolt holes are placed on the flange gage lines with subsequent holes spaced from 2¼ to 3¼″ on ¼″ increments. Gage lines for angle members are shown in Figure 11-4.

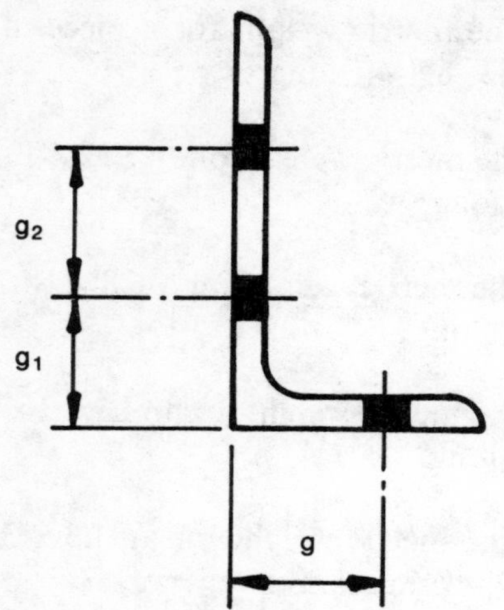

LEG SIZE, INCHES	8	7	6	5	4	3½	3	2½	2	1¾	1½	1⅜	1¼	1
g	4½	4	3½	3	2½	2	1¾	1⅜	1⅛	1	⅞	⅞	¾	⅝
g1	3	2½	2¼	2										
g2	3	3	2½	1¾										

Figure 11-4. Angle gage lines.

Drawing Steel Members

Design arrangement plan and elevation drawings use ⅜″ = 1′-0″ scale, if the overall dimensions allow it to be shown on a single drawing. For extremely large structures, arrangement drawings will use ¼″ = 1′-0″ scale, resulting in still smaller details. Because of these small scales it is impossible to draw members as double line showing flange and web thicknesses as they actually are. Current drawing methods call for these arrangement drawings to be drawn single line where all steel members are represented with a heavy single line. Usually some place on the heavy single line the member will be pictured double line to aid in angle or flange orientation. The structural steel detailer will prepare the detail drawings in full double line, supplying all details, to a much larger scale, usually 1″ = 1′-0″.

In single line drafting the single heavy line always represents the members *working line,* the line where locating dimensions are given on plan drawings, and elevations are given on elevation and sectional drawings. This means that the heavy single line indicates different parts of members on plans and elevations.

On plans the heavy single line represents:

1. Centerline of beam.
2. Back of channel.
3. Workline of angle or structure tee bracing which will be the gage line for bolts or gravity line for welding.
4. Back of angle when it is used as a beam or column.
5. Column end view is often shown single line.

On elevations the heavy single line represents:

1. Top of beam or channel (T.O.S.).
2. Work line of angle or structure tee bracing which will be the gage line for bolts or gravity line for welding.
3. Back of angle when it is used as a beam or column.
4. Centerline of wide flange columns.

Because there are many cases where the single heavy line may confuse or mislead the person reading the drawing, a short portion should be shown double line on each run at least once for like members. Figure 11-5 shows how to picture double line portions on single line drawings.

The orientation of angle bracing or struts of unequal legs is indicated by abbreviation LLV (long leg vertical) or by dimension. Figure 11-6 shows how this is done on a steel drawing.

Figure 11-7 shows how to draw steel members both double line and single line for plan drawings. Figure 11-8 provides the views when drawn on elevation or sectional drawings.

Steel Welding Symbols

Most structural steel welded joints utilize the fillet weld (see Chapter 6), and weld symbols are shown on the steel drawings. Weld sizes are calculated to withstand the required forces. Figure 11-9 pictures common fillet welds used on steel drawings and their proper weld symbol. Figure 11-10 shows bevel weld symbols.

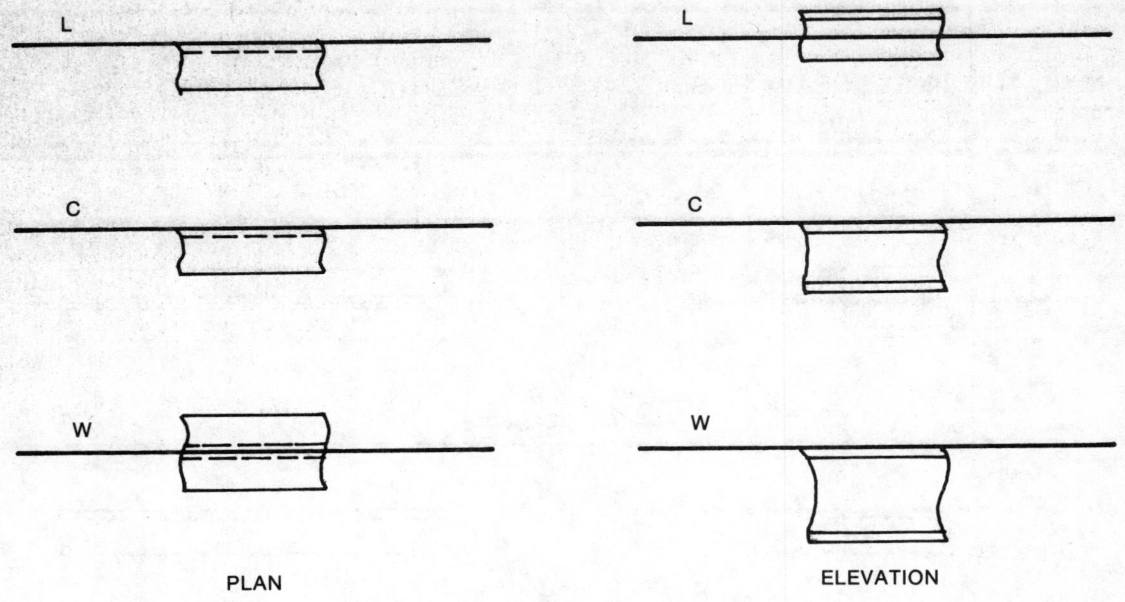

Figure 11-5. Single-line drawing with portions shown double-line.

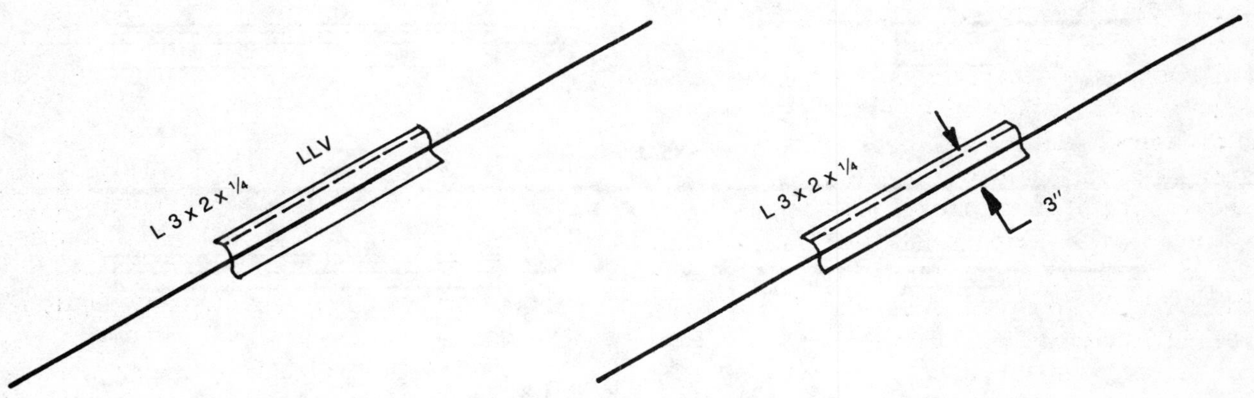

Figure 11-6. Unequal angle leg orientation methods.

Ladder and Platform Drawings

Walking Surfaces

Steel walking surfaces are eithert floor plate or grating. Grating is usually 1¼″ x 3⁄16″ in size, serrated and of the welded type. All drawings show the proper symbol for floor plate or grating per Figure 11-11. The line with arrowheads indicates the direction of bearing bars and is called a *span* symbol. Floor plate is usually tack-welded by the field to the supporting member to hold it in place. Grating is usually secured by saddle clip anchors so pieces may be removed.

Handrail

Handrail is made from structural steel and has four nomenclatures; top rail, midrail, toe plate and post. Figure 11-12 pictures these components. Posts are bolted to the supporting member, and the other components are welded to the post, unless the steel is galvanized. Welding is avoided on galvanized surfaces because it burns away the zinc protective coating. Bolted joints are used for galvanized steel.

(text continued on page 169)

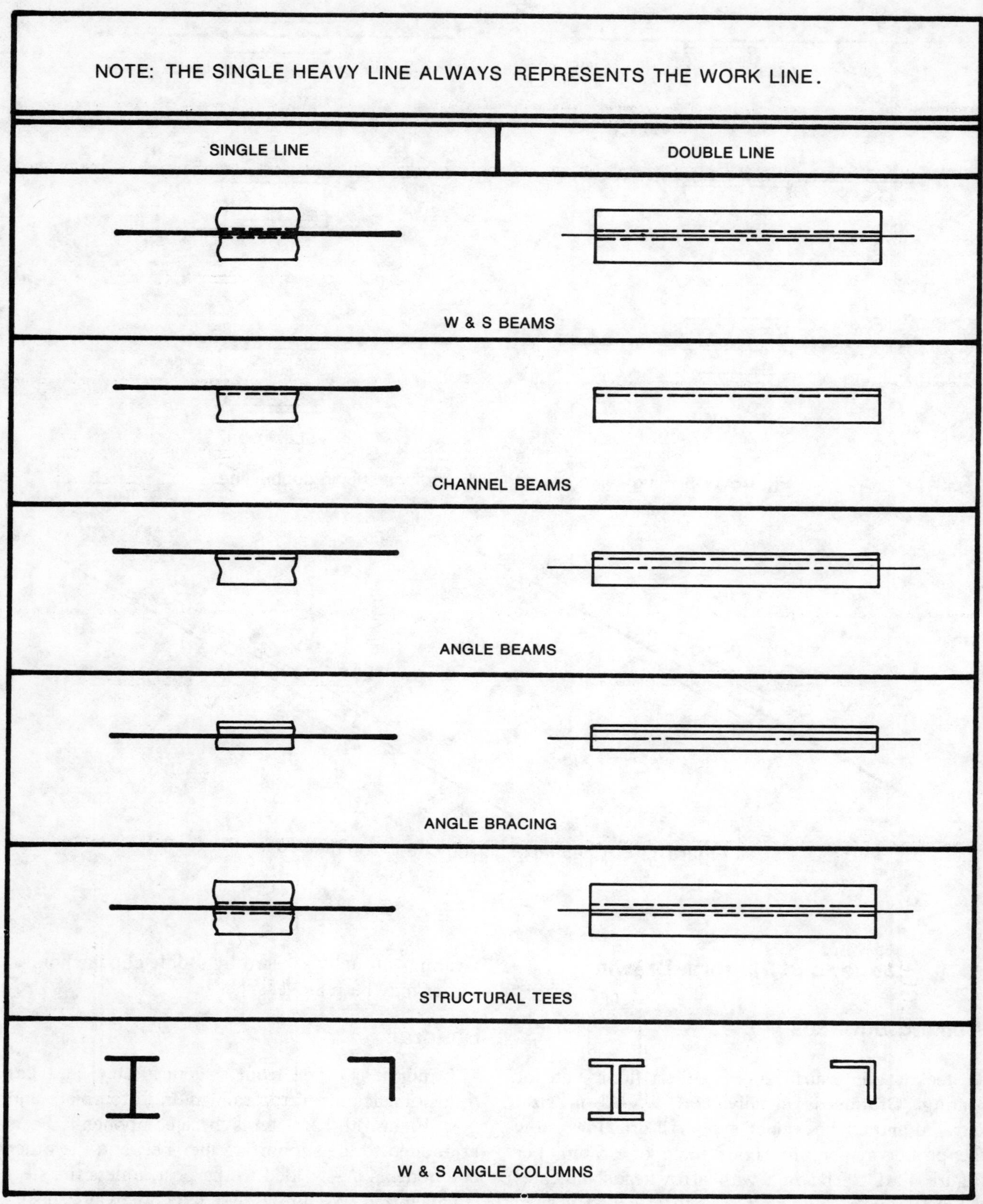

Figure 11-7. Steel member representations for plan drawings.

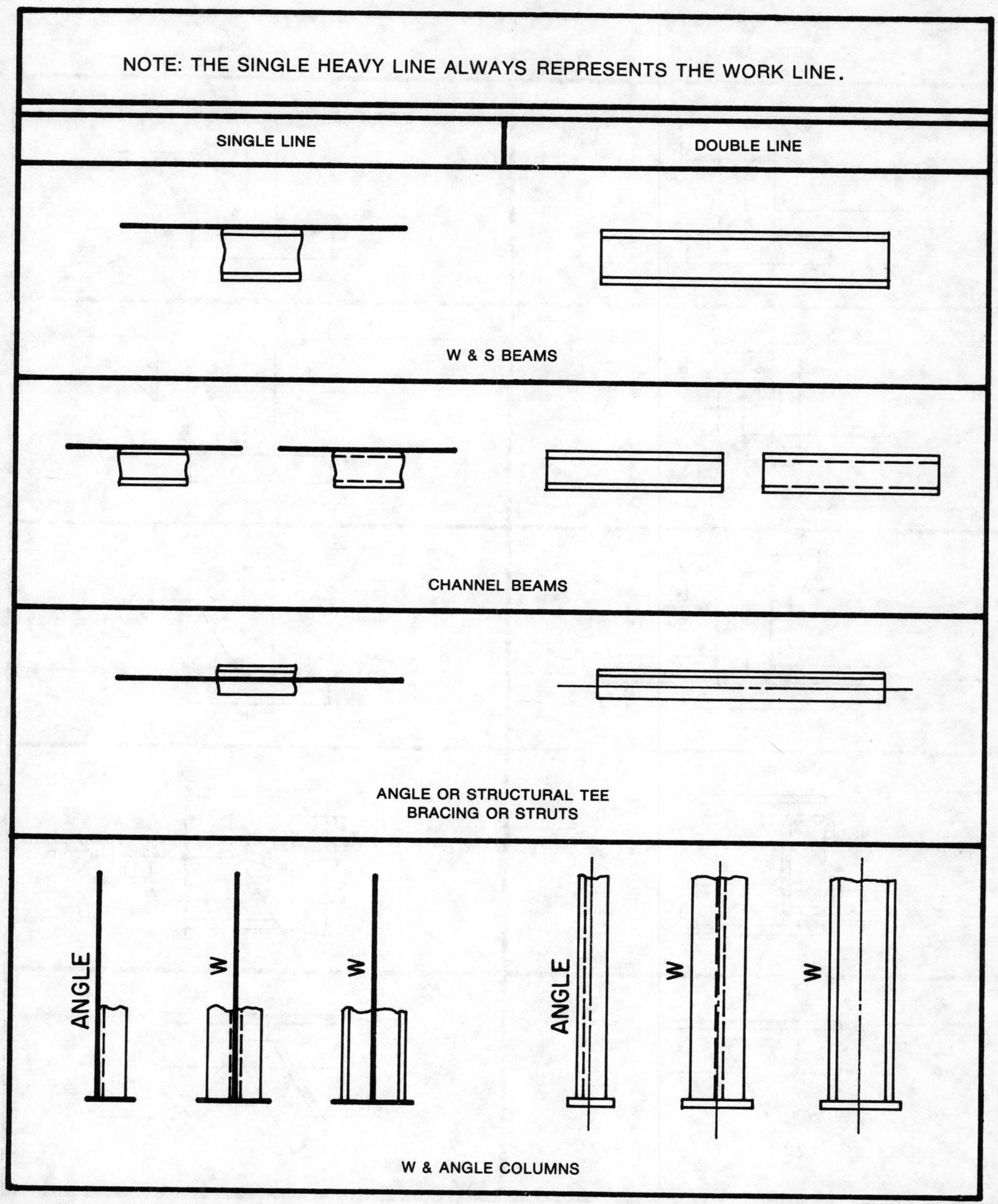

Figure 11-8. Steel member representations for elevation drawings.

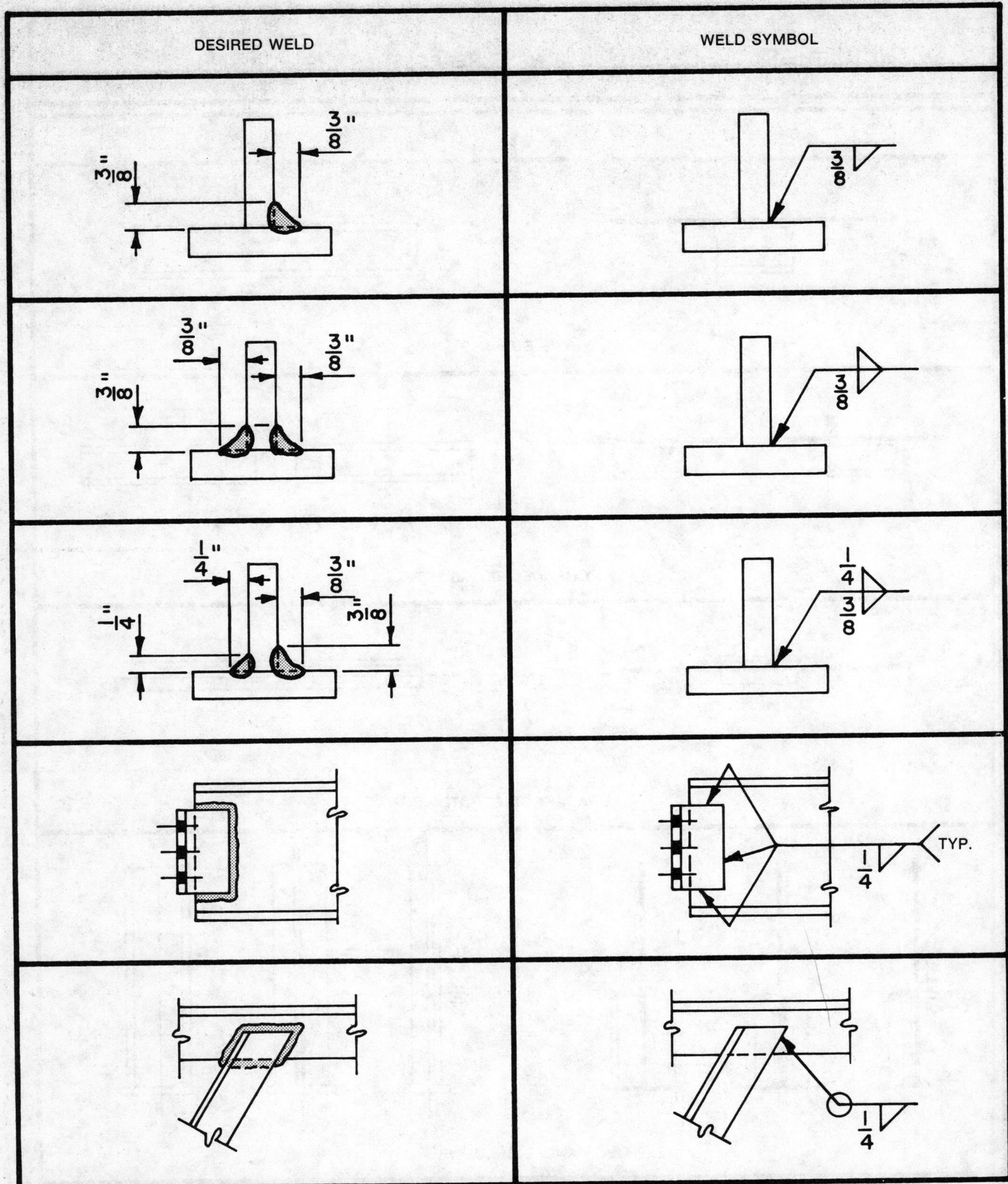

Figure 11-9. Fillet welds and weld symbols.

figure continued on next page

Figure 11-9 continued

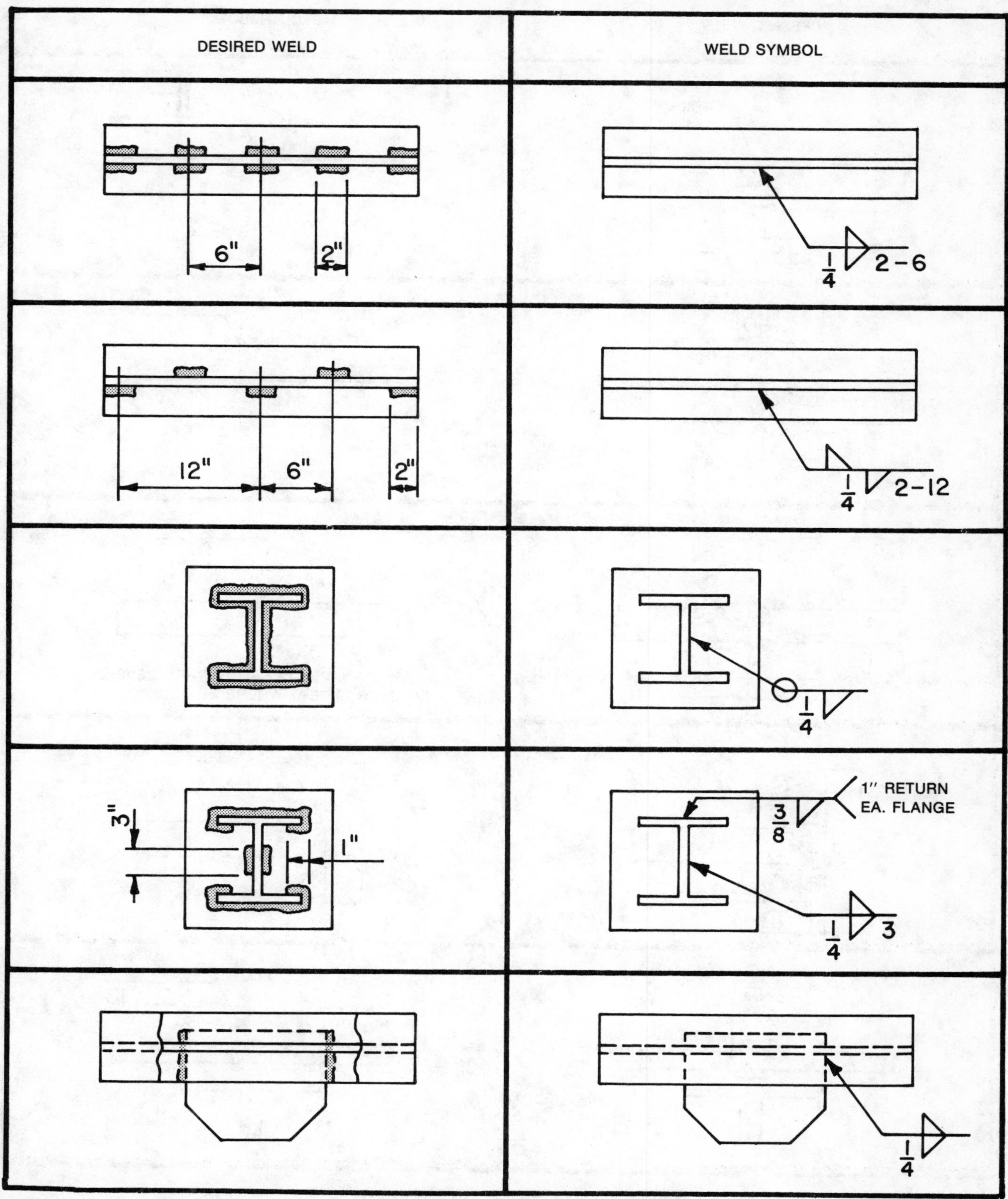

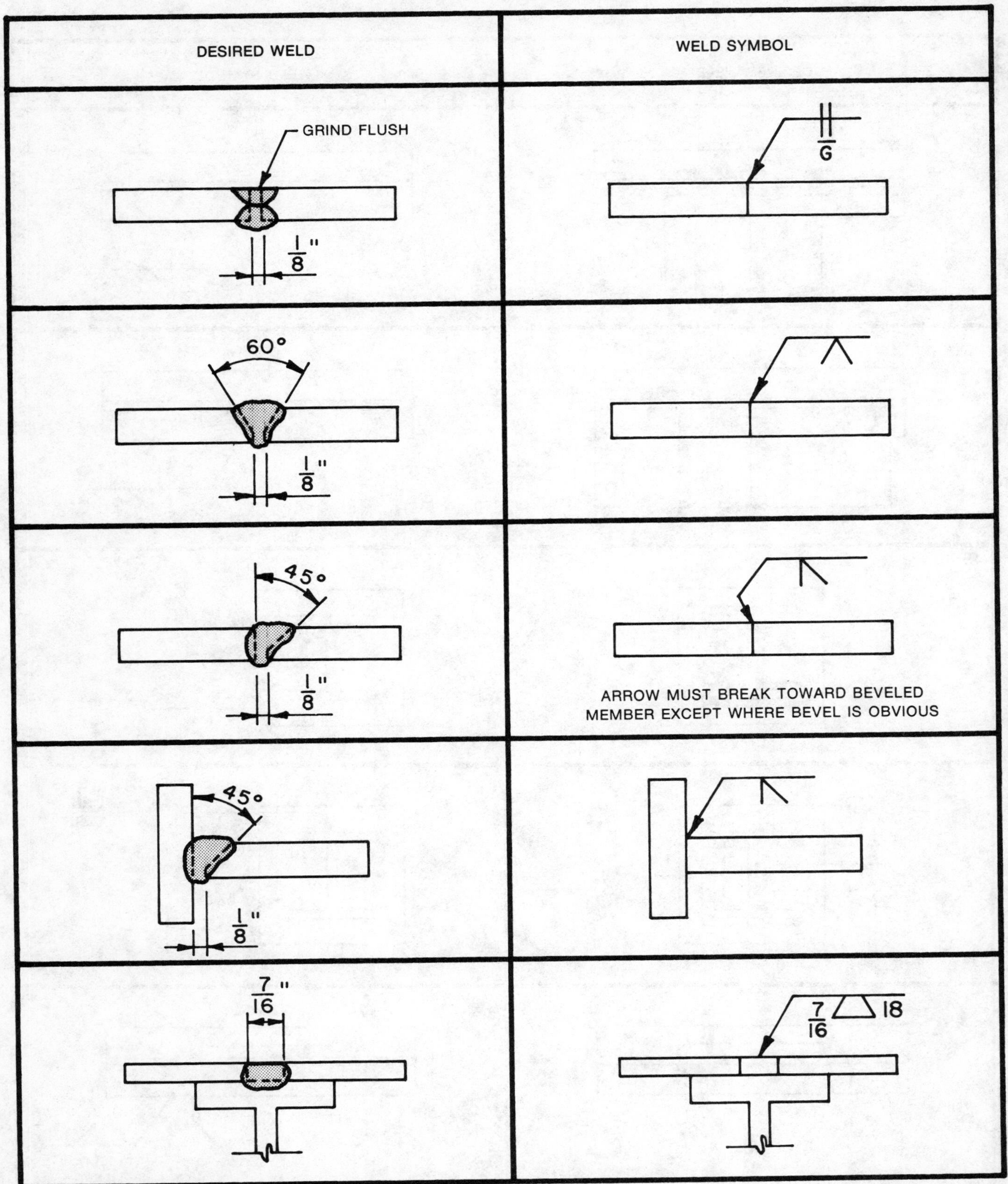

Figure 11-10. Bevel welding symbols.

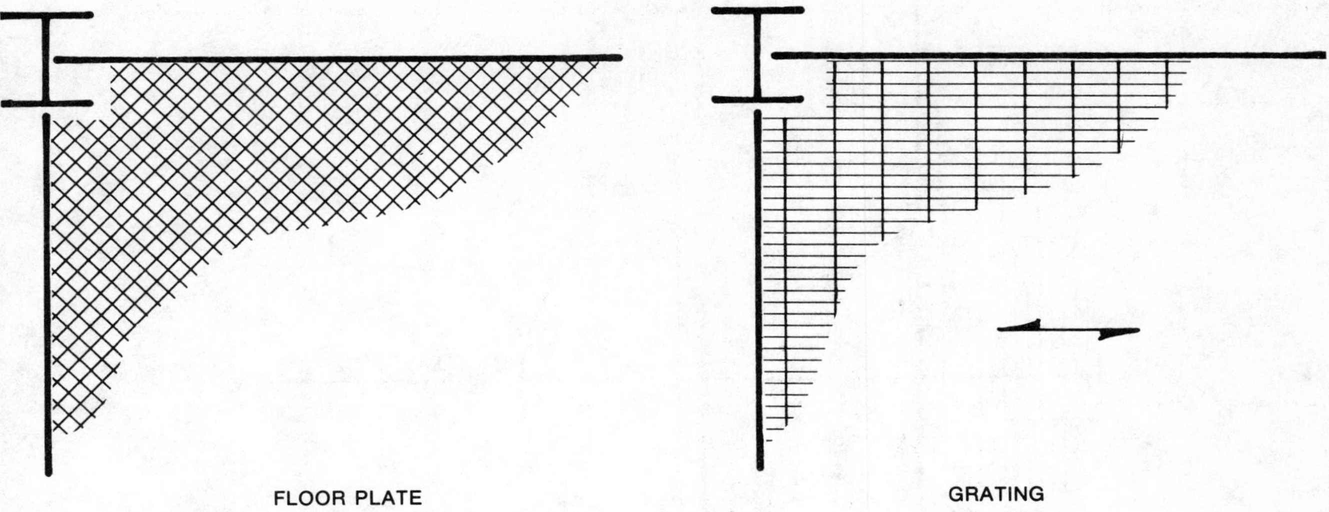

Figure 11-11. Floor plate and grating symbols.

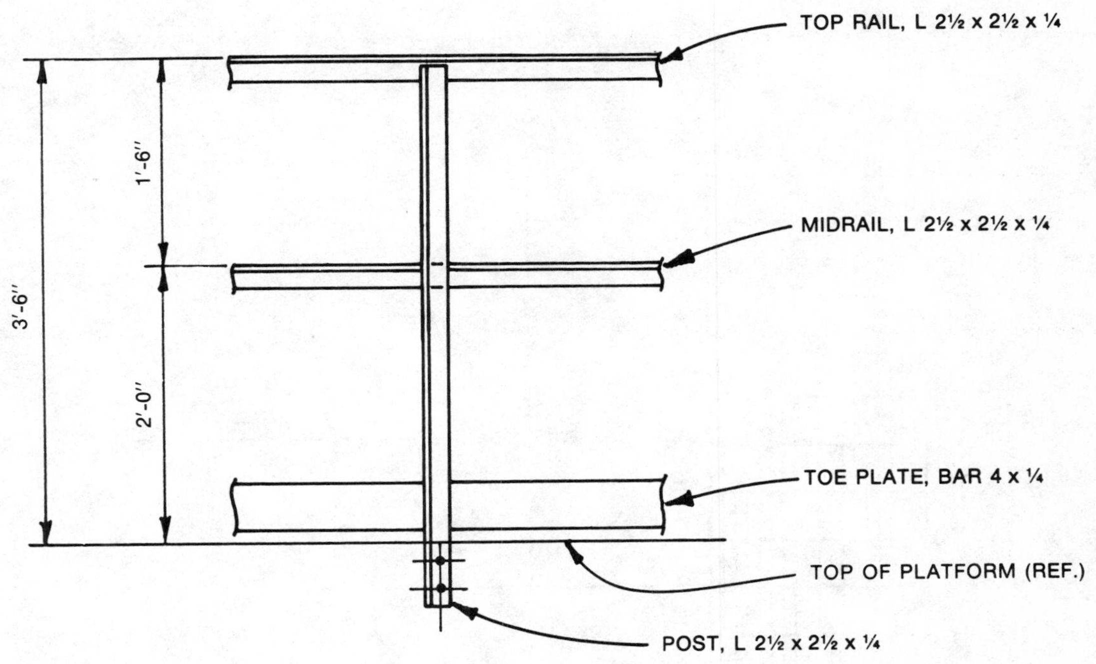

Figure 11-12. Handrail components.

(text continued from page 163)

Ladders

Access ladders are of two basic kinds, the front and the side approach. Figure 11-13 pictures these two ladders. On design drawings the ladders are shown on plans, and the centerlines are located by dimension. Rung spacing, elevation of ladder bottom and support clip locations are shown in the elevation view. Other details are shown on a typical detail drawing as shown in Figure 11-14.

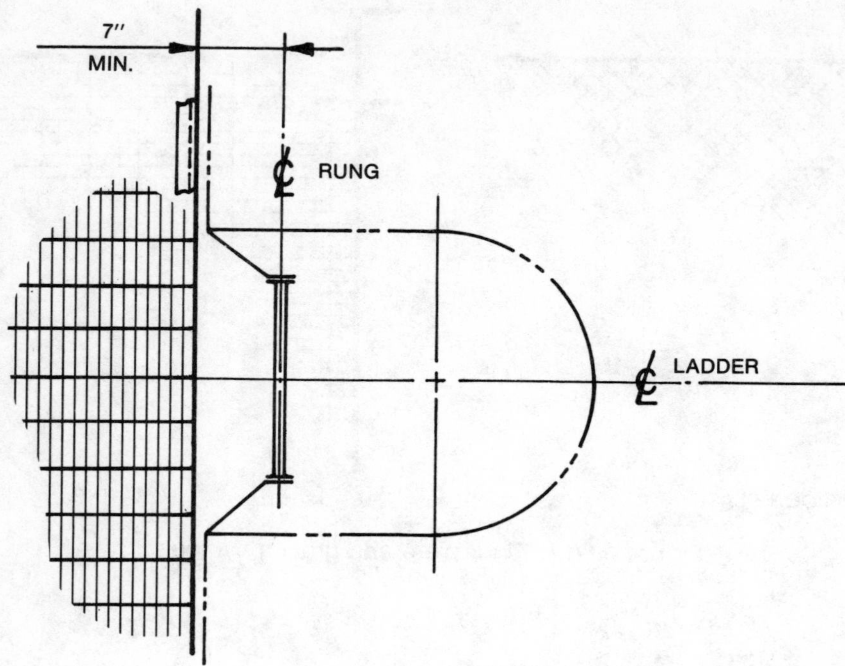

FRONT APPROACH LADDER

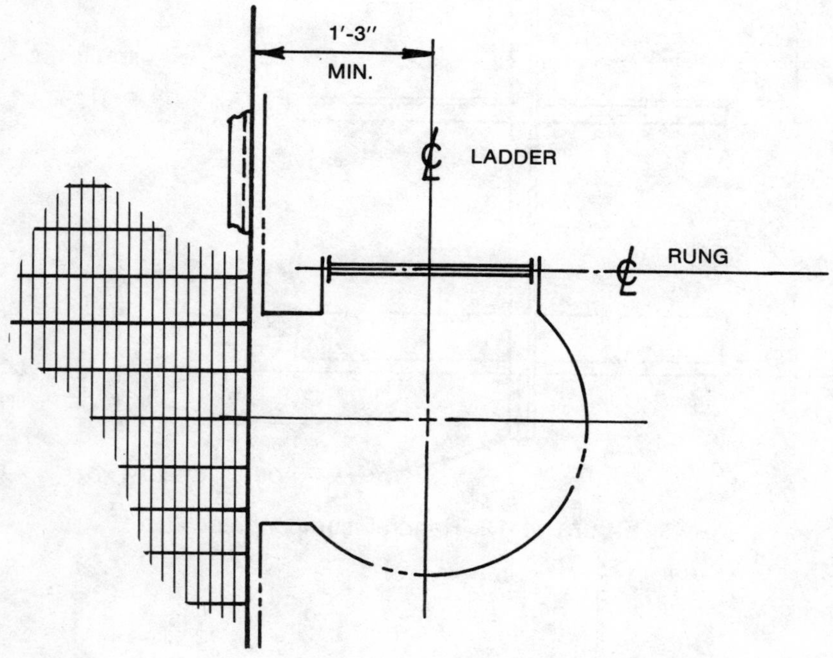

SIDE APPROACH LADDER

Figure 11-13. Front and side approach ladders.

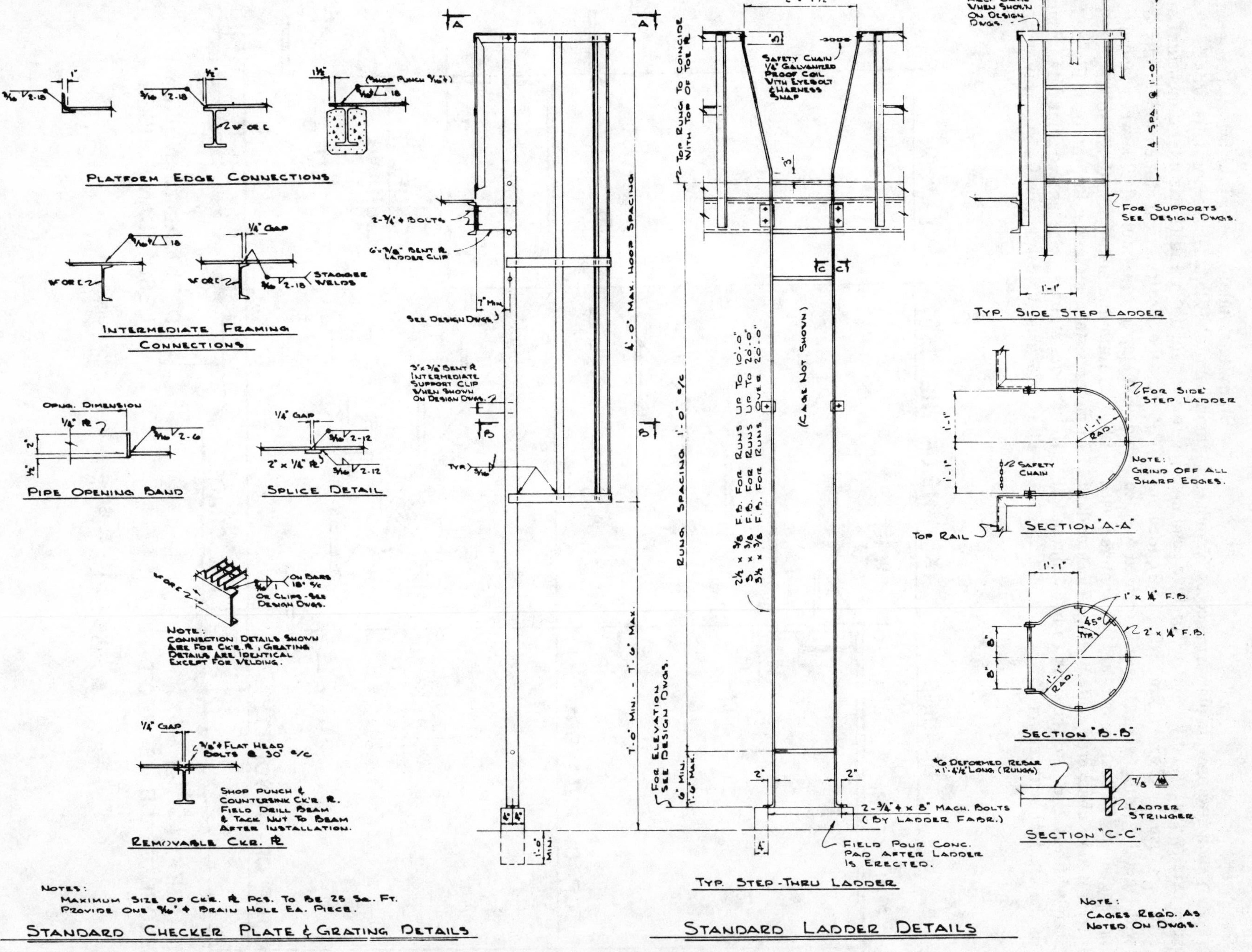

Figure 11-14. Typical ladder detail drawing.

Stairways

Stairways also provide platform access. Standard steel stairways usually consist of two channel stringers (usually C8 x 11.5 or C10 x 15.3), toed out with metal grating stair treads (serrated welded grating, 1″ x ⅛″ bearing bars at 13⁄16″ centers), with perforated or checkered plate nosing. The standard stair width is 2′-6″, so the back to back dimension of the channel stringers is 2′-6″.

Stairway standard rise and run is based on a maximum sum of 17½″. This is normally broken down into 9¾″ for the tread (horizontal dimension) and 7¾″ for the rise (vertical dimension or distance between steps). Maximum rise between landings is 12′-0″. Minimum landing length is 3′-0″. Figure 11-15 pictures stairway layout information. Figure 11-16 provides stairway clearance design data. Figure 11-17 provides typical details for stairway design.

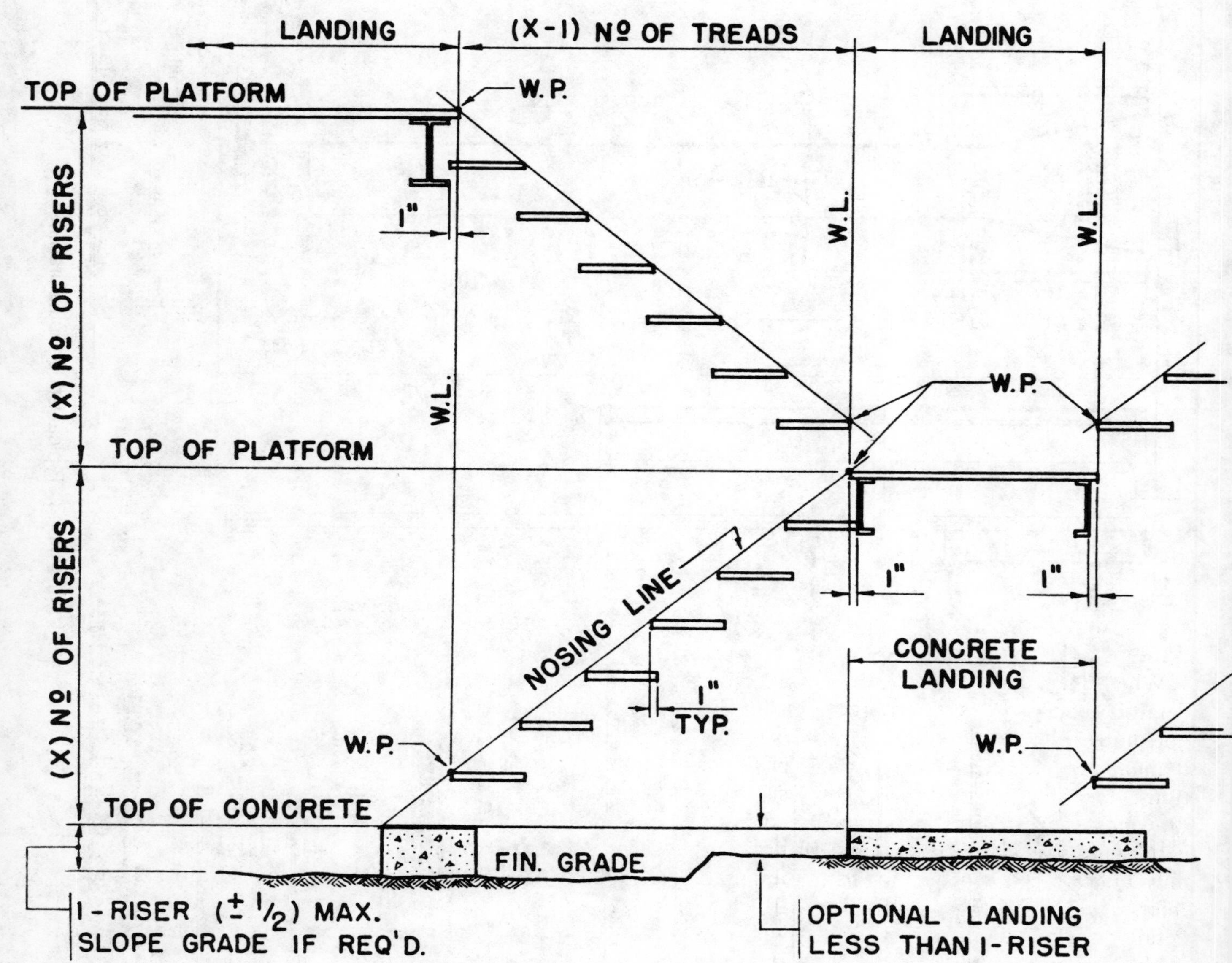

Figure 11-15. Stairway layout information.

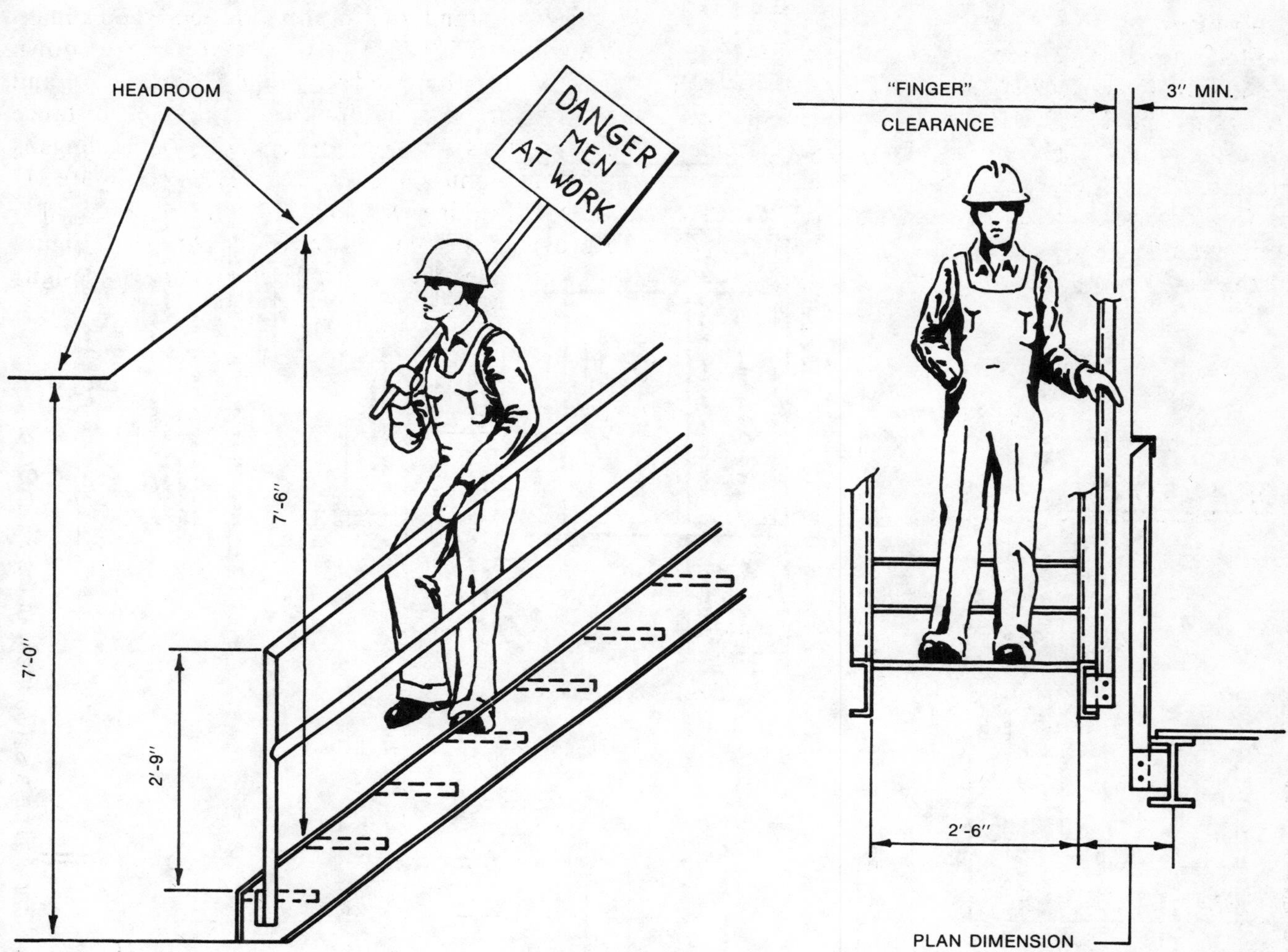

Figure 11-16. Stairway clearance design data.

Connection Details

When two steel members meet, a connection detail is required. Design arrangement drawings do not show this detail every time it occurs. The AISC manual provides many details for typical connections. The design contractor often issues a set of drawings showing typical connection details used with their drawings. The arrangement drawings will only indicate a number or letter at the joint. This refers the steel detailer to the typical connection detail drawing, which will show the joint in full detail. Figure 11-18 is an example of typical joint details for bolted bracing. Note that bolts are only indicated, not dimensioned. Using standard gage data, the detailer will assign bolt locations. Figure 11-19 explains how arrangement drawings specify bolt quantities.

Figure 11-20 is an example of how an engineering contractor's column and beam connection detail drawing might appear. Note the base plate detail; only bolt holes are shown. Bolt sizes appear on the concrete drawing, but the experienced detailer knows that base plate holes are punched 3/16" larger than bolt size so bolt size can be figured.

When the arrangement drawings do not refer to a typical connection detail, it becomes the steel

(text continued on page 175)

Stair Dimensions				
Bevel	Rise	Tread	A	B
9 : 12	7½"	10"	1'-1 3/16"	1'-3 9/16"
9 9/16 : 12	7¾"	9¾"	1'-0 9/16"	1'-2 9/16"
10 1/8 : 12	8"	9½"	11 7/8"	1'-2 1/8"
10 11/16 : 12	8¼"	9¼"	11 3/16"	1'-1 7/16"
11 5/16 : 12	8½"	9"	10 9/16"	1'-0 13/16"
12 : 12	8¾"	8¾"	10"	1'-0¼"

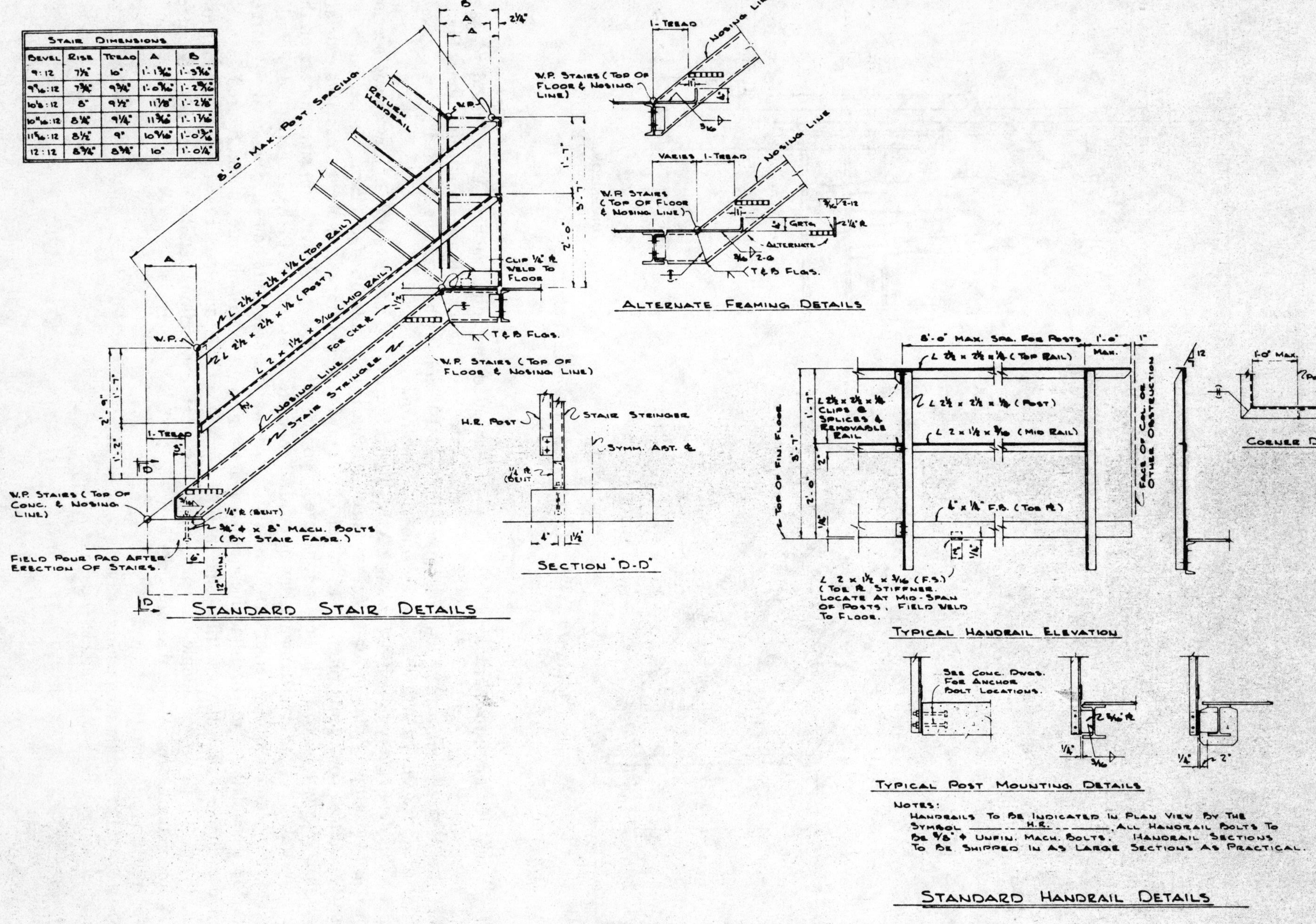

STANDARD STAIR DETAILS

SECTION "D-D"

ALTERNATE FRAMING DETAILS

TYPICAL HANDRAIL ELEVATION

CORNER DET.

TYPICAL POST MOUNTING DETAILS

NOTES:
HANDRAILS TO BE INDICATED IN PLAN VIEW BY THE SYMBOL ——— H.R. ———. ALL HANDRAIL BOLTS TO BE 5/8" ⌀ UNFIN. MACH. BOLTS. HANDRAIL SECTIONS TO BE SHIPPED IN AS LARGE SECTIONS AS PRACTICAL.

STANDARD HANDRAIL DETAILS

Figure 11-17. Standard stair and handrail details. (Courtesy of the Fluor Corp.)

detailers responsibility to select the proper joint detail from the AISC manual to withstand the proper loading. His responsibility doesn't stop there. The joint selected must be the most economical to do the job. Anyone can select the strongest possible joint detail, but why use twelve bolts when three will do the job? The detailer doing this will soon be out of a job.

Shop Drawings

Figure 11-21 is a design arrangement drawing issued to a steel fabrication company. The detailer is given the drawing with the instructions to make the shop details. To do this the detailer must:

1. Prepare orthographic plan and elevation drawing(s) showing all piece mark numbers.
2. Make necessary detail drawings for shop to fabricate.
3. Supply bill of material for all pieces including bolt count. Increase bolt quantity 15% to allow for field loss.
4. Supply total weight for each piece and total shipping weight for the job.

To analyze the problem the detailer reviews the isometric arrangement drawing. The easiest thing to start with is P.S.#5, a free-standing pipe support. P.S.#5 can be shipped in one piece so joint selection is Detail 13, Figure 11-20. Since the columns are W10, Figure 11-10 also supplies the base plate dimensions of 1′-5″ square, ⅞″ thick, 4 holes, 1 7/16″ diameter on 1′-1″ square centers and the base plate is welded to the column with ⅜″ fillet welds all around. The horizontal beam overhangs the column centerline by 2′-0″. Table 11-2 supplies dimensions for W10 x 33. Since this is so simple, the detailer might let his assistant prepare this detail. The student can do the above steps 2 and 4, since there are no bolts here.

While the detailer has his assistant working on P.S.#5, attention can be given to the other four pipe supports. Each P.S. can be shipped in one piece, but struts at elevation 18′-0″ and angle K-bracing will have to be shipped separately for field bolting. This means Figure 11-20, Detail 13 will be used for beam to column connections and Detail 14 will be rotated 90° and used for strut to column connections and for the K-bracing. In Figure 11-18, Detail 7 will be used at the strut and Detail 3 will be used at the base connections. A quick reference to Figure 11-19 shows proper number of bolts. Since no bolt size is specified on the isometric, ¾″ diameter high strength bolts will be used. Weld length of gusset is to be 6″, each side.

Now piece marks must be assigned. Since this drawing is of a pipe support and the drawing number is 11-21 (the figure number) all piece marks will be prefixed with PS-21-21. Make P.S.#5 piece number 1, so its mark number will be PS-21-21-1. To keep shop pieces from getting mixed with the wrong order, all pieces are prefixed with the sales order number, in this case 1234. So the full mark number for P.S.#5 becomes 1234-PS-21-21-1.

Now the detailer makes a piece mark index listing all items to be shipped, which will be issued to the shop and becomes the shipping list. A typical index for job 1234 would be:

Piece Mk	Description	Weight
1234-PS-21-21 -1	P.S.#5 bent	
" " " " -2	P.S.#1 bent	
" " " " -3	P.S.#2 bent	
" " " " -4	P.S.#3 bent	
" " " " -5	P.S.#4 bent	
" " " " -6	Strut between P.S.#1 and #2 (2 req'd)	
" " " " -7	Strut between P.S.#2 and #3 (2 req'd)	
" " " " -8	Strut between P.S.#3 and #4 (2 req'd)	
" " " " -9	L6 x 6 x ½	
" " " " -10	"	
" " " " -11	"	
" " " " -12	"	
――	¾″ Erection bolts	

After the detailer finishes the index the assistant detailer can be given the rest of this job. Students should prepare all necessary plans, elevations, shop details and other items noted above. Do a complete job and turn in all work for checking and grading.

(text continued on page 181)

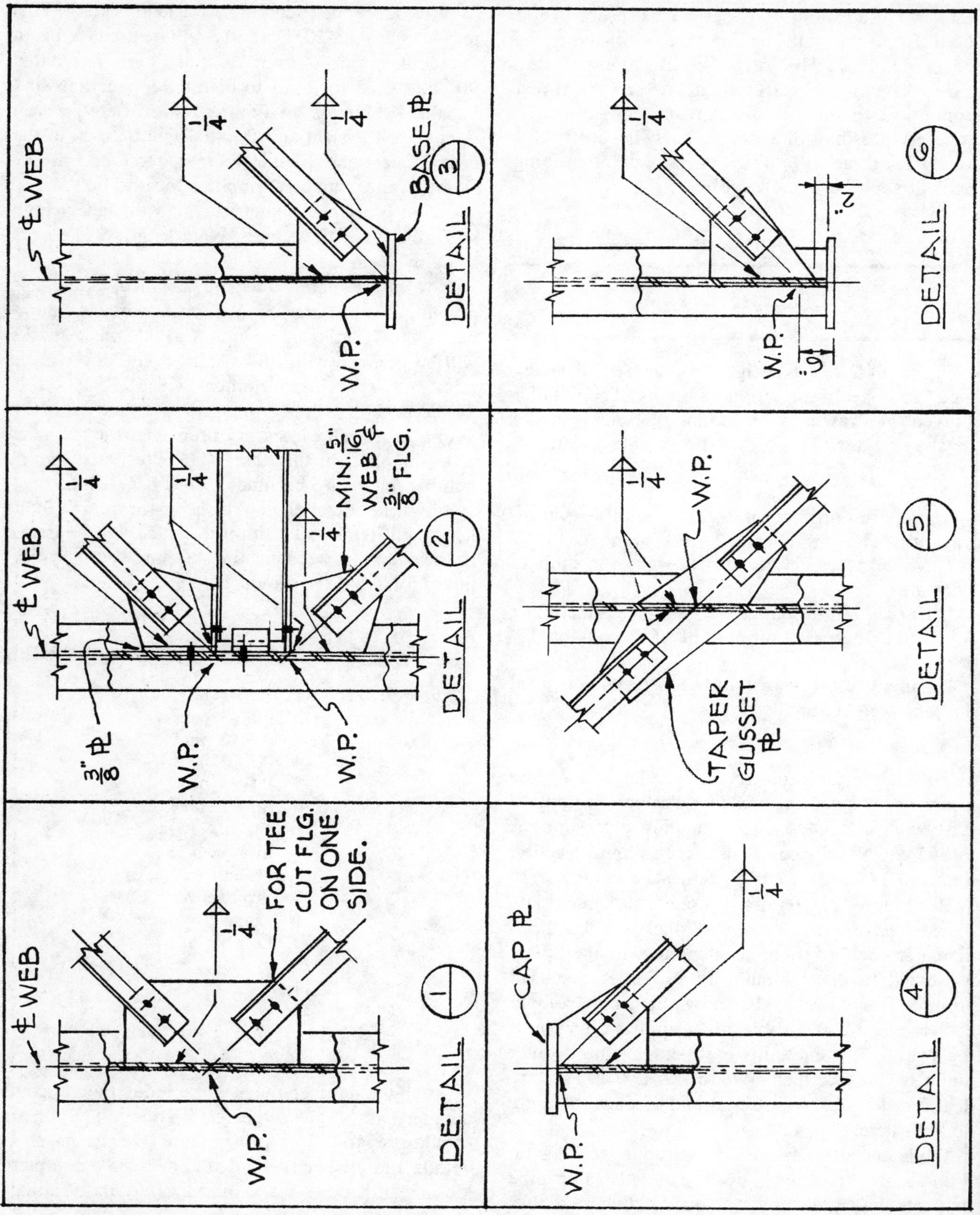

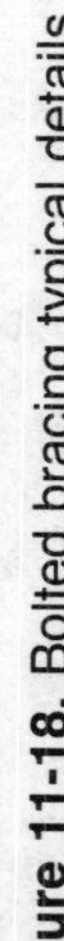

Figure 11-18. Bolted bracing typical details.

figure continued on next page

Figure 11-18 continued

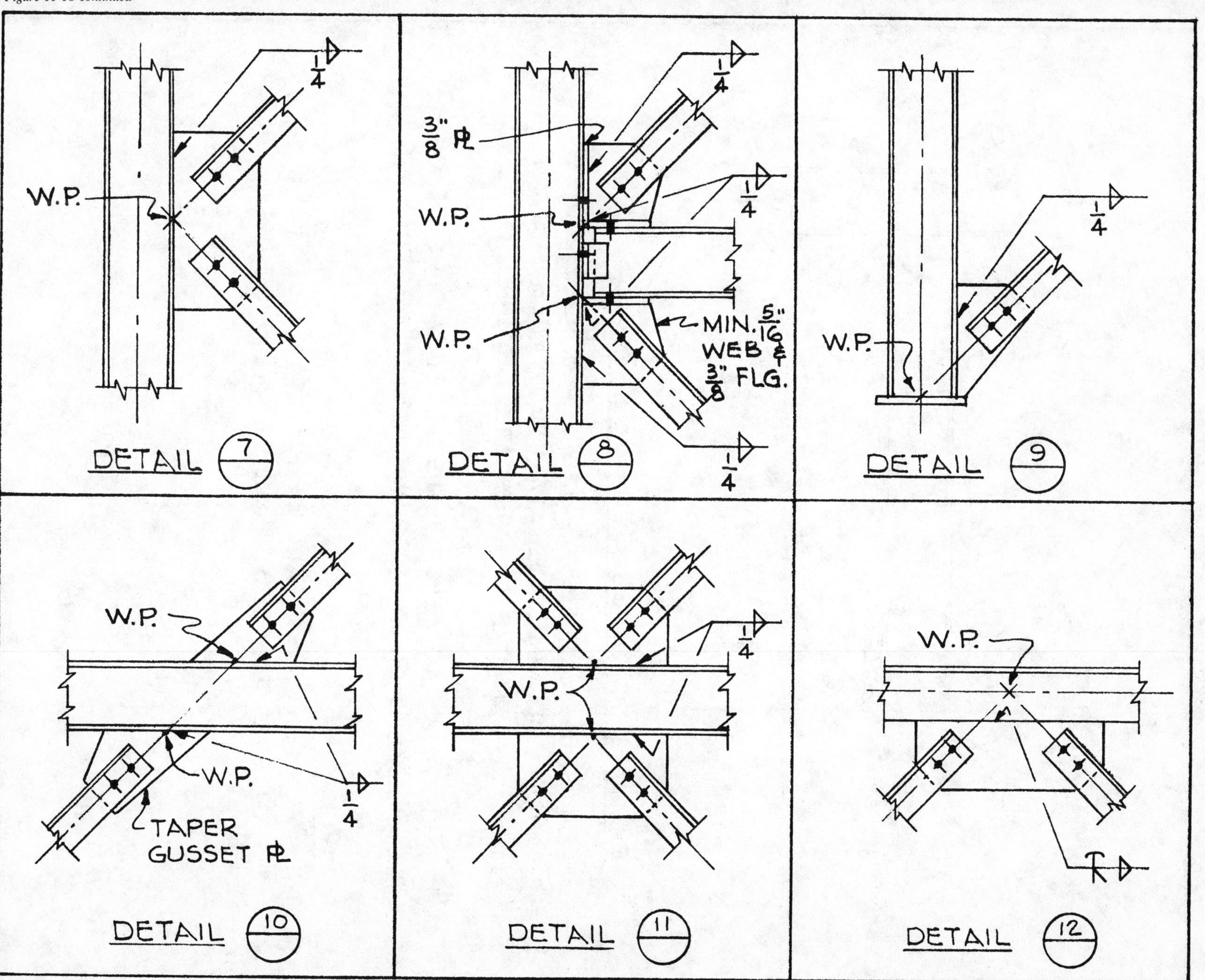

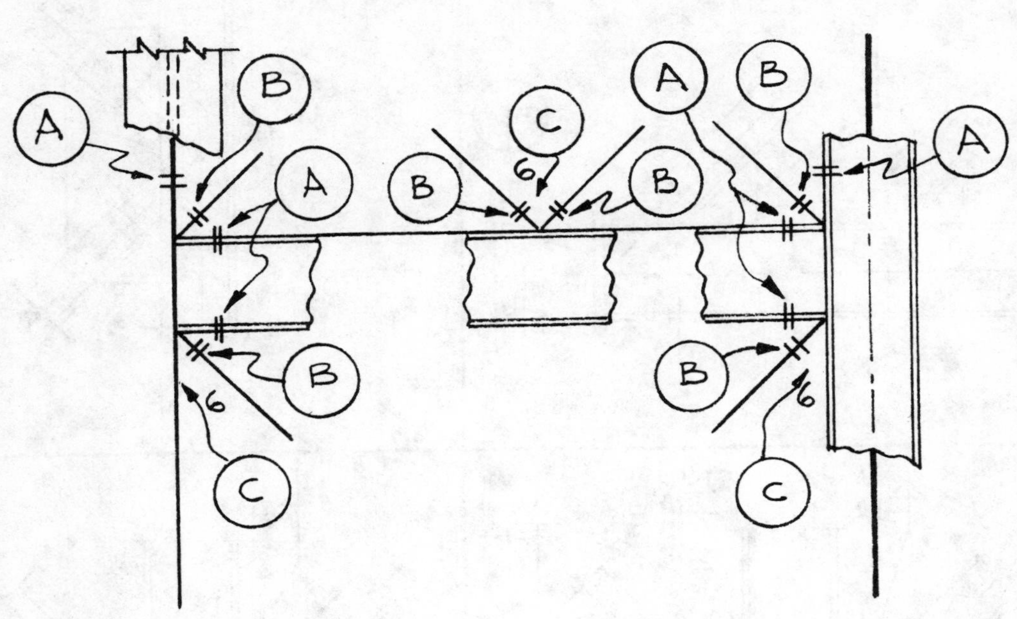

DESIGN DRAWING SYMBOL LEGEND

(A) – NUMBER OF BOLT PAIRS IN FLANGE OR WEB OF COLUMN OR BEAM. TWO PAIRS SHOWN.

(B) – TOTAL NUMBER OF BOLTS CONNECTING BRACING TO GUSSET PLATE. TWO SHOWN.

(C) – MINIMUM WELD LENGTH ON EACH SIDE OF GUSSET PLATE. SHOWN IS 6" WELD.

NOTES:

1. ALL BOLTS SHALL BE H.S. 3/4" φ MINIMUM
2. GUSSET PLATES SHALL BE 5/16" THK. MIN.

Figure 11-19. Design drawing bolted symbols.

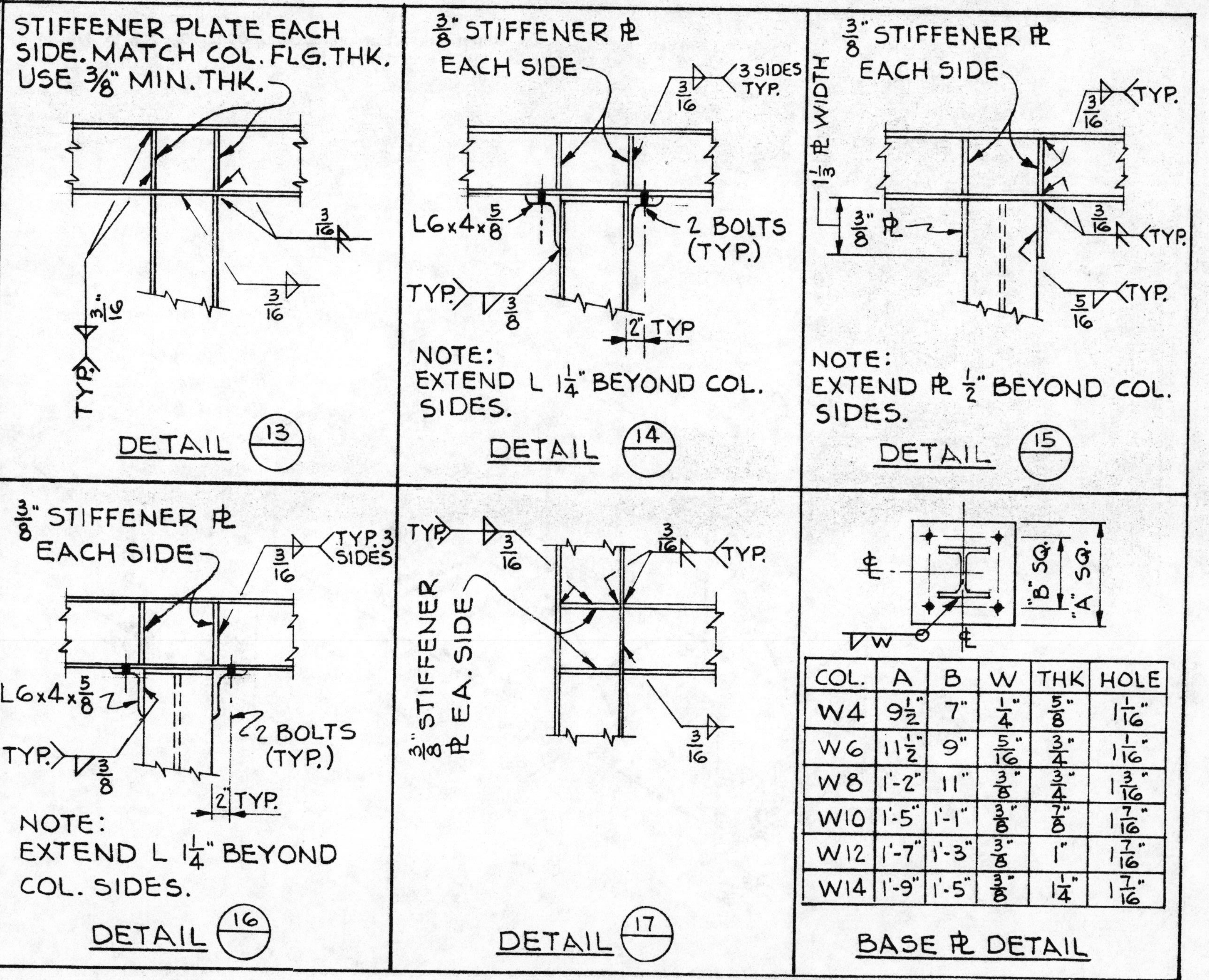

COL.	A	B	W	THK	HOLE
W4	9 1/2"	7"	1/4"	5/8"	1 1/16"
W6	11 1/2"	9"	5/16"	3/4"	1 1/16"
W8	1'-2"	11"	3/8"	3/4"	1 3/16"
W10	1'-5"	1'-1"	3/8"	7/8"	1 7/16"
W12	1'-7"	1'-3"	3/8"	1"	1 7/16"
W14	1'-9"	1'-5"	3/8"	1 1/4"	1 7/16"

Figure 11-20. Column and beam connection details.

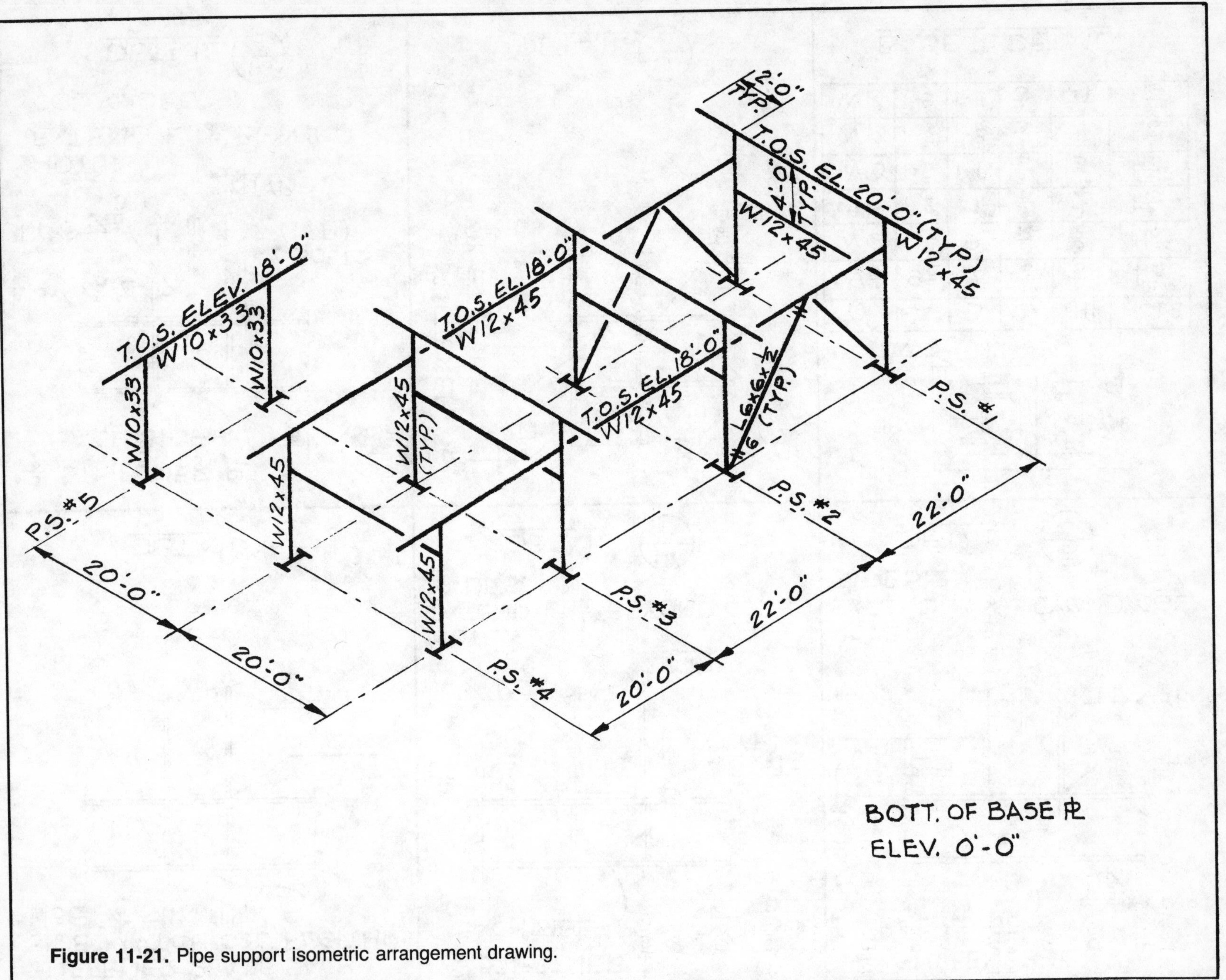

Figure 11-21. Pipe support isometric arrangement drawing.

Review Test

1. Name the two main types of structural steel draftsmen.
2. Define a fabrication subassembly.
3. What is a span symbol?
4. What is the metric weight for 50′ of W10 x 33?
5. What ASTM material specification is most commonly used for steel shapes?
6. Define AISC.
7. Give three uses for angle shapes.
8. Define strut.
9. Define bent.
10. What determines structural shape shipping length?
11. What is a kip?
12. Define cantilever beam.
13. Define dead load.
14. Define moment.
15. What is the total metric weight for P.S.#5 in Figure 11-21?
16. Define gage line.
17. Structural steel arrangement drawings usually use ________ = 1′-0″ scale.
18. Structural steel detail drawings usually are drawn to ________ = 1′-0″ scale.
19. What does LLV mean?
20. What is the normal floor grating size?

CHAPTER 12

Piping Drawings

The student is about to meet a completely new language. The piping draftsman must know the terms of the business. The student has heard some of these; many will be strange. But, even if he has heard the term, does he know what it means? First, the student will investigate some of these new terms. He must learn them well, for a professional knows his business; and this is the language of the professional piping draftsman.

Process Plant Terms

Refinery

A refinery is a plant that takes crude oil as its "feed" or "charge" stock and converts it into the many by-products that people use. Some of these are gasoline, jet fuel, kerosene, butane, propane, fuel oil and asphalt.

Gasoline Plant

The gasoline plant takes natural gas (a vapor) as its charge stock and separates the vapor's heavier products out and reinjects the lighter gas (methane) into a pipeline or perhaps into the gas field it came from. Again gasoline, propane and butane are extracted as products. But, since a gasoline plant starts with a vapor, the heavier hydrocarbons do not exist in its charge stock; so heavier products cannot be made. Asphalt is one of the products that is classified as a heavy hydrocarbon and is not produced in a gasoline plant.

Hydrocarbon

The hydrocarbon compound contains hydrogen and carbon. Hydrocarbon compounds are numerous and form the basis for petroleum products. They exist mostly as vapors and liquids but may also be solid. In general, piping systems in refineries and gasoline plants transport hydrocarbons or utilities.

Chemical Plant

The chemical plant takes semirefined products from refineries and gasoline plants and—by running them through their units, sometimes blending in other products—converts them into certain chemicals which may be sold as a finished consumer product. One such product widely demanded today is plastic. Chemical plants make many ingredients in modern medicines.

Tank Farm

The tank farm is the area that contains the huge storage tanks of the refinery and gasoline or chemical plants. The tanks are usually isolated from the main processing units in case of fire. They may be 200' or more in diameter and will contain the plant's charge stock for several days. The tanks also will store the plant's products, until the shipment goes to the consumer.

Process Plant Utilities

Utility

The utility is a refinery's service portion. While a home has water, gas and electricity, a refinery or other plant has many more, some of which are below.

Steam. Steam services many plant items. Heat generates steam in fired boilers or heaters, which will make many different steam pressures and temperatures. They apply heat and convert condensate (a pure water) to steam (a vapor). The steam then goes to the different plant units in the piping systems which use the steam.

Many students think they have seen steam, but they haven't. They cannot acutally see steam; it is invisible. What they have seen is the condensate condensing out of the steam. That is where the term "condensate" comes from.

Condensate. As the energy in steam is used, the steam turns to condensate. Another piping system collects this condensate, which is returned under a low pressure to a collection point and is pumped through the boiler tubing and converted to steam again. So the condensate is in a constant cycle from steam to condensate to steam.

Fuel Oil. Fuel oil is another utility that refineries make and partially consume. It is also sold as a product to heat homes and fire furnaces in private business.

Instrument Air. A utility that operates the plant instruments is instrument air. A piping system distributes this air, which has been compressed and dried to remove all its moisture, as the moisture would harm the instruments.

Utility Air. Utility air drives air motors and blows air on objects to clean them, such as some barbers blow cut hair off customers with air hoses.

Cooling Water. Cooling water cools various streams in a plant. The water starts at a cooling tower and is pumped through a piping system to exchangers, which exchange heat. It comes out hotter–much like water from a hot water heater in a home. This water then returns to the cooling tower, which cools the water, and then is ready for more circulation into the unit. Like the steam and condensate system above, this is a constantly circulating system.

Drains. An underground utility collects drains from funnels or catch basins and, in a separate piping system, transports them to a disposal point. Since no pressure is in this drain piping, the pipes must slope to cause flow. This slope is usually 1 foot per 100 feet of line or greater.

It can be very difficult to design drain systems. Since they run underground, they must miss all other underground items. As an example, a $25 million installation will use about 20,000 yards of concrete, most of which will be underground as foundations for the process equipment. The drainage system must twist and turn to miss all the foundations.

Most plants also have more than one drain system. They may have an oily water sewer, a storm water sewer and an acid sewer. The oily water sewer handles the oily drips and drains. The storm water sewer collects surface runoff from rains. The acid sewer collects acid drains and drips. There may be many other types of separate drain systems.

Flare System. The flare system transports vapors (via a piping system) to a flare stack which is very tall and has a flame burning at the top. This system burns waste gases and also collects and burns relief valve discharges. At night the flare stack usually stands out–sending flames high into the air. This is waste gas burning. If it did not burn, it would pollute the air.

These are some basic terms and piping systems you should learn completely. And you will be exposed to many more–beginning with the definition of piping.

Piping and Pipe Sizes

Piping transports a vapor or liquid and some solids. A familiar piping system is the gas and water pipes in a home. These are sized to flow sufficient products into the home. Most are ½" and ¾" and are usually screwed fittings because the pressures and temperatures are very low. Piping systems for refineries usually are 1"-24" with some special systems measuring several feet in diameter. A person could easily walk in some. Pressures and temperatures are very high, so these pipes and fittings are welded and not screwed.

Pipe sizes are calculated to flow a certain product at a set quantity at its pressure and temperature. The higher the pressure, the more pipe thickness required. As the system's temperature rises, it not only affects the thickness, but hot objects expand, and expanding pipe creates a force which must be considered.

Pipes are constructed of many materials–most commonly of carbon steel. They may also be of stainless steel, chrome steel, vitrified clay, cast iron, plastic, glass and many other materials.

Process Plant Equipment

It's important to remember the names and functions of equipment to which the piping draftsman will have to connect.

Vessel

A vessel is only a large diameter pipe which may have internals. Some are installed horizontally, like those at a service station. Most vessels there are underground to store regular and premium gasoline. Many are vertical and vary in size and shape. The tall ones, which have "fractionating trays" inside, are *fractionating towers*. They are 100' or more high.

Reactors are vertical vessels or spheres which contain a catalyst. A chemical reaction occurs in this vessel, changing the molecular structure of the fluid going through it.

No chemical change occurs in fractionating towers which separate the various compounds. The separation results from the different boiling points of the different products. The lighter the product, the lower the boiling point. The desired "cut," or product separation, is drawn off generally as a vapor from the top of the fractionating tower. This vapor is then cooled, usually by cooling water, and condensed to a liquid which the overhead accumulator (see below) then retains. The main point to remember is that the fractionating tower "fractionates," or separates, products. No chemical change occurs just like no chemical change occurs when cream is separated from whole milk. In a "reactor" a chemical change, or reaction, *does* occur.

Overhead accumulator is sometimes called "reflux accumulator." This is a horizontal vessel which collects the overhead product from a fractionating tower. It usually operates one-half full of liquid. These vessels usually have little, if any, internals.

Storage Tanks

Storage tanks still fall under the "vessel" category but are not in the process areas. They usually appear as bunches and are called a "tank farm." The tanks run more than 200' in diameter and are 40 to 60' high. They store the crude oil until the process units are ready for it, store all the various products until they are sold or the plant consumes them and also store "rerun" products, which have come from one unit and are held for further refinement in another.

These tanks have many types. Most are flat bottomed with a conical top. Some have a floating top which floats on top of the stored liquid. These tanks are used when the stored liquid has a high evaporation loss.

Most "light ends" products are stored under pressure so they won't evaporate. Some are propane and butane, which are stored in "bullets" or long horizontal vessels. Some of these lighter products are stored in "spheres," which legs support.

Exchangers

The "heat exchanger" gets its name from exchanging heat from one stream to another. Many methods accomplish this. A common exchanger is the car radiator. This heat exchange comes from water radiating heat through the metal of the radiator. Another common exchanger is the home hot water heater. This exchanges heat from the heating medium to the water. In most applications in process units, this exchange occurs between two process streams so that heat is not wasted. Heat is energy, and wasted energy costs money.

Exchangers also differ in size and shape. Most are the "shell and tube" type installed horizontally. Another is the "fin-fan" or air cooler type, which blows air over exposed tubes to cool the fluid, much like the car radiator works. While the car radiator is vertical, the "fin-fan" is usually horizontal. The "double pipe" exchanger is another type. It has pipe inside a larger pipe, transferring heat from one stream to the other stream. In an exchanger the two streams *never* mix. They exchange their heat through a pipe or "tube" just as the car radiator exchanges heat through the radiator. The water doesn't actually contact the air, or a leak would result, losing the water.

Pumps

Pumps increase the pressure of a *liquid* and cause circulation. The heart, for example, is a pump. The liquid comes to the pump at a low pressure and is discharged at a higher pressure, causing circulation.

Many different pumps exist. The most common is the "centrifugal," which uses a high speed impeller and centrifugal force to increase the pressure. A "reciprocating" pump's parts reciprocate and increase the pressure much like a car's pistons, which go back and forth. This type is often called a "positive displacement" pump.

Compressors

Compressors increase the pressure of a *vapor.* They also come as "reciprocating" and "centrifugal." Familiar compressors are the air compressors in a service station or a simple air pump that inflates a bicycle tire. Unlike liquids, vapors will compress. Car tires have compressed air. An inflated ballon must have compressed air.

Fired Heaters

Fired heaters are huge and are in most refineries, gasoline plants and chemical plants. They may be vertical like a hot water heater, or may be horizontal. They contain pipes, or "tubes."

A "vertical" heater is cylindrical and its diameter may be as much as 20'. The tubes or pipes will run vertically. Burners, firing fuel gas or fuel oil will be on the heater's bottom. Its bottom is usually 6'-7' from the ground.

The "horizontal" heater is shaped like a box and is often called a "box" heater. Its tubes run horizontally. The burners may be on the heater's bottom, ends or sides.

Vertical heaters generally operate for smaller duties, while the larger horizontal heaters carry out heavy duty services.

The heaters have two main sections–radient and convection. The radient section is the large part of the heater, where tubes receive heat radiating from the burners. The convection section of the heater is directly above the radient section and just below the stack. The inlet to heaters is usually in this convection section. The convection section of fired heaters often generates steam.

Boiler

The boiler is another fired heater. It takes a condensate and, by applying intense heat, converts it to steam. Fired heaters–instead of boilers–heat hydrocarbons. Boilers generate steam. Fired heaters may generate comparatively little steam in their convection section, but they mainly heat hydrocarbons.

Boilers and fired heaters have stacks. The stack is the large diameter pipe that carries off hot waste gases. The temperature of these gases in the stack runs from 700° F to 1000° F or more.

Valve Types and Uses

Valves

Valves stop or open and regulate flow. Some valves are huge and some are very small.

Gate valve is the most common type that plants use. It is usually manually operated and is designed for open or shut operation. It's not recommended for throttling.

Globe valve is for throttling. Good examples of globe valves are the faucets on a washbasin which throttle or adjust the flow to suit a person's needs.

Relief valve or *safety valve* is an automatic valve that opens when the pressure reaches "set pressure" on the relief valve. Without relief valves the plants could explode during periods of very high pressure. These valves have a spring that holds them shut. The spring holds until a set pressure is obtained; and, when the pressure is more than the set pressure, the spring "gives" and allows the fluid to escape, thereby relieving the pressure. As the pressure reduces, the spring closes and shuts off the flow.

Control Valve is usually an automatic valve built with a "globe valve" body and controls flow in a piping system. This valve opens, closes or throttles on a signal from an instrument. No manual operation is required, although some manual control valves are available. An example of an automatic control valve is the "Big Joe" type pressure regulating valve that controls a home's gas pressure. The gas line near the meter shows this "control valve." Control valves in a car, for instance, control water flow in the car heater.

Plug valve has a plug that rotates when turned and either lets flow pass through a hole in the plug or turns so that no flow is possible. This valve may be used for on-off service or for throttling. It has a more positive shutoff than the gate valve.

Ball valve uses a ball with a hole in it instead of a wedge-shaped plug, and the rotating ball opens or closes the flow. It also may be used for throttling. Ball valves are comparatively new and are gaining wide acceptance.

Check valve "checks" flow. It lets flow go one way and will not let it reverse. When you have a check valve in a line (or pipe), you have made a one-way street. The flow can only go one way. Many check valves are available. The common ones are *swing check,* in which a flapper lifts up to permit flow; the *piston check,* which has a piston in it that lifts to flow; the *ball check,* which has a ball in it which lifts; and the *butterfly check,* which has two vanes like a butterfly has wings. These "wings" fold back to permit flow but will close to stop backflow.

Plants are now using possibly a hundred other valves, but this book can't cover all of them. The student will be exposed to them as he gains actual experience. He should remember that these valves come in all sizes—from very small to sizes a person could walk through for special applications. The most common valving size is ½" through 24". Valves are expensive; their total cost is approximately 20-25% of the piping system in most plants. Like pipe, they are manufactured to all material specifications. The most common one is carbon steel.

Piping Terms

Flanges

Flanges make a bolted joint. This book will show later that most valves have flanged ends and must have a companion or matching flange attached. A gasket is then inserted between them, and the bolts are tightened to form a flanged joint.

Fittings

Fittings are many and varied. Some are elbows, tees, reducers, reducing ells and caps.

Instruments

Instruments tell the operator what is happening inside a vessel or pipe. A *pressure gage* tells him the pressure like an oil gage on a car tells the oil pressure in its piping system. A *gage glass* connected to a vessel tells the operator what the vessel's liquid level is. A *level indicator* tells him what the level is from a remote location. A car's gas gage is a level indicator because it is not hooked to the tank but is remotely located on the dashboard. Gage glasses on large coffee urns in restaurants show how much coffee is in the urns by the level in the gage glass itself.

Temperature indicators tell the fluid temperature in the pipe or vessel. They can be remotely located like a car's "temperature indicator." They also can be connected directly to the pipe or vessel.

This book will cover other instruments later.

Fluid

Most students may think of fluid as a liquid, but it can also be a vapor. Fluid means something that will flow—something not solid. Piping directs fluid flow.

Basic Piping Data

Many people have never seen pipe like the process plants use. The pipe is dimensionally set by the ANSI (American National Standards Institute) code. Wall thickness varies with the "schedule" number, but the *outside* diameter remains constant for the various sizes. As the thickness changes, the inside diameter changes. A schedule number and "standard weight" and "extra strong" list pipe and fittings. Several schedule numbers are available. This book deals with standard weight and extra strong. In Table 12-1 the (—) thickness does not match a schedule number.

An example of pipe is in Figure 12-1.

Table 12-1. Pipe Diameter, Wall Thickness, and Schedule Number

Nominal Size Inches	Outside Diameter Inches	Std Weight Thickness and Schedule Inches		Extra Strong Thickness and Schedule Inches	
1/8	.405	.068	(40)	.095	(80)
1/4	.540	.088	(40)	.119	(80)
3/8	.675	.091	(40)	.126	(80)
1/2	.840	.109	(40)	.147	(80)
3/4	1.050	.113	(40)	.154	(80)
1	1.315	.133	(40)	.179	(80)
1-1/2	1.900	.145	(40)	.200	(80)
2	2.375	.154	(40)	.218	(80)
3	3.500	.216	(40)	.300	(80)
4	4.500	.237	(40)	.337	(80)
6	6.625	.280	(40)	.432	(80)
8	8.625	.322	(40)	.500	(80)
10	10.750	.365	(40)	.500	(60)
12	12.750	.375	(–)	.500	(–)
14	14.000	.375	(30)	.500	(–)
16	16.000	.375	(30)	.500	(40)
18	18.000	.375	(–)	.500	(–)
20	20.000	.375	(20)	.500	(30)
24	24.000	.375	(20)	.500	(–)

Several methods of manufacturing pipe exist. The most common method makes "seamless," where the pipe is smooth with no seam or joint on the longitudinal axis. "Welded" pipe has a weld lengthwise. This may be "buttwelded" or "ERW" (electric resistance welded) or may be "spiral welded," which has a weld spiraling around the pipe.

Pipe is manufactured in "random length," which is ± 20'-0", and in "double random length," which is ± 40'-0". Unless double random is specified, the draftsmen will get single random lengths.

Pipe is single line and double line on piping drawings. Most companies have converted to single line piping drawings because they take less manhours to draw. Single line piping uses the heavy line for the pipe's center line. Whether single or double line, the OD (outside diameter) is always drawn to scale. For pipe sizes 1½" and below, this would be very difficult; so in both cases these are single line. (See Figures 12-2a and 12-2b.)

Fittings are welded, screwed and socketwelded. The student will find many other types as he learns the business.

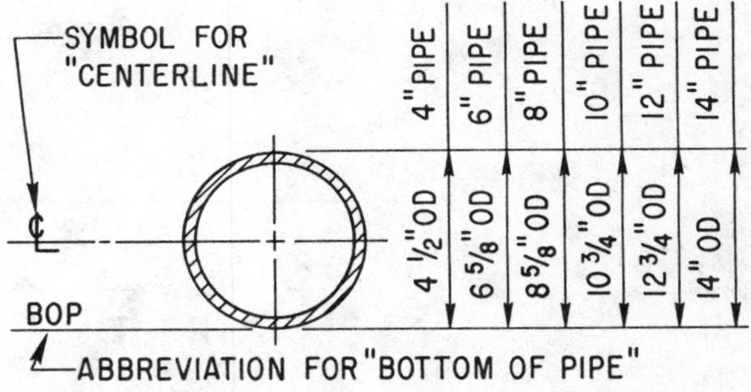

Figure 12-1. Pipe nominal versus actual outside diameter.

The elbow makes turns. It is commonly called an ell and mainly makes 90° and 45° turns. The 90° elbow comes in "long radius" and "short radius." The 45° elbow comes only in long radius. A reducing elbow is also widely used today. One end of it is larger than the other. This is for 90° turns only, and the radius is 1½ times the large-end size. Long radius is "LR" and short radius is "SR." (See Figure 12-3).

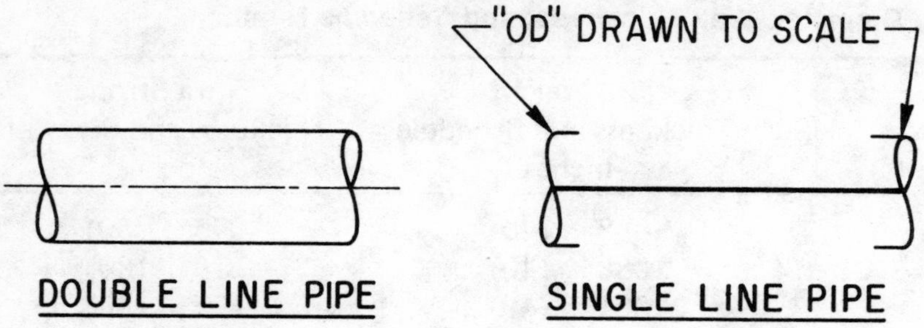

"OD" DRAWN TO SCALE

DOUBLE LINE PIPE

SINGLE LINE PIPE

Figure 12-2a. Double-line pipe.

Figure 12-2b. Single-line pipe.

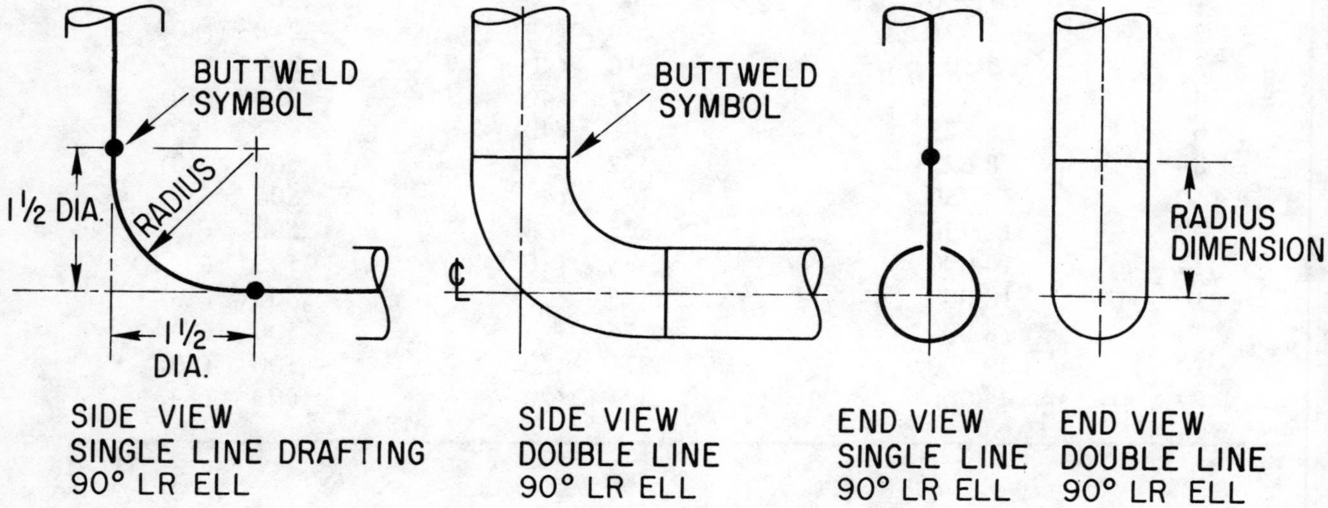

BUTTWELD SYMBOL

RADIUS

1 1/2 DIA.

1 1/2 DIA.

SIDE VIEW
SINGLE LINE DRAFTING
90° LR ELL

BUTTWELD SYMBOL

SIDE VIEW
DOUBLE LINE
90° LR ELL

END VIEW
SINGLE LINE
90° LR ELL

RADIUS DIMENSION

END VIEW
DOUBLE LINE
90° LR ELL

Figure 12-3. 90° elbows.

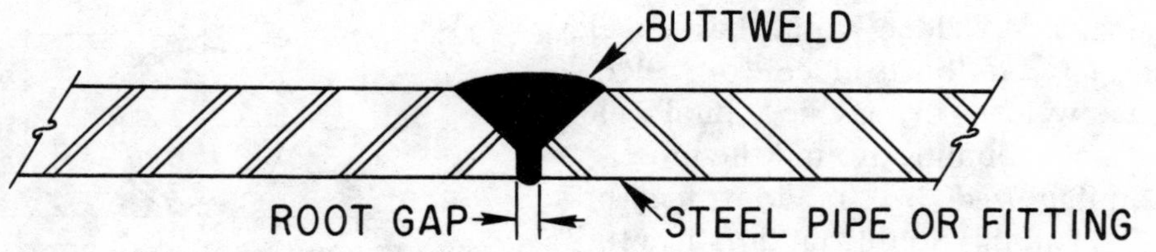

Figure 12-4. Example of a buttweld.

The LR ell's radius is 1½ times the nominal size. A 4″ LR ell has a 6″ radius on the centerline. (See Figure 12-3). A 6″ LR elbow has a 9″ radius. The SR ell's radius is only 1 time the nominal size. A 4″ SR ell has a 4″ radius. The 6″ SR ell has a 6″ radius, et cetera. The radius is to be drawn to scale.

The buttweld (Figure 12-4) is a weld that connects pipe or attaches pipe to fittings or fittings to fittings. It is called a buttweld because two beveled ends are butted and welded together. The symbol for a buttweld is a simple dot in single line piping. The symbol for a buttweld in double line piping is a heavy line (Figure 12-3).

The 45° ell is in Figure 12-5. Radius is 1½ x pipe size.

Figure 12-6 shows the reducing ell. Its radius is 1½ x the large end pipe size.

Whether drawing single line or double line, the piping draftsman will *always* draw the pipe size and fittings to scale. On single line drawings, pipe size is

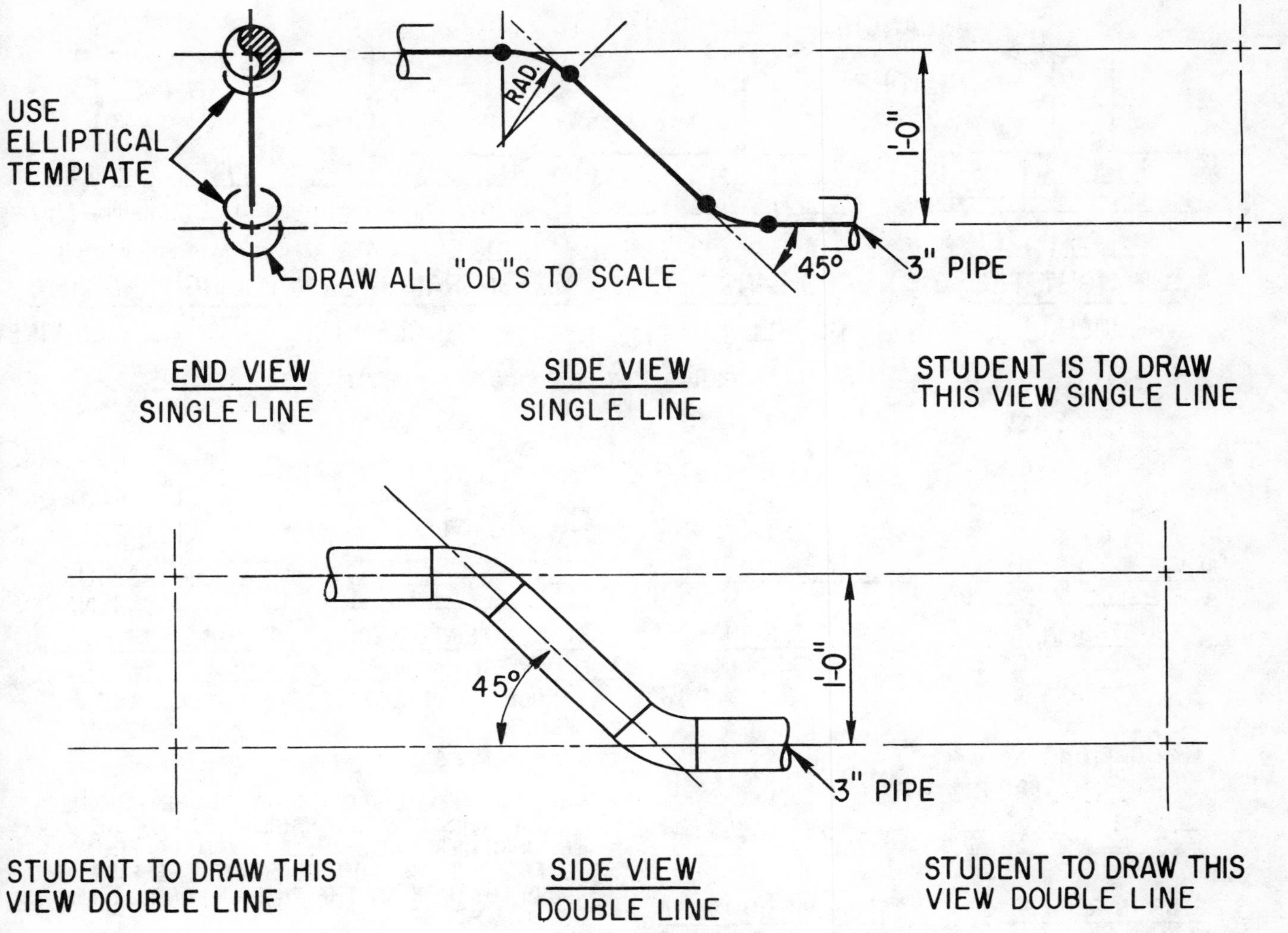

USE
ELLIPTICAL
TEMPLATE
DRAW ALL "OD"S TO SCALE
RAD.
1'-0"
45°
3" PIPE
END VIEW
SINGLE LINE
SIDE VIEW
SINGLE LINE
STUDENT IS TO DRAW
THIS VIEW SINGLE LINE
45°
1'-0"
3" PIPE
STUDENT TO DRAW THIS
VIEW DOUBLE LINE
SIDE VIEW
DOUBLE LINE
STUDENT TO DRAW THIS
VIEW DOUBLE LINE

Figure 12-5. 45° elbows.

PIPE CUT-DRAW ALL CUTS TO SCALE
A
NOTE: ALL REDUCING FITTINGS ARE
DRAWN DOUBLE LINE EVEN
IN SINGLE LINE PIPING.
NOTE SINGLE LINE
BUTTWELD SYMBOL.
DOUBLE LINE
SINGLE LINE

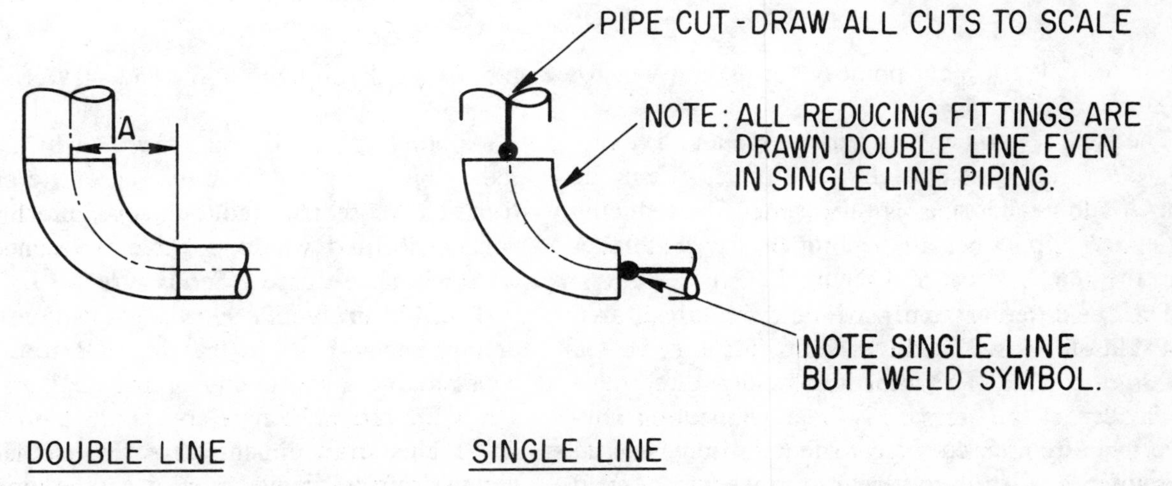

Figure 12-6. Reducing elbows.

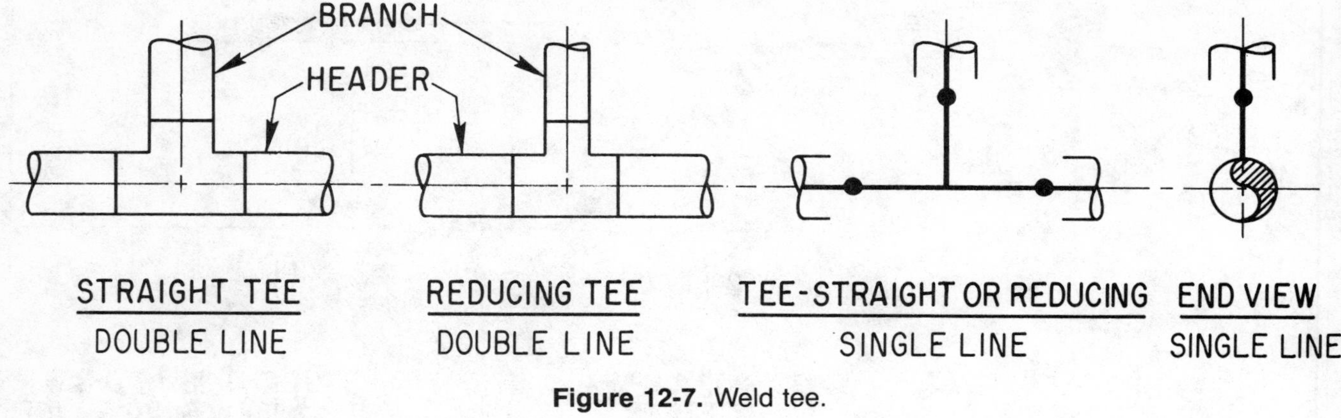

Figure 12-7. Weld tee.

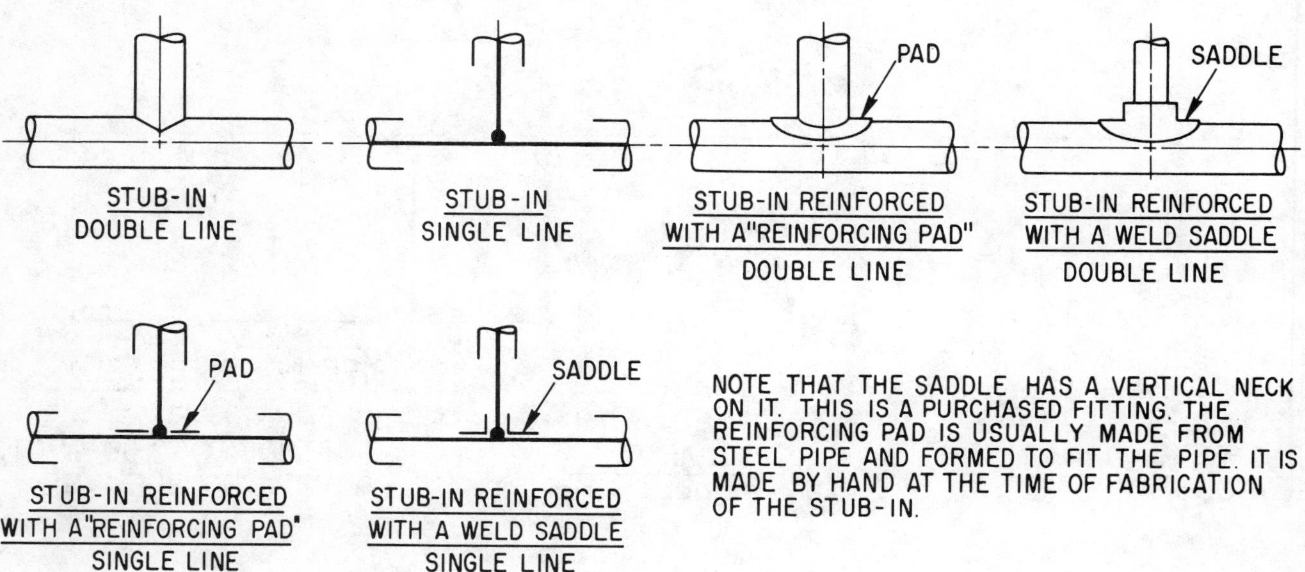

Figure 12-8. Stub-in—reinforced and nonreinforced.

drawn to scale for a cut point or for an end view of an elbow or pipe.

The weld tee comes as a *straight tee* and a *reducing outlet tee*. The straight tee applies when the branch and header sizes are the same. The reducing outlet tee applies when the branch size is smaller than the header size. (See Figure 12-7.)

The weld tee is expensive and requires three buttwelds for installation. A substitute for the tee is a *stub-in* where the branch is welded directly to the header. If the pressure is great, the stub-in may have to be reinforced. Many methods can be used to reinforce a stub-in. Some of these are "weld saddle," "reinforcing pad," "weldolet," or other fittings used throughout the industry. (See Figure 12-8.)

Another fitting is the *reducer* which makes a reduction in a pipe. It's either eccentric or concentric. The eccentric reducer costs more than the concentric, and where possible the concentric reducer should be used. (See Figure 12-9.)

The *eccentric reducer* is used mostly in pipeways or pipe racks. It keeps the BOP (Bottom of Pipe) constant for self draining or for resting on a constant support. The offset appears in the two centerlines. This offset dimension is ½ the difference of the two *inside* diameters. For a 6″ x 4″ eccentric reducer, the eccentricity is 1″.

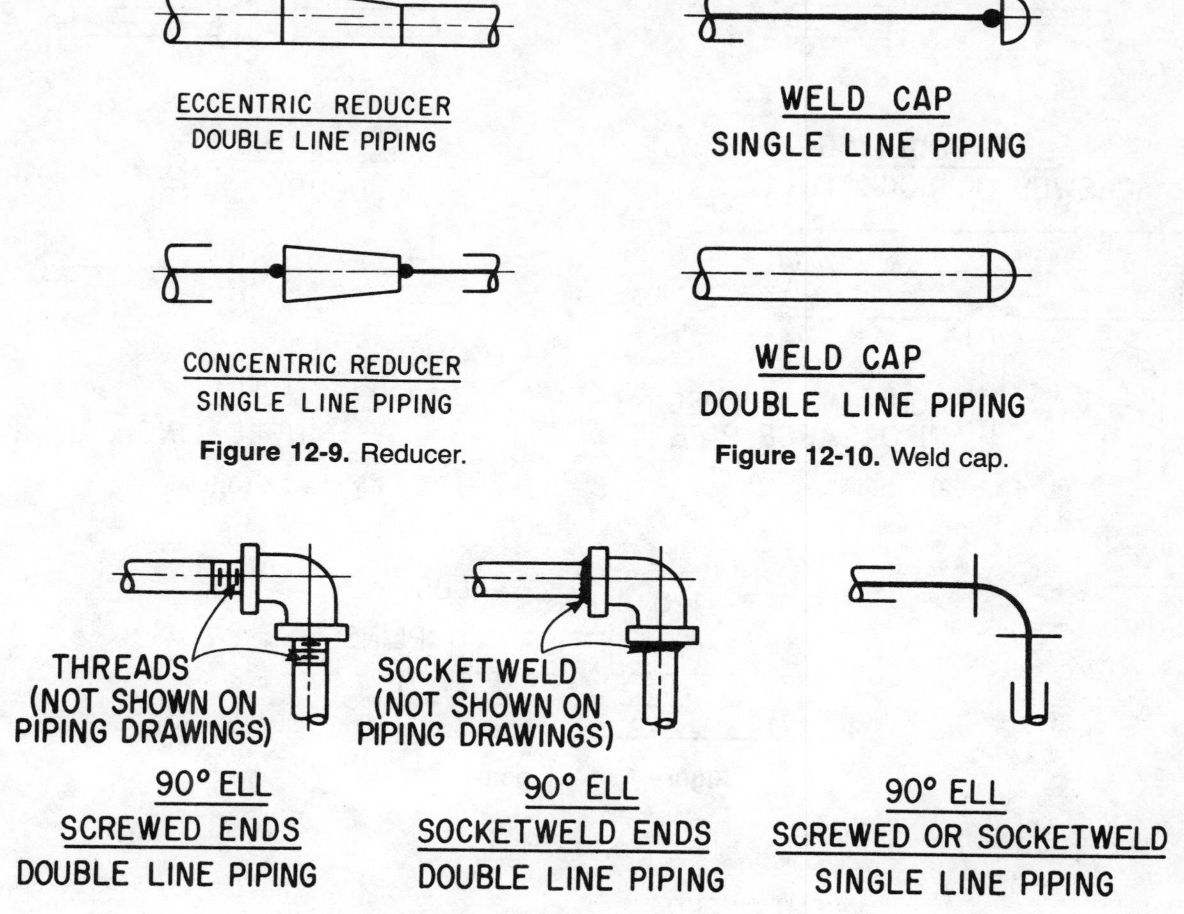

Figure 12-9. Reducer.

Figure 12-10. Weld cap.

Figure 12-11. Screwed or socketweld elbows.

The *concentric reducer* has the same centerline as shown.

Whether drawing single or double line piping, reducers are double line and drawn to scale by pipe size and end-to-end length.

The weld cap terminates a pipe. This fitting should *always* be drawn double line. (See Figure 12-10.)

Screwed and socketweld fittings are used in sizes 2″ and below. Some companies use them for 1½″ and below. Figure 12-11 shows them in double line for clarity and single line as they are usually drawn.

Screwed and socketweld fittings come in the same types as those welded fittings already shown. They have 90° ells, 45° ells, tees, caps, et cetera.

Some different fittings are described as follows:

Swage nipple is a reducer, except it is much longer. It, too, is eccentric or concentric and comes in the welded sizes. It is most commonly used in sizes 2″ and smaller instead of using reducers.

Coupling joins two pieces of pipe or male connections. It also stubs-in a small pipe or connection into a large one. (See Figure 12-12.)

Union joins screwed and socketweld pipe and make connections when they may need to be broken apart in the future. The union is a possible leak point and should only be used where the "break-away" feature is necessary. (See Figure 12-13.)

Plug plugs or stops flow in a screwed fitting at the end. (See Figure 12-14.)

Flanges and Flange Facings

Flanges come in all sizes and materials. The forged steel flange comes in seven basic ratings which the ANSI set: 150#, 300#, 400#, 600#, 900#, 1500# and 2500#. The student must remember these.

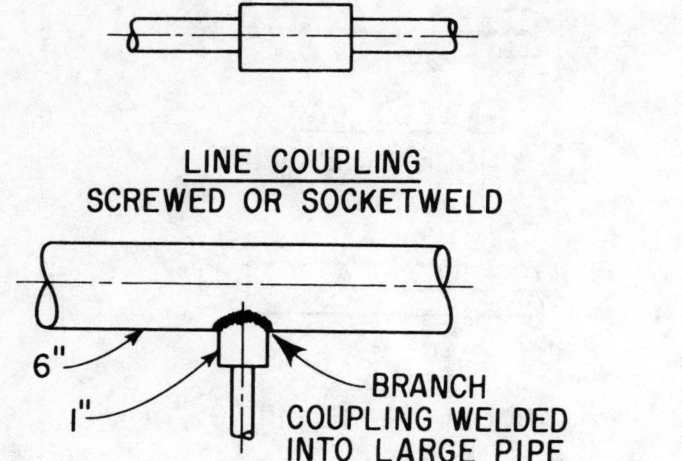

Figure 12-12. Coupling.

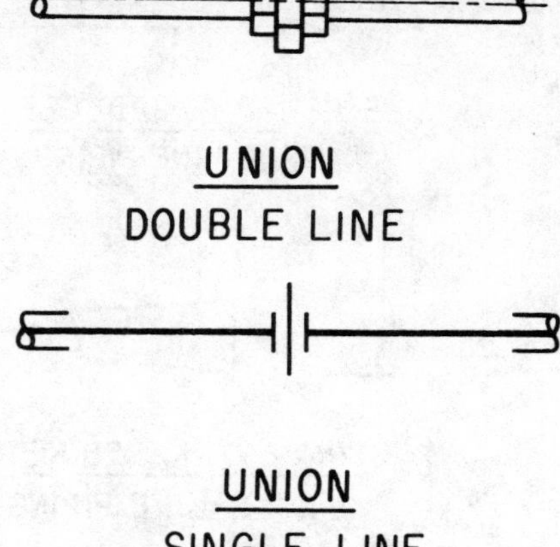

Figure 12-13. Union.

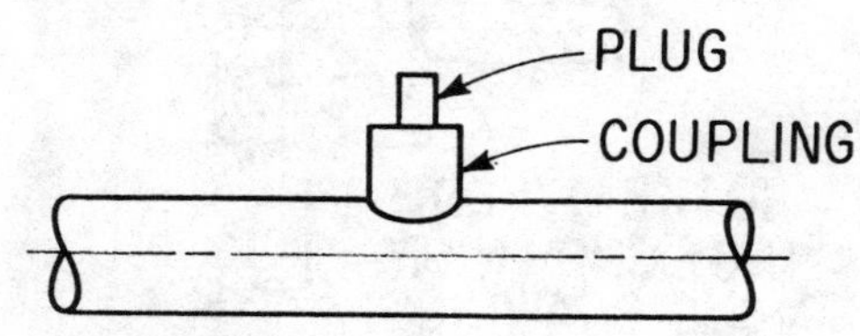

Figure 12-14. Pipe plug.

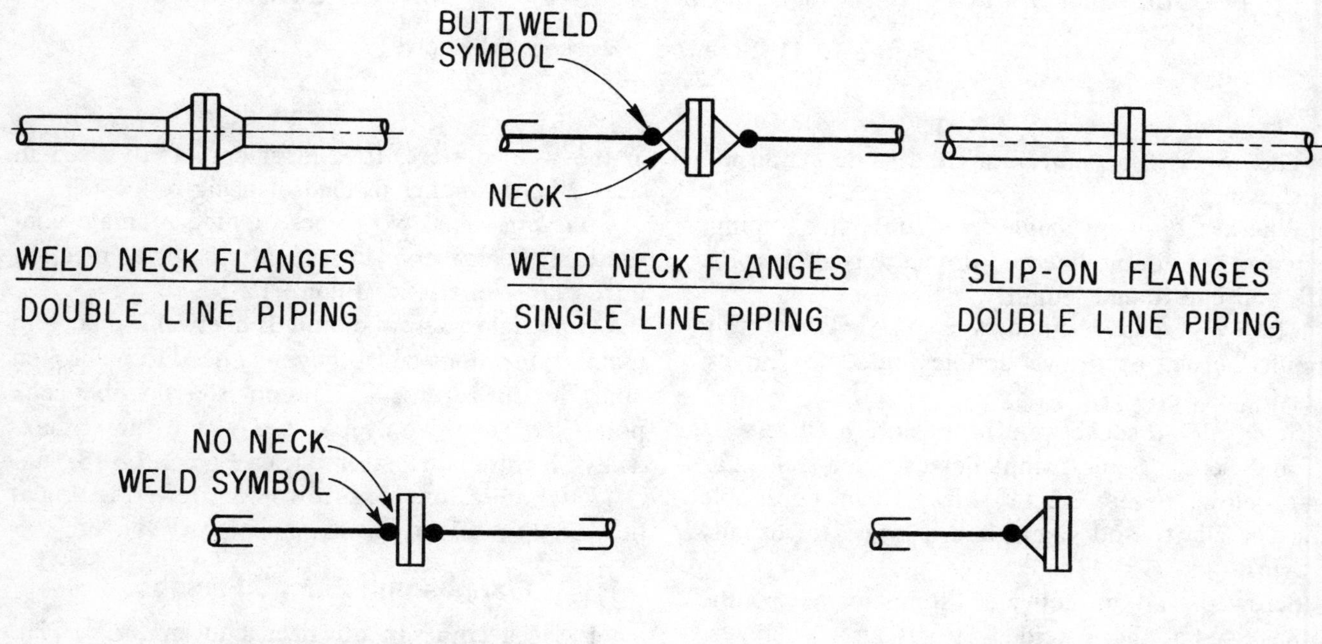

Figure 12-15. Basic flange types.

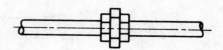

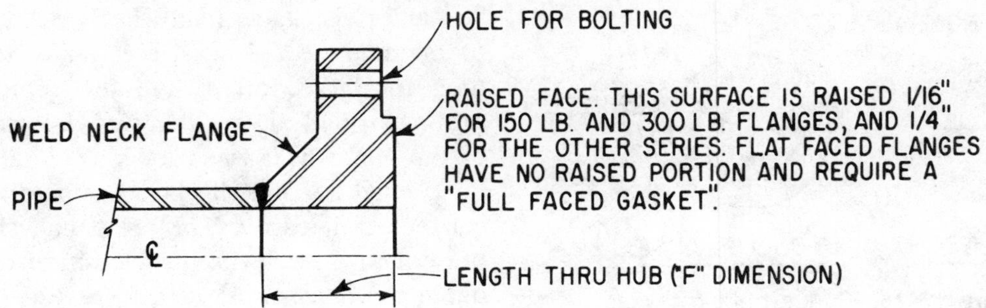
HOLE FOR BOLTING
RAISED FACE. THIS SURFACE IS RAISED 1/16" FOR 150 LB. AND 300 LB. FLANGES, AND 1/4" FOR THE OTHER SERIES. FLAT FACED FLANGES HAVE NO RAISED PORTION AND REQUIRE A "FULL FACED GASKET".
WELD NECK FLANGE
PIPE
℄
LENGTH THRU HUB ("F" DIMENSION)
1/2 RAISED FACE FLANGE
SHOWN IS A WELD NECK FLANGE BUT THIS RAISED FACE COULD ALSO BE PUT ON A SLIP-ON, BLIND OR OTHER TYPES OF FLANGES.

Figure 12-16. Weld neck raised face flange.

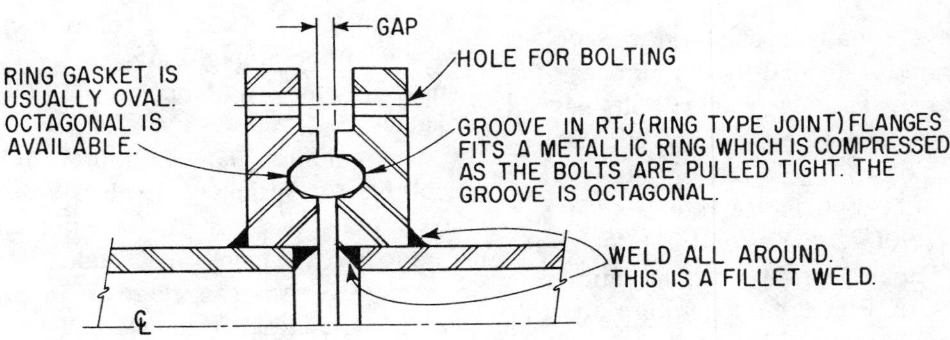
GAP
HOLE FOR BOLTING
RING GASKET IS USUALLY OVAL. OCTAGONAL IS AVAILABLE.
GROOVE IN RTJ (RING TYPE JOINT) FLANGES FITS A METALLIC RING WHICH IS COMPRESSED AS THE BOLTS ARE PULLED TIGHT. THE GROOVE IS OCTAGONAL.
WELD ALL AROUND. THIS IS A FILLET WELD.
℄
1/2 RING JOINT FLANGES
SHOWN IS A PAIR OF SLIP-ON FLANGES. RING JOINT FACING COULD ALSO BE SPECIFIED FOR WELD NECK AND OTHER TYPES. THE SLIP-ON FLANGE IS SHOWN SO YOU CAN SEE THE INTERNAL AND EXTERNAL WELDS REQUIRED FOR THIS TYPE OF FLANGED JOINT.

Figure 12-17. Slip-on RTJ flange.

Cast iron flanges come in two ratings. The 125# rating has a flat face, while the 250# rating usually has a raised face.

All flanges make a bolted joint, which can be broken away. Flanges are expensive, and the good "piper" will use them only when necessary. Piping specifications, which the student will see later, will indicate when to use flanges.

The student must also remember the basic flange types. They are in Figure 12-15.

Flange bolting changes with each size and rating. The gasket, placed between flanges, changes with the flange facing. Many flange facings exist. This text will deal with three basic types: raised face, flat face and ring joint (See Figures 12-16 and 12-17.)

When drawing flanges the length through hub, "F" dimension, is drawn to scale. Flange thickness may be shown double line to scale as depicted in Figure 12-15 or may be shown as a single line (see Figure 12-23). For this class, draw flanges with double lines and to scale for OD and thickness.

Flange Bolting

Process pressure piping is under pressure of varying degrees. The more pressure existing, the more

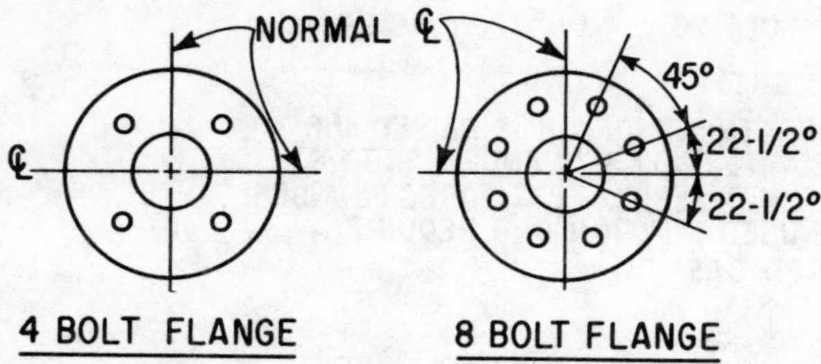

Figure 12-18. Flange bolting.

bolts required. ANSI has defined bolting requirements. Bolts, regardless of number, will straddle the flange's "normal centerline." Normal centerline is on the North-South, East-West or Up-Down axis. On this basis a flange has four quadrants. Bolt holes are always added in quantities of four: 4, 8, 12, 16, 20, 24, et cetera.

Bolting is always equally spaced on the bolt circle. The 360° circle is divided by the number of bolts to determine their spacing. Twelve bolts have 30° spacing; 16 bolts have 22½° spacing, et cetera. (See Figure 12-18.)

The seven series of ANSI flange ratings (150#, 300#, 400#, 600#, 900#, 1500# and 2500#) are a nominal rating. It does not mean that the rating is the maximum pressure. Pressure-temperature rating determines the flange rating used. For instance, the 150# flange is good for 275# at 100° F. As the temperature rises, allowable pressure goes down. At 500° F the maximum allowable pressure is 150#. At 750° F the 150# flange is good for only 100#. All listed ratings are for carbon steel. Alloy materials will have different ratings. All pressure-temperature ratings are in the ANSI standard on flanges and fittings.

Valves

Valves stop or control flow. The most common refinery-chemical plant valve is the flanged-end valve. They also have screwed ends, socketweld ends, buttwelded ends and other special types.

The most important part of the valve is the body type. The gate body is for on-off service. A gate body is not for throttling service; the globe body is. The student sees globe valves every day. He calls them faucets and controls the water flow in the washbasin or bathtub with them. These are globe body valves.

Valves also have handwheels, which are turned to operate the valve. These handwheels are very long for large valves, and the competent piping draftsman will always think of the proper location of the handwheel when locating a valve. It should be located so that the person operating the valve can get to it easily. The draftsman should always draw it to scale in its open position to insure adequate clearance.

Other valves are check valves, motor operated control valves, angle valves, plug valves, butterfly valves, ball valves, relief valves and diaphragm valves. These all have specific uses and symbols which the piping draftsman will know.

Symbols

Tables 12-2 and 12-3 summarize symbols, which the student *must* memorize before going on. These symbols are the piping dratfsman's ABC's. Without completely knowing them, he can't "write" piping.

Tables 12-4 and 12-5 are dimensional tables for the most common flanges, fittings and valves. For special flanges, fittings and valves a draftsman may face on a high pressure job, he will want to refer to manufacturer's dimensional catalogs. What is high pressure? Everything is relative. In process plants 1000 psi (pounds per square inch) is not considered real high. Some plants run several thousand psi. The student should compare this to the 25-30 psi pressure in his automobile tire.

Projection Exercise

Now it is time for the piping drafting student to do some projection exercises. Figures 12-19 through 12-32 show various piping configurations. These are all single line exercises. Use ⅜″ = 1′0″ for scale with 10″ and 6″ pipe sizes.

Figure 12-19 shows a simple 90° ell's side and plan views. The student is to draw the two end views and bottom view.

Figure 12-20 shows a straight tee in elevation and top plan. The student is to complete the end views and bottom view.

Figure 12-20 shows a straight tee in elevation and top plan. The student is to complete the end views and bottom view.

Figure 12-21 has an elevation of two 90° ells welded together. The student is to draw all four projections.

(text continued on page 199)

Table 12-2. Symbols of Fittings and Flanges for Piping Drawings

TYPE OF FITTING		SCREWED OR SOCKET WELD	WELDED		FLANGED	
		SINGLE LINE	DOUBLE LINE	SINGLE LINE	DOUBLE LINE	SINGLE LINF
90° ELL	TOP					
	SIDE					
	BOTTOM					
45° ELL	TOP					
	SIDE					
	BOTTOM					
TEE	TOP					
	SIDE					
	BOTTOM					
LATERAL	TOP					
	SIDE					
	BOTTOM					
REDUCERS	CONC.					
	ECC.					

FLANGES	SINGLE LINE				T.G.		
	DOUBLE LINE				T.G.		
		SLIP-ON	WELD NECK	LAPPED	TONGUE & GROOVE	ORIFICE	BLIND
MISC.	SINGLE LINE						
	DOUBLE LINE						
		PIPE WELD	STUB IN	WELD SADDLE	WELDED PIPE CAP	UNION	PLUG

Courtesy of Fluor Corporation

Table 12-3. Symbols of Valves and Fittings for Piping Drawings

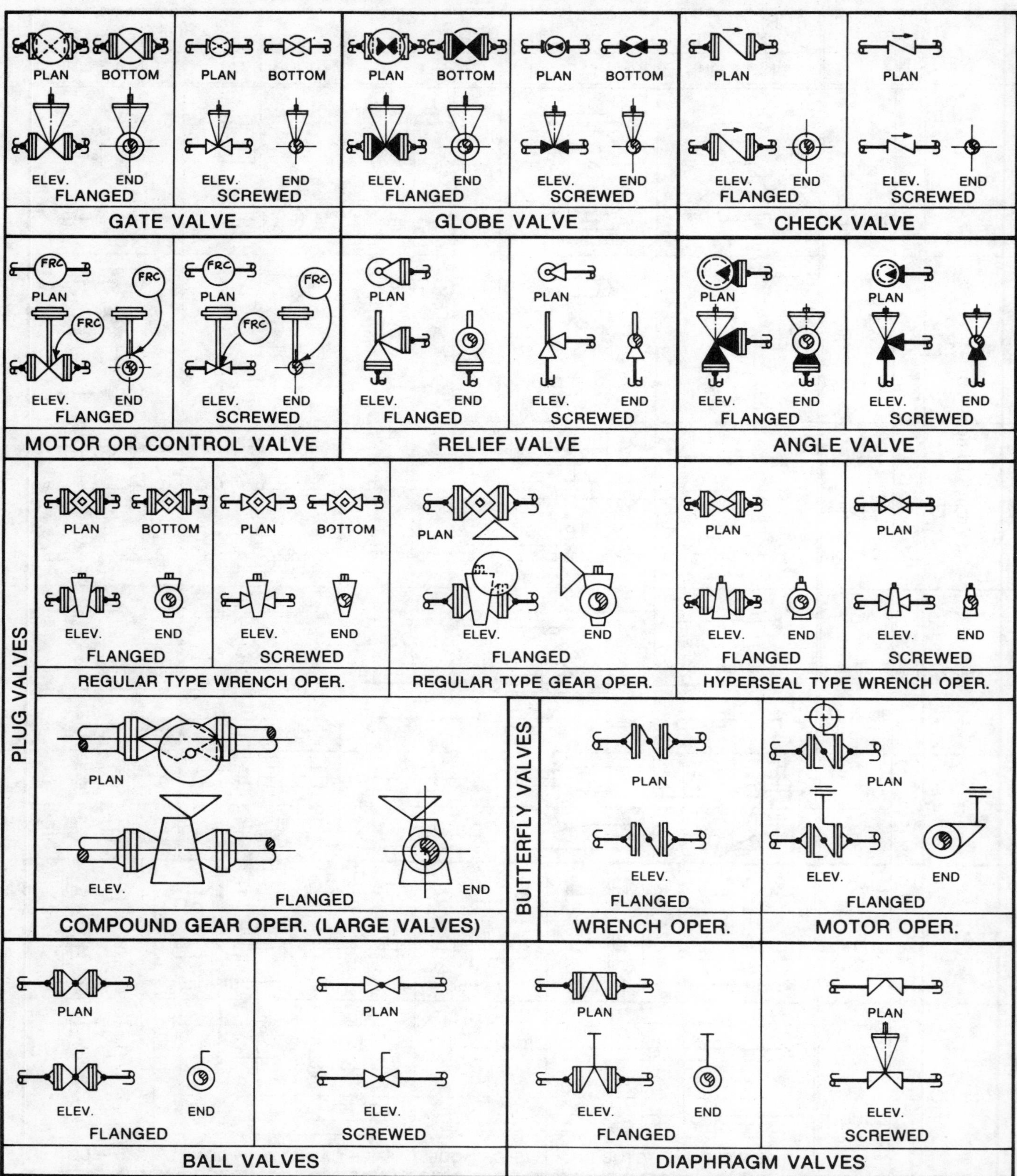

Courtesy of Fluor Corporation

Table 12-4. Flange and Fitting Dimensions

NOMINAL PIPE SIZE	WELD FITTING			WELD NECK						WELD NECK AND SLIP-ON									SWAGE NIPPLE		SCREWED FITTING			UNION			COUPLING	
	CENTER TO END		END TO END	TOTAL LENGTH R.F.			TOTAL LENGTH R.T.J.			TOTAL THICKNESS R.F.			TOTAL THICKNESS R.T.J.			OUTSIDE DIAMETER					TEE AND ELL CENTER TO END			END TO END			6000# 1/2", 3/4", 1" / 3000# 1 1/2", 2"	
																					M.I.	STL.	STL.	M.I.	F.S.	F.S.		
				150	300	600	150	300	600	150	300	600	150	300	600	150	300	600	REDUCTION	LGTH	150	2000	3000	250	2000 3000	6000	LG.	O.D.
1/2	5/8	1		1 7/8	2 1/16	2 5/16		2 5/16	2 5/16	7/16	9/16	13/16		13/16	13/16	3 1/2	3 3/4	3 3/4	1/4 TO 3/8	2 3/4	1 1/8	1 1/8	1 5/16	1 7/8	1 15/16	2 7/8	1 5/8	1 1/2
3/4	7/16	1 1/8	1 1/2	2 1/16	2 1/4	2 1/2		2 1/2	2 1/2	1/2	5/8	7/8		7/8	7/8	3 7/8	4 5/8	4 5/8	1/4 TO 1/2	3	1 5/16	1 5/16	1 1/2	2 1/8	2 1/4	3 3/8	2	1 3/4
1	7/8	1 1/2	2	2 3/16	2 7/16	2 11/16	2 7/16	2 11/16	2 11/16	9/16	11/16	15/16	13/16	15/16	15/16	4 1/4	4 7/8	4 7/8	1/4 TO 3/4	3 1/2	1 1/2	1 1/2	1 3/4	2 3/8	2 7/16	3 3/4	2 5/8	2 1/4
1 1/2	1 1/8	2 1/4	2 1/2	2 7/16	2 11/16	3	2 11/16	2 15/16	3	11/16	13/16	1 1/8	15/16	1 1/16	1 1/8	5	6 1/8	6 1/8	1/4 TO 1 1/4	4 1/2	1 15/16	2	2 3/8	2 15/16	3	4 1/4	3 1/8	2 1/2
2	1 3/8	2 1/2	3	2 1/2	2 3/4	3 1/8	2 3/4	3 1/16	3 3/16	3/4	7/8	1 1/4	1	1 3/16	1 5/16	6	6 1/2	6 1/2	1/4 TO 1 1/2	6 1/2	2 1/4	2 3/8	2 1/2	3 1/4	3 3/8	4 5/8	3 3/8	3 1/4
3	2	3 3/8	3 1/2	2 3/4	3 1/8	3 1/2	3	3 7/16	3 9/16	15/16	1 1/8	1 1/2	1 3/16	1 7/16	1 9/16	7 1/2	8 1/4	8 1/4	1/4 TO 2 1/2	8								
4	2 1/2	4 1/8	4	3	3 3/8	4 1/4	3 1/4	3 11/16	4 5/16	15/16	1 1/4	1 3/4	1 3/16	1 9/16	1 13/16	9	10	10 3/4	1/4 TO 3 1/2	9								
6	3 3/4	5 5/8	5 1/2	3 1/2	3 7/8	4 7/8	3 3/4	4 3/16	4 15/16	1	1 7/16	2 1/8	1 1/4	1 3/4	2 3/16	11	12 1/2	14	1/2 TO 5	12								
8	5	7	6	4	4 3/8	5 1/2	4 1/4	4 11/16	5 9/16	1 1/8	1 5/8	2 7/16	1 3/8	1 15/16	2 1/2	13 1/2	15	16 1/2	2 TO 6	13								
10	6 1/4	8 1/2	7	4	4 5/8	6 1/4	4 1/4	4 15/16	6 5/16	1 3/16	1 7/8	2 3/4	1 7/16	2 3/16	2 15/16	16	17 1/2	20										
12	7 1/2	10	8	4 1/2	5 1/8	6 3/8	4 3/4	5 7/16	6 7/16	1 1/4	2	2 7/8	1 1/2	2 5/16	2 15/16	19	20 1/2	22										
14	8 3/4	11	13	5	5 5/8	6 3/4	5 1/4	5 15/16	6 13/16	1 3/8	2 1/8	3	1 5/8	2 7/16	3 1/16	21	23	23 3/4										
16	10	12	14	5	5 3/4	7 1/4	5 1/4	6 1/16	7 5/16	1 7/16	2 1/4	3 1/4	1 11/16	2 9/16	3 5/16	23 1/2	25 1/2	27										
18	11 1/4	13 1/2	15	5 1/2	6 1/4	7 1/2	5 3/4	6 9/16	7 9/16	1 9/16	2 3/8	3 1/2	1 13/16	2 11/16	3 9/16	25	28	29 1/4										
20	12 1/2	15	20	5 11/16	6 3/8	7 3/4	5 15/16	6 3/4	7 7/8	1 11/16	2 1/2	3 3/4	1 15/16	2 7/8	3 7/8	27 1/2	30 1/2	32										
24	15	17	20	6	6 5/8	8 1/4	6 1/4	7 1/16	8 7/16	1 7/8	2 3/4	4 1/4	2 1/8	3 3/16	4 7/16	32	36	37										

Courtesy of Fluor Corporation

Table 12-5. Valve Dimensions

NOMINAL PIPE SIZE	FLANGED GATE VALVE											
	150				300				600			
	R.F.	R.T.J.			R.F.	R.T.J.			R.F.	R.T.J.		
	A	A	D	E	A	A	D	E	A	A	D	E
1/2												
3/4												
1	5	5 1/2	10 11/16	5					8 1/2	8 1/2	10 11/16	5
1 1/2	6 1/2	7	13 7/16	6	7 1/2	8	13 7/16	6	9 1/2	9 1/2	13 7/16	6
2	7	7 1/2	16 1/2	8	8 1/2	9 1/8	18 1/2	8	11 1/2	11 5/8	18 1/4	8
3	8	8 1/2	20 3/4	9	11 1/8	11 3/4	23 1/4	9	14	14 1/8	25 3/4	10
4	9	9 1/2	25 3/4	10	12	12 5/8	28 1/4	10	17	17 1/8	32	14
6	10 1/2	11	35 1/4	14	15 7/8	16 1/2	38 1/2	14	22	22 1/8	42 3/4	20
8	11 1/2	12	43	14	16 1/2	17 1/8	47	16	26	26 1/8	52 1/4	24
10	13	13 1/2	52 1/2	18	18	18 5/8	56 1/2	20	31	31 1/8	62 1/4	27
12	14	14 1/2	60 1/2	18	19 3/4	20 3/8	64 1/4	20	33	33 1/8	70	27
14	15	15 1/2	70 1/4	22	30	30 5/8	74 3/4	27	35	35 1/8	77 1/4	30
16	16	15 1/2	79 3/4	24	33	33 5/8	80 1/2	27	39	39 1/8	83 3/4	30
18	17	17 1/2	89	27	36	36 5/8	91	30	43	43 1/8	93 3/4	36
20	18	18 1/2	97 1/4	30	39	39 3/4	100 1/2	36	47	47 1/4	104 1/2	36
24	20	20 1/2	112 3/4	30	45	45 7/8	120 1/2	36	55	55 3/8	126	42

NOMINAL PIPE SIZE	FLANGED GLOBE VALVE											
	150				300				600			
	R.F.	R.T.J.			R.F.	R.T.J.			R.F.	R.T.J.		
	A	A	D	E	A	A	D	E	A	A	D	E
1/2												
3/4												
1	5 1/8	5 1/2	8 5/16	3 5/8	8	8 1/2	8 3/4	4 1/2	8 1/2	8 1/2	9 3/4	5
1 1/2	6 5/8	7	10	4 5/8	9	9 1/2	12	7	9 1/2	9 1/2	12 3/16	7
2	8	8 1/2	13 3/4	8	10 1/2	11 1/8	17 3/4	9	11 1/2	11 5/8	19	10
3	9 1/2	10	16 1/2	9	12 1/2	13 1/8	20 1/2	10	14	14 1/8	23 1/2	12
4	11 1/2	12	19 3/4	10	14	14 5/8	24 3/4	14	17	17 1/8	27 1/2	18
6	16	16 1/2	24 1/2	12	17 1/2	18 1/8	29 3/4	18	22	22 1/8	35	24
8	19 1/2	20	26	16	22	22 5/8	36 1/2	24				
10												
12												
14												
16												
18												
20												
24												

NOMINAL PIPE SIZE	FLANGED CONTROL VALVE — FACE TO FACE					
	150		300		600	
	R.F.	R.T.J.	R.F.	R.T.J.	R.F.	R.T.J.
1/2			7 1/2	7 15/16	8	7 15/16
3/4	7 1/4		7 5/8	8 1/8	8 1/8	8 1/8
1	7 1/4	7 3/4	7 3/4	8 1/4	8 1/4	8 1/4
1 1/2	8 3/4	9 1/4	9 1/4	9 3/4	9 7/8	9 7/8
2	10	10 1/2	10 1/2	11 1/8	11 1/4	11 3/8
3	11 3/4	12 1/4	12 1/2	13 1/8	13 1/4	13 3/8
4	13 7/8	14 3/8	14 1/2	15 1/8	15 1/2	15 5/8
6	17 3/4	18 1/4	18 5/8	19 1/4	20	20 1/8
8	21 3/8	21 7/8	22 3/8	23	24	24 1/8
10						
12						
14						
16						
18						
20						
24						

Courtesy of Fluor Corporation

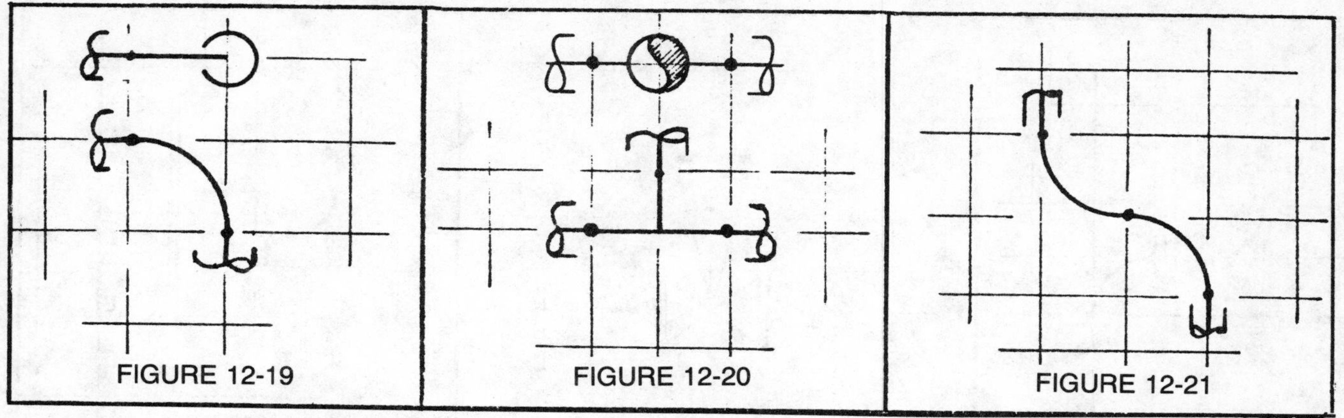

Figures 12-19 through 12-21. Various piping configurations using single-line symbols. (Courtesy of the Fluor Corp.)

(text continued from page 194)

Figure 12-22 has a concentric reducer with ells welded to both ends. The student is to draw the top, bottom and both end views.

Figure 12-23 shows a flanged gate valve with flanges bolted to both ends. The student is to project the top, bottom and both end views.

Figure 12-24 shows a flanged check valve, flanges, 90° elbows and some pipe welded to the ells; the student is to draw the other four views.

Figure 12-25 gets a little more complex. It combines fittings in Figures 12-19 through 12-24. The student is to draw the top, bottom and end projections.

Figure 12-26 has two 45° ells. The student musι use the ellipse template for these projections, all four of which he is to draw.

The student should note that, on all of the above figures, the piping in buttwelded. He should show every weld in all views as a dot or an ellipitical line for 45° ells.

The next projection exercises will be for screwed or socketweld piping. They do not show the welds, but a short dash across the piping shows the end of the fitting.

Figure 12-27 has a straight tee, two 90° ells and some pipe. The student is to draw the upper plan, the bottom and two end views.

Figure 12-28 has the same fittings as Figure 12-27 but also has a concentric swage and a union. The student is to draw the other four views.

Figure 12-29 has two 45° ells plus a tee and a union. Using the elliptical template, the student is to draw the other four views.

Figure 12-30 has only the end view. The student is to draw the side elevation and the other three views.

Figure 12-31 requires some imagination on the student's part. He doesn't know how long the pipe that he can't see is. He is to assume it's a short piece and is to draw the other four views.

Figure 12-32 shows an elevation and three views of some piping. Here the student sees how he is to use the elliptical template to draw projections. For practice he is to duplicate this study, drawing the full elevation first. Then, he is to extend his own projection lines (lightly) and draw the other three views.

After the student has completed this study, he is to draw the side elevation again. But this time he is to draw the left projection and the bottom view, which Figure 12-32 does not show.

In Figure 12-33 the student is to complete four views of each picture.

Abbreviations

Many abbreviations, like those below, are common language for the piping draftsman. The memory must be "put in gear" for the first group and the second group need not be memorized. The text has some of the first group but repeats them so the student will always have them together for ready reference.

Group 1

ANSI American National Standards Institute
ASME American Society of Mechanical Engineers
ASTM American Society for Testing and Materials

(text continued on page 204)

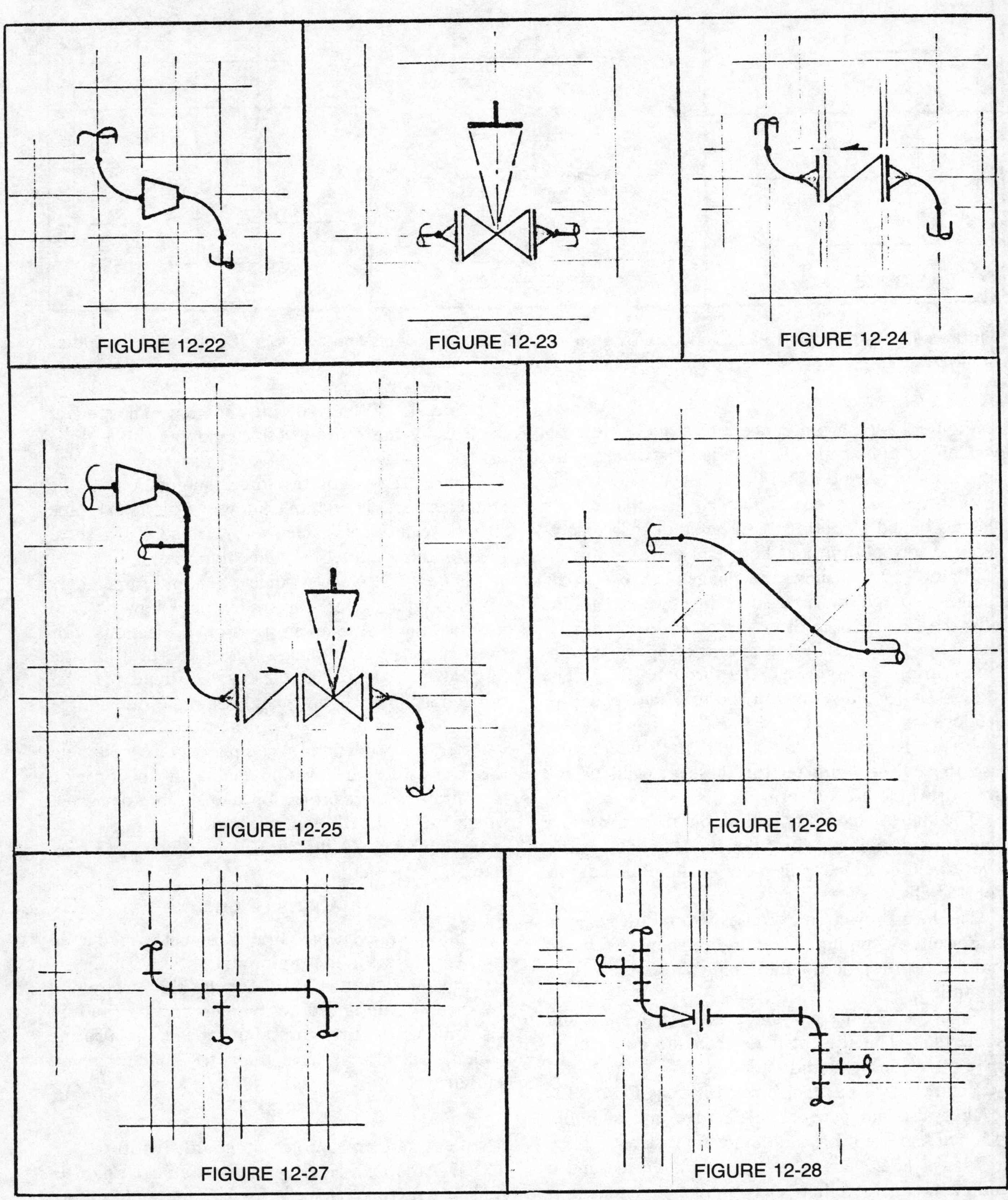

Figures 12-22 through 12-28. Various piping configurations using single-line symbols. (Courtesy of the Fluor Corp.)

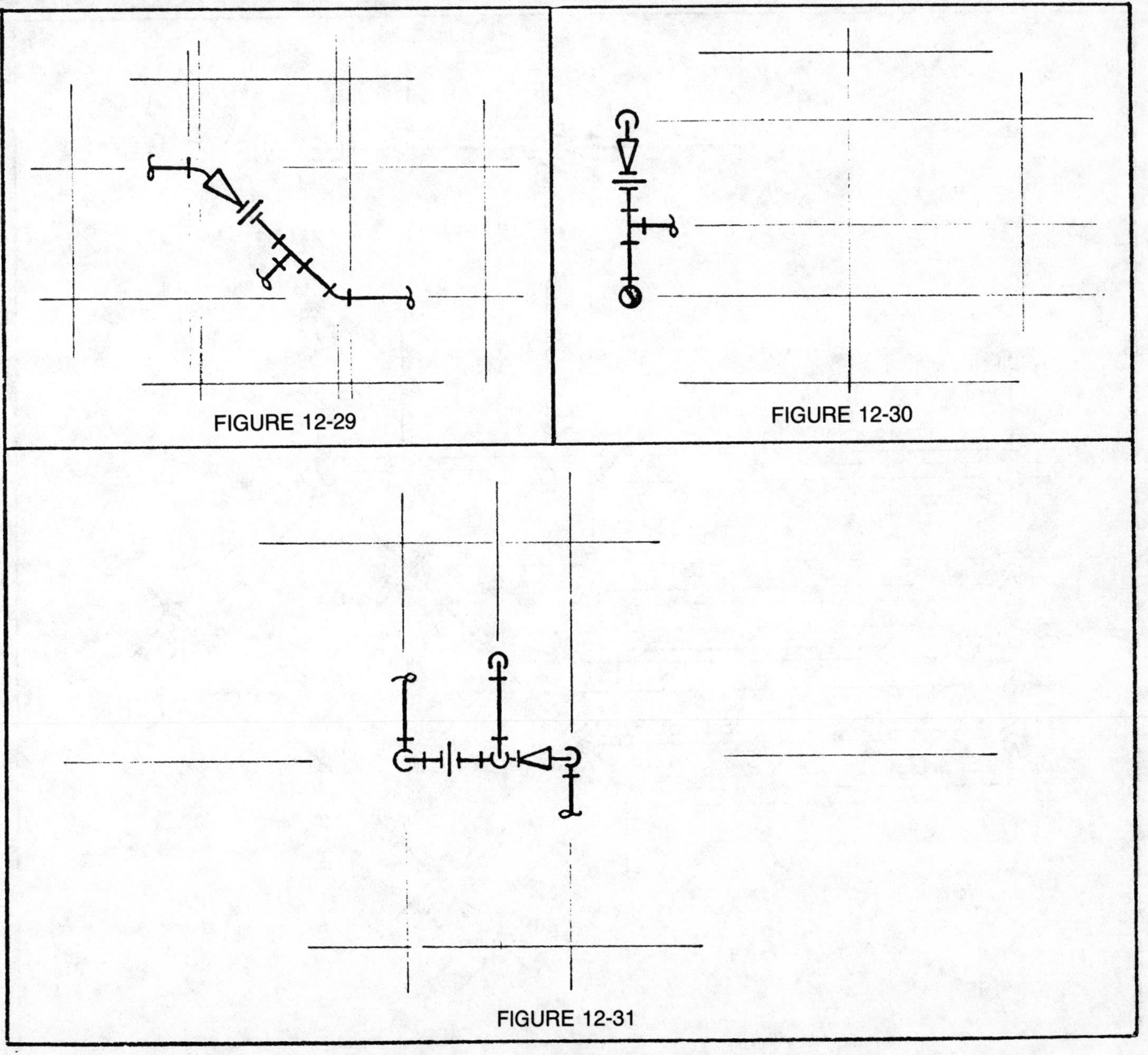

Figures 12-29 through 12-31. Various piping configurations using single-line symbols. (Courtesy of Fluor Corp.)

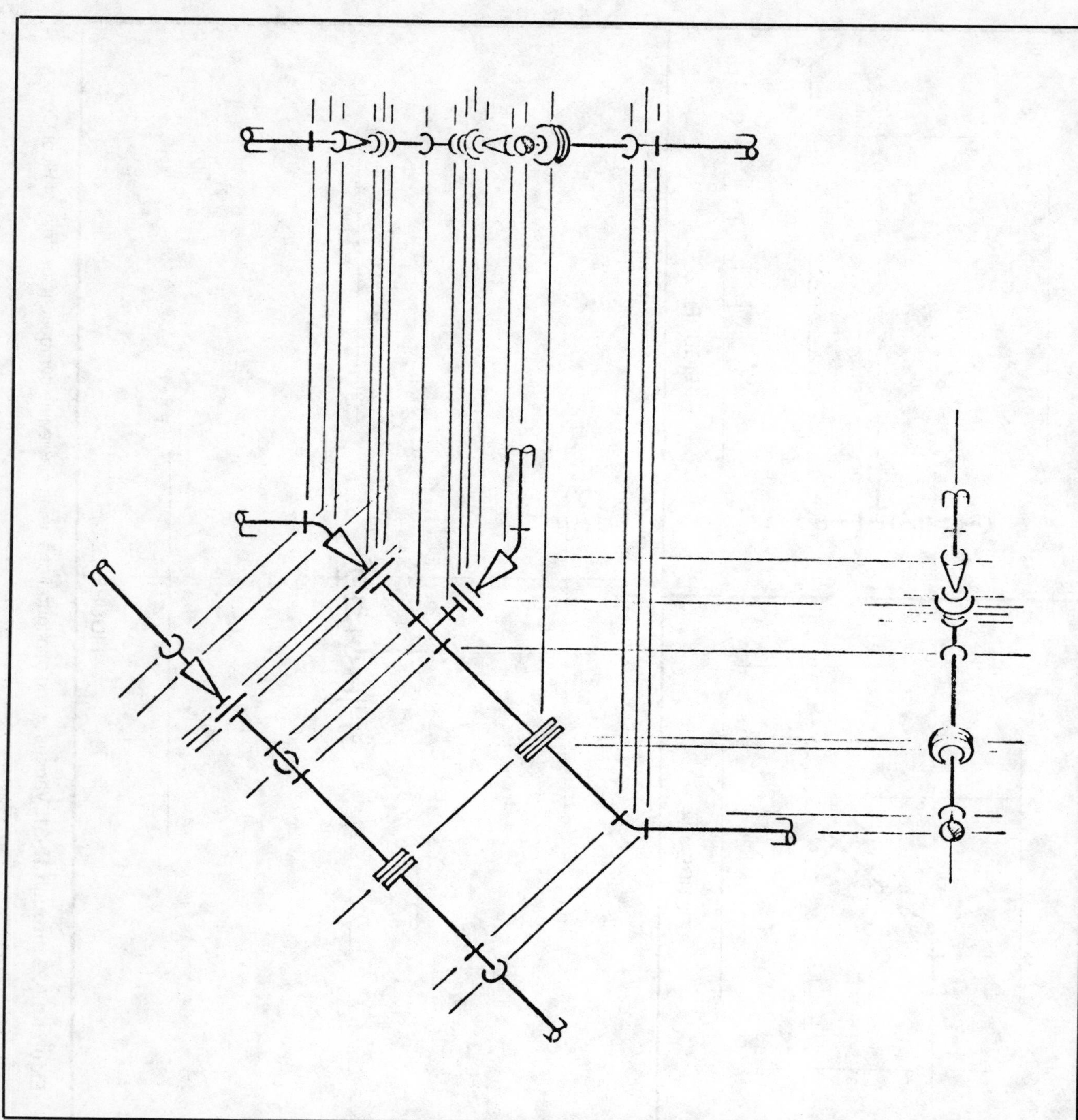

Figure 12-32. Projection study sheet. (Courtesy of the Fluor Corp.)

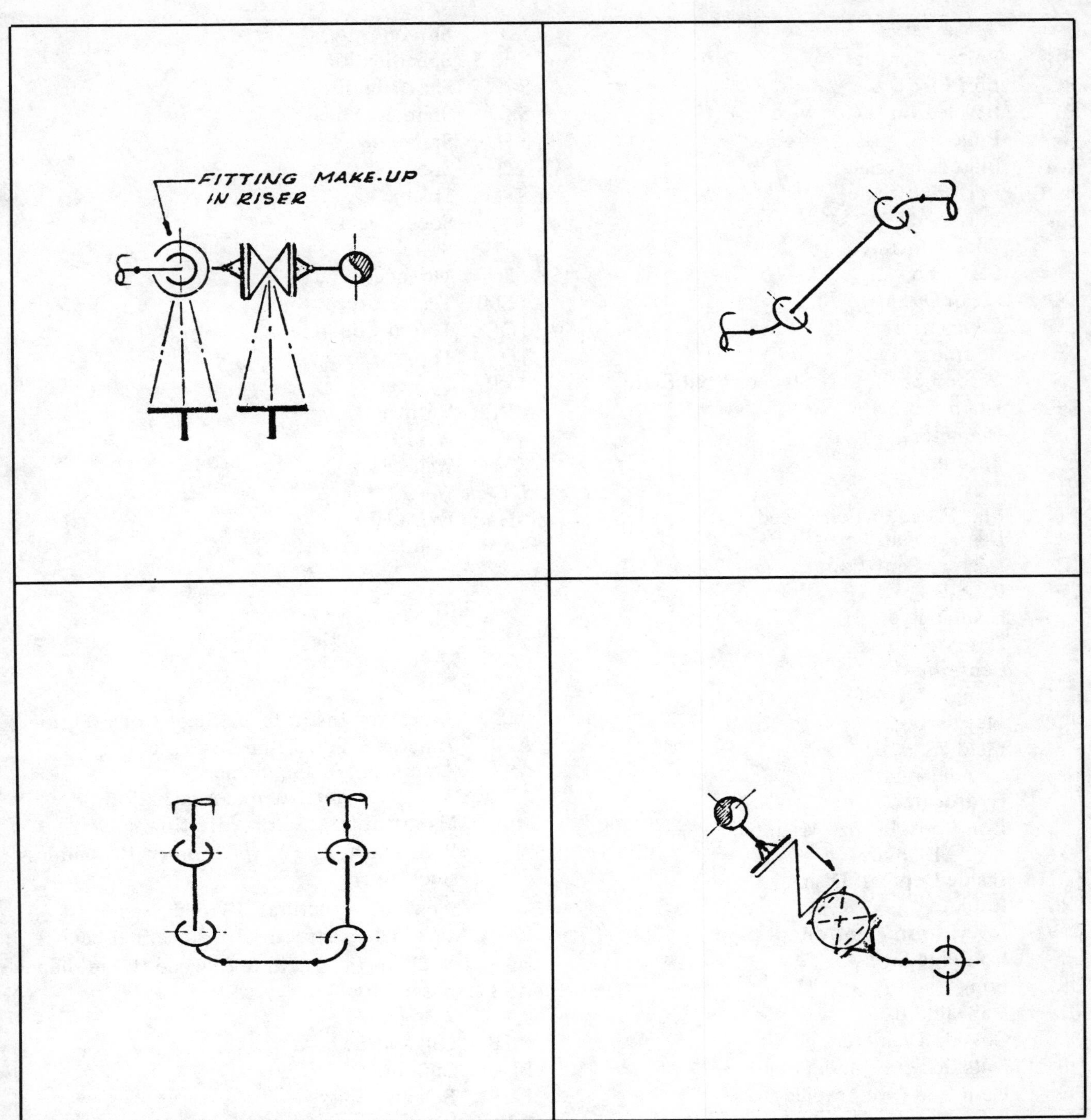

Figure 12-33. Classroom study sheet. (Courtesy of the Fluor Corp.)

(text continued from page 199)

BB Bolted Bonnet
BC Bolt Circle
BE Beveled Ends (for welding)
BF Blind Flange
BM Bill of Material
BOP Bottom of Pipe
BW Buttweld
Ch. Op. Chain Operated
CI Cast Iron
CO Clean Out
CONC. Concentric
CPLG. Coupling
CS Carbon Steel, Cast Steel or Cold Spring
DF Drain Funnel
DIA. Diameter
ECC Eccentric
ELEV. Elevation
FF Flat Faced or Full Faced
°F Degrees Fahrenheit
°C Degrees Centigrade
LBS Pounds
ø Diameter
℘ Plate
℄ Centerline
FLG. Flange
FOB Flat on Bottom
FW Field Weld
GJ Ground Joint
HC Hydrocarbon
IBBM Iron Body Bronze Mounted
ID Inside Diameter
IDD Inside Depth of Dish
INS Insulate
INV Invert (*inside* bottom of pipe)
IPS Iron Pipe Size
LR Long Radius
MI Malleable Iron
OD Outside Diameter
OS & Y Outside Screw and Yoke
PE Plain End (not beveled)
PR Pair
PSIA Pounds Per Square Inch Absolute
PSIG Pounds Per Square Inch Gage
RED Reducer
RF Raised Face
RTJ Ring Type Joint (sometimes just designated RJ)
SCH Schedule
SCRD Screwed
SF Semi-finished
SMLS Seamless
SO Slip-on
SPEC Specification
SR Short Radius
SS Stainless Steel
STD Standard
STL Steel
STM Steam
SW Socketweld
SWG Swage
TE Threaded End
TEMP Temperature
TOC Top of Concrete
TOS Top of Steel
TYP Typical
VERT Vertical
WE Weld End
WN Weld Neck
WT Weight
XH Extra Heavy
XXH Double Extra Heavy

Group 2

AISC American Institute of Steel Construction
API American Petroleum Institute
AWS American Welding Society
AWWA American Water Works Association
MSS Manufacturers Standardization Society
W/ With–such as, W/SS trim (with stainless steel trim)
L Angle (a structural 4" angle shape)
C Channel (a structural 4" channel shape)
W W Shape (a structural 8" wide flange shape)
ASSY. Assembly
AVG Average
B&B Bell and Bell
BLDG Building
B&S Bell and Spigot
BWG Birmingham Wire Gage
CAS Cast Alloy Steel
CO_2 Carbon Dioxide
COND Condensate
CORR. Corrosion
DWG Drawing
EF Electric Furnace
EFW Electric Fusion Welded
ERW Electric Resistance Welded
FIG Figure or Figure Number
FS Forged Steel

FSS Forged Stainless Steel
FT Feet or Foot
GALV Galvanized
GR Grade
H_2 Hydrogen
HDR Header
LC Lock Closed
LO Lock Open
LW Lap Weld
M Miscellaneous Shapes, Steel
M&F Male and Female
MFG Manufacture
MIN Minimum
MW Miter Weld
NI Nickel
NC Normally Closed
NO Normally Open
OH Open Hearth
REINF Reinforce
S Standard Beams, usually called I-Beams
S.O. Steam Out
SQ Square
SWP Standard Working Pressure
S.C. Sample Connection
T.C. Test Connection
T&C Thread and Coupled
T&G Tongue and Groove
VC Vitrified Clay
WB Welded Bonnet

Piping Plans and Elevations or Sections

Figure 12-34 is an actual piping plan. It was initially drawn on 24″ × 36″ tracing paper and has been reduced. Many construction companies use a camera to microfilm all their drawings. A copy of this microfilm is then sent to the field construction office. They have a machine that takes the microfilm and produces them a large working print.

Any light lines on the original tracing, small lettering or generally poor draftsmanship will not reproduce on the microfilm and therefore will not reproduce on the field print. So again, *linework and lettering must be good and sharp*. And, if one works for a company that uses microfilm, his lettering should be a minimum of ⅛″ high.

Now, in reading this piping plan, students will read it in the general order that the good piping draftsman might draw it. Also, they'll consider where he would have to go to get his information to make the finished piping plan. The draftsman's supervisor might give him a copy of the piping drawing index, which would show him his drawing area, or his match lines. It also would show him what equipment, vessels, exchangers, reboilers, pumps, et cetera were in his area.

To learn the coordinates of his equipment, he must refer to the foundation location plan. There he also can learn the coordinates and spacing of his pipe rack. If he is given this project to do before the foundation location plan is drawn—which sometimes happens—then the piping layout, done by an experienced piping designer, will establish the coordinates. Should this occur, the piping draftsman is to check and agree with the layout coordinates.

The piping draftsman will need the mechanical flow diagram and the utility flow diagram. If a sewer flow diagram is made, he will need a copy of it to know where to locate funnels. If the underground piping drawings have been made, this will also tell him this data. He also needs a copy of the line list.

Then, to draw the piping sections (Figures 12-35 and 12-36) he must refer to the foundation drawings for the equipment elevation and the structural steel drawings for the pipe rack elevations.

He also must review the underground electrical drawings to determine if there are underground electrical trenches, electrical conduit or electrical manholes. In general, the piping draftsman should review all possible information about his drawing area. He should know more about what is in the area than anyone else. And he should know this *before* he makes his *first* line on his drawing.

The author has picked a very easy piping plan and sections for the student to study. Many actual areas are so full of piping that not much white is on the drawing. The author is telling the student this to impress on him the need for accuracy in this business. There's only one way to pipe up an area—the accurate way. The area he is looking at will cost about $250,000. If the draftsman is thinking of something else while doing piping, a small error could cost $5,000 to fix in the field. He must learn to concentrate on what he is doing and do it right.

The first thing to do in starting the piping drawing is to lightly draw in the four drawing limit lines. Three are "match lines," matching another drawing area. The fourth is the "area limits,"

(text continued on page 210)

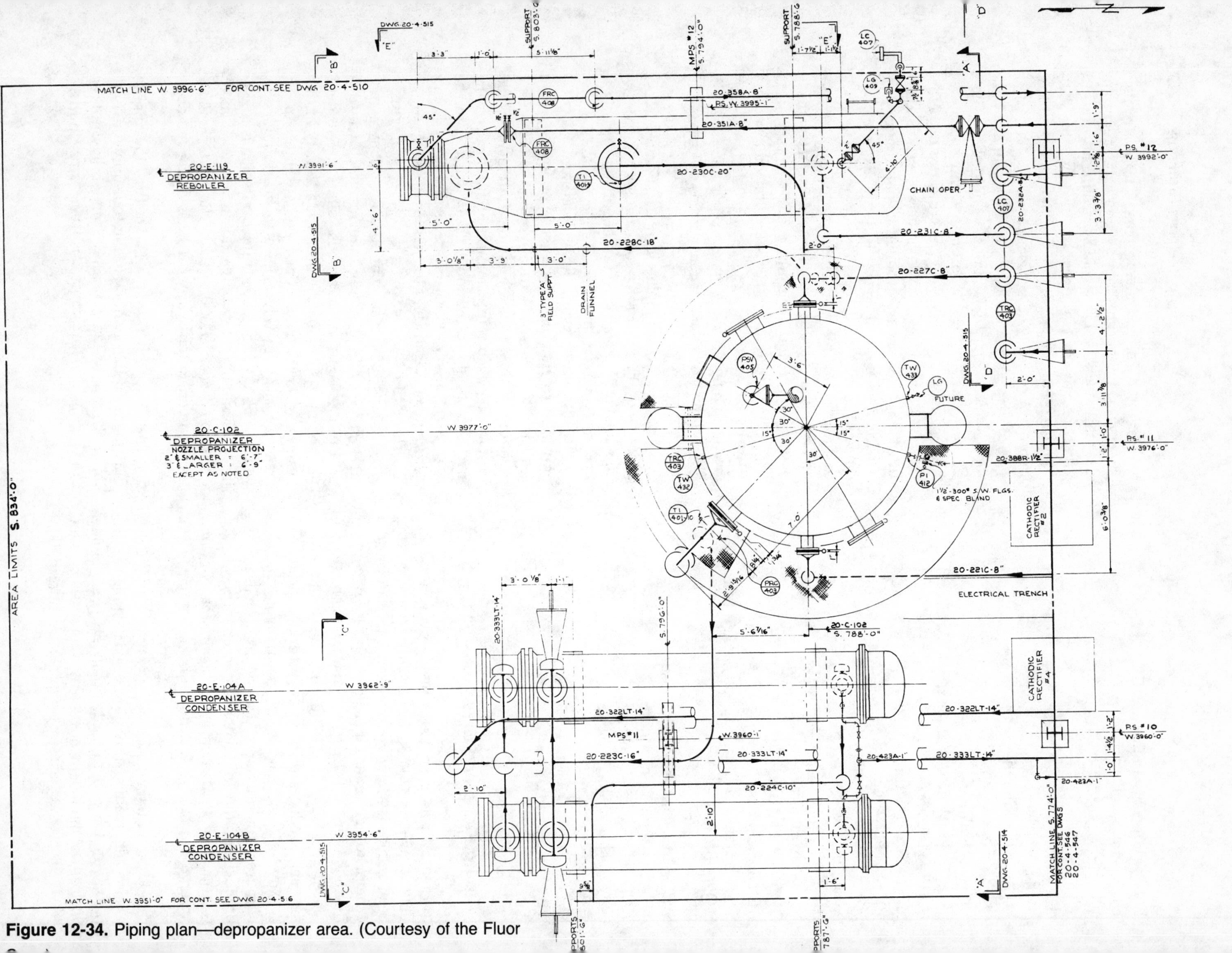

Figure 12-34. Piping plan—depropanizer area. (Courtesy of the Fluor

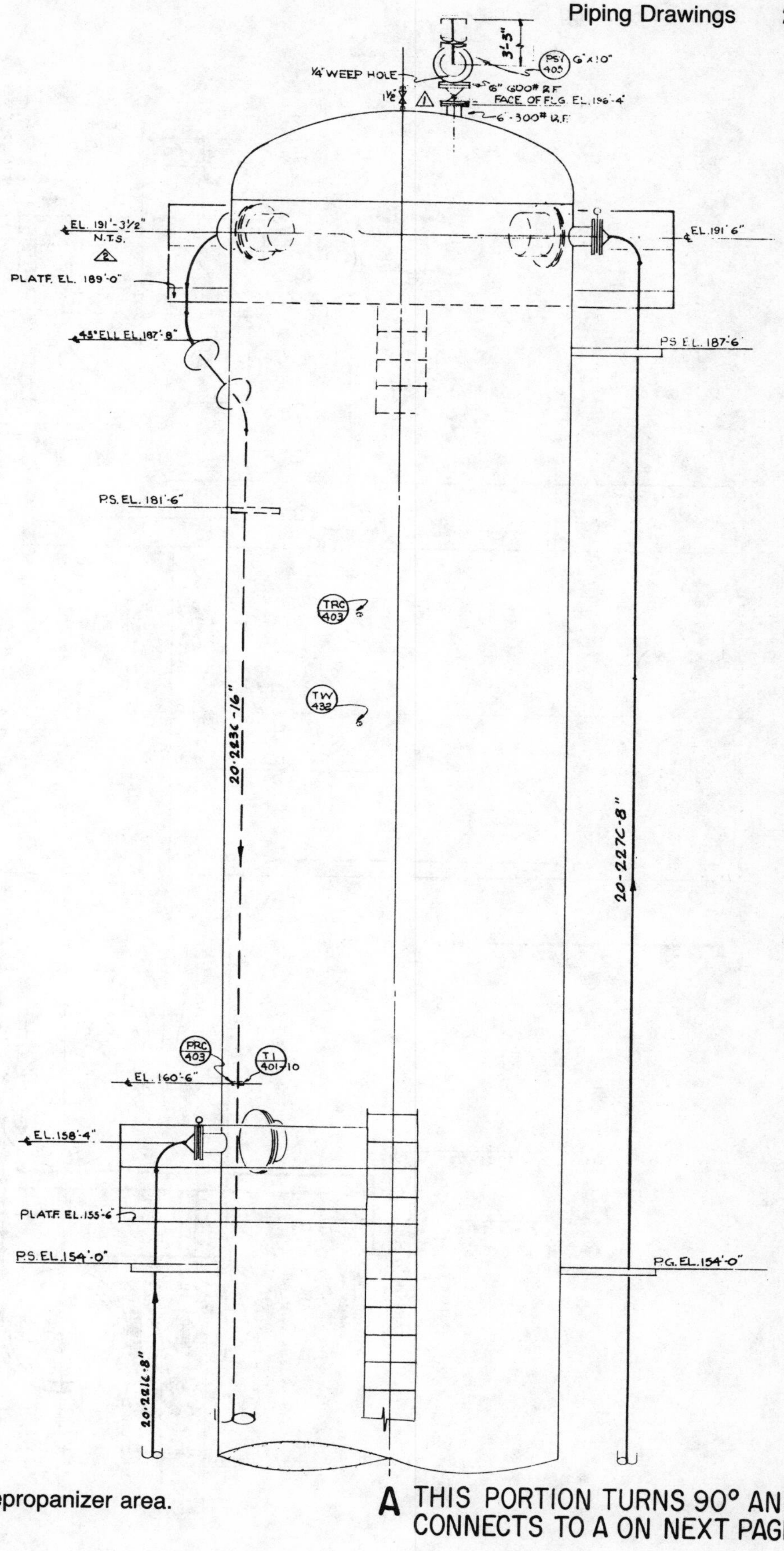

Figure 12-35. Piping sections—depropanizer area. (Courtesy of the Fluor Corp.)

A THIS PORTION TURNS 90° AND CONNECTS TO A ON NEXT PAGE.

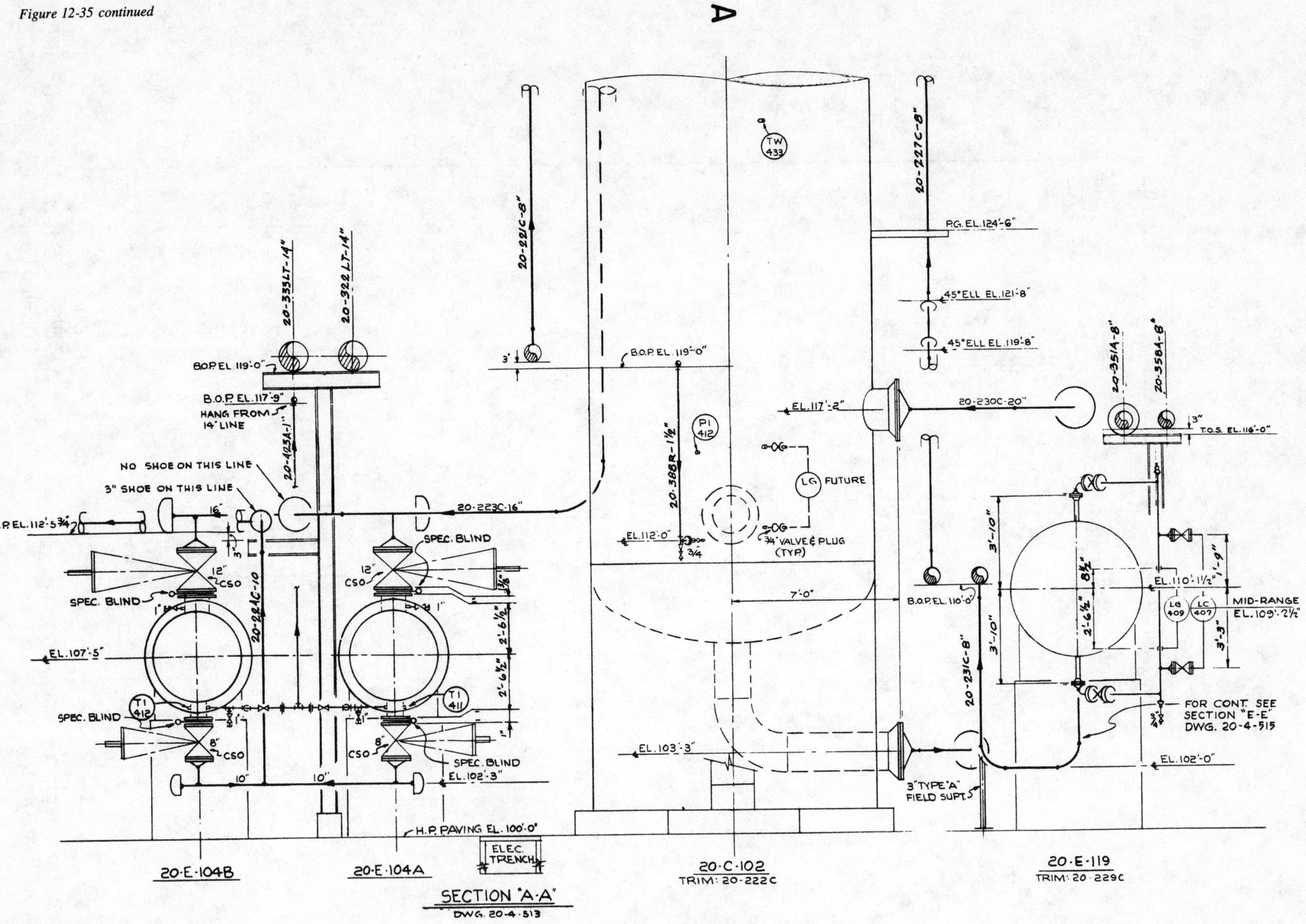

Figure 12-35 continued

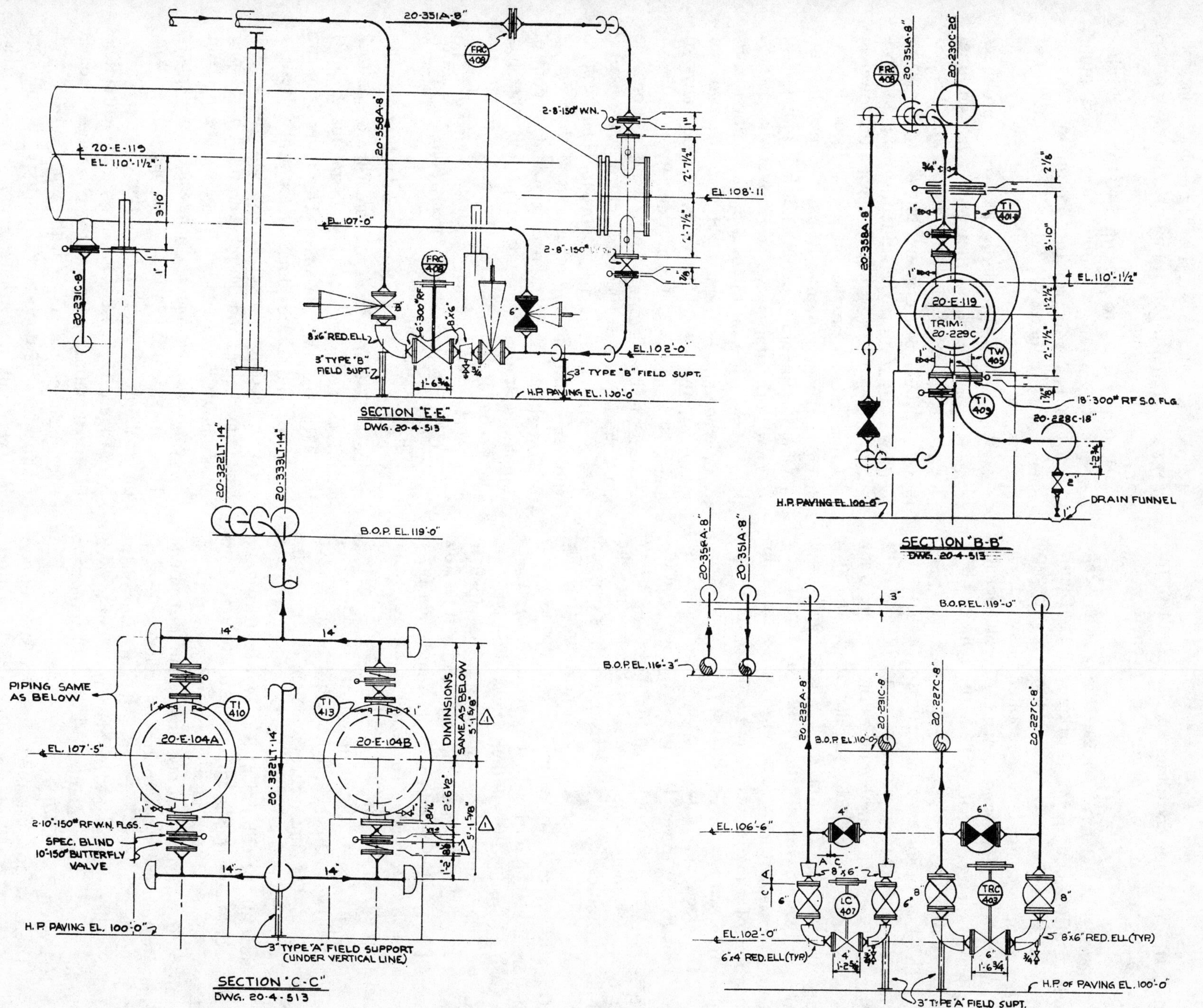

Figure 12-36. Piping sections—depropanizer area. (Courtesy of the Fluor Corp.)

continued from page 205

where the point or coordinate is the extent of the unit being drawn. Coordinates are given for each line.

Next, the student is to locate the crosshair centerlines of the equipment, starting with the Depropanizer. Then, he is to draw in the centerlines of the exchangers; locate the reboiler; draw in the foundations of all this equipment. After he pencils in lightly the coordinates of all these lines, he is ready to draw the equipment.

The Depropanizer is his main piece of equipment in this drawing area. So, he is to draw it in first. The vessel drawing indicates the size and what nozzles are required. The vessel drawing will also give the "nozzle orientation," where nozzles are located around the circumference of the tower. The vessel ladder and platform drawings show the location of these items. One must always orient himself in relation to the North arrow.

The mechanical flow diagram will show all connections to the tower. It may even differ from the vessel drawing as to location in relation to a tray number for some connections. In that event, the student should ask his supervisor which is correct.

The Depropanizer is now drawn in to scale. The student is to draw in the Depropanizer Reboiler, E-119, referring to the vendor's print and drawing it in lightly. He then draws in E-104A and B, two shell and tube exchangers to be piped in parallel. Had threy been piped in series, the flow would have gone through E-104A and then B for the shell or tube side. The other side flow would have been E-104B first; then, E-104A.

Exchangers are piped up with the fluid being cooled flowing down and the fluid being heated flowing up. One must always remember to reverse flow exchangers.

Now all the equipment is drawn in lightly and the coordinates are shown lightly. It's time to draw the piping in. The student must always start at the top of the tower and draw down. The instrument specification sheet determines the relief valve size. The student draws it in to scale and shows the tailpipe discharging up. He then draws in the overhead vapor line, number 223C-16" going to E-104A and B shellside.

The next nozzle is the reflux. The student connects line 227C-8" to the reflux nozzle and goes right on down the tower, showing each connection on the vessel and piping it up or showing the proper instrument bubble.

Now he is ready to pipe up E-104A and B tubeside, drawing in the 14" cooling water lines, number 322LT-14" going in the bottom of the exchanger and number 333LT-14" leaving the top of the tubeside.

As the student probably has some problems trying to visualize how that pipe twists around, now is the time for him to start drawing his sections. His supervisor may tell him where he wants sections cut. If not, the student should cut elevation A-A for all process units, showing the tower in elevation and all nozzles on the shellside for exchangers. Then, section B-B and C-C can be cut, showing the tubeside nozzles and their piping.

On a horizontal vessel or reboiler like E-119, the student will want to cut a partial section, E-E, to show details he cannot show in the two end elevations. But he is not to draw side views of simple exchangers like E-104A and B.

As he is piping up his plans and sections, he must check the flow sheet for control valve assemblies and meter runs and follow the line back to whoever is drawing the other part of that line. If the flow sheet calls for a meter run and/or control station, the student must not assume the other person is showing it in his area. One of the two draftsmen must make sure he is. The author once saw one line going through two areas, *both* having the control valve station.

A good spot at which to locate control valve stations is 2′-0″ out from the centerline of the rack columns. In Figure 12-34 the stations LC-407 and TRC-403 are examples. The latter is board mounted. The student notes the control valve-temperature recorder. He also notes that the by-pass valve, it's flanges and the related piping in the plan aren't shown. Section D-D will show all of this. But, the control valve block's handwheels do appear. They are oriented North, so the operator walking under the pipe rack can reach them.

In Figure 12-35, section A-A, the student notices the fractionating tower is cut in elevation. When he does this, he must also cut the piping in the same relative area as done here. The tower is cut when it is very tall, and a long section has no nozzles.

Basically, there's nothing to show. If one did have some connections, he could not cut the tower.

Also, at the top of the vessel, the 6″ × 10″ PSV has a 6″-600# RF flange on it. The nozzle on C-102 is a 6″ 300# RF. Since the draftsman cannot put a 600# flange against a 300# flange, a pair of flanges has to be welded hub to hub to make this fitup.

All elevations of nozzles are shown. Line 227C-8″ shows the reflux liquid going up to the top tray. At elevation 191′-6″ is the vessel nozzle. Platform elevation is 189′-0″. But what is the PS at elevation 187′-6″? That is a pipe support needed to support the weight of the line to keep this "dead weight" off the nozzle. It also supports "live weight," which would be the weight of the liquid in the line.

Down the tower on the same line is a P.G. at elevation 154′-0″ and 124′-6″? These are pipe guides. They do not actually connect to the pipe, as a pipe support does, but encircle the pipe to keep it from swaying, putting forces on the nozzle.

Pipe supports and guides are drawn on the vessel ladder and platform detail drawing. The piping designer establishes them as he orients the tower and lays out the piping area.

The lines are all single line. Many companies draw all piping as single line. It is quicker. But, for line sizes 14″ and up, double line should be used. It is easier to notice any interference.

Figure 12-36 shows sections B-B, C-C, D-D and E-E of the Depropanizer area. As subtracting coordinates on plans gives the distances between two or more points, subtracting elevations gives the distance between those two elevation points.

In section C-C, the centerline elevation of E-104 A and B is 107′-5″. Adding 5′-1⅝″ to this gets to the 14″ header at elevation 112′-6⅝″. The 90° ell is at BOP El. 119′-0″. A BOP elevation appears only where supports are located. Dimensioning pipe always use centerlines. One-half the OD, or 7″, must be added to the BOP to get centerline elevation 119′-7″. Then, by subtracting 112′-6⅝″, the dimension between the centerline of the 90° ell and that of the centerline of the 14″ header is 7′-0⅜″.

Figures 12-37 and 12-38 are plans of the upper and lower levels in the pipe rack. These are typical of pipe racks–i.e., utilities are in the top level and the process piping is in the bottom level.

Figures 12-39 and 12-40 show the piping plan and sections for Depropanizer Reflux Drums and their related pumps. Piping here gets a little more congested. Pumps G-104A and G-105 are the normally operating pumps. Piping at G-104B can take suction from either Reflux Drum. It is called a "common spare." The student is to study this piping carefully, for common spares are used quite often.

One can order pumps with a top suction instead of an end suction, like those here. The discharge remains on the top in either case. Piping for common spares is usually simpler if all three pumps have top suctions.

Piping Isometrics (Spools)

Piping isometrics are the end product of all those expended manhours. Figure 12-41 is an isometric of line 226C, the 12″ suction line to G-104A and B. An isometric is drawn for *each line* and is sent to a fabricator or to the field if they are field fabricating.

The author likes to think of these process units as erector sets. Each unit is custom designed. Isometric detail sheets of each piece of piping are made. The isometric shows all flanges, valves, instrumentation connecting to it and every weld as well as the equipment to which it connects. It also shows all dimensions needed to fabricate the pipe. Dimensions are to the closest 1/16″. After fabrication, field forces will erect the pieces, which must fit.

Assembly material–such as valves, bolts, gaskets, spectacle blinds and start-up strainers–are bought and sent directly to the job site. The pipe fabricator will only furnish pieces that require welding.

This line would be made and shipped in three pieces. Piece 1 would be from the weld neck flange at C-106 nozzle and down to the two 12" block valves. Piece 2 might be the small piece at G-104A suction which would be two 12" weld necks, one 90° ell, a short piece of 12" pipe and a ¾" coupling for a PI. Piece 3 would be the suction piece at G-104B.

Supports under the ells at the pump suction are called out as 3" Type "A" FS (field support), and no dimension is shown. The field makes these. After the 12" line is in place, the field will measure the support length needed and then fabricate and install this support.

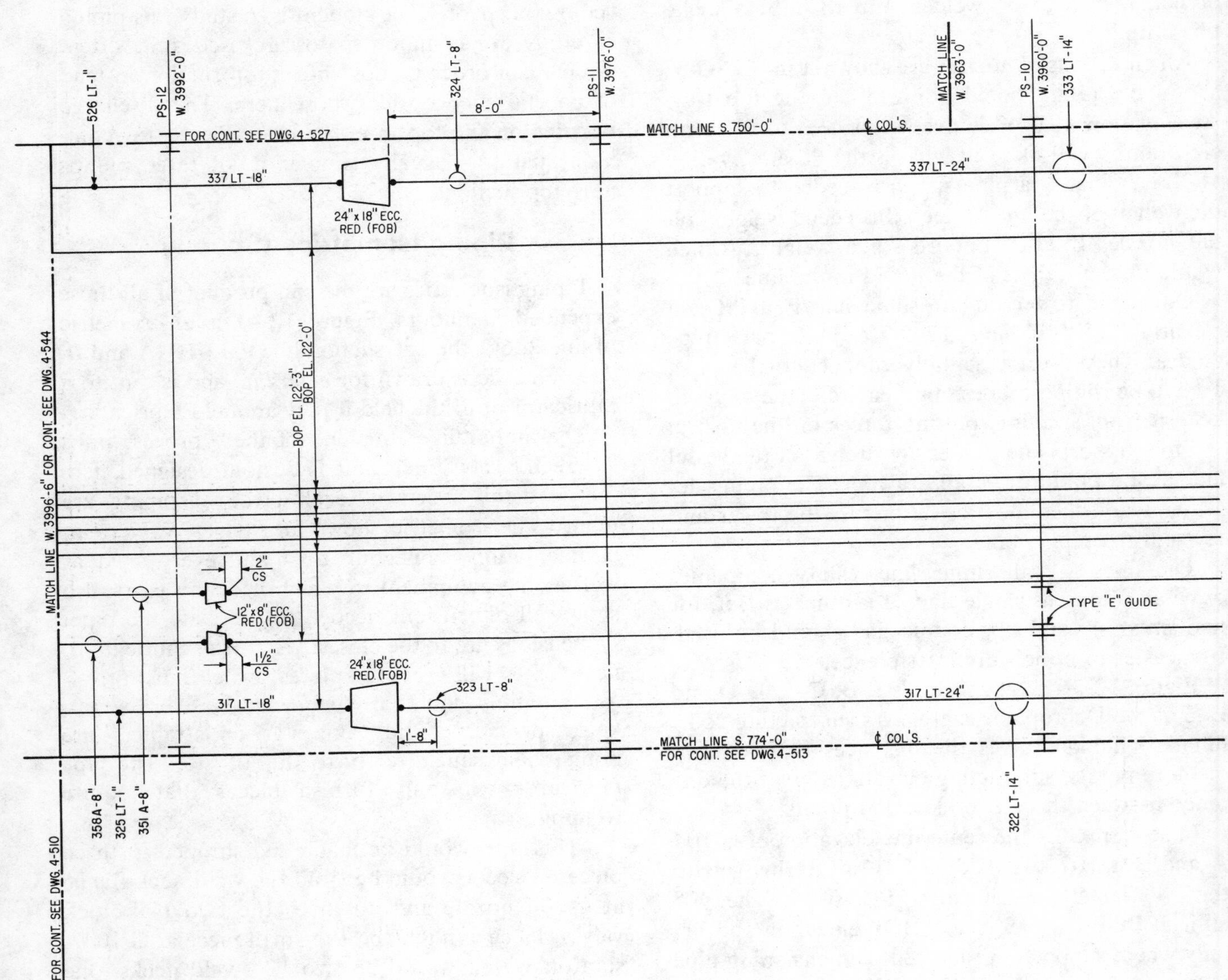

Figure 12-37. Piping plan—upper pipe rack—fractionation unit. (Courtesy of the Fluor Corp.)

figure continued on next page

Figure 12-37 continued

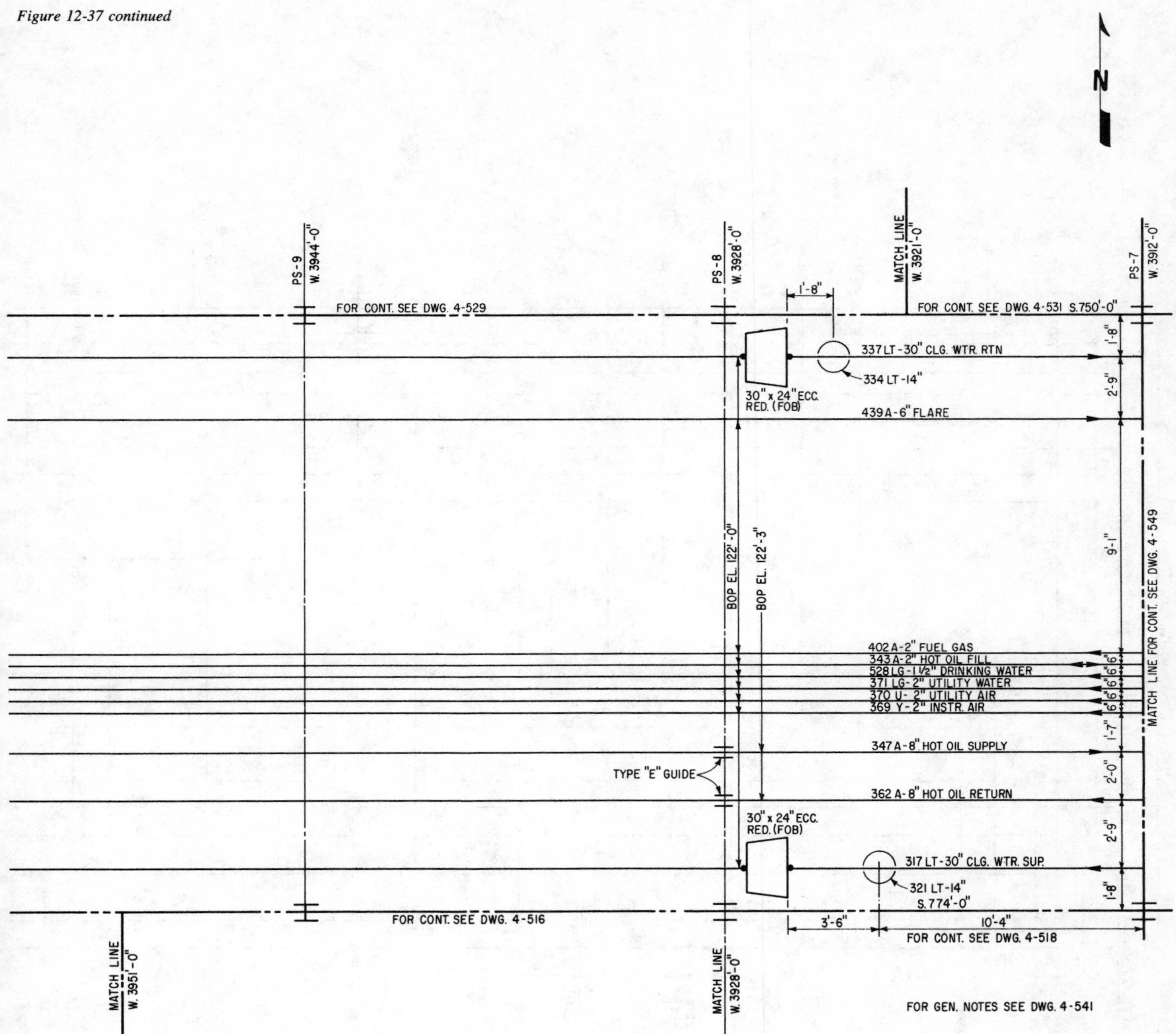
N
PS-9
W. 3944'-0"
PS-8
W. 3928'-0"
MATCH LINE
W. 3921'-0"
PS-7
W. 3912'-0"
1'-8"
FOR CONT. SEE DWG. 4-529
FOR CONT. SEE DWG. 4-531 S.750'-0"
337 LT-30" CLG. WTR. RTN
334 LT-14"
30" x 24" ECC. RED. (FOB)
439 A-6" FLARE
2'-9"
9'-1"
BOP EL. 122'-0"
BOP EL. 122'-3"
MATCH LINE FOR CONT. SEE DWG. 4-549
402 A-2" FUEL GAS
343 A-2" HOT OIL FILL
528 LG-1 1/2" DRINKING WATER
371 LG-2" UTILITY WATER
370 U-2" UTILITY AIR
369 Y-2" INSTR. AIR
6" 6" 6" 6" 6"
1'-7"
347 A-8" HOT OIL SUPPLY
2'-0"
TYPE "E" GUIDE
362 A-8" HOT OIL RETURN
30" x 24" ECC. RED. (FOB)
317 LT-30" CLG. WTR. SUP.
321 LT-14"
S.774'-0"
FOR CONT. SEE DWG. 4-516
3'-6"
10'-4"
FOR CONT. SEE DWG. 4-518
MATCH LINE
W. 3951'-0"
MATCH LINE
W. 3928'-0"
FOR GEN. NOTES SEE DWG. 4-541

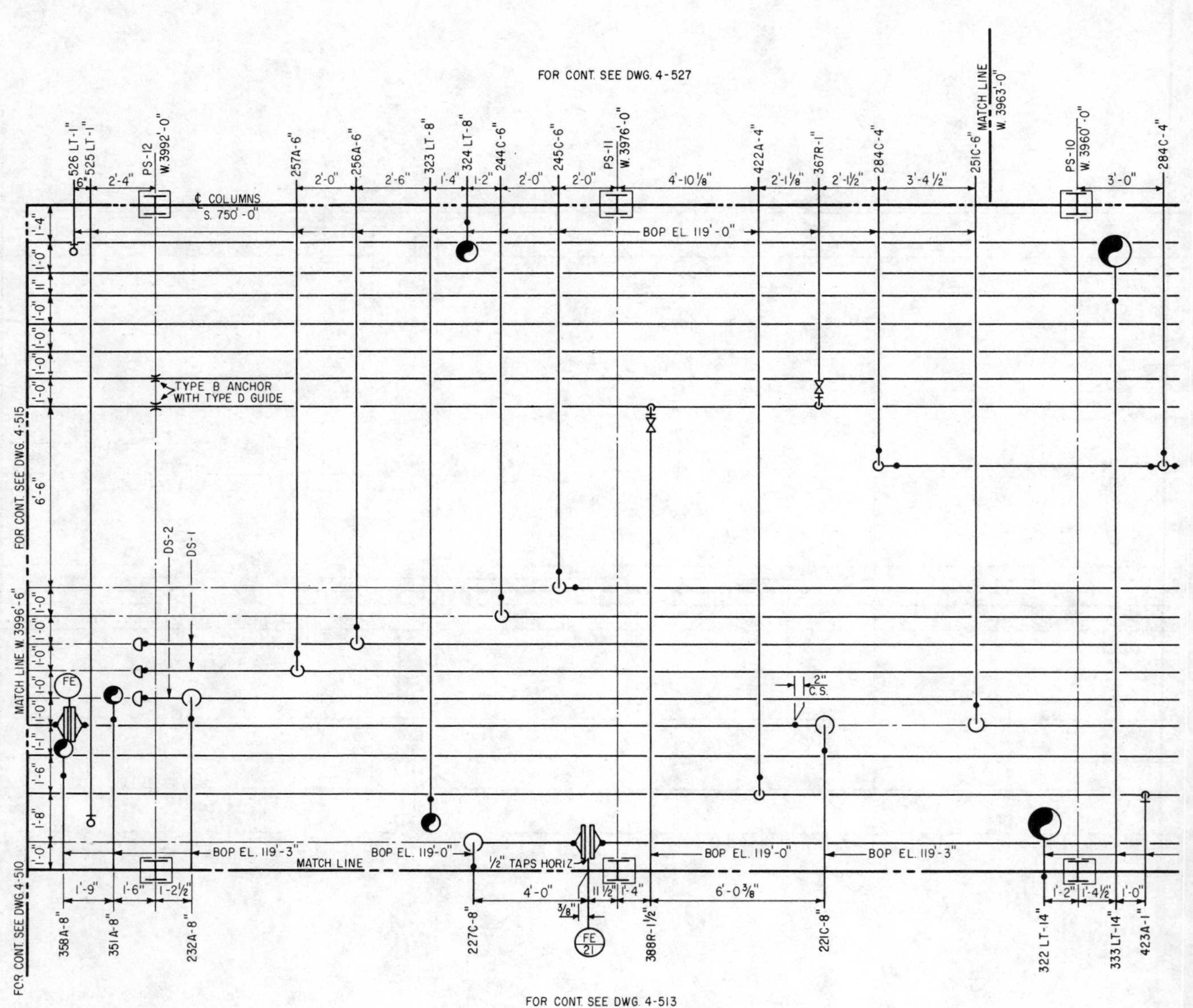

Figure 12-38. Piping plan—lower pipe rack—fractionation unit. (Courtesy of the Fluor Corp.) *figure continued on next page*

Figure 12-38 continued

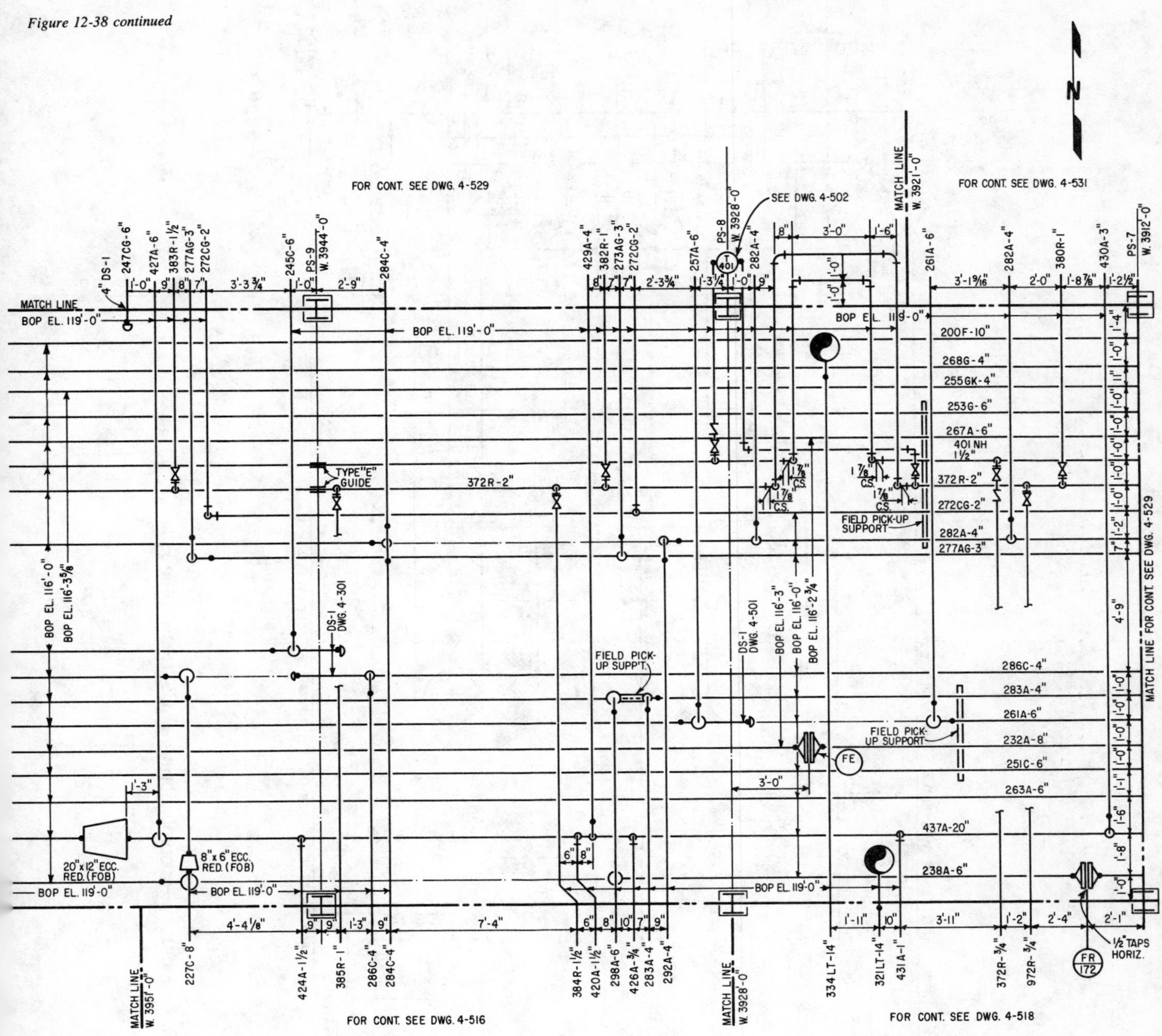

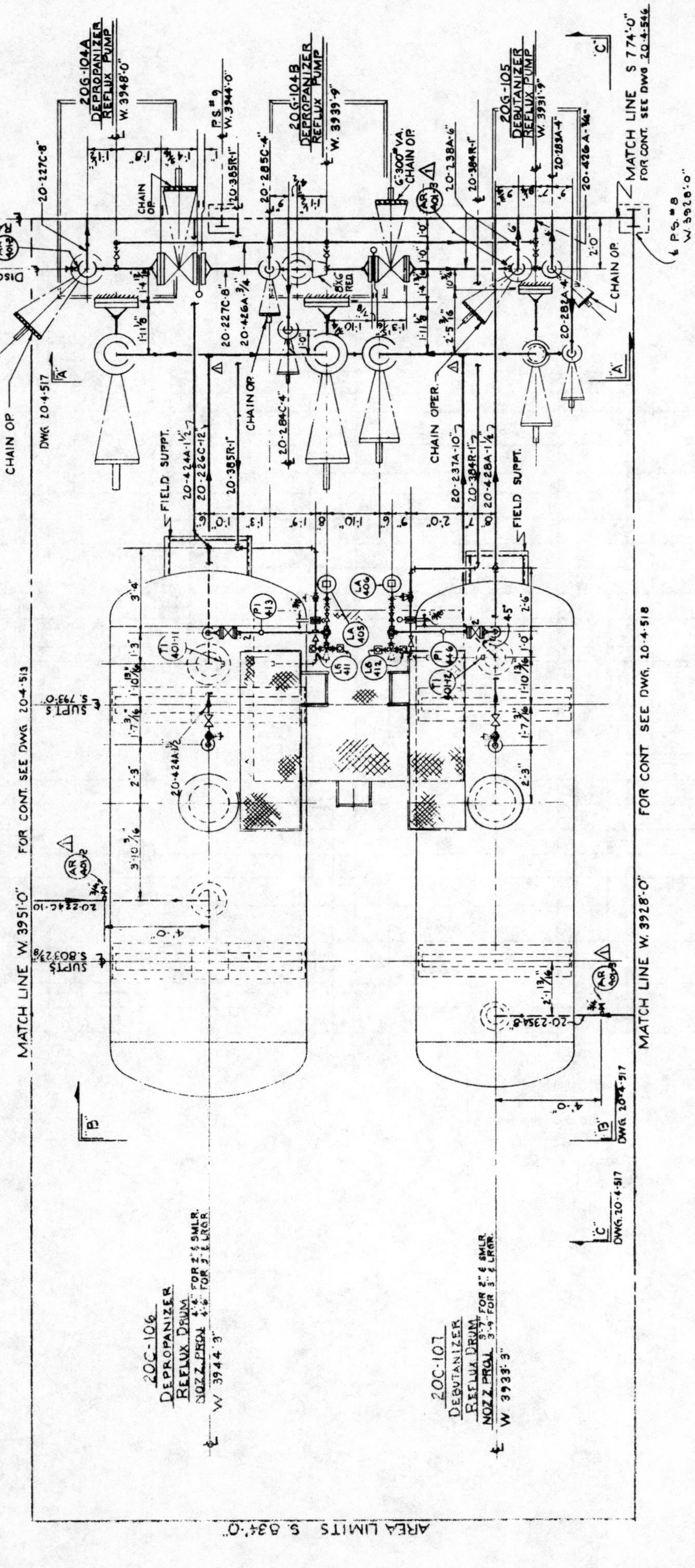

Figure 12-39. Piping plan—accumulator area. (Courtesy of the Fluor Corp.)

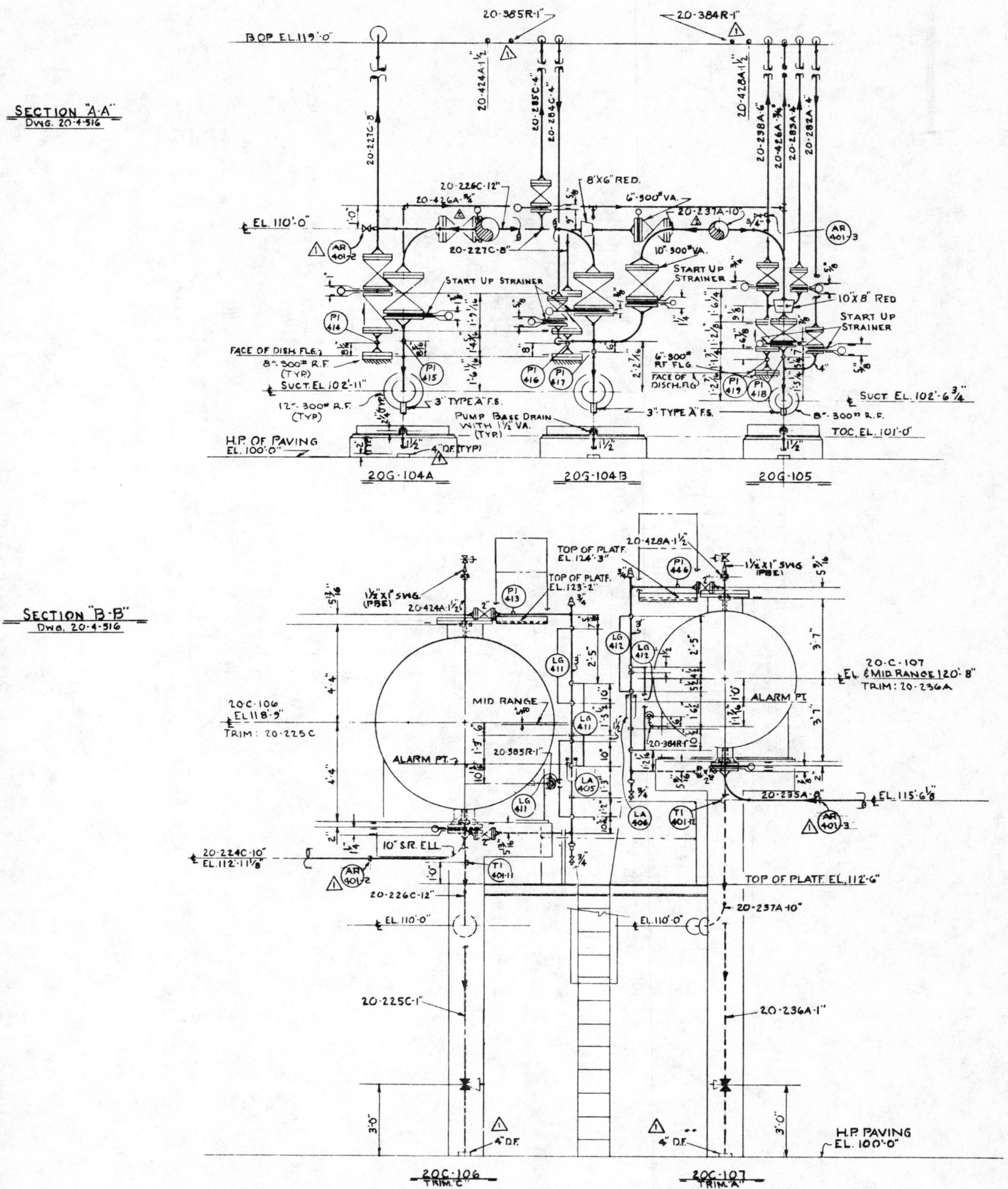

Figure 12-40. Piping sections—accumulator area. (Courtesy of the Fluor Corp.)

Figure 12-40 continued

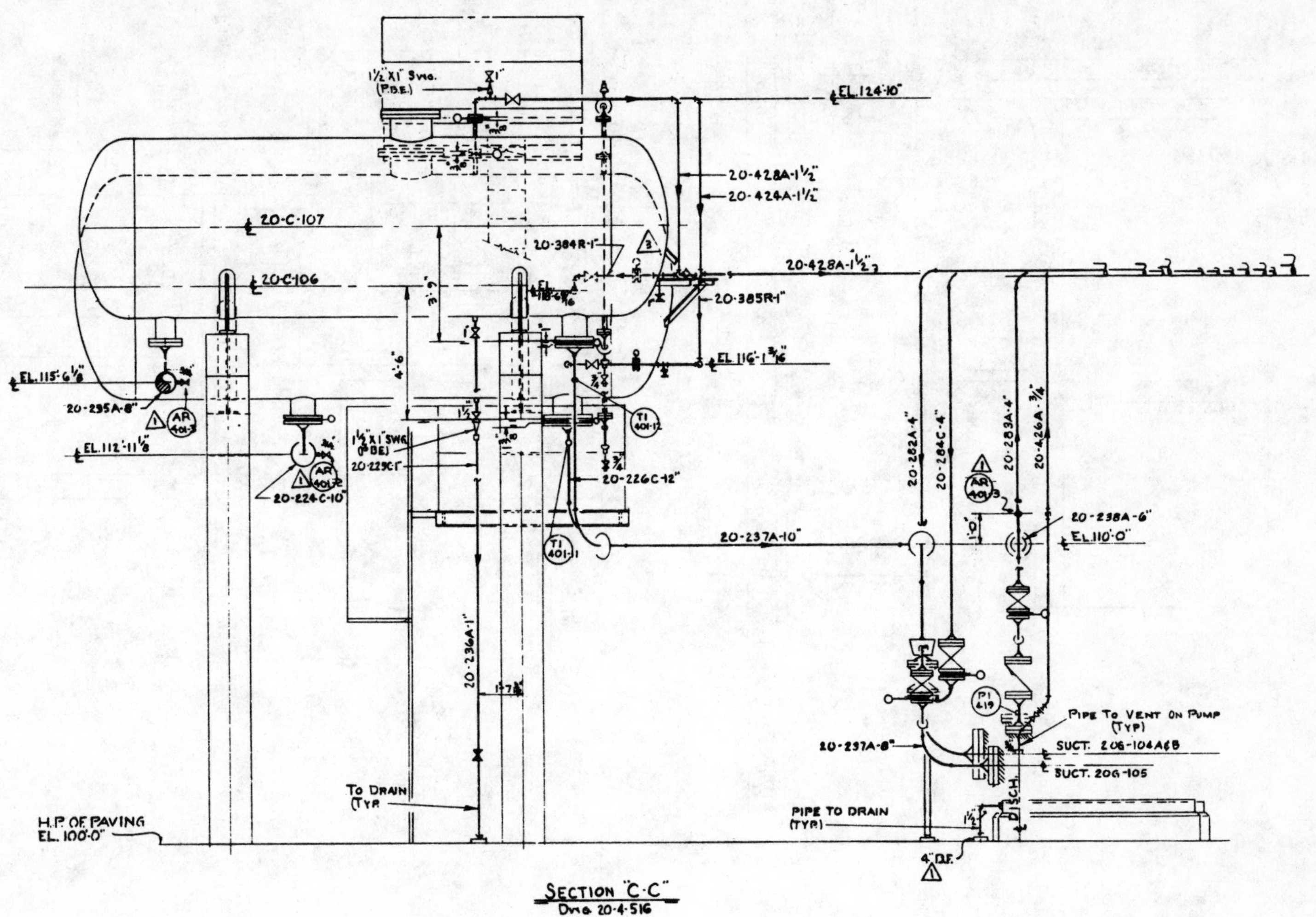

DATE ____ CHKD. ____ DATE ____ BY ____

"S" DENOTES "SCREWED"
"SW" DENOTES "SOCKET WELD"
PLAIN BOTH END NIPPLES CUT FROM PIPE (LENGTH INCL. W/PIPE)

SHOP FABRICATED MATERIAL

REQ'D.	SIZE	WT.	ITEM	REQ'D.	SIZE	WT.	ITEM	REQ'D.	SIZE	WT.	ITEM	REQ'D.	SIZE	ASA RATING	FLANGES
			90°W.ELL				W.TEE		x		W.RED.				# S.O.
			90°W.ELL				W.TEE		x		W.RED.				# S.O.
			90°W.ELL				W.TEE		x		W.RED.				
			90°W.ELL												
			SR W.ELL				W.CAP		x		Sw-				# W.NK. BORE
							W.CAP		x		Sw-				# W.NK.
			45°W.ELL				W.CAP								# W.NK.
			45°W.ELL								# FULL COUP.				# W.NK.
											# FULL COUP.				# W.NK.
											# FULL COUP.	SET			# ORIFICE W.NK.

FEET REQ.	SIZE	SCHED.	PIPE: WELDED	SMLS.	MAT.	A.S.T.M.	GR.	WALL

FIELD FABRICATED MATERIAL

REQ'D	SIZE	WT.	ITEM	REQ'D	SIZE	WT.	ITEM	REQ'D	SIZE	WT.	ITEM	REQ'D	SIZE	ASA RATING	FLANGES
2	4"	STD	90°W.ELL	0+	12"	STD	W.TEE		x		W.RED.				# S.O.
1	10"	STD	90°W.ELL	△1			W.TEE		x		W.RED.				# S.O.
5	12"	STD	90°W.ELL				W.TEE		x		W.RED.				
			90°W.ELL												
							W,CAP		x		Sw-	1	4"	300 RF	# W.NK BORE 4.026
			SR W.ELL				W.CAP		x		Sw-	1	10"	300 RF	# W.NK 10.020
							W.CAP					7	12"	300 RF	# W.NK 12.000
			45°W.ELL					2	3/4"	3000 # SCRD	FULL COUP.				# W.NK.
			45°W.ELL					1	1"	3000 # SCRD	FULL COUP.	SET			# ORIFICE W.NK.

FEET REQ.	SIZE	SCHED.	PIPE: WELDED	SMLS.	MAT.	A.S.T.M.	GR.	WALL
1	4"	40		✓	CS	A-53	B	
1	10"	30		✓	✓	✓	✓	
17'	12"	✓		✓	✓	✓	✓	

FIELD ASSEMBLY, SCREWED & SOCKET WELD MATERIAL

ITEM	Q	S	FIG. NO.	Q	S	FIG. NO.	Q	S	FIG. NO.	Q	S	FIG. NO.	Q	S	FIG. NO.
GATE	2	1/2"	VOGT # 12113	4	3/4"	VOGT # 12114	1	4"	OIC # 3002-C	1	10"	OIC # 3002-C	2	12"	OIC # 3002-C
GLOBE															
PLUG															
CHECK															

Q	S		ITEM	Q	S	ITEM	Q	S	ITEM	Q	S	ITEM
2	3/4" x 1/2"	S/160	Sw.-		x	Sw.-			# BLIND FLG.			# SCRD.FLG.
	x		Sw.-		x	Sw.-						
2	1 1/2"	3000# SW	90 ELL	2	1/2"	2000# SCRD TEE			# UNION	3	12"	300# SPEC. BLD.
			# 90 ELL			# TEE			# UNION			
			# 90 ELL			# TEE			# UNION			XH PIPE NIP.-TBE
							2	1/2"	SOLID STL.PLUG	1	4"	START UP STR.
			# 45 ELL			# CAP			STL.PLUG	1	10"	START UP STR
			# 45 ELL			# CAP			STL.PLUG	2	12"	START UP STR.

ITEM	Q	S		Q	S		Q	S		Q	S		Q	S		Q	S	
1/16" THK. JM 60 ASB. GASKETS	3	4"	300#	3	10"	300#	12	12"	300#			#			#			#
GASKETS			#			#			#			#			#			#
MACHINE BOLTS			x			x			x			x			x			x
GR "B 7" STUD BOLTS	8	3/4" x 4 1/4		8	3/4" x 5"		16	1" x 6"		16	1" x 7 1/4		64	1 1/8" x 6 1/2		48	1 1/8 x 8"	

FEET REQ.	SIZE	SCHED.	PIPE: WELDED	SMLS.	MAT.	A.S.T.M.	GR.	WALL	Q	S	ITEM
1'	3/4"	80		✓	CS	A-106	B		2	3"	TYPE A FS
3'	1 1/2"	✓		✓	✓	✓	✓		1	4"	300# SPEC BLD
									1	10"	300# SPEC BLD.

Figure 12-41. Piping isometric. (Courtesy of the Fluor Corp.)

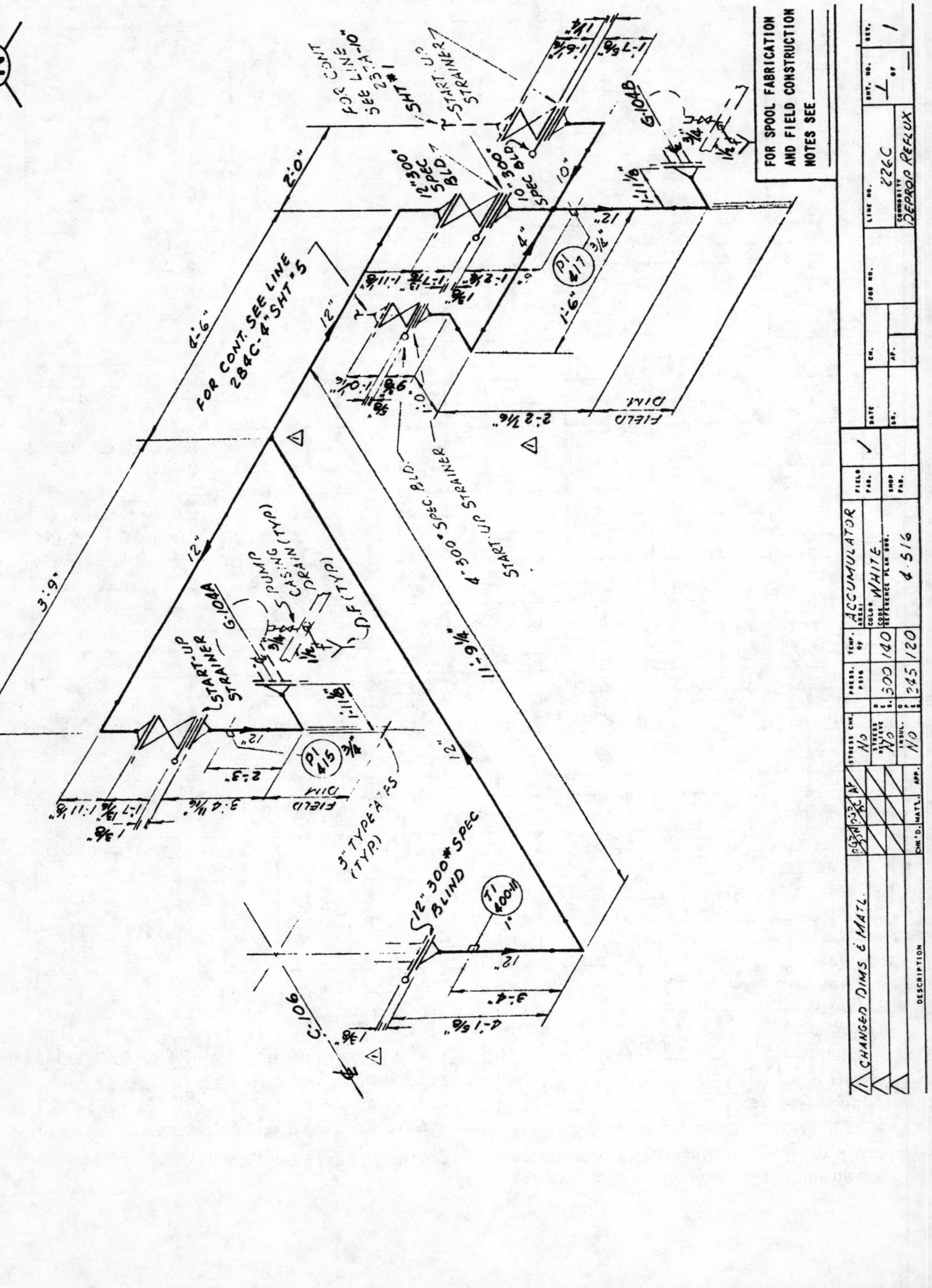

Figure 12-41 continued

CHAPTER 13

Pipe Fabrication Drawings

Fabrication is the assembling and attaching of component pieces to make a completed item. Fabrication of piping, the joining together of weldable pipe and fittings, is done by field personnel at the job site and by "shop fabricators," a shop located at a metropolitan area with access to qualified personnel and all materials on a large scale. Shop fabrication is usually less costly than field fabrication due to modern assembly line techniques and access to the latest fabrication equipment. This savings will average 10-15% over the usual field fabrication. Because shops have instant communication with steel mills and large suppliers, they can locate special materials or fittings for their customers' needs. And as they buy huge quantities of piping materials yearly, they can get the best possible prices. Shop fabrication thus saves money on both labor and materials.

Most pipe shops employ all union personnel. The national union agreement calls for all pipe 2″ and smaller to be fabricated in the field. By arrangement with the local union, smaller pipe may be shop fabricated. They usually agree to this if they cannot supply enough people to field fabricate the work.

Normally, pipe 3″ and larger is shop fabricated. Long straight runs are not supplied by shops. This pipe is sent directly to the field. Underground pipe is field fabricated.

Shops fabricate pipe for all companies. Each company's drawings and specifications are sent to the shop in many different kinds of forms. Half of their drawings are supplied as fully dimensioned plans and elevations. The other half are line isometrics, referred to as "spools" by contractors. Shops do not call isometrics spools. Shops prepare a drawing of a piece that is shippable and they call this piece a spool. A line isometric may produce a dozen shop spools.

Welding

Shops have welders and engineers fully qualified to make almost any type of weld necessary. Most large contractors, such as Fluor, have complete welding specifications and procedures which shops use. However, for customers who do not furnish welding specifications, shops have their own specifications.

Welding symbols have been standardized by AWS, the American Welding Society. Figure 13-1, welding symbols, shows these standards.

Shop Details

The shop's drafting room prepares drawings of shop spools, showing every piece and every detail dimension necessary for fabrication. With each shop piece drawing a list of material is shown with cut lengths for each piece of pipe to the closest 1/16″.

When plans and elevations are supplied, a "take-off" man is given the job of preparing a

(text continued on page 224)

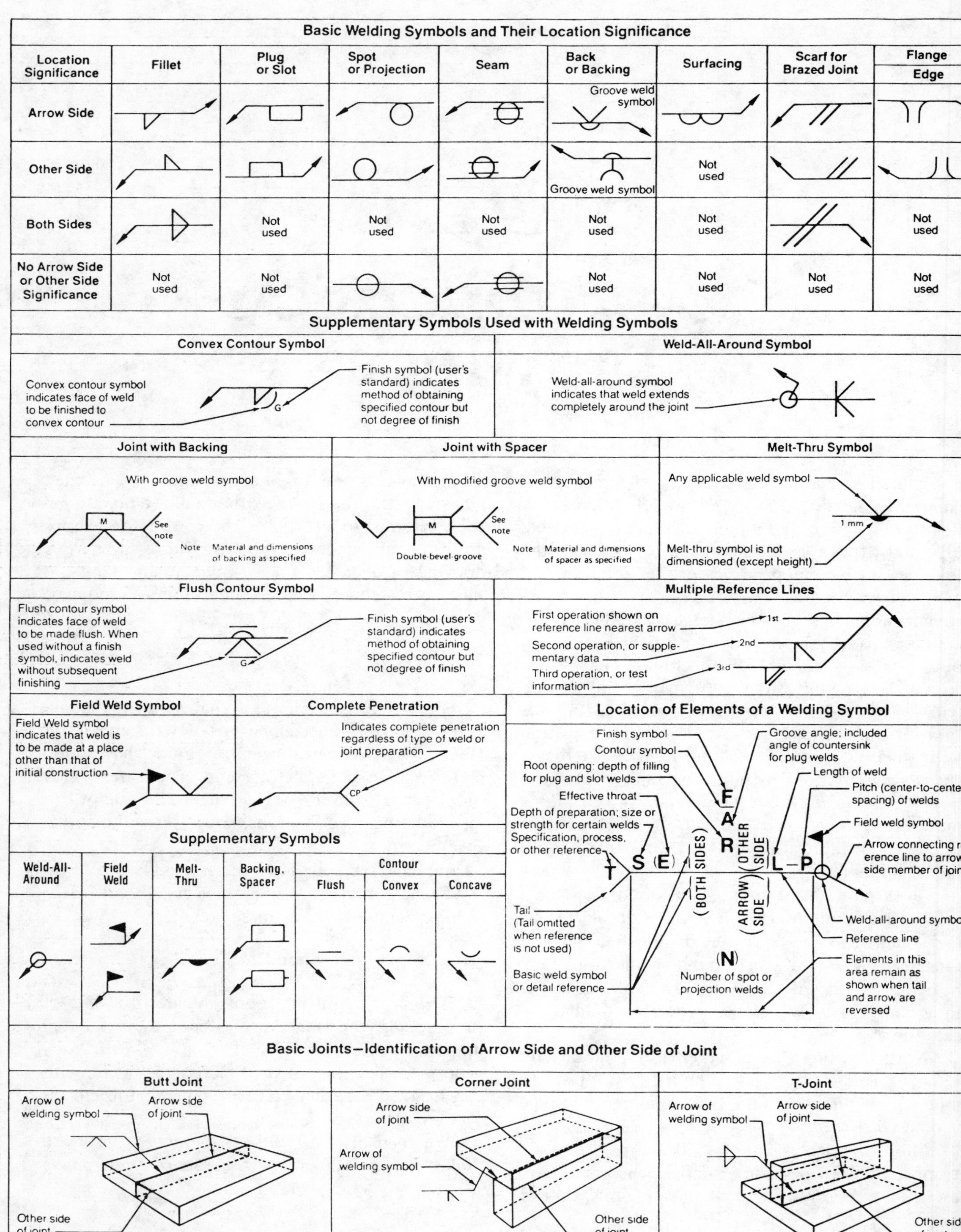

Figure 13-1. Welding symbols. (Courtesy of Texas Pipe Bending Company, Inc.)

Figure 13-1 continued

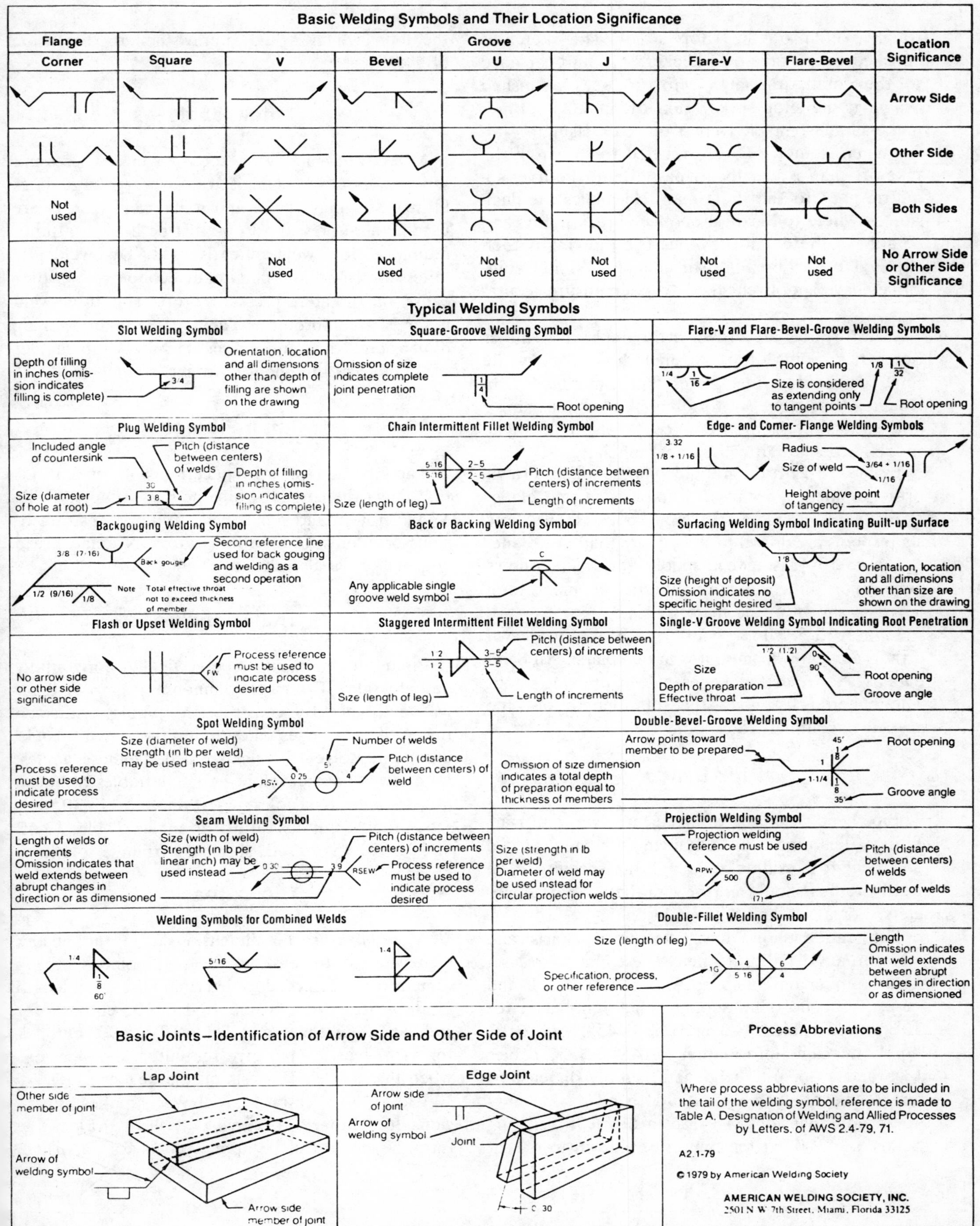

rough sketch of the shop spool. The sketch is checked by a "take-off checker." Then it is passed to the detail draftsman who produces a finished drawing, the shop spool. Take-off men sometimes make a line isometric to pass on to draftsmen.

When isometrics are furnished to the shop, the take-off man is usually by-passed and the line isometric goes to the draftsman. He makes the decision of where to locate shop piece break points and completes the detail drawing. The drawing is then checked by a "break-down checker." Some companies have spools checked twice to ensure accuracy.

Shop spools are drawn orthographically and as isometrics. Some shops do single line spools while some do double line. Texas Pipe Bending Co. in Houston has done all kinds and now does single line spools in isometric except foı single plane spools, which are drawn orthographically.

Figure 13-2, orthographic pipe spool, is drawn orthographically because the pipe is in one plane.

Figure 13-3, isometric pipe spool, shows how shop spools are drawn to show more than one plane.

Figure 13-4, isometric spool with miters, shows detail miter dimensioning. Note that the pipe's total length is shown in the material list.

Figure 13-5, shop spool with bends, is drawn orthographically because it is in one plane. All bend data is shown in the boxes at the drawing's top. The material list gives the total length of pipe needed to complete the spool.

Pipe Bends

Pipe bends are used to make turns without using fittings. The pipe is usually filled with sand, heated and bent to a radius and angle as specified. The bend's radius should not be less than 5 pipe diameters. For a 12″ line this would be 5′.

Fabrication shops have developed charts and tables to aid detail draftsmen. These charts are useful to piping designers as well. Tables 13-1 and 13-2, 30° bend data, supply dimensional data for 30° bends. Tables 13-3 and 13-4, 45° bend data, supply the same information for 45° bends. Tables 13-5 and 13-6, 60° bend data, supply dimensions for 60° bends. Table 13-7, 90° bend data, shows 90° bend dimensions. To read this table, for a 3′-6″ radius, see 3′-0″ at the top and go down to 6″ (at the left) which would supply the arc dimension of 5′-6″.

Miter Welds

Miter welds are often specified in low pressure services as elbow substitutes. In very large lines fittings are unavailable and miter weld elbows are used. Where pressure drop must be held to a minimum, the four-weld miter is used. Two-weld 90° miters are used for maximum economy, but they cause the greatest pressure drop. The three-weld miter is a compromise. Table 13-8, miter welding dimensions, gives full details on miters. For angles of 45° or less the one-weld miter is common.

Small Fittings

The pipe fabricator is concerned with dimensions that affect the length of pipe he must supply. Table 13-9, screwed and socketweld fittings, gives dimensions of interest to pipe fabricators. Normal thread engagement is also shown.

The Triangle

Piping designers run pipe vertically, horizontally and at angles. The most common angles formed are 30° and 45°. By construction, piping draftsmen make 90° triangles and apply their math background to solve triangles formed by these angles. Pipe shops have developed triangle tables to aid in quick solutions. Tables 13-10 and 13-11, 30° offsets, supply solutions for 30° triangles. Tables 13-12 and 13-13, 45° offsets, show 45° triangle solutions.

The Cutback

A *cutback* is the dimension from the header centerline to the nozzle's nearest point. Cutback dimensions are needed to determine the exact length of the nozzle pipe. Table 13-14, 90° cutback for standard weight pipe and Table 13-15, 90° cutback for extra heavy pipe, give the cutback dimensions when the nozzle's ID rests on the header's OD.

Table 13-16, cutback at elbows, supplies dimensions for cutbacks occurring at 90° elbows. The formula shown can be applied for sizes not listed.

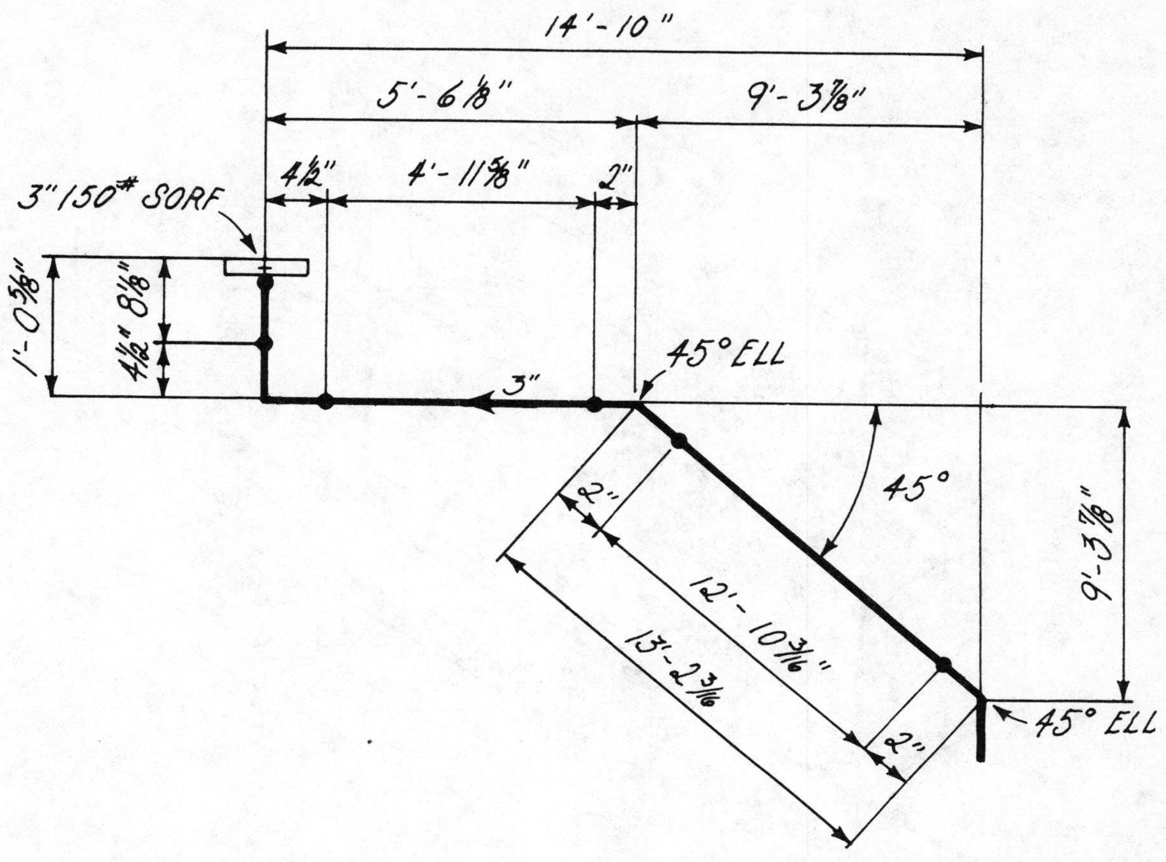

Material List

Quantity	Size	Description
1	3"	Std. Wt. LR 90° Ell
1	3"	Std. Wt. 45° Ell
1	3"	150 lb. SO RF Flg.

All pipe to be ASTM A-106 Gr. B Smls. Sch 40

1	3"	0' – 7-3/4"	IPE 1BE
1	3"	4' – 11-5/8"	2BE
1	3"	12' – 10-3/16	2BE

Total length 18' – 7"

Figure 13-2. Orthographic pipe spool. (Courtesy of Texas Pipe Bending Company, Inc.)

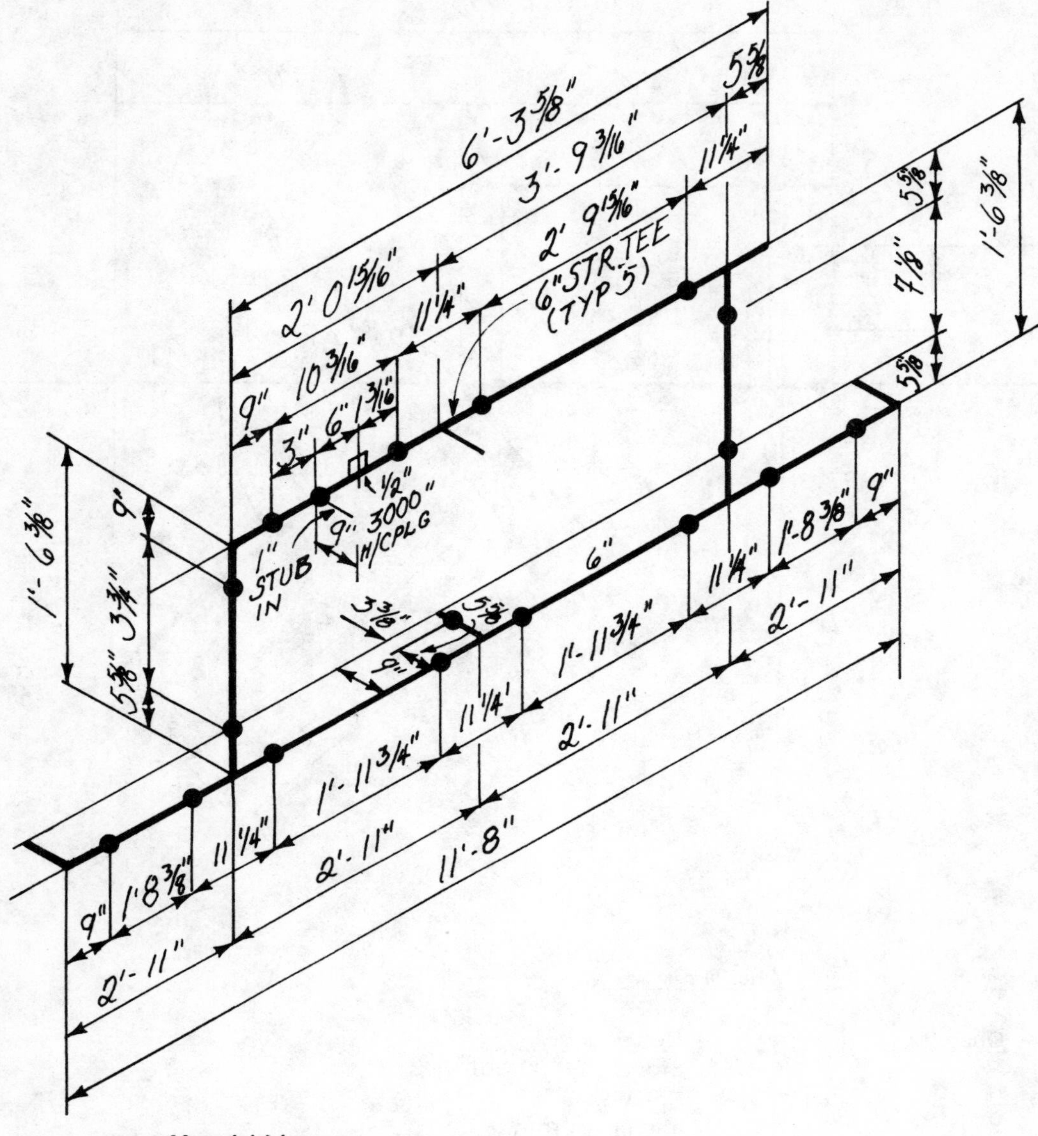

Material List

Quantity	Size	Description	
3	6"	Std. Wt. LR 90° Ell	
5	6"	Std. Wt. Str. Tee	
1	1/2"	3000 lb. Thd'd Half Cplg.	
All pipe to be ASTM A-106 Gr B. Smls Sch. 40			
1	6"	0' – 3-3/8"	2BE
1	6"	0' – 3-3/4"	2BE
1	6"	0' – 7-1/8"	2BE
1	6"	0' – 10-3/16"	2BE
2	6"	1' – 8-3/8"	2BE
2	6"	1' – 11-3/4"	2BE
1	6"	2' – 9-15/16"	2BE
Total length 12' – 11"			

Figure 13-3. Isometric pipe spool. (Courtesy of Texas Pipe Bending Company, Inc.)

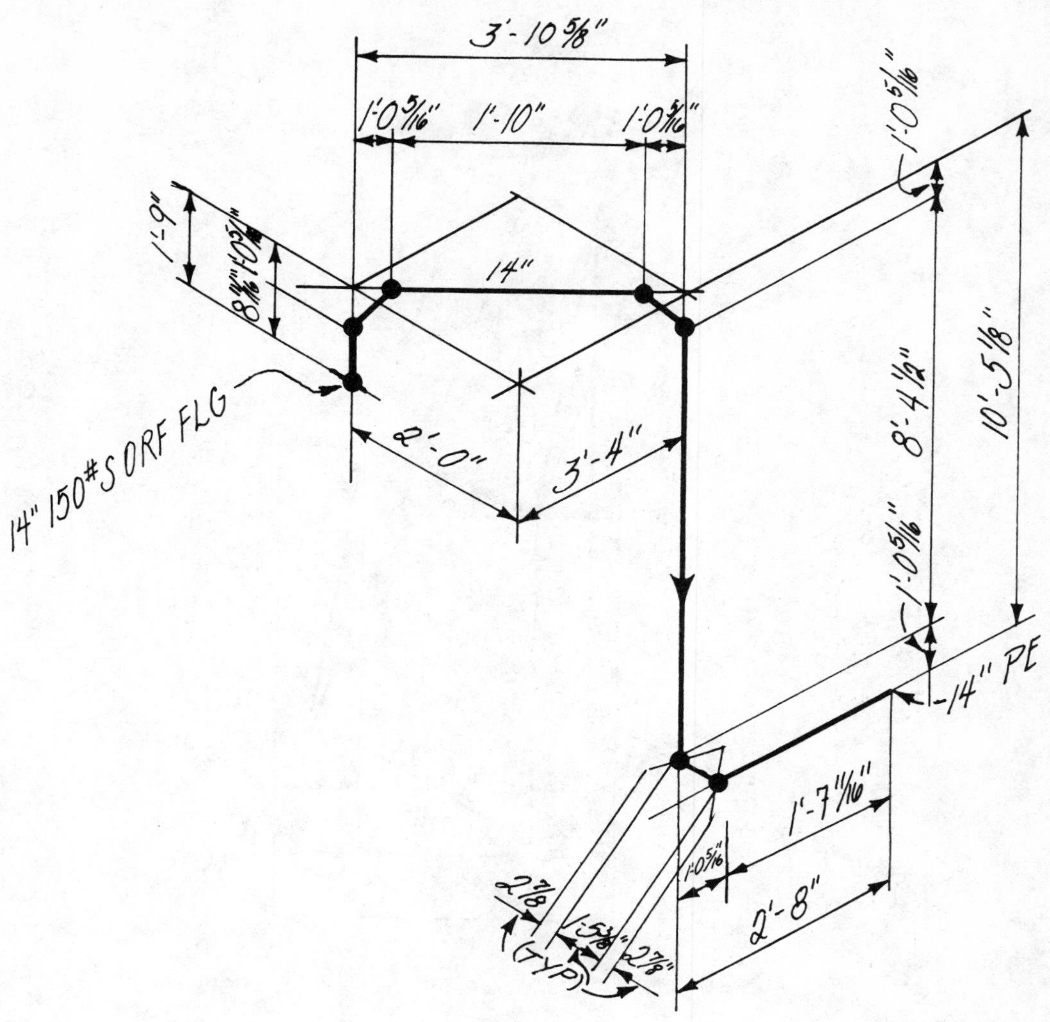

Material List

Quantity	Size	Description
1	14"	150 lb. SO RF Flg.

All pipe to be ASTM A-106 Gr. B. Smls. 0.375" Wall

1	14"	16' – 11-1/8"	2PE

Figure 13-4. Isometric spool with mitres. (Courtesy of Texas Pipe Bending Company, Inc.)

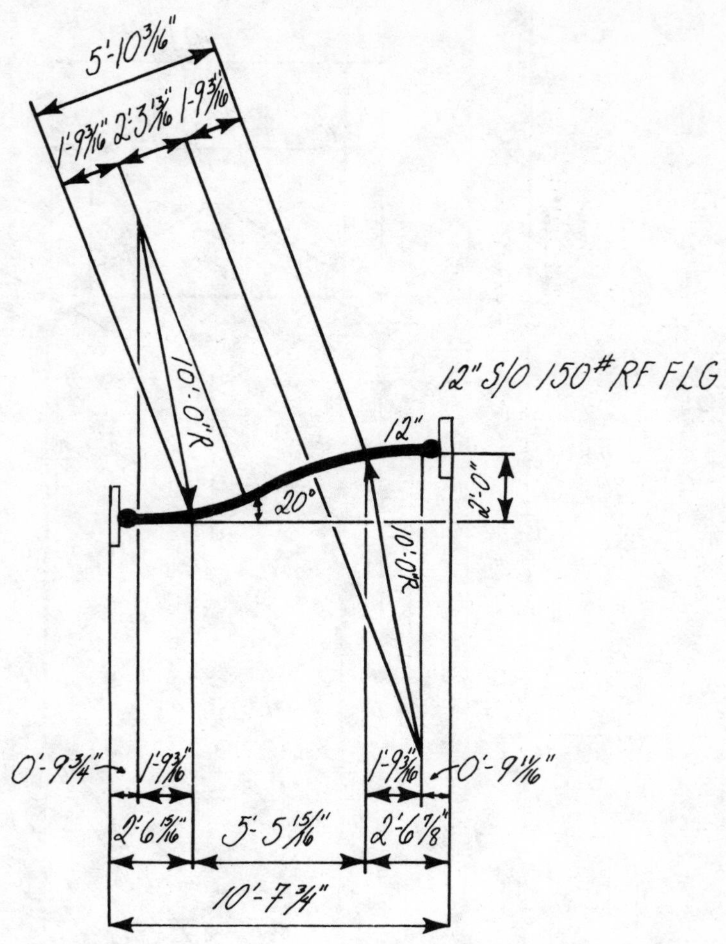

Material List

Quantity	Size	Description
2	12"	150 lb. SO RF Flg.

All pipe to be ASTM A-106 Gr. B. Smls 0.375" Wall

1	12"	10' – 11-1/4"	2 PE

Figure 13-5. Shop spool with bends. (Courtesy of Texas Pipe Bending Company, Inc.)

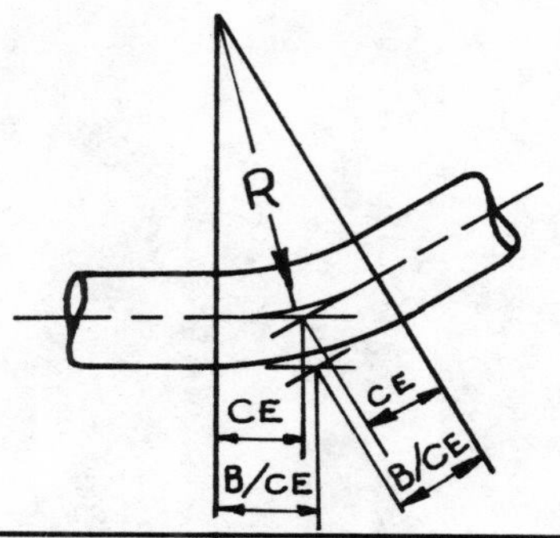

Table 13-1. Center-to-End (CE), Back Center-to-End (B/CE), and Arc Length for 30° Bends of Varying Radii and Pipe Sizes

Feet — Inches			(B/CE) Nominal Pipe Sizes (inches)							
Rad.	Arc	(CE)	2	3	4	5	6	8	10	12
6	3-1/8	1-5/8	1-15/16							
7	3-11/16	1-7/8	2-3/16							
8	4-3/16	2-1/8	2-7/16							
9	4-11/16	2-7/16	2-3/4	2-15/16						
10	5-1/4	2-11/16	3	3-3/16						
11	5-3/4	2-15/16	3-1/4	3-7/16						
1 – 0	6-5/16	3-3/16	3-1/2	3-11/16	3-13/16					
1 – 1	6-13/16	3-1/2	3-13/16	4	4-1/8					
1 – 2	7-5/16	3-3/4	4-1/16	4-1/4	4-3/8					
1 – 3	7-7/8	4	4-5/16	4-1/2	4-5/8	4-3/4				
1 – 4	8-3/8	4-5/16	4-5/8	4-13/16	4-15/16	5-1/16				
1 – 5	8-7/8	4-9/16	4-7/8	5-1/16	5-3/16	5-5/16				
1 – 6	9-7/16	4-13/16	5-1/8	5-5/16	5-7/16	5-9/16	5-11/16			
1 – 7	9-15/16	5-1/16	5-3/8	5-9/16	5-11/16	5-13/16	5-15/16			
1 – 8	10-1/2	5-3/8	5-11/16	5-7/8	6	6-1/8	6-1/4			
1 – 9	11	5-5/8	5-15/16	6-1/8	6-1/4	6-3/8	6-1/2			
1 – 10	11-1/2	5-7/8	6-3/16	6-3/8	6-1/2	6-5/8	6-3/4			
1 – 11	1 – 0-1/16	6-3/16	6-1/2	6-11/16	6-13/16	6-15/16	7-1/16			
2 – 0	1 – 0-9/16	6-7/16	6-3/4	6-15/16	7-1/16	7-3/16	7-5/16	7-9/16		
2 – 1	1 – 1-1/16	6-11/16	7	7-3/16	7-5/16	7-7/16	7-9/16	7-13/16		
2 – 2	1 – 1-5/8	6-15/16	7-1/4	7-7/16	7-9/16	7-11/16	7-13/16	8-1/16		
2 – 3	1 – 2-1/8	7-1/4	7-9/16	7-3/4	7-7/8	8	8-1/8	8-3/8		
2 – 4	1 – 2-5/8	7-1/2	7-13/16	8	8-1/8	8-1/4	8-3/8	8-5/8		
2 – 5	1 – 3-3/16	7-3/4	8-1/16	8-1/4	8-3/8	8-1/2	8-5/8	8-7/8		
2 – 6	1 – 3-11/16	8-1/16	8-3/8	8-9/16	8-11/16	8-13/16	8-15/16	9-3/16	9-1/2	
2 – 7	1 – 4-1/4	8-5/16	8-5/8	8-13/16	8-15/16	9-1/16	9-3/16	9-7/16	9-3/4	
2 – 8	1 – 4-3/4	8-9/16	8-7/8	9-1/16	9-3/16	9-5/16	9-7/16	9-11/16	10	
2 – 9	1 – 5-1/4	8-13/16	9-1/8	9-5/16	9-7/16	9-9/16	9-11/16	9-15/16	10-1/4	
2 – 10	1 – 5-13/16	9-1/8	9-7/16	9-5/8	9-3/4	9-7/8	10	10-1/4	10-9/16	
2 – 11	1 – 6-5/16	9-3/8	9-11/16	9-7/8	10	10-1/8	10-1/4	10-1/2	10-13/16	
3 – 0	1 – 6-7/8	9-5/8	9-15/16	10-1/8	10-1/4	10-3/8	10-1/2	10-3/4	11-1/16	11-5/16
3 – 1	1 – 7-3/8	9-15/16	10-1/4	10-7/16	10-9/16	10-11/16	10-13/16	11-1/16	11-3/8	11-5/8
3 – 2	1 – 7-7/8	10-3/16	10-1/2	10-11/16	10-13/16	10-15/16	11-1/16	11-5/16	11-5/8	11-7/8
3 – 3	1 – 8-7/16	10-7/16	10-3/4	10-15/16	11-1/16	11-3/16	11-5/16	11-9/16	11-7/8	1 – 0-1/8

Courtesy of Texas Pipe Bending Co., Inc.

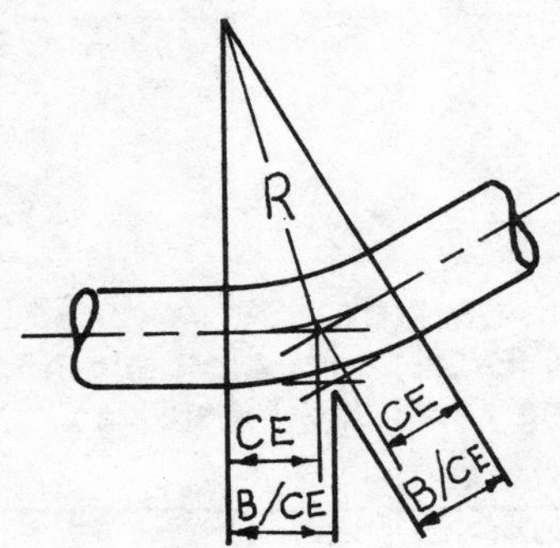

Table 13-2. Center-to-End (CE), Back Center-to-End (B/CE), and Arc Length for 30° Bends of Varying Radii and Pipe Sizes

Feet – Inches			(B/CE) Nominal Pipe Sizes							
Rad.	Arc	(CE)	2	3	4	5	6	8	10	12
3 – 4	1 – 8-15/16	10-11/16	11	11-3/16	11-5/16	11-7/16	11-9/16	11-13/16	1 – 0-1/8	1 – 0-3/8
3 – 5	1 – 9-1/2	11	11-5/16	11-1/2	11-5/8	11-3/4	11-7/8	1 – 0-1/8	1 – 0-7/16	1 – 0-11/16
3 – 6	1 – 10	11-1/4	11-9/16	11-3/4	11-7/8	1 – 0	1 – 0-1/8	1 – 0-3/8	1 – 0-11/16	1 – 0-15/16
3 – 7	1 – 10-1/2	11-1/2	11-13/16	1 – 0	1 – 0-1/8	1 – 0-1/4	1 – 0-3/8	1 – 0-5/8	1 – 0-15/16	1 – 1-3/16
3 – 8	1 – 11-1/16	11-13/16	1 – 0-1/8	1 – 0-5/16	1 – 0-7/16	1 – 0-9/16	1 – 0-11/16	1 – 0-15/16	1 – 1-1/4	1 – 1-1/2
3 – 9	1 – 11-9/16	1 – 0-1/16	1 – 0-3/8	1 – 0-9/16	1 – 0-11/16	1 – 0-13/16	1 – 0-15/16	1 – 1-3/16	1 – 1-1/2	1 – 1-3/4
3 – 10	2 – 0-1/8	1 – 0-5/16	1 – 0-5/8	1 – 0-13/16	1 – 0-15/16	1 – 1-1/16	1 – 1-3/16	1 – 1-7/16	1 – 1-3/4	1 – 2
3 – 11	2 – 0-5/8	1 – 0-5/8	1 – 0-15/16	1 – 1-1/8	1 – 1-1/4	1 – 1-3/8	1 – 1-1/2	1 – 1-3/4	1 – 2-1/16	1 – 2-5/16
4 – 0	2 – 1-1/8	1 – 0-7/8	1 – 1-3/16	1 – 1-3/8	1 – 1-1/2	1 – 1-5/8	1 – 1-3/4	1 – 2	1 – 2-5/16	1 – 2-9/16
4 – 1	2 – 1-11/16	1 – 1-1/8	1 – 1-7/16	1 – 1-5/8	1 – 1-3/4	1 – 1-7/8	1 – 2	1 – 2-1/4	1 – 2-9/16	1 – 2-13/16
4 – 2	2 – 2-3/16	1 – 1-3/8	1 – 1-11/16	1 – 1-7/8	1 – 2	1 – 2-1/8	1 – 2-1/4	2 – 2-1/2	1 – 2-13/16	1 – 3-1/16
4 – 3	2 – 2-11/16	1 – 1-11/16	1 – 2	1 – 2-3/16	1 – 2-5/16	1 – 2-7/16	1 – 2-9/16	1 – 2-13/16	1 – 3-1/8	1 – 3-3/8
4 – 4	2 – 3-1/4	1 – 1-15/16	1 – 2-1/4	1 – 2-7/16	1 – 2-9/16	1 – 2-11/16	1 – 2-13/16	1 – 3-1/16	1 – 3-3/8	1 – 3-5/8
4 – 5	2 – 3-3/4	1 – 2-3/16	1 – 2-1/2	1 – 2-11/16	1 – 2-13/16	1 – 2-15/16	1 – 3-1/16	1 – 3-5/16	1 – 3-5/8	1 – 3-7/8
4 – 6	2 – 4-1/4	1 – 2-1/2	1 – 2-13/16	1 – 3	1 – 3-1/8	1 – 3-1/4	1 – 3-3/8	1 – 3-5/8	1 – 3-15/16	1 – 4-3/16
4 – 7	2 – 4-13/16	1 – 2-3/4	1 – 3-1/16	1 – 3-1/4	1 – 3-3/8	1 – 3-1/2	1 – 3-5/8	1 – 3-7/8	1 – 4-3/16	1 – 4-7/16
4 – 8	2 – 5-5/16	1 – 3	1 – 3-5/16	1 – 3-1/2	1 – 3-5/8	1 – 3-3/4	1 – 3-7/8	1 – 4-1/8	1 – 4-7/16	1 – 4-11/16
4 – 9	2 – 5-7/8	1 – 3-1/4	1 – 3-9/16	1 – 3-3/4	1 – 3-7/8	1 – 4	1 – 4-1/8	1 – 4-3/8	1 – 4-11/16	1 – 4-15/16
4 – 10	2 – 6-3/8	1 – 3-9/16	1 – 3-7/8	1 – 4-1/16	1 – 4-3/16	1 – 4-1/4	1 – 4-3/8	1 – 4-5/8	1 – 4-15/16	1 – 5-3/16
4 – 11	2 – 6-7/8	1 – 3-13/16	1 – 4-1/8	1 – 4-5/16	1 – 4-7/16	1 – 4-9/16	1 – 4-11/16	1 – 4-15/16	1 – 5-1/4	1 – 5-1/2
5 – 0	2 – 7-7/16	1 – 4-1/16	1 – 4-3/8	1 – 4-9/16	1 – 4-11/16	1 – 4-13/16	1 – 4-15/16	1 – 5-3/16	1 – 5-1/2	1 – 5-3/4
5 – 1	2 – 7-13/16	1 – 4-3/8	1 – 4-11/16	1 – 4-7/8	1 – 5	1 – 5-1/8	1 – 5-1/4	1 – 5-1/2	1 – 5-13/16	1 – 6-1/16
5 – 2	2 – 8-7/16	1 – 4-5/8	1 – 4-15/16	1 – 5-1/8	1 – 5-1/4	1 – 5-3/8	1 – 5-1/2	1 – 5-3/4	1 – 6-1/16	1 – 6-5/16
5 – 3	2 – 9	1 – 4-7/8	1 – 5-3/16	1 – 5-3/8	1 – 5-1/2	1 – 5-5/8	1 – 5-3/4	1 – 6	1 – 6-5/16	1 – 6-9/16
5 – 4	2 – 9-1/2	1 – 5-1/8	1 – 5-7/16	1 – 5-5/8	1 – 5-3/4	1 – 5-7/8	1 – 6	1 – 6-1/4	1 – 6-9/16	1 – 6-13/16
5 – 5	2 – 10-1/16	1 – 5-7/16	1 – 5-3/4	1 – 5-15/16	1 – 6-1/16	1 – 6-3/16	1 – 6-5/16	1 – 6-9/16	1 – 6-7/8	1 – 7-1/8
5 – 6	2 – 10-9/16	1 – 5-11/16	1 – 6	1 – 6-3/16	1 – 6-5/16	1 – 6-7/16	1 – 6-9/16	1 – 6-13/16	1 – 7-1/8	1 – 7-3/8
5 – 7	2 – 11-1/16	1 – 5-15/16	1 – 6-1/4	1 – 6-7/16	1 – 6-9/16	1 – 6-11/16	1 – 6-13/16	1 – 7-1/16	1 – 7-3/8	1 – 7-5/8
5 – 8	2 – 11-5/8	1 – 6-1/4	1 – 6-9/16	1 – 6-3/4	1 – 6-7/8	1 – 7	1 – 7-1/8	1 – 7-3/8	1 – 7-11/16	1 – 7-15/16
5 – 9	3 – 0-1/8	1 – 6-1/2	1 – 6-13/16	1 – 7	1 – 7-1/8	1 – 7-1/4	1 – 7-3/8	1 – 7-5/8	1 – 7-15/16	1 – 8-3/16
5 – 10	3 – 0-5/8	1 – 6-3/4	1 – 7-1/16	1 – 7-1/4	1 – 7-3/8	1 – 7-1/2	1 – 7-5/8	1 – 7-7/8	1 – 8-3/16	1 – 8-7/16
5 – 11	3 – 1-3/16	1 – 7	1 – 7-5/16	1 – 7-1/2	1 – 7-5/8	1 – 7-3/4	1 – 7-7/8	1 – 8-1/8	1 – 8-7/16	1 – 8-11/16
6 – 0	3 – 1-11/16	1 – 7-5/16	1 – 7-5/8	1 – 7-13/16	1 – 7-15/16	1 – 8-1/16	1 – 8-3/16	1 – 8-7/16	1 – 8-3/4	1 – 9

Courtesy of Texas Pipe Bending Co., Inc.

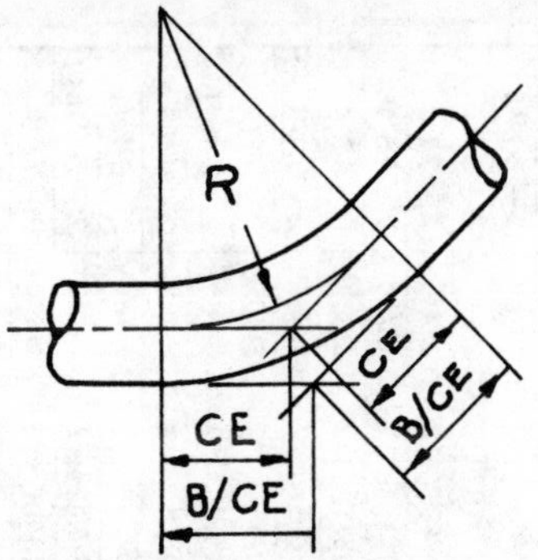

Table 13-3. Center-to-End (CE), Back Center-to-End (B/CE), and Arc Length for 45° Bends of Varying Radii and Pipe Sizes

Feet — Inches			(B/CE) Nominal Pipe Sizes							
Rad.	Arc	CE	2	3	4	5	6	8	10	12
6	4-11/16	2-1/2	3							
7	5-1/2	2-7/8	3-3/8							
8	6-5/16	3-5/16	3-13/16							
9	7-1/16	3-3/4	4-1/4	4-1/2						
10	7-7/8	4-1/8	4-5/8	4-7/8						
11	8-5/8	4-9/16	5-1/16	5-5/16						
1 — 0	9-7/16	5	5-1/2	5-3/4	5-15/16					
1 — 1	10-3/16	5-3/8	5-7/8	6-1/8	6-5/16					
1 — 2	11	5-13/16	6-5/16	6-9/16	6-3/4					
1 — 3	11-3/4	6-3/16	6-11/16	6-15/16	7-1/8	7-5/16				
1 — 4	1 — 0-9/16	6-5/8	7-1/8	7-3/8	7-9/16	7-3/4				
1 — 5	1 — 1-3/8	7-1/16	7-9/16	7-13/16	8	8-3/16				
1 — 6	1 — 2-1/8	7-7/16	7-15/16	8-3/16	8-3/8	8-9/16	8-13/16			
1 — 7	1 — 2-15/16	7-7/8	8-3/8	8-5/8	8-13/16	9	9-1/4			
1 — 8	1 — 3-11/16	8-5/16	8-13/16	9-1/16	9-1/4	9-7/16	9-11/16			
1 — 9	1 — 4-1/2	8-11/16	9-3/16	9-7/16	9-5/8	9-13/16	10-1/16			
1 — 10	1 — 5-1/4	9-1/8	9-5/8	9-7/8	10-1/16	10-1/4	10-1/2			
1 — 11	1 — 6-1/16	9-1/2	10	10-1/4	10-7/16	10-11/16	10-15/16			
2 — 0	1 — 6-7/8	9-15/16	10-7/16	10-11/16	10-7/8	11-1/16	11-5/16	11-3/4		
2 — 1	1 — 7-5/8	10-3/8	10-7/8	11-1/8	11-5/16	11-1/2	11-3/4	1 — 0-3/16		
2 — 2	1 — 8-7/16	10-3/4	11-1/4	11-1/2	11-11/16	11-7/8	1 — 0-1/8	1 — 0-9/16		
2 — 3	1 — 9-3/16	11-3/16	11-11/16	11-15/16	1 — 0-1/8	1 — 0-5/16	1 — 0-9/16	1 — 1		
2 — 4	1 — 10	11-5/8	1 — 0-1/8	1 — 0-3/8	1 — 0-9/16	1 — 0-3/4	1 — 1	1 — 1-7/16		
2 — 5	1 — 10-3/4	1 — 0	1 — 0-1/2	1 — 0-3/4	1 — 0-15/16	1 — 1-1/8	1 — 1-3/8	1 — 1-13/16		
2 — 6	1 — 11-9/16	1 — 0-7/16	1 — 0-15/16	1 — 1-3/16	1 — 1-3/8	1 — 1-9/16	1 — 1-13/16	1 — 2-1/4	1 — 2-11/16	
2 — 7	2 — 0-3/8	1 — 0-13/16	1 — 1-5/16	1 — 1-9/16	1 — 1-3/4	1 — 1-15/16	1 — 2-3/16	1 — 2-5/8	1 — 3-1/16	
2 — 8	2 — 1-1/8	1 — 1-1/4	1 — 1-3/4	1 — 2	1 — 2-3/16	1 — 2-3/8	1 — 2-5/8	1 — 3-1/16	1 — 3-1/2	
2 — 9	2 — 1-15/16	1 — 1-11/16	1 — 2-3/16	1 — 2-7/16	1 — 2-5/8	1 — 2-13/16	1 — 3-1/16	1 — 3-1/2	1 — 3-15/16	
2 — 10	2 — 2-11/16	1 — 2-1/16	1 — 2-9/16	1 — 2-13/16	1 — 3	1 — 3-3/16	1 — 3-7/16	1 — 3-7/8	1 — 4-5/16	
2 — 11	2 — 3-1/2	1 — 2-1/2	1 — 3	1 — 3-1/4	1 — 3-7/16	1 — 3-5/8	1 — 3-7/8	1 — 4-5/16	1 — 4-3/4	
3 — 0	2 — 4-1/4	1 — 2-15/16	1 — 3-7/16	1 — 3-11/16	1 — 3-7/8	1 — 4-1/16	1 — 4-5/16	1 — 4-3/4	1 — 5-3/16	1 — 5-9/16
3 — 1	2 — 5-1/16	1 — 3-5/16	1 — 3-13/16	1 — 4-1/16	1 — 4-1/4	1 — 4-7/16	1 — 4-11/16	1 — 5-1/8	1 — 5-9/16	1 — 5-15/16
3 — 2	2 — 5-7/8	1 — 3-3/4	1 — 4-1/4	1 — 4-1/2	1 — 4-11/16	1 — 4-7/8	1 — 5-1/8	1 — 5-9/16	1 — 6	1 — 6-3/8
3 — 3	2 — 6-5/8	1 — 4-1/8	1 — 4-5/8	1 — 4-7/8	1 — 5-1/16	1 — 5-1/4	1 — 5-1/2	1 — 5-15/16	1 — 6-3/8	1 — 6-3/4

Courtesy of Texas Pipe Bending Co., Inc.

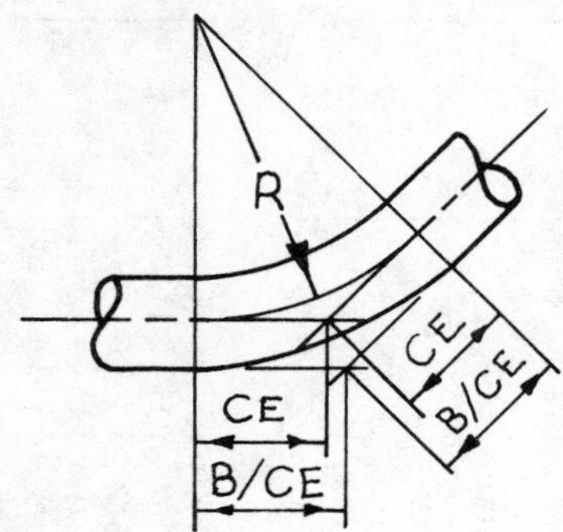

Table 13-4. Center-to-End (CE), Back Center-to-End (B/CE), and Arc Length for 45° Bends of Varying Radii and Pipe Sizes

Feet — Inches			(B/CE) Nominal Pipe Sizes							
Rad.	Arc	(CE)	2	3	4	5	6	8	10	12
3 – 4	2 – 7-7/16	1 – 4-9/16	1 – 5-1/16	1 – 5-5/16	1 – 5-1/2	1 – 5-11/16	1 – 5-15/16	1 – 6-3/8	1 – 6-13/16	1 – 7-3/16
3 – 5	2 – 8-3/16	1 – 5	1 – 5-1/2	1 – 5-3/4	1 – 5-15/16	1 – 6-1/8	1 – 6-3/8	1 – 6-13/16	1 – 7-1/4	1 – 7-5/8
3 – 6	2 – 9	1 – 5-3/8	1 – 5-7/8	1 – 6-1/8	1 – 6-5/16	1 – 6-1/2	1 – 6-3/4	1 – 7-3/16	1 – 7-5/8	1 – 8
3 – 7	2 – 9-3/4	1 – 5-13/16	1 – 6-5/16	1 – 6-9/16	1 – 6-3/4	1 – 6-15/16	1 – 7-3/16	1 – 7-5/8	1 – 8-1/16	1 – 8-7/16
3 – 8	2 – 10-9/16	1 – 6-1/4	1 – 6-3/4	1 – 7	1 – 7-3/16	1 – 7-3/8	1 – 7-5/8	1 – 8-1/16	1 – 8-1/2	1 – 8-7/8
3 – 9	2 – 11-5/16	1 – 6-5/8	1 – 7-1/8	1 – 7-3/8	1 – 7-9/16	1 – 7-3/4	1 – 8	1 – 8-7/16	1 – 8-7/8	1 – 9-1/4
3 – 10	3 – 0-1/8	1 – 7-1/16	1 – 7-9/16	1 – 7-13/16	1 – 8	1 – 8-3/16	1 – 8-7/16	1 – 8-7/8	1 – 9-5/16	1 – 9-11/16
3 – 11	3 – 0-15/16	1 – 7-7/16	1 – 7-15/16	1 – 8-3/16	1 – 8-3/8	1 – 8-9/16	1 – 8-13/16	1 – 9-1/4	1 – 9-11/16	1 – 10-1/16
4 – 0	3 – 1-11/16	1 – 7-7/8	1 – 8-3/8	1 – 8-5/8	1 – 8-13/16	1 – 9	1 – 9-1/4	1 – 9-11/16	1 – 10-1/8	1 – 10-1/2
4 – 1	3 – 2-1/2	1 – 8-5/16	1 – 8-13/16	1 – 9-1/16	1 – 9-1/4	1 – 9-7/16	1 – 9-11/16	1 – 10-1/8	1 – 10-9/16	1 – 10-15/16
4 – 2	3 – 3-1/4	1 – 8-11/16	1 – 9-3/16	1 – 9-7/16	1 – 9-5/8	1 – 9-13/16	1 – 10-1/16	1 – 10-1/2	1 – 1-015/16	1 – 11-5/16
4 – 3	3 – 4-1/16	1 – 9-1/8	1 – 9-5/8	1 – 9-7/8	1 – 10-1/16	1 – 10-1/4	1 – 10-1/2	1 – 10-15/16	1 – 11-3/8	1 – 11-3/4
4 – 4	3 – 4-13/16	1 – 9-9/16	1 – 10-1/16	1 – 10-5/16	1 – 10-1/2	1 – 10-11/16	1 – 10-15/16	1 – 11-3/8	1 – 11-13/16	2 – 0-3/16
4 – 5	3 – 5-5/8	1 – 9-15/16	1 – 10-7/16	1 – 10-11/16	1 – 10-7/8	1 – 11-1/16	1 – 11-5/16	1 – 11-3/4	2 – 0-3/16	2 – 0-9/16
4 – 6	3 – 6-7/16	1 – 10-3/8	1 – 10-7/8	1 – 11-1/8	1 – 11-5/16	1 – 11-1/2	1 – 11-3/4	2 – 0-3/16	2 – 0-5/8	2 – 1
4 – 7	3 – 7-3/16	1 – 10-13/16	1 – 11-5/16	1 – 11-9/16	1 – 11-3/4	1 – 11-15/16	2 – 0-3/16	2 – 0-5/8	2 – 1-1/16	2 – 1-7/16
4 – 8	3 – 8	1 – 11-3/16	1 – 11-11/16	1 – 11-15/16	2 – 0-1/8	2 – 0-5/16	2 – 0-9/16	2 – 1	2 – 1-7/16	2 – 1-13/16
4 – 9	3 – 8-3/4	1 – 11-5/8	2 – 0-1/8	2 – 0-3/8	2 – 0-9/16	2 – 0-3/4	2 – 1	2 – 1-7/16	2 – 1-7/8	2 – 2-1/2
4 – 10	3 – 9-9/16	2 – 0	2 – 0-1/2	2 – 0-3/4	2 – 0-15/16	2 – 1-1/8	2 – 1-3/8	2 – 1-13/16	2 – 2-1/4	2 – 2-5/8
4 – 11	3 – 10-5/16	2 – 0-7/16	2 – 0-15/16	2 – 1-3/16	2 – 1-3/8	2 – 1-9/16	2 – 1-13/16	2 – 2-1/4	2 – 2-11/16	2 – 3-1/16
5 – 0	3 – 11-1/8	2 – 0-7/8	2 – 1-3/8	2 – 1-5/8	2 – 1-13/16	2 – 2	2 – 2-1/4	2 – 2-11/16	2 – 3-1/8	2 – 3-1/2
5 – 1	3 – 11-15/16	2 – 1-1/4	2 – 1-3/4	2 – 2	2 – 2-3/16	2 – 3-3/8	2 – 2-5/8	2 – 3-1/16	2 – 3-1/2	2 – 3-7/8
5 – 2	4 – 0-11/16	2 – 1-11/16	2 – 2-3/16	2 – 2-7/16	2 – 2-5/8	2 – 2-13/16	2 – 3-1/16	2 – 3-1/2	2 – 3-15/16	2 – 4-5/16
5 – 3	4 – 1-1/2	2 – 2-1/8	2 – 2-5/8	2 – 2-7/8	2 – 3-1/16	2 – 3-1/4	2 – 3-1/2	2 – 3-15/16	2 – 4-3/8	2 – 4-3/4
5 – 4	4 – 2-1/4	2 – 2-1/2	2 – 3	2 – 3-1/4	2 – 3-7/16	2 – 3-5/8	2 – 3-7/8	2 – 4-5/16	2 – 4-3/4	2 – 5-1/8
5 – 5	4 – 3-1/16	2 – 2-15/16	2 – 3-7/16	2 – 3-11/16	2 – 3-7/8	2 – 4-1/16	2 – 4-5/16	2 – 4-3/4	2 – 5-3/16	2 – 5-9/16
5 – 6	4 – 3-13/16	2 – 3-5/16	2 – 3-13/16	2 – 4-1/16	2 – 4 1/4	2 – 4-7/16	2 – 4-11/16	2 – 5-1/8	2 – 5-9/16	2 – 6-15/16
5 – 7	4 – 4-5/8	2 – 3-3/4	2 – 4-1/4	2 – 4-1/2	2 – 4-11/16	2 – 4-7/8	2 – 5-1/8	2 – 5-9/16	2 – 6	2 – 6-3/8
5 – 8	4 – 5-7/16	2 – 4-3/16	2 – 4-11/16	2 – 4-15/16	2 – 5-1/8	2 – 5-5/16	2 – 5-9/16	2 – 6	2 – 6-7/16	2 – 6-13/16
5 – 9	4 – 6-3/16	2 – 4-9/16	2 – 5-1/16	2 – 5-5/16	2 – 5-1/2	2 – 5-11/16	2 – 5-15/16	2 – 6-3/8	2 – 6-13/16	2 – 7-3/16
5 – 10	4 – 7	2 – 5	2 – 5-1/2	2 – 5-3/4	2 – 5-15/16	2 – 6-1/8	2 – 6-3/8	2 – 6-13/16	2 – 7-1/4	2 – 7-5/8
5 – 11	4 – 7-3/4	2 – 5-3/8	2 – 5-7/8	2 – 6-1/8	2 – 6-5/16	2 – 6-1/2	2 – 6-3/4	2 – 7-3/16	2 – 7-5/8	2 – 8
6 – 0	4 – 8-9/16	2 – 5-13/16	2 – 6-5/16	2 – 6-9/16	2 – 6-3/4	2 – 6-15/16	2 – 7-3/16	2 – 7-5/8	2 – 8-1/16	2 – 8-7/16

Courtesy of Texas Pipe Bending Co., Inc.

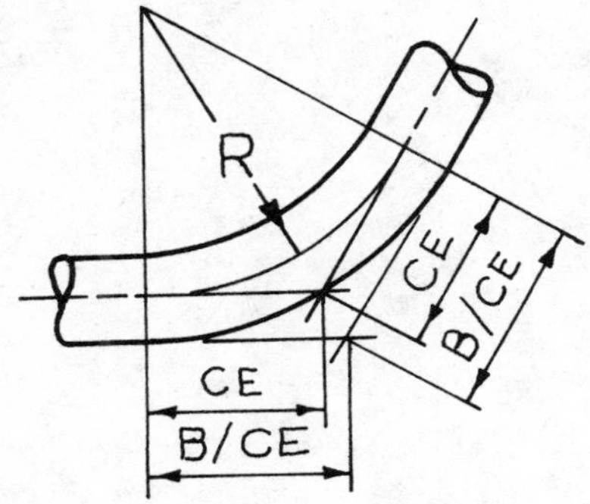

Table 13-5. Center-to-End (CE), Back Center-to-End (B/CE), and Arc Length for 60° Bends of Varying Radii and Pipe Sizes

Feet – Inches			(B/CE) Nominal Pipe Sizes							
Rad.	Arc	(CE)	2	3	4	5	6	8	10	12
6	6-5/16	3-7/16	4-1/8							
7	7-5/16	4-1/16	4-3/4							
8	8-3/8	4-5/8	5-5/16							
9	9-7/16	5-3/16	5-7/8	6-3/16						
10	10-1/2	5-3/4	6-7/16	6-3/4						
11	11-1/2	6-3/8	7-1/16	7-3/8						
1 – 0	1 – 0-9/16	6-15/16	7-5/8	7-15/16	8-1/4					
1 – 1	1 – 1-5/8	7-1/2	8-3/16	8-1/2	8-13/16					
1 – 2	1 – 2-5/8	8-1/16	8-3/4	9-1/16	9-3/8					
1 – 3	1 – 3-11/16	8-11/16	9-3/8	9-11/16	10	10-5/16				
1 – 4	1 – 4-3/4	9-1/4	9-15/16	10-1/4	10-9/16	10-7/8				
1 – 5	1 – 5-13/16	9-13/16	10-1/2	10-13/16	11-1/8	11-7/16				
1 – 6	1 – 6-7/8	10-3/8	11-1/16	11-3/8	11-11/16	1 – 0	1 – 0-5/16			
1 – 7	1 – 7-7/8	11	11-11/16	1 – 0	1 – 0-5/16	1 – 0-5/8	1 – 0-15/16			
1 – 8	1 – 8-15/16	11-9/16	1 – 0-1/4	1 – 0-9/16	1 – 0-7/8	1 – 1-3/16	1 – 1-1/2			
1 – 9	1 – 10	1 – 0-1/8	1 – 0-13/16	1 – 1-1/8	1 – 1-7/16	1 – 1-3/4	1 – 2-1/16			
1 – 10	1 – 11-1/16	1 – 0-11/16	1 – 1-3/8	1 – 1-11/16	1 – 2	1 – 2-5/16	1 – 2-5/8			
1 – 11	2 – 0-1/8	1 – 1-1/4	1 – 1-15/16	1 – 2-1/4	1 – 2-9/16	1 – 2-7/8	1 – 3-3/16			
2 – 0	2 – 1-1/8	1 – 1-7/8	1 – 2-9/16	1 – 2-7/8	1 – 3-3/16	1 – 3-1/2	1 – 3-13/16	1 – 4-3/8		
2 – 1	2 – 2-3/16	1 – 2-7/16	1 – 3-1/16	1 – 3-7/16	1 – 3-3/4	1 – 4-1/16	1 – 4-3/8	1 – 4-15/16		
2 – 2	2 – 3-1/4	1 – 3	1 – 3-11/16	1 – 4	1 – 4-5/16	1 – 4-5/8	1 – 4-15/16	1 – 5-1/2		
2 – 3	2 – 4-1/4	1 – 3-9/16	1 – 4-1/4	1 – 4-9/16	1 – 4-7/8	1 – 5-3/16	1 – 5-1/2	1 – 6-1/16		
2 – 4	2 – 5-5/16	1 – 4-3/16	1 – 4-7/8	1 – 5-3/16	1 – 5-1/2	1 – 5-13/16	1 – 6-1/8	1 – 6-11/16		
2 – 5	2 – 6-3/8	1 – 4-3/4	1 – 5-7/16	1 – 5-3/4	1 – 6-1/16	1 – 6-3/8	1 – 6-11/16	1 – 7-1/4		
2 – 6	2 – 7-7/16	1 – 5-5/16	1 – 6	1 – 6-5/16	1 – 6-5/8	1 – 6-15/16	1 – 7-1/4	1 – 7-13/16	1 – 8-7/16	
2 – 7	2 – 8-7/16	1 – 5-7/8	1 – 6-9/16	1 – 6-7/8	1 – 7-3/16	1 – 7-1/2	1 – 7-13/16	1 – 8-3/8	1 – 9	
2 – 8	2 – 9-1/2	1 – 6-1/2	1 – 7-3/16	1 – 7-1/2	1 – 7-13/16	1 – 8-1/8	1 – 8-7/16	1 – 9	1 – 9-5/8	
2 – 9	2 – 10-9/16	1 – 7-1/16	1 – 7-3/4	1 – 8-1/16	1 – 8-3/8	1 – 8-11/16	1 – 9	1 – 9-9/16	1 – 10-3/16	
2 – 10	2 – 11-5/8	1 – 7-5/8	1 – 8-5/16	1 – 8-5/8	1 – 8-15/16	1 – 9-1/4	1 – 9-9/16	1 – 10-1/8	1 – 10-3/4	
2 – 11	3 – 0-5/8	1 – 8-3/16	1 – 8-7/8	1 – 9-3/16	1 – 9-1/2	1 – 9-13/16	1 – 10-1/8	1 – 10-11/16	1 – 11-5/16	
3 – 0	3 – 1-11/16	1 – 8-13/16	1 – 9-1/2	1 – 9-13/16	1 – 10-1/8	1 – 10-7/16	1 – 10-3/4	1 – 11-5/16	1 – 11-15/16	2 – 0-1/2
3 – 1	3 – 2-3/4	1 – 9-3/8	1 – 10-1/16	1 – 10-3/8	1 – 10-11/16	1 – 11	1 – 11-5/16	1 – 11-7/8	2 – 0-1/2	2 – 1-1/16
3 – 2	3 – 3-13/16	1 – 9-15/16	1 – 10-5/8	1 – 10-15/16	1 – 11-1/4	1 – 11-9/16	1 – 11-7/8	2 – 0-7/16	2 – 1-1/16	2 – 1-5/8
3 – 3	3 – 4-13/16	1 – 10-1/2	1 – 11-3/16	1 – 11-1/2	1 – 11-13/16	2 – 0-1/8	2 – 0-7/16	2 – 1	2 – 1-5/8	2 – 2-3/16

Courtesy of Texas Pipe Bending Co., Inc.

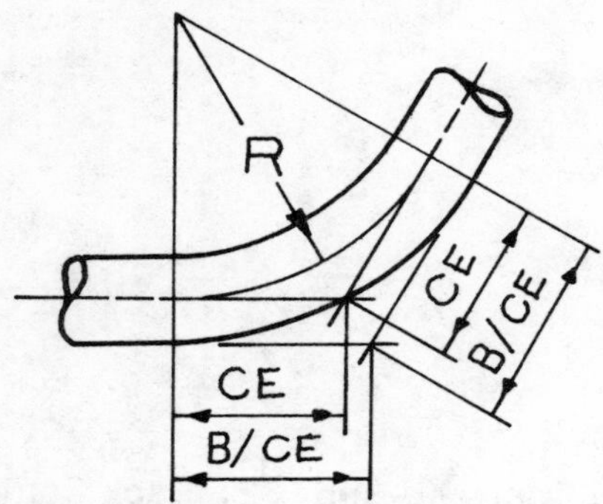

Table 13-6. Center-to-End (CE), Back Center-to-End (B/CE), and Arc Length for 60° Bends of Varying Radii and Pipe Sizes

Feet — Inches			(B/CE) Nominal Pipe Sizes							
Rad.	Arc	(CE)	2	3	4	5	6	8	10	12
3 – 4	3 – 5-7/8	1 – 11-1/8	1 – 11-13/16	2 – 0-1/8	2 – 0-7/16	2 – 0-3/4	2 – 1-1/16	2 – 1-5/8	1 – 2-1/4 2 –	2-13/16
3 – 5	3 – 6-15/16	1 – 11-11/16	2 – 0-3/8	2 – 0-11/16	2 – 1	2 – 1-5/16	2 – 1-5/8	2 – 2-3/16	2 – 2-13/16	2 – 3-3/8
3 – 6	3 – 8	2 – 0-1/4	2 – 0-15/16	2 – 1-1/4	2 – 1-9/16	2 – 1-7/8	2 – 2-3/16	2 – 2-3/4	2 – 3-3/8	2 – 3-15/16
3 – 7	3 – 9	2 – 0-13/16	2 – 1-1/2	2 – 1-13/16	2 – 2-1/8	2 – 2-7/16	2 – 2-3/4	2 – 3-5/16	2 – 3-15/16	2 – 4-1/2
3 – 8	3 – 10-1/16	2 – 1-3/8	2 – 2-1/16	2 – 2-3/8	2 – 2-11/16	2 – 3	2 – 3-5/16	2 – 3-7/8	2 – 4-1/2	2 – 5-1/16
3 – 9	3 – 11-1/8	2 – 2	2 – 2-11/16	2 – 3	2 – 3-5/16	2 – 3-5/8	2 – 3-15/16	2 – 4-1/2	2 – 5-1/8	2 – 5-11/16
3 – 10	4 – 0-3/16	2 – 2-9/16	2 – 3-1/4	2 – 3-9/16	2 – 3-7/8	2 – 4-3/16	2 – 4-1/2	2 – 5-1/16	2 – 5-11/16	2 – 6-1/4
3 – 11	4 – 1-3/16	2 – 3-1/8	2 – 3-13/16	2 – 4-1/8	2 – 4-7/16	2 – 4-3/4	2 – 5-1/16	2 – 5-5/8	2 – 6-1/4	2 – 6-13/16
4 – 0	4 – 2-1/4	2 – 3-11/16	2 – 4-3/8	2 – 4-11/16	2 – 5	2 – 5-5/16	2 – 5-5/8	2 – 6-3/16	2 – 6-13/16	2 – 7-3/8
4 – 1	4 – 3-5/16	2 – 4-5/16	2 – 5	2 – 5-5/16	2 – 5-5/8	2 – 5-15/16	2 – 6-1/4	2 – 6-13/16	2 – 7-7/16	2 – 8
4 – 2	4 – 4-3/8	2 – 4-7/8	2 – 5-9/16	2 – 5-7/8	2 – 6-3/16	2 – 6-1/2	2 – 6-13/16	2 – 7-3/8	2 – 8	2 – 8-9/16
4 – 3	4 – 5-7/16	2 – 5-7/16	2 – 6-1/8	2 – 6-7/16	2 – 6-3/4	2 – 7-1/16	2 – 7-3/8	2 – 7-15/16	2 – 8-9/16	2 – 9-1/8
4 – 4	4 – 6-7/16	2 – 6	2 – 6-11/16	2 – 7	2 – 7-5/16	2 – 7-5/8	2 – 7-15/16	2 – 8-1/2	2 – 9-1/8	2 – 9-11/16
4 – 5	4 – 7-1/2	2 – 6-5/8	2 – 7-5/16	2 – 7-5/8	2 – 7-15/16	2 – 8-1/4	2 – 8-9/16	2 – 9-1/8	2 – 9-3/4	2 – 10-5/16
4 – 6	4 – 8-9/16	2 – 7-3/16	2 – 7-7/8	2 – 8-3/16	2 – 8-1/2	2 – 8-13/16	2 – 9-1/8	2 – 9-11/16	2 – 10-5/16	2 – 10-7/8
4 – 7	4 – 9-5/8	2 – 7-3/4	2 – 8-7/16	2 – 8-3/4	2 – 9-1/16	2 – 9-3/8	2 – 9-11/16	2 – 10-1/4	2 – 10-7/8	2 – 11-7/16
4 – 8	4 – 10-5/8	2 – 8-5/16	2 – 9	2 – 9-5/16	2 – 9-5/8	2 – 9-15/16	2 – 10-1/4	2 – 10-13/16	2 – 11-7/16	3 – 0
4 – 9	4 – 11-11/16	2 – 8-15/16	2 – 9-5/8	2 – 9-15/16	2 – 10-1/4	2 – 10-9/16	2 – 10-7/8	2 – 11-7/16	2 – 0-1/16	3 – 0-5/8
4 – 10	5 – 0-3/4	2 – 9-1/2	2 – 10-3/16	2 – 10-1/2	2 – 10-13/16	2 – 11-1/8	2 – 11-7/16	3 – 0	3 – 0-5/8	3 – 1-3/16
4 – 11	5 – 1-13/16	2 – 10-1/16	2 – 10-3/4	2 – 11-1/16	2 – 11-3/8	2 – 11-11/16	3 – 0	3 – 0-9/16	3 – 1-3/16	3 – 1-3/4
5 – 0	5 – 2-13/16	2 – 10-5/8	2 – 11-5/16	2 – 11-5/8	2 – 11-15/16	3 – 0-1/4	3 – 0-9/16	3 – 1-1/8	3 – 1-3/4	3 – 2-5/16
5 – 1	5 – 3-7/8	2 – 11-1/4	2 – 11-15/16	3 – 0-1/4	3 – 0-9/16	3 – 0-7/8	3 – 1-3/16	3 – 1-3/4	3 – 2-3/8	3 – 2-15/16
5 – 2	5 – 4-15/16	2 – 11-13/16	3 – 0-1/2	3 – 0-13/16	3 – 1-1/8	3 – 1-7/16	3 – 1-3/4	3 – 2-5/16	3 – 2-15/16	3 – 3-1/2
5 – 3	5 – 6	3 – 0-3/8	3 – 1-1/16	3 – 1-3/8	3 – 1-11/16	3 – 2	3 – 2-5/16	3 – 2-7/8	3 – 3-1/2	3 – 4-1/16
5 – 4	5 – 7	3 – 0-15/16	3 – 1-5/8	3 – 1-15/16	3 – 2-1/4	3 – 2-9/16	3 – 2-7/8	3 – 3-7/16	3 – 4-1/16	3 – 4-5/8
5 – 5	5 – 8-1/16	3 – 1-1/2	3 – 2-3/16	3 – 2-1/2	2 – 2-13/16	3 – 3-1/8	3 – 3-7/16	3 – 4	3 – 4-5/8	3 – 5-3/16
5 – 6	5 – 9-1/8	3 – 2-1/8	3 – 2-13/16	3 – 3-1/8	3 – 3-7/16	3 – 3-3/4	3 – 4-1/16	3 – 4-5/8	3 – 5-1/4	3 – 5-13/16
5 – 7	5 – 10-3/16	3 – 2-11/16	3 – 3-3/8	3 – 3-11/16	3 – 4	3 – 4-5/16	3 – 4-5/8	3 – 5-3/16	3 – 5-13/16	3 – 6-3/8
5 – 8	5 – 11-3/16	3 – 3-1/4	3 – 3-15/16	3 – 4-1/4	3 – 4-9/16	3 – 4-7/8	3 – 5-3/16	3 – 5-3/4	3 – 6-3/8	3 – 6-15/16
5 – 9	6 – 0-1/4	3 – 3-13/16	3 – 4-1/2	3 – 4-13/16	3 – 5-1/8	3 – 5-7/16	3 – 5-3/4	3 –6-5/16	3 – 6-15/16	3 – 7-1/2
5 – 10	6 – 1-5/16	3 – 4-7/16	3 – 5-1/8	3 – 5-7/16	3 – 5-3/4	3 – 6-1/16	3 – 6-3/8	3 – 6-15/16	3 – 7-9/16	3 – 8-1/8
5 – 11	6 – 2-3/8	3 – 5	3 – 5-11/16	3 – 6	3 – 6-5/16	3 – 6-5/8	3 – 6-15/16	3 – 7-1/2	3 – 8-1/8	3 – 8-11/16
6 – 0	6 – 3-3/8	3 – 5-9/16	3 – 6-1/4	3 – 6-9/16	3 – 6-7/8	3 – 7-3/16	3 – 7-1/2	3 – 8-1/16	3 – 8-11/16	3 – 9-1/4

Courtesy of Texas Pipe Bending Co., Inc.

Table 13-7. Arc for 90° Bends

Radius (inches)	Radius (feet) 0 – 0	1 – 0	2 – 0	3 – 0	4 – 0	5 – 0	6 – 0	7 – 0	8 – 0		(inches) Rad.	Arc
0		1 – 6-7/8	3 – 1-11/16	4 – 8-9/16	6 – 3-3/8	7 – 10-1/4	9 – 5-1/8	10 – 11-15/16	12 – 6-13/16	0	1-1/4	1-15/16
1	1-9/16	1 – 8-7/16	3 – 3-1/4	4 – 10-1/8	6 – 5	7 – 11-13/16	9 – 6-11/16	11 – 1-1/2	12 – 8-3/8	1	1-1/2	2-3/8
2	3-1/8	1 – 10	3 – 4-13/16	4 – 11-11/16	6 – 6-9/16	8 – 1-3/8	9 – 8-1/4	11 – 3-1/16	12 – 9-15/16	2	1-3/4	2-3/4
3	4-11/16	1 – 11-9/16	3 – 6-7/16	5 – 1-1/4	6 – 8-1/8	8 – 2-15/16	9 – 9-13/16	11 – 4-11/16	12 – 11-1/2	3	2-1/4	3-9/16
4	6-5/16	2 – 1-1/8	3 – 8	5 – 2-13/16	6 – 9-11/16	8 – 4-1/2	9 – 11-3/8	11 – 6-1/4	13 – 1-1/16	4	2-1/2	3-15/16
5	7-7/8	2 – 2-11/16	3 – 9-9/16	5 – 4-3/8	6 – 11-1/4	8 – 6-1/8	10 – 0-15/16	11 – 7-13/16	13 – 2-5/8	5	2-3/4	4-5/16
6	9-7/16	2 – 4-1/4	3 – 11-1/8	5 – 6	7 – 0-13/16	8 – 7-11/16	10 – 2-1/2	11 – 9-3/8	13 – 4-1/4	6	3-1/4	5-1/8
7	11	2 – 5-7/8	4 – 0-11/16	5 – 7-9/16	7 – 2-3/8	8 – 9-1/4	10 – 4-1/16	11 – 10-15/16	13 – 5-13/16	7	3-1/2	5-1/2
8	1 – 0-9/16	2 – 7-7/16	4 – 2-1/4	5 – 9-1/8	7 – 3-15/16	8 – 10-13/16	10 – 5-11/16	12 – 0-1/2	13 – 7-3/8	8	3-3/4	5-7/8
9	1 – 2-1/8	2 – 9	4 – 3-13/16	5 – 10-11/16	7 – 5-9/16	9 – 0-3/8	10 – 7-1/4	12 – 2-1/16	13 – 8-15/16	9	4-1/4	6-11/16
10	1 – 3-11/16	2 – 10-9/16	4 – 5-7/16	6 – 0-1/4	7 – 7-1/8	9 – 1-15/16	10 – 8-13/16	12 – 3-5/8	13 – 10-1/2	10	4-1/2	7-1/16
11	1 – 5-1/4	3 – 0-1/8	4 – 7	6 – 1-13/16	7 – 8-11/16	9 – 3-1/2	10 – 10-3/8	12 – 5-1/4	14 – 0-1/16	11	4-3/4	7-7/16
	9 – 0	10 – 0	11 – 0	12 – 0	13 – 0	14 – 0	15 – 0	16 – 0	17 – 0		5-1/4	8-1/4
0	14 – 1-5/8	15 – 8-1/2	17 – 3-3/8	18 – 10-3/16	20 – 5-1/16	21 – 11-7/8	23 – 6-3/4	25 – 1-5/8	26 – 8-7/16	0	5-1/2	8-5/8
1	14 – 3-3/16	15 – 10-1/16	17 – 4-15/16	18 – 11-3/4	20 – 6-5/8	22 – 1-7/16	23 – 8-5/16	25 – 3-3/16	26 – 10	1	5-3/4	9-1/16
2	14 – 4-13/16	15 – 11-5/8	17 – 6-1/2	19 – 1-5/16	20 – 8-3/16	22 – 3-1/16	23 – 9-7/8	25 – 4-3/4	26 – 11-9/16	2	6-1/2	10-3/16
3	14 – 6-3/8	16 – 1-3/16	17 – 8-1/16	19 – 2-15/16	20 – 9-3/4	22 – 4-5/8	23 – 11-7/16	25 – 6-5/16	27 – 1-1/8	3	7-1/2	11-3/4
4	14 – 7-15/16	16 – 2-3/4	17 – 9-5/8	19 – 4-1/2	20 – 11-5/16	22 – 6-3/16	24 – 1	25 – 7-7/8	27 – 2-3/4	4	8-1/2	13-3/8
5	14 – 9-1/2	16 – 4-3/8	17 – 11-3/16	19 – 6-1/16	21 – 0-7/8	22 – 7-3/4	24 – 2-5/8	25 – 9-7/16	27 – 4-5/16	5	9-1/2	14-15/16
6	14 – 11-1/16	16 – 5-15/16	18 – 0-3/4	19 – 7-5/8	21 – 2-1/2	22 – 9-5/16	24 – 4-3/16	25 – 11	27 – 5-7/8	6	10-1/2	16-1/2
7	15 – 0-5/8	16 – 7-1/2	18 – 2-5/16	19 – 9-3/16	21 – 4-1/16	22 – 10-7/8	24 – 5-3/4	26 – 0-9/16	27 – 7-7/16	7	11-1/2	18-1/16
8	15 – 2-3/16	16 – 9-1/16	18 – 3-15/16	19 – 10-3/4	21 – 5-5/8	23 – 0-7/16	24 – 7-5/16	26 – 2-3/16	27 – 9	8	12-1/2	19-5/8
9	15 – 3-13/16	16 – 10-5/8	18 – 5-1/2	20 – 0-5/16	21 – 7-3/16	23 – 2-1/16	24 – 8-7/8	26 – 3-3/4	27 – 10-9/16	9	13-1/2	21-3/16
10	15 – 5-3/8	17 – 0-3/16	18 – 7-1/16	20 – 1-7/8	21 – 8-3/4	23 – 3-5/8	24 – 10-7/16	26 – 5-5/16	28 – 0-1/8	10	14-1/2	22-3/4
11	15 – 6-15/16	17 – 1-3/4	18 – 8-5/8	20 – 3-1/2	21 – 10-5/16	23 – 5-3/16	25 – 0	26 – 6-7/8	28 – 1-3/4	11	15-1/2	24-3/8

Courtesy of Texas Pipe Bending Co., Inc.

Table 13-8A. Miter Welding

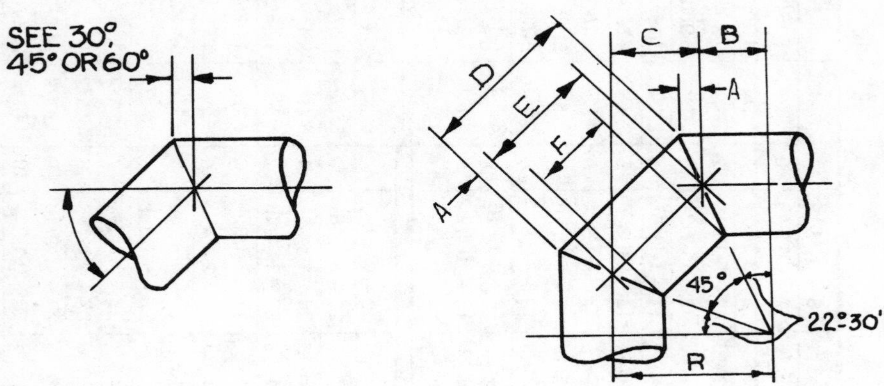

Size	30°	45°	60°	R	A	B	C	D	E	F
3	1/2	3/4	1	4-1/2	3/4	1-7/8	2-5/8	5-1/4	3-3/4	2-1/4
4	5/8	15/16	1-5/16	6	15/16	2-1/2	3-1/2	6-7/8	5	3-1/8
6	7/8	1-3/8	1-15/16	9	1-3/8	3-3/4	5-1/4	10-3/16	7-7/16	4-11/16
8	1-1/8	1-13/16	2-1/2	1 – 0	1-13/16	5	7	1 – 1-9/16	9-15/16	6-5/16
10	1-7/16	2-1/4	3-1/8	1 – 3	2-1/4	6-3/16	8-13/16	1 – 4-15/16	1 – 0-7/16	7-15/16
12	1-11/16	2-5/8	3-11/16	1 – 6	2-5/8	7-7/16	10-9/16	1 – 8-3/16	1 – 2-15/16	9-11/16
14	1-7/8	2-7/8	4-1/16	1 – 9	2-7/8	8-11/16	1 – 0-5/16	1 – 11-1/8	1 – 5-3/8	11-5/8
16	2-1/8	3-5/16	4-5/8	2 – 0	3-5/16	9-15/16	1 – 2-1/16	2 – 2-1/2	1 – 7-7/8	1 – 1-1/4
18	2-7/16	3-3/4	5-3/16	2 – 3	3-3/4	11-3/16	1 – 3-13/16	2 – 5-7/8	1 – 10-3/8	1 – 2-7/8
20	2-11/16	4-1/8	5-3/4	2 – 6	4-1/8	1 – 0-7/16	1 – 5-9/16	2 – 9-1/8	2 – 0-7/8	1 – 4-5/8
22	2-15/16	4-9/16	6-3/8	2 – 9	4-9/16	1 – 1-11/16	1 – 7-5/16	3 – 0-7/16	2 – 3-5/16	1 – 6-3/16
24	3-3/16	5	6-15/16	3 – 0	5	1 – 2-15/16	1 – 9-1/16	3 – 3-13/16	2 – 5-13/16	1 – 7-13/16
26	3-1/2	5-3/8	7-1/2	3 – 3	5-3/8	1 – 4-1/8	1 – 10-7/8	3 – 7-1/16	2 – 8-5/16	1 – 9-9/16
28	3-3/4	5-13/16	8-1/16	3 – 6	5-13/16	1 – 5-3/8	2 – 0-5/8	3 – 10-7/16	2 – 10-13/16	1 – 11-3/16
30	4	6-3/16	8-5/8	3 – 9	6-3/16	1 – 6-5/8	2 – 2-3/8	4 – 1-5/8	3 – 1-1/4	2 – 0-7/8
32	4-5/16	6-5/8	9-1/4	4 – 0	6-5/8	1 – 7-7/8	2 – 4-1/8	4 – 5	3 – 3-3/4	2 – 2-1/2
34	4-9/16	7-1/16	9-13/16	4 – 3	7-1/16	1 – 9-1/8	2 – 5-7/8	4 – 8-3/8	3 – 6-1/4	2 – 4-1/8
36	4-13/16	7-7/16	10-3/8	4 – 6	7-7/16	1 – 10-3/8	2 – 7-5/8	4 – 11-5/8	3 – 8-3/4	2 – 5-7/8
38	5-1/16	7-7/8	11	4 – 9	7-7/8	1 – 11-5/8	2 – 9-3/8	5 – 3	3 – 11-1/4	2 – 7-1/2
40	5-3/8	8-5/16	11-9/16	5 – 0	8-5/16	2 – 0-7/8	2 – 11-1/8	5 – 6-5/16	4 – 1-11/16	2 – 9-1/16
42	5-5/8	8-11/16	1 – 0-1/8	5 – 3	8-11/16	2 – 2-1/8	3 – 0-7/8	5 – 9-9/16	4 – 4-3/16	2 – 10-13/16
48	6-7/16	9-15/16	1 – 1-7/8	6 – 0	9-15/16	2 – 5-13/16	3 – 6-3/16	6 – 7-1/2	4 – 11-5/8	3 – 3-3/4
54	7-1/4	11-3/16	1 – 3-9/16	6 – 9	11-3/16	2 – 9-9/16	3 – 11-7/8	7 – 5-1/2	5 – 7-1/8	3 – 8-3/4
60	8-1/16	1 – 0-7/16	1 – 5-5/16	7 – 6	1 – 0-7/16	3 – 1-1/4	4 – 4-3/4	8 – 3-7/16	6 – 2-9/16	4 – 1-11/16
72	9-5/8	1 – 2-15/16	1 – 8-13/16	9 – 0	1 – 2-15/16	3 – 8-3/4	5 – 3-1/4	9 – 11-3/8	7 – 5-1/2	4 – 11-5/8

Courtesy of Texas Pipe Bending Co., Inc.

Table 13-8B. Miter Welding Dimensions

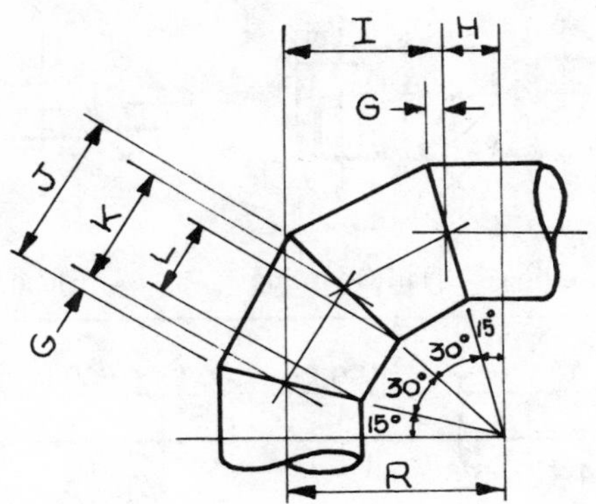

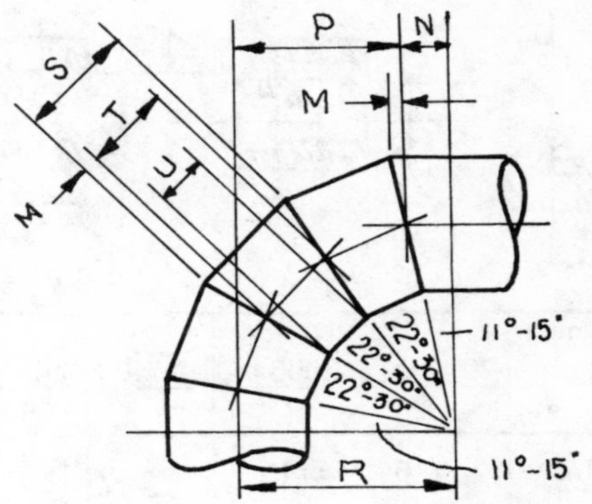

G	H	I	J	K	L	M	N	P	S	T	U
1/2	1-3/16	3-5/16	3-7/16	2-7/16	1-7/16	3/8	7/8	3-5/8	2-9/16	1-13/16	1-1/16
5/8	1-5/8	4-3/8	4-7/16	3-3/16	1-15/16	7/16	1-3/16	4-13/16	3-1/4	2-3/8	1-1/2
7/8	2-7/16	6-9/16	6-9/16	4-13/16	3-1/16	11/16	1-13/16	7-3/16	4-15/16	3-9/16	2-3/16
1-1/8	3-3/16	8-13/16	8-11/16	6-7/16	4-3/16	7/8	2-3/8	9-5/8	6-1/2	4-3/4	3
1-7/16	4	11	10-15/16	8-1/16	5-3/16	1-1/16	3	1 – 0	8-1/16	5-15/16	3-13/16
1-11/16	4-13/16	1 – 1-3/16	1 – 1	9-5/8	6-1/4	1-1/4	3-9/16	1 – 2-7/16	9-11/16	7-3/16	4-11/16
1-7/8	5-5/8	1 – 3-3/8	1 – 3	11-1/4	7-1/2	1-3/8	4-3/16	1 – 4-13/16	11-1/8	8-3/8	5-5/8
2-1/8	6-7/16	1 – 5-9/16	1 – 5-1/8	1 – 0-7/8	8-5/8	1-9/16	4-3/4	1 – 7-1/4	1 – 0-11/16	9-9/16	6-7/16
2-7/16	7-1/4	1 – 7-3/4	1 – 7-5/16	1 – 2-7/16	9-9/16	1-13/16	5-3/8	1 – 9-5/8	1 – 2-3/8	10-3/4	7-1/8
2-11/16	8-1/16	1 – 9-15/16	1 – 9-7/16	1 – 4-1/16	10-11/16	2	5-15/16	2 – 0-1/16	1 – 3-15/16	11-15/16	7-15/16
2-15/16	8-13/16	2 – 0-3/16	1 – 11-9/16	1 – 5-11/16	11-13/16	2-3/16	6-9/16	2 – 2-7/16	1 – 5-1/2	1 – 1-1/8	8-3/4
3-3/16	9-5/8	2 – 2-3/8	2 – 1-11/16	1 – 7-5/16	1 – 0-15/16	2-3/8	7-3/16	2 – 4-13/16	1 – 7-1/16	1 – 2-5/16	9-9/16
3-1/2	10-7/16	2 – 4-9/16	2 – 3-7/8	1 – 8-7/8	1 – 1-7/8	2-9/16	7-3/4	2 – 7-1/4	1 – 8-5/8	1 – 3-1/2	10-3/8
3-3/4	11-1/4	2 – 6-3/4	2 – 6	1 – 10-1/2	1 – 3	2-13/16	8-3/8	2 – 9-5/8	1 – 10-5/16	1 – 4-11/16	11-1/16
4	1 – 0-1/16	2 – 8-15/16	2 – 8-1/8	2 – 0-1/8	1 – 4-1/8	3	8-15/16	3 – 0-1/16	1 – 11-7/8	1 – 5-7/8	11-7/8
4-5/16	1 – 0-7/8	2 – 11-1/8	2 – 10-5/16	2 – 1-11/16	1 – 5-1/16	3-3/16	9-9/16	3 – 2-7/16	2 – 1-1/2	1 – 7-1/8	1 – 0-3/4
4-9/16	1 – 1-11/16	3 – 1-5/16	3 – 0-7/16	2 – 3-5/16	1 – 6-3/16	3-3/8	10-1/8	3 – 4-7/8	2 – 3-1/16	1 – 8-5/16	1 – 1-9/16
4-13/16	1 – 2-7/16	3 – 3-9/16	3 – 2-9/16	2 – 4-15/16	1 – 7-5/16	3-9/16	10-3/4	3 – 7-1/4	2 – 4-5/8	1 – 9-1/2	1 – 2-3/8
5-1/16	1 – 3-1/4	3 – 5-3/4	3 – 4-11/16	2 – 6-9/16	1 – 8-7/16	3-3/4	11-5/16	3 – 9-11/16	2 – 6-3/16	1 – 10-11/16	1 – 3-3/16
5-3/8	1 – 4-1/16	3 – 7-15/16	3 – 6-7/8	2 – 8-1/8	1 – 9-3/8	4	11-15/16	4 – 0-1/16	2 – 7-7/8	1 – 11-7/8	1 – 3-7/8
5-5/8	1 – 4-7/8	3 – 10-1/8	3 – 9	2 – 9-3/4	1 – 10-1/2	4-3/16	1 – 0-1/2	4 – 2-1/2	2 – 9-7/16	2 – 1-1/16	1 – 4-11/16
6-7/16	1 – 7-5/16	4 – 4-11/16	4 – 3-7/16	3 – 2-9/16	2 – 1-11/16	4-3/4	1 – 2-5/16	4 – 9-11/16	3 – 2-1/8	2 – 4-5/8	1 – 7-1/8
7-1/4	1 – 9-11/16	4 – 11-5/16	4 – 9-7/8	3 – 7-3/8	2 – 4-7/8	5-3/8	1 – 4-1/8	5 – 4-7/8	3 – 7	2 – 8-1/4	1 – 9-1/2
8-1/16	2 – 0-1/8	5 – 5-7/8	5 – 4-3/8	4 – 0-1/4	2 – 8-1/8	5-15/16	1 – 5-7/8	6 – 0-1/8	3 – 11-5/8	2 – 11-3/4	1 – 11-7/8
9-5/8	2 – 5	6 – 7	6 – 5-1/4	4 – 10	3 – 2-3/4	7-3/16	1 – 9-1/2	7 – 2-1/2	4 – 9-3/8	3 – 7	2 – 4-5/8

Courtesy of Texas Pipe Bending Co., Inc.

Table 13-9A. Socket Weld Fittings

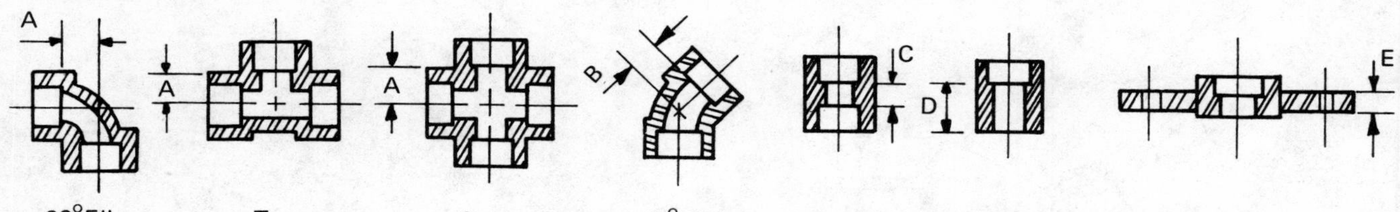

90° Ell | Tee | Cross | 45° Ell | Coupling | Half Coupling | Flange

Nominal Pipe Size	2000 lb. & 3000 lb. A	2000 lb. & 3000 lb. B	4000 lb. A	4000 lb. B	6000 lb. A	6000 lb. B	All Wts. C	All Wts. D	150 lb. E	300 lb. E	600 lb. E	Nominal Pipe Size
1/8	7/16	5/16					1/4	5/8				1/8
1/4	7/16	5/16					1/4	5/8	1/4	1/2	3/4	1/4
3/8	9/16	5/16			9/16	3/8	1/4	11/16	1/4	1/2	3/4	3/8
1/2	5/8	7/16	3/4	1/2	5/8	3/8	3/8	7/8	1/4	1/2	3/4	1/2
3/4	3/4	1/2	7/8	9/16	3/4	7/16	3/8	15/16	3/16	9/16	13/16	3/4
1	7/8	9/16	1-1/16	11/16	7/8	1/2	1/2	1-1/8	3/16	9/16	13/16	1
1-1/4	1-1/16	11/16	1-1/4	13/16	1-1/16	5/8	1/2	1-3/16	1/4	1/2	13/16	1-1/4
1-1/2	1-1/4	13/16	1-1/2	1	1-1/4	5/8	1/2	1-1/4	1/4	9/16	7/8	1-1/2
2	1-1/2	1	1-5/8	1-1/8	1-1/2	7/8	3/4	1-5/8	5/16	5/8	1	2

Courtesy of Texas Pipe Bending Co., Inc.

Table 13-9B. Threaded Fittings

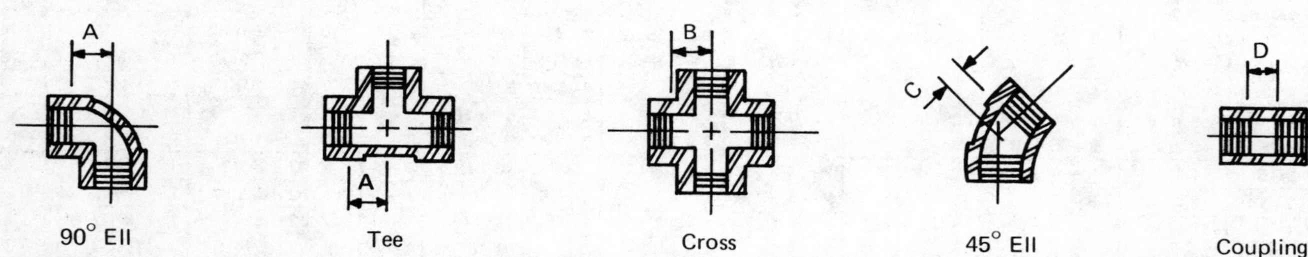

Nominal Pipe Size	2000 lb. A	2000 lb. B	2000 lb. C	3000 lb. A	3000 lb. B	3000 lb. C	6000 lb. A	6000 lb. B	6000 lb. C	3000 lb. 6000 lb. D	Normal Thread Engagement	Nominal Pipe Size
1/8	9/16	3/4	7/16	9/16	3/4	7/16	3/4	3/4	1/2	3/4	1/4	1/8
1/4	7/16	5/8	5/16	5/8	5/8	3/8	3/4	3/4	1/2	5/8	3/8	1/4
3/8	5/8	5/8	3/8	3/4	3/4	1/2	15/16	15/16	5/8	3/4	3/8	3/8
1/2	5/8	5/8	3/8	13/16	13/16	1/2	1	1	5/8	7/8	1/2	1/2
3/4	3/4	3/4	7/16	15/16	15/16	9/16	1-3/16	1-3/16	3/4	7/8	9/16	3/4
1	13/16	13/16	7/16	1-1/16	1-1/16	5/8	1-5/16	1-5/16	11/16	1	11/16	1
1-1/4	1-1/16	1-1/16	5/8	1-5/16	1-5/16	11/16	1-11/16	1-11/16	1	1-1/4	11/16	1-1/4
1-1/2	1-5/16	1-5/16	11/16	1-11/16	1-11/16	1	1-13/16	1-13/16	1-1/16	1-3/4	11/16	1-1/2
2	1-5/8	1-5/8	15/16	1-3/4	1-3/4	1	2-1/2	2-1/2	1-5/16	1-7/8	3/4	2

Courtesy of Texas Pipe Bending Co., Inc.

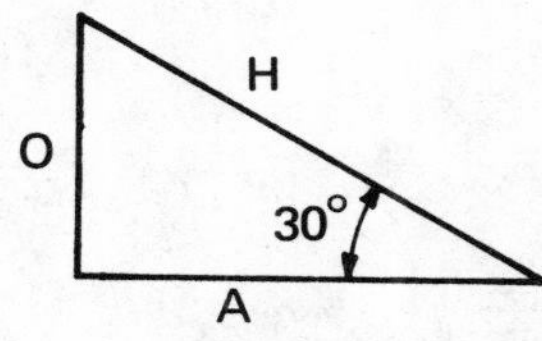

Table 13-10. Table of 30° Offsets

0 ft. – 0-1/4 in. to 0 ft. – 11-3/4 in.

O	H	A	O	H	A	O	H	A
			0 – 4	0 – 8	0 – 6-15/16	0 – 8	1 – 4	1 – 1-15/16
0 – 0-1/4	0 – 0-1/2	0 – 0-7/16	0 – 4-1/4	0 – 8-1/2	0 – 7-3/8	0 – 8-1/4	1 – 4-1/2	1 – 2-5/16
0 – 0-1/2	0 – 1	0 – 0-7/8	0 – 4-1/2	0 – 9	0 – 7-13/16	0 – 8-1/2	1 – 5	1 – 2-3/4
0 – 0-3/4	0 – 1-1/2	0 – 1-5/16	0 – 4-3/4	0 – 9-1/2	0 – 8-1/4	0 – 8-3/4	1 – 5-1/2	1 – 3-1/8
0 – 1	0 – 2	0 – 1-3/4	0 – 5	0 – 10	0 – 8-11/16	0 – 9	1 – 6	1 – 3-9/16
0 – 1-1/4	0 – 2-1/2	0 – 2-3/16	0 – 5-1/4	0 – 10-1/2	0 – 9-1/16	0 – 9-1/4	1 – 6-1/2	1 – 4
0 – 1-1/2	0 – 3	0 – 2-5/8	0 – 5-1/2	0 – 11	0 – 9-1/2	0 – 9-1/2	1 – 7	1 – 4-7/16
0 – 1-3/4	0 – 3-1/2	0 – 3	0 – 5-3/4	0 – 11-1/2	0 – 9-15/16	0 – 9-3/4	1 – 7-1/2	1 – 4-7/8
0 – 2	0 – 4	0 – 3-7/16	0 – 6	1 – 0	0 – 10-3/8	0 – 10	1 – 8	1 – 5-5/16
0 – 2-1/4	0 – 4-1/2	0 – 3-7/8	0 – 6-1/4	1 – 0-1/2	0 – 10-13/16	0 – 10-1/4	1 – 8-1/2	1 – 5-3/4
0 – 2-1/2	0 – 5	0 – 4-5/16	0 – 6-1/2	1 – 1	0 – 11-1/4	0 – 10-1/2	1 – 9	1 – 6-3/16
0 – 2-3/4	0 – 5-1/2	0 – 4-3/4	0 – 6-3/4	1 – 1-1/2	0 – 11-11/16	0 – 10-3/4	1 – 9-1/2	1 – 6-5/8
0 – 3	0 – 6	0 – 5-3/16	0 – 7	1 – 2	1 – 0-1/8	0 – 11	1 – 10	1 – 7-1/16
0 – 3-1/4	0 – 6-1/2	0 – 5-5/8	0 – 7-1/4	1 – 2-1/2	1 – 0-9/16	0 – 11-1/4	1 – 10-1/2	1 – 7-1/2
0 – 3-1/2	0 – 7	0 – 6-1/16	0 – 7-1/2	1 – 3	1 – 1	0 – 11-1/2	1 – 11	1 – 7-15/16
0 – 3-3/4	0 – 7-1/2	0 – 6-1/2	0 – 7-3/4	1 – 3-1/2	1 – 1-7/16	0 – 11-3/4	1 – 11-1/2	1 – 8-3/8

Table 13-10 Continued

1 ft. – 0 in. to 1 ft. – 11-3/4 in.

O	H	A	O	H	A	O	H	A
1 – 0	2 – 0	1 – 8-13/16	1 – 4	2 – 8	2 – 3-11/16	1 – 8	3 – 4	2 – 10-5/8
1 – 0-1/4	2 – 0-1/2	1 – 9-3/16	1 – 4-1/4	2 – 8-1/2	2 – 4-1/8	1 – 8-1/4	3 – 4-1/2	2 – 11-1/16
1 – 0-1/2	2 – 1	1 – 9-5/8	1 – 4-1/2	2 – 9	2 – 4-9/16	1 – 8-1/2	3 – 5	2 – 11-1/2
1 – 0-3/4	2 – 1-1/2	1 – 10-1/16	1 – 4-3/4	2 – 9-1/2	2 – 5	1 – 8-3/4	3 – 5-1/2	2 – 11-15/16
1 – 1	2 – 2	1 – 10-1/2	1 – 5	2 – 10	2 – 5-7/16	1 – 9	3 – 6	3 – 0-3/8
1 – 1-1/4	x2 – 2-1/2	1 – 10-15/16	1 – 5-1/4	2 – 10-1/2	2 – 5-7/8	1 – 9-1/4	3 – 6-1/2	3 – 0-13/16
1 – 1-1/2	2 – 3	1 – 11-3/8	1 – 5-1/2	2 – 11	2 – 6-5/16	1 – 9-1/2	3 – 7	3 – 1-1/4
1 – 1-3/4	2 – 3-1/2	1 – 11-13/16	1 – 5-3/4	2 – 11-1/2	2 – 6-3/4	1 – 9-3/4	3 – 7-1/2	3 – 1-11/16
1 – 2	2 – 4	2 – 0-1/4	1 – 6	3 – 0	3 – 7-3/16	1 – 10	3 – 8	3 – 2-1/8
1 – 2-1/4	2 – 4-1/2	2 – 0-11/16	1 – 6-1/4	3 – 0-1/2	2 – 7-5/8	1 – 10-1/4	3 – 8-1/2	3 – 2-9/16
1 – 2-1/2	2 – 5	2 – 1-1/8	1 – 6-1/2	3 – 1	2 – 8-1/16	1 – 10-1/2	3 – 9	3 – 3
1 – 2-3/4	2 – 5-1/2	2 – 1-9/16	1 – 6-3/4	3 – 1-1/2	2 – 8-1/2	1 – 10-3/4	3 – 9-1/2	3 – 3-3/8
1 – 3	2 – 6	2 – 2	1 – 7	3 – 2	2 – 8-15/16	1 – 11	3 – 10	3 – 3-13/16
1 – 3-1/4	2 – 6-1/2	2 – 2-7/16	1 – 7-1/4	2 – 2-1/2	2 – 9-5/16	1 – 11-1/4	3 – 10-1/2	3 – 4-1/4
1 – 3-1/2	2 – 7	2 – 2-7/8	1 – 7-1/2	3 – 3	2 – 9-3/4	1 – 11-1/2	3 – 11	3 – 4-11/16
1 – 3-3/4	2 – 7-1/2	2 – 3-1/4	1 – 7-3/4	3 – 3-1/2	2 – 10-3/16	1 – 11-3/4	3 – 11-1/2	3 – 5-1/8

Courtesy of Texas Pipe Bending Co., Inc.

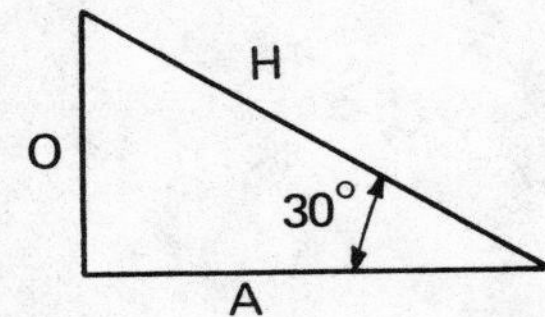

Table 13-11. Table of 30° Offsets

2′ – 0″ to 2′ – 11-3/4″

O	H	A	O	H	A	O	H	A
2 – 0	4 – 0	3 – 5-9/16	2 – 4	4 – 8	4 – 0-1/2	2 – 8	5 – 4	4 – 7-7/16
2 – 0-1/4	4 – 0-1/2	3 – 6	2 – 4-1/4	4 – 8-1/2	4 – 0-15/16	2 – 8-1/4	5 – 4-1/2	4 – 7-7/8
2 – 0-1/2	4 – 1	3 – 6-7/16	2 – 4-1/2	4 – 9	4 – 1-3/8	2 – 8-1/2	5 – 5	4 – 8-5/16
2 – 0-3/4	4 – 1-1/2	3 – 6-7/8	2 – 4-3/4	4 – 9-1/2	4 – 1-13/16	2 – 8-3/4	5 – 5-1/2	4 – 8-3/4
2 – 1	4 – 2	3 – 7-5/16	2 – 5	4 – 10	4 – 2-1/4	2 – 9	5 – 6	4 – 9-3/16
2 – 1-1/4	4 – 2-1/2	3 – 7-3/4	2 – 5-1/4	4 – 10-1/2	4 – 2-11/16	2 – 9-1/4	5 – 6-1/2	4 – 9-9/16
2 – 1-1/2	4 – 3	3 – 8-3/16	2 – 5-1/2	4 – 11	4 – 3-1/8	2 – 9-1/2	5 – 7	4 – 10
2 – 1-3/4	4 – 3-1/2	3 – 8-5/8	2 – 5-3/4	4 – 11-1/2	4 – 3-1/2	2 – 9-3/4	5 – 7-1/2	4 – 10-7/16
2 – 2	4 – 4	3 – 9-1/16	2 – 6	5 – 0	4 – 3-15/16	2 – 10	5 – 8	4 – 10-7/8
2 – 2-1/4	4 – 4-1/2	3 – 9-1/2	2 – 6-1/4	5 – 0-1/2	4 – 4-3/8	2 – 10-1/4	5 – 8-1/2	4 – 11-5/16
2 – 2-1/2	4 – 5	3 – 9-7/8	2 – 6-1/2	5 – 1	4 – 4-13/16	2 – 10-1/2	5 – 9	4 – 11-3/4
2 – 2-3/4	4 – 5-1/2	3 – 10-5/16	2 – 6-3/4	5 – 1-1/2	4 – 5-1/4	2 – 10-3/4	5 – 9-1/2	5 – 0-3/16
2 – 3	4 – 6	3 – 10-3/4	2 – 7	5 – 2	4 – 5-11/16	2 – 11	5 – 10	5 – 0-5/8
2 – 3-1/4	4 – 6-1/2	3 – 11-3/16	2 – 7-1/4	5 – 2-1/2	4 – 6-1/8	2 – 11-1/4	5 – 10-1/2	5 – 1-1/16
2 – 3-1/2	4 – 7	3 – 11-5/8	2 – 7-1/2	5 – 3	4 – 6-9/16	2 – 11-1/2	5 – 11	5 – 1-1/2
2 – 3-3/4	4 – 7-1/2	4 – 0-1/16	2 – 7-3/4	5 – 3-1/2	4 – 7	2 – 11-3/4	5 – 11-1/2	5 – 1-13/16

Table 13-11 Continued

3′ – 0″ to 3′ – 11-3/4″

O	H	A	O	H	A	O	H	A
3 – 0	6 – 0	5 – 2-3/8	3 – 4	6 – 8	5 – 9-5/16	3 – 8	7 – 4	6 – 4-3/16
3 – 0-1/4	6 – 0-1/2	5 – 2-13/16	3 – 4-1/4	6 – 8-1/2	5 – 9-11/16	3 – 8-1/4	7 – 4-1/2	6 – 4-5/8
3 – 0-1/2	6 – 1	5 – 3-1/4	3 – 4-1/2	6 – 9	5 – 10-1/8	3 – 8-1/2	7 – 5	6 – 5-1/16
3 – 0-3/4	6 – 1-1/2	5 – 3-5/8	3 – 4-3/4	6 – 9-1/2	5 – 10-9/16	3 – 8-3/4	7 – 5-1/2	6 – 6-1/2
3 – 1	6 – 2	5 – 4-1/16	3 – 5	6 – 10	5 – 11	3 – 9	7 – 6	6 – 5-15/16
3 – 1-1/4	6 – 2-1/2	5 – 4-1/2	3 – 5-1/4	6 – 10-1/2	5 – 11-7/16	3 – 9-1/4	7 – 6-1/2	6 – 6-3/8
3 – 1-1/2	6 – 3	5 – 4-15/16	3 – 5-1/2	6 – 11	5 – 11-7/8	3 – 9-1/2	7 – 7	6 – 6-13/16
3 – 1-3/4	6 – 3-1/2	5 – 5-3/8	3 – 5-3/4	6 – 11-1/2	6 – 0-5/16	3 – 9-3/4	7 – 7-1/2	6 – 7-1/4
3 – 2	6 – 4	5 – 5-13/16	3 – 6	7 – 0	6 – 0-3/4	3 – 10	7 – 8	6 – 7-11/16
3 – 2-1/4	6 – 4-1/2	5 – 6-1/4	3 – 6-1/4	7 – 0-1/2	6 – 1-3/16	3 – 10-1/4	7 – 8-1/2	6 – 8-1/8
3 – 2-1/2	6 – 5	5 – 6-11/16	3 – 6-1/2	7 – 1	6 – 1-5/8	3 – 10-1/2	7 – 9	6 – 8-9/16
3 – 2-3/4	6 – 5-1/2	5 – 7-1/8	3 – 6-3/4	7 – 1-1/2	6 – 2-1/16	3 – 10-3/4	7 – 9-1/2	6 – 9
3 – 3	6 – 6	5 – 7-9/16	3 – 7	7 – 2	6 – 2-1/2	3 – 11	7 – 10	6 – 9-7/16
3 – 3-1/4	6 – 6-1/2	5 – 8	3 – 7-1/4	7 – 2-1/2	6 – 2-15/16	3 – 11-1/4	7 – 10-1/2	6 – 9-13/16
3 – 3-1/2	6 – 7	5 – 8-7/16	3 – 7-1/2	7 – 3	6 – 3-3/8	3 – 11-1/2	7 – 11	6 – 10-1/4
3 – 3-3/4	6 – 7-1/2	5 – 8-7/8	3 – 7-3/4	7 – 3-1/2	6 – 3-3/4	3 – 11-3/4	7 – 11-1/2	6 – 10-11/16

Courtesy of Texas Pipe Bending Co., Inc.

Table 13-12A. 45° Offsets

	0	1	2	3	4	5	6	7	8	9	10	11	
0 1/16	0 1/16	1-7/16 1-1/2	2-13/16 2-7/8	4-1/4 4-5/16	5-11/16 5-3/4	7-1/16 7-3/16	8-1/2 8-9/16	9-7/8 10	11-5/16 11-3/8	12-3/4 12-13/16	14-1/8 14-1/4	15-9/16 15-5/8	0 1/16
1/8 3/16	3/16 1/4	1-9/16 1-11/16	3 3-1/16	4-7/16 4-1/2	5-13/16 5-7/8	7-1/4 7-5/16	8-11/16 8-3/4	10-1/16 10-3/16	11-1/2 11-9/16	12-7/8 13	14-5/16 14-7/16	15-3/4 15-13/16	1/8 3/16
1/4 5/16	3/8 7/16	1-3/4 1-7/8	3-3/16 3-1/4	4-5/8 4-11/16	6 6-0/8	7-7/16 7-1/2	8-13/16 8-15/16	10-1/4 10-3/8	11-11/16 11-3/4	13-1/16 13-3/16	14-1/2 14-9/16	15-15/16 16	1/4 5/16
3/8 7/16	1/2 5/8	1-15/16 2-1/16	3-3/8 3-7/16	4-3/4 4-7/8	6-3/16 6-1/4	7-5/8 7-11/16	9 9-1/8	10-7/16 10-1/2	11-7/8 11-15/16	13-1/4 13-3/8	14-11/16 14-3/4	16-1/16 16-3/16	3/8 7/16
1/2 9/16	11/16 13/16	2-1/8 2-3/16	3-9/16 3-5/8	4-15/16 5-1/16	6-3/8 6-7/16	7-3/4 7-7/8	9-3/16 9-1/4	10-5/8 10-11/16	12 12-1/8	13-7/16 13-9/16	14-7/8 14-15/16	16-1/4 16-3/8	1/2 9/16
5/8 11/16	7/8 1	2-5/16 2-3/8	3-11/16 3-13/16	5-1/8 5-3/16	6-9/16 6-5/8	7-15/16 8-1/16	9-3/8 9-7/16	10-13/16 10-7/8	12-3/16 12-5/16	13-5/8 13-11/16	15 15-1/8	16-7/16 16-1/2	5/8 11/16
3/4 13/16	1-1/16 1-1/8	2-1/2 2-9/16	3-7/8 4	5-5/16 5-3/8	6-11/16 6-13/16	8-1/8 8-1/4	9-9/16 9-5/8	11 11-1/16	12-3/8 12-7/16	13-13/16 13-7/8	15-3/16 15-5/16	16-5/8 16-11/16	3/4 13/16
7/8 15/16	1-1/4 1-5/16	2-5/8 2-3/4	4-1/16 4-1/8	5-1/2 5-9/16	6-7/8 7	8-5/16 8-3/8	9-11/16 9-13/16	11-1/8 11-1/4	12-9/16 12-5/8	13-15/16 14-1/16	15-3/8 15-7/16	16-13/16 16-7/8	7/8 15/16

Courtesy of Texas Pipe Bending Co., Inc.

Table 13-12B. 45° Triangles - Base to Hypotenuse

	12	13	14	15	16	17	18	19	20	21	22	23	
0 1/16	17 17-1/16	18-3/8 18-1/2	19-13/16 19-7/8	21-3/16 21-5/16	22-5/8 22-11/16	2 – 0-1/16 2 – 0-1/8	2 – 1-7/16 2 – 1-9/16	2 – 2-7/8 2 – 2-15/16	2 – 4-5/16 2 – 4-3/8	2 – 5-11/16 2 – 5-13/16	2 – 7-1/8 2 – 7-3/16	2 – 8-1/2 2 – 8-5/8	0 1/16
1/8 3/16	17-1/8 17-1/4	18-9/16 18-5/8	20 20-1/16	21-3/8 21-1/2	22-13/16 22-7/8	2 – 0-1/4 2 – 0-5/16	2 – 1-5/8 2 – 1-3/4	2 – 3-1/16 2 – 3-1/8	2 – 4-1/2 2 – 4-9/16	2 – 5-7/8 2 – 5-15/16	2 – 7-5/16 2 – 7-3/8	2 – 8-11/16 2 – 8-13/16	1/8 3/16
1/4 5/16	17-5/16 17-7/8	18-3/4 18-13/16	20-1/8 20-1/4	21-9/16 21-5/8	23 23-1/16	2 – 0-3/8 2 – 0-1/2	2 – 1-13/16 2 – 1-7/8	2 – 3-1/4 2 – 3-5/16	2 – 4-5/8 2 – 4-3/4	2 – 6-1/16 2 – 6-1/8	2 – 7-7/16 2 – 7-9/16	2 – 8-7/8 2 – 9	1/4 5/16
3/8 7/16	17-1/2 17-9/16	18-7/8 19	20-5/16 20-7/16	21-3/4 21-13/16	23-3/16 23-1/4	2 – 0-9/16 2 – 0-11/16	2 – 2 2 – 2-1/16	2 – 3-3/8 2 – 3-1/2	2 – 4-13/16 2 – 4-7/8	2 – 6-1/4 2 – 6-5/16	2 – 7-5/8 2 – 7-3/4	2 – 9-1/16 2 – 9-1/8	3/8 7/16
1/2 9/16	17-11/16 17-3/4	19-1/8 19-3/16	20-1/2 20-5/8	21-15/16 22	23-5/16 23-7/16	2 – 0-3/4 2 – 0-13/16	2 – 2-3/16 2 – 2-1/4	2 – 3-9/16 2 – 3-11/16	2 – 5 2 – 5-1/16	2 – 6-3/8 2 – 6-7/16	2 – 7-13/16 2 – 7-15/16	2 – 9-1/4 2 – 9-5/16	1/2 9/16
5/8 11/16	17-7/8 17-15/16	19-1/4 19-3/8	20-11/16 20-3/4	22-1/8 22-3/16	23-1/2 23-5/8	2 – 0-15/16 2 – 1	2 – 2-5/16 2 – 2-7/16	2 – 3-3/4 2 – 3-13/16	2 – 5-3/16 2 – 5-1/4	2 – 6-9/16 2 – 6-11/16	2 – 8 2 – 8-1/16	2 – 9-7/16 2 – 9-1/2	5/8 11/16
3/4 13/16	18-1/16 18-1/8	19-7/16 19-9/16	20-7/8 20-15/16	22-1/4 22-3/8	23-11/16 22-3/4	2 – 1-1/16 2 – 1-3/16	2 – 2-1/2 2 – 2-5/8	2 – 3-15/16 2 – 4	2 – 5-3/8 2 – 5-7/16	2 – 6-3/4 2 – 6-7/8	2 – 8-3/16 2 – 8-1/4	2 – 9-9/16 2 – 9-11/16	3/4 13/16
7/8 15/16	18-3/16 18-5/16	19-5/8 19-11/16	21-1/16 21-1/8	22-7/16 22-9/16	23-7/8 23-15/16	2 – 1-1/4 2 – 1-3/8	2 – 2-11/16 2 – 2-13/16	2 – 4-1/8 2 – 4-3/16	2 – 5-1/2 2 – 5-5/8	2 – 6-15/16 2 – 7	2 – 8-3/8 2 – 8-7/16	2 – 9-3/4 2 – 9-7/8	7/8 15/16

Courtesy of Texas Pipe Bending Co., Inc.

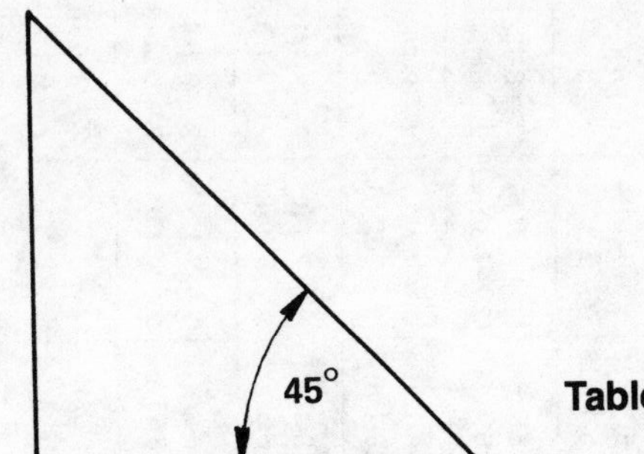

Table 13-13. 45° Offsets

	2 – 4	2 – 1	2 – 2	2 – 3	2 – 4	2 – 5	2 – 6	2 – 7	2 – 8	2 – 9	2 – 10	2 – 11	
0 1/16	2 – 9-15/16 2 – 10	2 – 11-3/8 2 – 11-7/16	3 – 0-3/4 3 – 0-7/8	3 – 2-3/16 3 – 2-1/4	3 – 3-5/8 3 – 3-11/16	3 – 5 3 – 5-1/8	3 – 6-7/16 3 – 6-1/2	3 – 7-13/16 3 – 7-15/16	3 – 9-1/4 3 – 9-5/16	3 – 10-11/16 3 – 10-3/4	4 – 0-1/16 4 – 0-3/16	4 – 1-1/2 4 – 1-9/16	0 1/16
1/8 3/16	2 – 10-1/8 2 – 10-3/16	2 – 11-9/16 2 – 11-5/8	3 – 0-15/16 3 – 1-1/16	3 – 2-3/8 3 – 2-7/16	3 – 3-3/4 3 – 3-7/8	3 – 5-3/16 3 – 5-1/4	3 – 6-5/8 3 – 6-11/16	3 – 8 3 – 8-1/8	3 – 9-7/16 3 – 9-1/2	3 – 10-7/8 3 – 10-15/16	4 – 0-1/4 4 – 0-3/8	4 – 4-11/16 4 – 1-3/4	1/8 3/16
1/4 3/16	2 – 10-5/16 2 – 10-3/8	2 – 11-11/16 2 – 11-13/16	3 – 1-1/8 3 – 1-3/16	3 – 2-9/16 3 – 2-5/8	3 – 3-15/16 3 – 4-1/16	3 – 5-3/8 3 – 5-7/16	3 – 6-3/4 3 – 6-7/8	3 – 8-3/16 3 – 8-5/16	3 – 9-5/8 3 – 9-11/16	3 – 11 3 – 11-1/8	4 – 0-7/16 4 – 0-1/2	4 – 1-7/8 4 – 1-15/16	1/4 3/16
3/8 7/16	2 – 10-1/2 2 – 10-9/16	2 – 11-7/8 3 – 0	3 – 1-5/16 3 – 1-3/8	3 – 2-11/16 3 – 2-13/16	3 – 4-1/8 3 – 4-3/16	3 – 5-9/16 3 – 5-5/8	3 – 6-15/16 3 – 7-1/16	3 – 8-3/8 3 – 8-7/16	3 – 9-3/16 3 – 9-7/8	3 – 11-3/16 3 – 11-5/16	4 – 0-5/8 4 – 0-11/16	4 – 2 4 – 2-1/16	3/8 7/16
1/2 9/16	2 – 10-5/8 2 – 10-3/4	3 – 0-1/16 3 – 0-1/8	3 – 1-1/2 3 – 1-9/16	3 – 2-7/8 3 – 3	3 – 4-5/16 3 – 4-3/8	3 – 5-3/4 3 – 5-13/16	3 – 7-1/8 3 – 7-1/4	3 – 8-9/16 3 – 8-5/8	3 – 9-15/16 3 – 10-1/16	3 – 11-3/8 3 – 11-7/16	4 – 0-13/16 4 – 0-7/8	4 – 2-3/16 4 – 2-5/16	1/2 9/16
3/8 11/16	2 – 10-13/16 2 – 10-15/16	3 – 0-1/4 3 – 0-5/16	3 – 1-11/16 3 – 1-3/4	3 – 3-1/16 3 – 3-1/8	3 – 4-1/2 3 – 4-9/16	3 – 5-7/8 3 – 6	3 – 7-5/16 3 – 7-3/8	3 – 8-3/4 3 – 8-13/16	3 – 10-1/8 3 – 10-1/4	3 – 11-9/16 3 – 11-5/8	4 – 0-15/16 4 – 1-1/16	4 – 2-3/8 4 – 2-1/2	3/8 11/16
3/4 13/16	2 – 11 2 – 11-1/16	3 – 0-7/16 3 – 0-1/2	3 – 1-13/16 3 – 1-15/16	3 – 3-1/4 3 – 3-5/16	3 – 4-11/16 3 – 4-3/4	3 – 6-1/16 3 – 6-3/16	3 – 7-1/2 3 – 7-9/16	3 – 8-15/16 3 – 9	3 – 10-5/16 3 – 10-3/8	3 – 11-11/16 3 – 11-13/16	4 – 1-1/8 4 – 1-1/4	4 – 2-9/16 4 – 2-5/8	3/4 13/16
7/8 15/16	2 – 11-3/16 2 – 11-1/4	3 – 0-9/16 3 – 0-11/16	3 – 2 3 – 2-1/8	3 – 3-7/16 3 – 3-1/2	3 – 4-13/16 3 – 4-15/16	3 – 6-1/4 3 – 6-5/16	3 – 7-11/16 3 – 7-3/4	3 – 9-1/16 3 – 9-3/16	3 – 10-1/2 3 – 10-9/16	3 – 11-13/16 4 – 0	4 – 1-3/16 4 – 1-7/16	4 – 2-3/4 4 – 2-13/16	7/8 15/16

Courtesy of Texas Pipe Bending Co., Inc.

table continued on next page

Table 13-13 continued

	3 – 0	3 – 1	3 – 2	3 – 3	3 – 4	3 – 5	3 – 6	3 – 7	3 – 8	3 – 9	3 – 10	3 – 11	
0 1/16	4 – 2-13/16 4 – 3	4 – 4-3/16 4 – 4-7/16	4 – 5-3/4 4 – 5-13/16	4 – 7-1/8 4 – 7-1/4	4 – 8-9/16 4 – 8-5/8	4 – 10 4 – 10-1/16	4 – 11-3/8 4 – 11-1/2	5 – 0-13/16 5 – 0-7/8	5 – 2-1/4 5 – 2-5/16	5 – 3-5/8 5 – 3-3/4	5 – 5-1/16 5 – 5-1/8	5 – 6-7/16 5 – 6-9/16	0 1/16
1/8 3/16	4 – 3-1/16 4 – 3-3/16	4 – 4-1/2 4 – 4-9/16	4 – 5-15/16 4 – 6	4 – 7-5/16 4 – 7-7/16	4 – 8-3/4 4 – 8-13/16	4 – 10-3/16 4 – 10-1/4	4 – 11-9/16 4 – 11-11/16	5 – 1 5 – 1-1/16	5 – 2-3/8 5 – 2-1/2	5 – 3-13/16 5 – 3-7/8	5 – 5-3/16 5 – 5-5/16	5 – 6-5/8 5 – 6-3/4	1/8 3/16
1/4 5/16	4 – 3-1/4 4 – 3-3/8	4 – 4-11/16 4 – 4-3/4	4 – 6-1/8 4 – 6-3/8	4 – 7-1/2 4 – 7-5/8	4 – 8-13/16 4 – 9	4 – 10-5/16 4 – 10-7/16	4 – 11-3/4 4 – 11-13/16	5 – 1-3/16 5 – 1-1/4	5 – 2-9/16 5 – 2-11/16	5 – 4 5 – 4-1/16	5 – 5-7/16 5 – 5-1/2	5 – 6-13/16 5 – 6-15/16	1/4 5/16
3/8 7/16	4 – 3-7/16 4 – 3-1/2	4 – 4-7/8 4 – 4-13/16	4 – 6-1/4 4 – 6-3/8	4 – 7-11/16 4 – 7-3/4	4 – 9-1/8 4 – 9-3/16	4 – 10-1/2 4 – 10-5/8	4 – 11-13/16 5 – 0	5 – 1-5/16 5 – 1-7/16	5 – 2-3/4 5 – 2-7/8	5 – 4-3/16 5 – 4-1/4	5 – 5-9/16 5 – 5-11/16	5 – 7 5 – 7-1/16	3/8 7/16
1/2 9/16	4 – 3-5/8 4 – 3-11/16	4 – 5-1/16 4 – 5-1/8	4 – 6-7/16 4 – 6-9/16	4 – 7-7/8 4 – 7-15/16	4 – 9-1/4 4 – 9-3/8	4 – 10-11/16 4 – 10-3/4	5 – 0-1/8 5 – 0-3/16	5 – 1-1/2 5 – 1-5/8	5 – 2-15/16 5 – 3	5 – 4-3/8 5 – 4-7/16	5 – 5-3/4 5 – 5-7/8	5 – 7-3/16 5 – 7-1/4	1/2 9/16
5/8 11/16	4 – 3-13/16 4 – 3-7/8	4 – 5-3/16 4 – 5-5/16	4 – 6-3/8 4 – 6-11/16	4 – 8-1/16 4 – 8-1/8	4 – 9-7/16 4 – 9-9/16	4 – 10-7/8 4 – 10-15/16	5 – 0-1/4 5 – 0-3/8	5 – 1-11/16 5 – 1-13/16	5 – 3-1/8 5 – 3-3/16	5 – 4-1/2 5 – 4-5/8	5 – 5-15/16 5 – 6	5 – 7-3/8 5 – 7-7/16	5/8 11/16
3/4 13/16	4 – 4 4 – 4-1/16	4 – 5-3/8 4 – 5-1/2	4 – 6-13/16 4 – 6-7/8	4 – 8-3/16 4 – 8-3/16	4 – 9-5/8 4 – 9-11/16	4 – 11-1/16 4 – 11-1/8	5 – 0-7/16 5 – 0-9/16	5 – 1-7/8 5 – 1-15/16	5 – 3-5/16 5 – 3-3/8	5 – 4-11/16 5 – 4-13/16	5 – 6-1/8 5 – 6-3/16	5 – 7-1/2 5 – 7-5/16	3/4 13/16
7/8 15/16	4 – 4-1/8 4 – 4-1/4	4 – 5-9/16 4 – 5-5/8	4 – 7 4 – 7-1/16	4 – 8-3/8 4 – 8-1/2	4 – 9-13/16 4 – 9-7/8	4 – 11-1/4 4 – 11-5/16	5 – 0-5/8 5 – 0-3/4	5 – 2-1/16 5 – 2-1/8	5 – 3-7/16 5 – 3-9/16	5 – 4-7/8 5 – 4-15/16	5 – 6-5/16 5 – 6-3/8	5 – 7-11/16 5 – 7-13/16	7/8 15/16

Courtesy of Texas Pipe Bending Co., Inc.

Table 13-14. Cutback at 90° for ID of Nozzle to OD of Header Standard Weight Pipe

Header ↓ / Nozzle →	3/4	1	1-1/2	2	2-1/2	3	3-1/2	4	5	6	8	10	12	14	16	18	20	22	24	30
3/4	0	1/2	7/8	1-1/8	1-3/8	1-11/16	1-15/16	2-3/16	2-3/4	3-5/16	4-5/16	5-3/8	6-3/8	7	8	9	10	11	12	15
1		0	13/16	1-1/16	1-5/16	1-11/16	1-15/16	2-3/16	2-3/4	3-1/4	4-1/4	5-3/8	6-3/8	7	8	9	10	11	12	15
1-1/2			0	7/8	1-3/16	1-9/16	1-13/16	2-1/8	2-11/16	3-3/16	4-1/4	5-5/16	6-5/16	6-15/16	7-15/16	8-15/16	9-15/16	11	12	15
2				0	1	1-7/16	1-11/16	2	2-9/16	3-1/8	4-3/16	5-1/4	6-5/16	6-15/16	7-15/16	8-15/16	9-15/16	10-15/16	11-15/16	14-15/16
2-1/2					0	1-1/4	1-9/16	1-7/8	2-1/2	3-1/16	4-1/8	5-1/4	6-1/4	6-7/8	7-7/8	8-15/16	9-15/16	10-15/16	11-15/16	14-15/16
3						0	1-5/16	1-5/8	2-5/16	2-15/16	4-1/16	5-1/8	6-3/16	6-13/16	7-7/8	8-7/8	9-7/8	10-7/8	11-7/8	14-15/16
3-1/2							0	1-3/8	2-1/8	2-13/16	3-15/16	5-1/16	6-1/8	6-3/4	7-13/16	8-13/16	9-13/16	10-7/8	11-7/8	14-7/8
4								0	1-15/16	2-5/8	3-13/16	5	6-1/16	6-11/16	7-3/4	8-3/4	9-13/16	10-13/16	11-13/16	14-7/8
5									0	2-1/8	3-1/2	4-3/4	5-7/8	6-1/2	7-9/16	8-5/8	9-11/16	10-11/16	11-3/4	14-13/16
6										0	3-1/16	4-7/16	5-5/8	6-5/16	7-3/8	8-1/2	9-1/2	10-9/16	11-5/8	14-11/16
8											0	3-9/16	4-15/16	5-3/4	6-15/16	8-1/16	9-3/16	10-1/4	11-5/16	14-7/16
10												0	3-15/16	4-7/8	6-1/4	7-1/2	8-11/16	9-13/16	10-15/16	14-1/8
12													0	3-5/8	5-5/16	6-11/16	8	9-1/2	10-3/8	13-3/4
14														0	4-1/2	6-1/16	7-1/2	8-3/16	10	13-7/16
16															0	4-3/4	6-1/2	7-15/16	9-1/4	12-15/16
18																0	5-1/16	6-13/16	8-5/16	12-1/4
20																	0	5-5/16	7-3/16	11-1/2
22																		0	5-9/16	10-9/16
24																			0	9-1/2
30																				0

CUTBACK

Courtesy of Texas Pipe Bending Co., Inc.

Table 13-15. Cutback at 90° for ID of Nozzle to OD of Header Extra Heavy Pipe

	1/2	3/4	1	1-1/2	2	2-1/2	3	3-1/2	4	5	6	8	10	12	14	16	18	20	24	30
1/2	0	7/16	9/16	15/16	1-1/8	1-7/16	1-3/4	2	2-1/4	2-3/4	3-5/16	4-5/16	5-3/8	6-3/8	7	8	9	10	12	15
3/4		0	9/16	7/8	1-1/8	1-3/8	1-11/16	1-15/16	2-3/16	2-3/4	3-5/16	4-5/16	5-3/8	6-3/8	7	8	9	10	12	15
1			0	13/16	1-1/16	1-3/8	1-11/16	1-15/16	2-3/16	2-3/4	3-1/4	4-5/16	5-3/8	6-3/8	7	8	9	10	12	15
1-1/2				0	15/16	1-1/4	1-9/16	1-7/8	2-1/8	2-11/16	3-1/4	4-1/4	5-5/16	6-5/16	6-15/16	7-15/16	8-15/16	10	12	15
2					0	1-1/16	1-7/16	1-3/4	2	2-5/8	3-3/16	4-3/16	5-3/16	6-5/16	6-15/16	7-15/16	8-15/16	9-15/16	11-15/16	14-15/16
2-1/2						0	1-5/16	1-5/8	1-15/16	2-1/2	3-1/8	4-1/8	5-1/4	6-1/4	6-7/8	7-15/16	8-15/16	9-15/16	11-15/16	14-15/16
3							0	1-3/8	1-11/16	2-3/8	3	4-1/16	5-3/16	6-3/16	6-7/8	7-7/8	8-7/8	9-7/8	11-15/16	14-15/16
3-1/2								0	1-1/2	2-3/16	2-7/8	3-15/16	5-1/8	6-1/8	6-13/16	7-13/16	8-13/16	9-7/8	11-7/8	14-7/8
4									0	2	2-11/16	3-7/8	5	6-1/16	6-3/4	7-3/4	8-13/16	9-13/16	11-7/8	14-7/8
5										0	2-1/4	3-9/16	4-13/16	5-7/8	6-9/16	7-5/8	8-11/16	9-11/16	11-3/4	14-13/16
6											0	3-3/16	4-9/16	5-11/16	6-3/8	7-7/16	8-1/2	9-9/16	11-5/8	14-3/4
8												0	3-13/16	5-1/8	5-7/8	7-1/16	8-1/8	9-1/4	11-3/8	14-1/2
10													0	4-1/8	5	6-5/16	7-9/16	8-3/4	10-15/16	14-3/16
12														0	3-13/16	5-7/16	6-13/16	8-1/16	10-7/16	13-13/16
14															0	4-11/16	6-1/4	7-5/8	10-1/16	13-1/2
16																0	5	6-5/8	9-3/8	13
18																	0	5-1/4	8-1/2	12-3/8
20																		0	7-5/16	11-5/8
24																			0	9-5/8

CUTBACK

Courtesy of Texas Pipe Bending Co., Inc.

Table 13-16. ID of Nozzle to OD of LR Ell on Centerline Standard Weight Pipe

HEADER ELL

1/2	3/4	1	1-1/2	2	2-1/2	3	3-1/2	4	5	6	8	10	12	14	16	18	20	24	NOZZLE
13/16	5/16	5/16	5/16	7/16	1/2	1/2	5/8	11/16	13/16	15/16	1-5/16	1-9/16	1-15/16	2-13/16	3-3/16	3-9/16	3-7/8	4-5/8	1/2
	1/2	1/2	7/16	9/16	5/8	5/8	11/16	13/16	15/16	1-1/16	1-7/16	1-11/16	2	2-15/16	3-5/16	3-11/16	4	4-11/16	3/4
		9/16	11/16	3/4	13/16	13/16	7/8	15/16	1-1/16	1-3/16	1-9/16	1-13/16	2-3/16	3-1/8	3-7/16	3-13/16	4-3/16	4-7/8	1
			1-5/16	1-1/4	1-5/16	1-3/16	1-1/4	1-3/8	1-7/16	1-9/16	1-7/8	2-1/8	2-1/2	3-7/16	3-13/16	4-1/8	4-1/2	5-3/16	1-1/2
				1-7/8	1-3/4	1-9/16	1-5/8	1-11/16	1-3/4	1-7/8	2-3/16	2-7/16	2-3/4	3-3/4	4-1/16	4-7/16	4-3/4	5-7/16	2
					2-5/16	2	2	2-1/16	2-1/16	2-1/8	2-7/16	2-11/16	3-1/16	4	4-5/16	4-11/16	5	5-3/4	2-1/2
						2-7/8	2-11/16	2-5/8	2-9/16	2-5/8	2-7/8	3-1/16	3-7/16	4-3/8	4-11/16	5-1/16	5-3/8	6-1/16	3
							3-1/2	3-1/4	3-1/16	3-1/16	3-1/4	3-7/16	3-3/4	4-3/4	5-1/16	5-3/8	5-11/16	6-3/8	3-1/2
								4	3-9/16	3-7/16	3-5/8	3-3/4	4-1/16	5	5-3/8	5-11/16	6	6-11/16	4
									5-3/16	4-5/8	4-9/16	4-5/8	4-13/16	5-3/4	6-1/16	6-3/8	6-11/16	7-5/16	5
										6-3/8	5-11/16	5-1/2	5-11/16	6-5/8	6-7/8	7-1/8	7-7/16	8-1/16	6
											8-13/16	7-5/8	7-1/2	8-3/8	8-1/2	8-11/16	8-15/16	9-7/16	8
												11-1/8	9-15/16	10-5/8	10-1/2	10-1/2	10-5/8	11-1/16	10
													13-3/4	13-9/16	12-7/8	12-5/8	12-9/16	12-3/4	12
														16-7/16	14-3/4	14-1/8	13-15/16	13-15/16	14
															19-1/8	17-1/8	16-7/16	16	16
																21-13/16	19-5/8	18-5/16	18
																	2 – 0-9/16	21-1/16	20
																		2 – 6	24

FORMULA:

$$Z = R + \frac{\text{O.D. Of Ell}}{2}$$

$$S = 1/2 \text{ I.D. Nozzle}$$

$$Z^2 - (R + S)^2 = Y^2$$

$$R - Y = X$$

Courtesy of Texas Pipe Bending Co., Inc.

1. Define pipe "fabrication." ____

____

2. Normally, pipe_______ inch and larger is shop fabricated.

3. Define AWS. ____

4. Shop spools are dimensioned to the closest_____ inch.

5. A pipe bends radius should not be less than_______ pipe diameters.

6. Miter welds are specified in place of_______ for low pressure services and for use in_________.

7. For minimum pressure drop, a________ weld miter is used.

8. Branches and nozzles are specified as OD to ID and ID to OD types. Prepare a cross-sectional view explaining these two types.

9. What is a "cutback?" ____

____

10. Define how a shop spool differs from a contractor's spool. ______________________

____

CHAPTER 14

Machine Drawings

This chapter discusses the common terms, symbols, and conventions used in connection with machine drawings. A detailed machine drawing is described, and engine drawings, boiler drawings, and patternmaker's prints are briefly discussed.

After studying this chapter, you should be able to define the common terms and recognize the common symbols and conventions used with machine drawings. You should also be able to describe the patternmaker's print and compare it to the machinist's print.

Common Terms and Symbols

In learning to read machine drawings you must first become familiar with the common terms, symbols, and conventions as discussed in the following paragraphs.

Tolerances—Because engineers realize that absolute accuracy is impossible, they figure how much variation is permissible. This leeway is known as tolerance, and is stated on the drawing as ± (plus or minus) a certain amount, either by fraction or decimal.

Limits are the maximum and minimum values prescribed for a specific dimension, whereas tolerance represents the total amount by which a specific dimension may vary.

Tolerances may be shown on drawings by several different methods. The unilateral method (Figure 14-1A) is used when variation from the design size is permissible in one direction only. In the bilateral method (Figure 14-1B), the dimension figure indicates the plus or minus variation that is acceptable. In the limit dimensioning method (Figure 14-1C), the maximum and minimum measurements are both stated.

The surfaces being toleranced have geometrical characteristics, such as roundness or perpendicularity to another surface. Typical geometrical characteristic symbols are shown in Figure 14-2. A datum is a surface or line or point from which a geometric position is to be determined or from

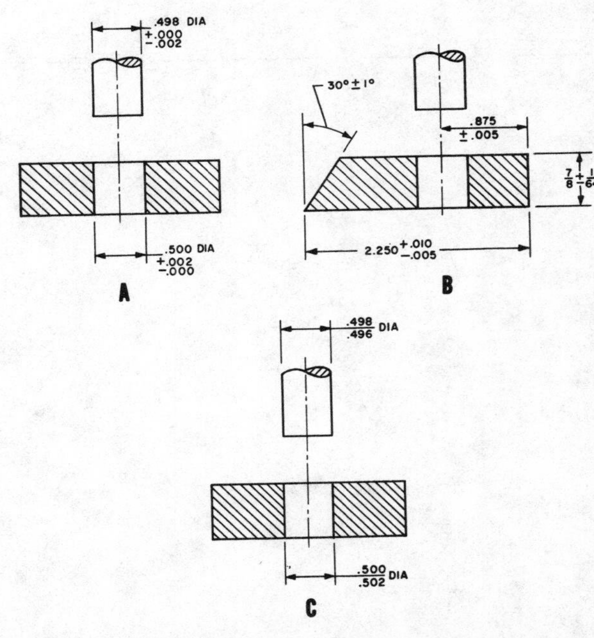

Figure 14-1. Methods of indicating tolerance.

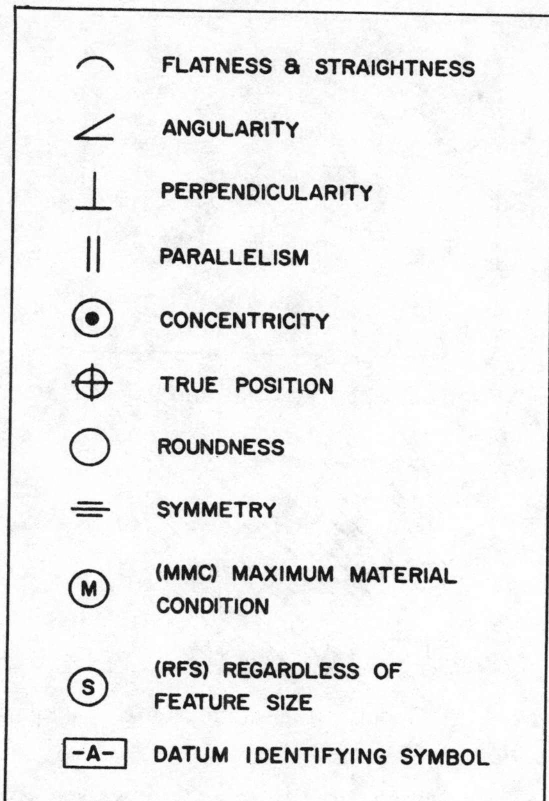

Figure 14-2. Geometric characteristic symbols.

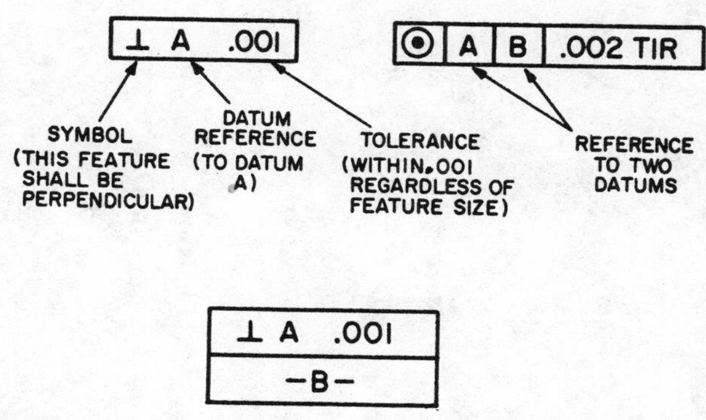

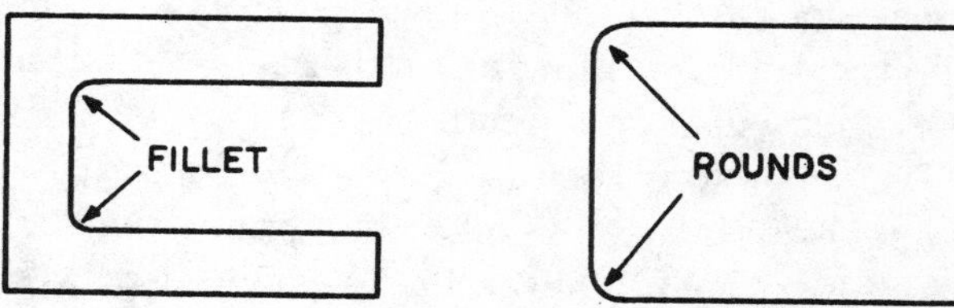

Figure 14-3. Feature control symbol incorporating datum reference.

Figure 14-4. Fillets and rounds.

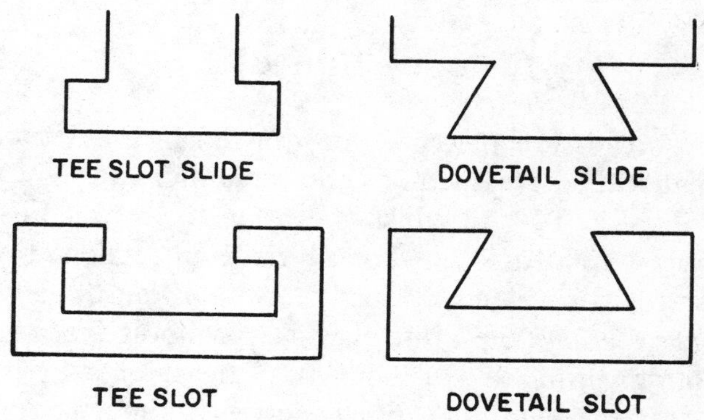

Figure 14-5. Slots and slides.

which a distance is to be measured. Any letter of the alphabet except I, O, and Q may be used as a datum identifying symbol. A feature control symbol is made of geometric symbols and tolerances. As shown in Figure 14-3, the feature control system may also include datum references.

Fillets and rounds—Fillets are concave metal corner (inside) surfaces. In a casting, a fillet normally increases the strength of a metal corner because a rounded corner cools more evenly than a sharp corner, thereby reducing the possibility of a break. Rounds or radii are edges or outside corners that have been rounded to prevent chipping, and to avoid sharp cutting edges. Fillets and rounds are illustrated in Figure 14-4.

Slots and slides—Slots and slides are used for the mating of two specially shaped pieces of material to securely hold them together, yet allow them to move, or slide. Illustrated in Figure 14-5 are two types, the tee slot, and the dovetail slot. A tee slot arrangement is used on a milling machine table. The dovetail slot is used on the cross slide assembly of a metal lathe.

Casting—A casting is an object made by pouring molten metal into a mold (normally of sand) of the desired shape, and allowing it to cool.

Forging—Forging is a process of shaping metal while it is hot or pliable by a hammering or forging process either manually (blacksmith) or by machine.

Key—A key is a small wedge or rectangular piece of metal inserted in a slot or groove between a shaft and a hub to prevent slippage (Figure 14-6).

Keyseat—A keyseat is a slot or groove into which the key fits (Figure 14-7A).

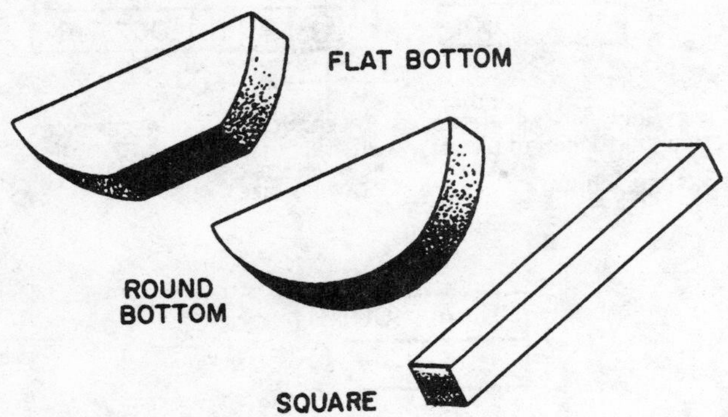

Figure 14-6. Three types of keys.

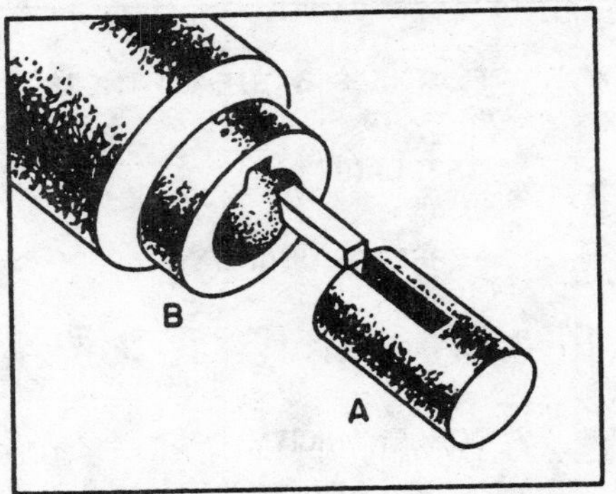

Figure 14-7. A keyseat and keyway.

Figure 14-8. Simplified all-purpose material symbol.

Keyway—A keyway is a slot or groove within a cylindrical tube or pipe into which a key fitted into a key seat will slide (Figure 14-7B).

Temper—To harden steel by heating and sudden cooling by immersion in oil, water, or other coolant.

Common metal conventions—Previous to MIL-STD-100, various symbols were used on machine drawings to indicate types of metals. The simplified all-purpose symbol shown in Figure 14-8 is now used to indicate materials of all types.

Screw Threads

Threads are presented on drawings by different methods. Several methods of presentation are shown in Figures 14-9 through 14-12. On the left in Figure 14-13 you see a thread profile in section. On the right is a common method of showing threads. To save time the draftsman uses symbols that are not drawn to scale. The length of the threaded part is dimensioned, but other necessary information appears in the Note, which in this case is ¼-20 UNC-2.

The first number of the note, ¼, indicates the nominal size which is the outside diameter. The number after the first dash, 20, shows that there are 20 threads per inch. The letters UNC indicate the thread series, Unified National Coarse. The last number, 2, indicates the class of thread and tolerance, commonly called the "fit." If it is a left-hand thread, a dash and the letters LH will follow the class of thread. Threads without the LH are right-hand threads.

Specifications necessary for the manufacture of screws include thread diameter, number of threads per inch, thread series, and class of thread. According to thread pitch, the two most widely used screw-thread series are the Unified or National Form Threads, National Coarse, or NC, and National Fne, or NF threads. The NF threads have more threads per inch of screw length than the NC.

Classes of thread are distinguished from each other by the amount of tolerance and/or allowance specified. Classes of thread was formerly known as "class of fit," a term which will probably remain in use for many years. The new term, "class of thread," was established by the National Bureau of Standards in the *Screw-Thread Standards for Federal Services*, Handbook H-28.

The terminology used in referring to screw threads as illustrated in Figure 14-14. Definitions are as follows:

Helix—The curve formed on any cylinder by a straight line in a plane that is wrapped around the cylinder with a forward progression.

External thread—An external thread is a thread on the outside of a member. Example, a thread of a bolt.

Internal thread—A thead on the inside of a member. Example, the thread inside a nut.

Major diameter—The largest diameter of an internal or external thread.

(text continued on page 253)

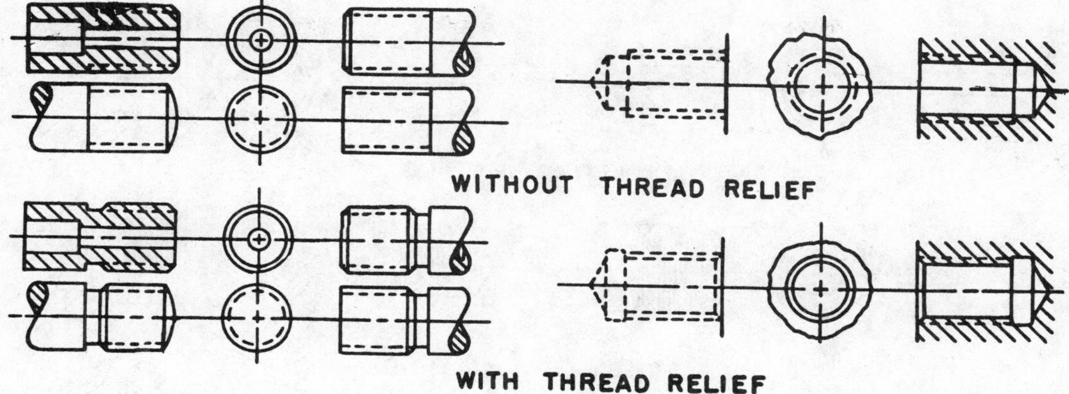

Figure 14-9. Simplified method of thread representation.

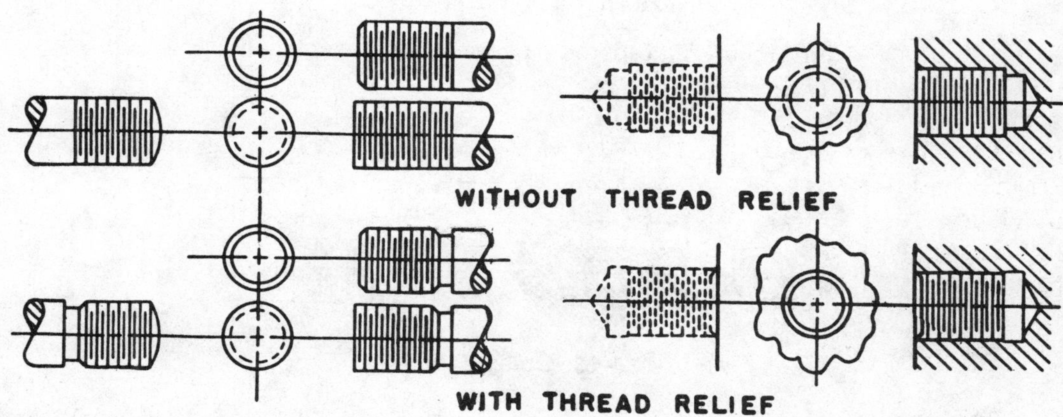

Figure 14-10. Schematic method of thread representation.

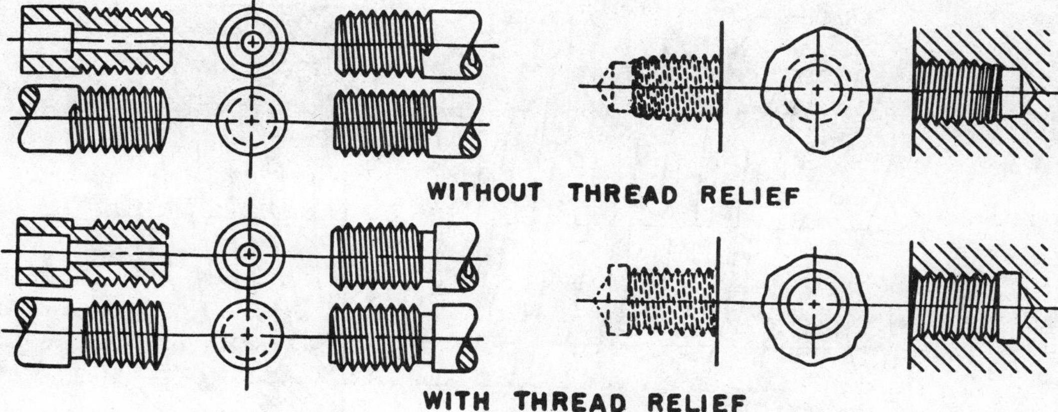

Figure 14-11. Detailed method of thread representation.

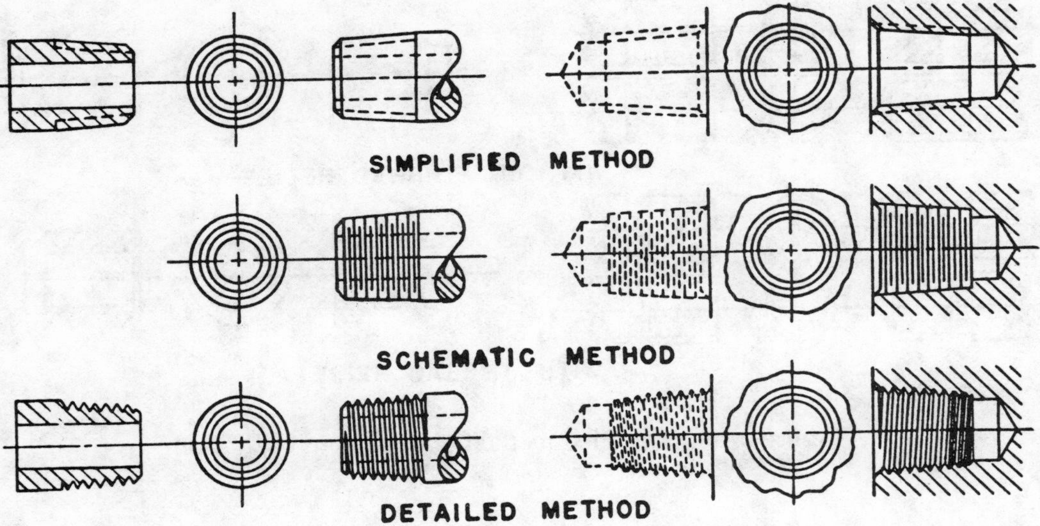

Figure 14-12. Tapered pipe thread representation.

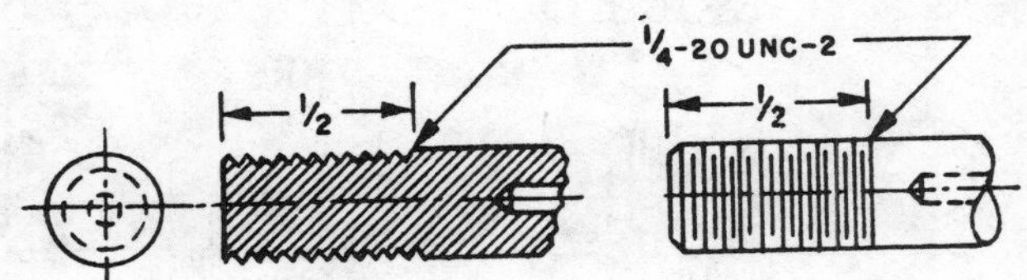

Figure 14-13. Outside threads.

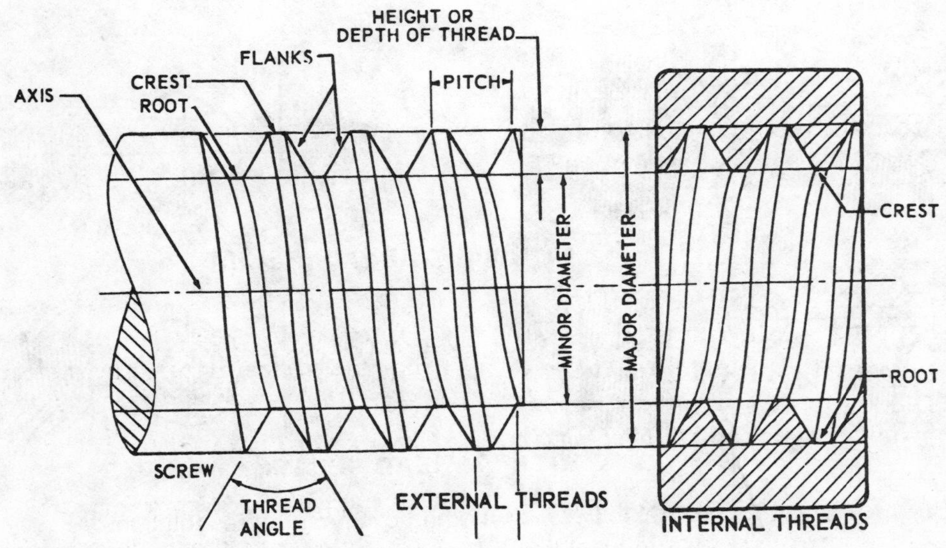

Figure 14-14. Screw-thread terminology.

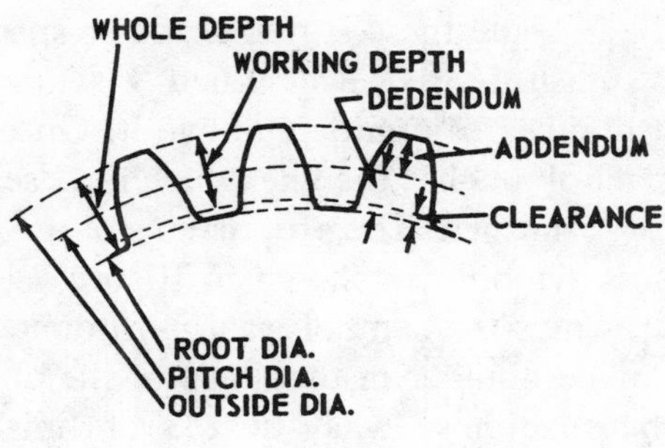

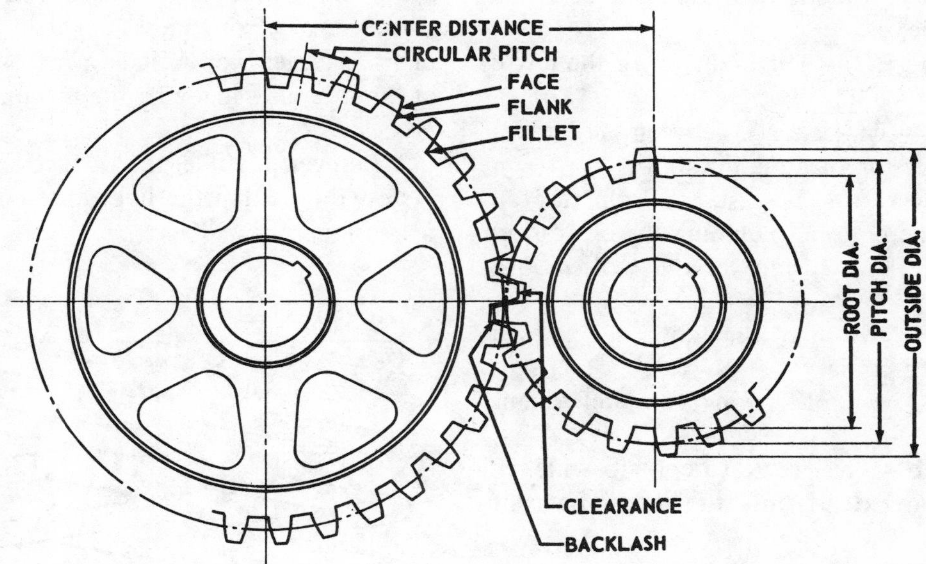

Figure 14-15. Gear nomenclature.

(text continued from page 250)

Axis—The center line running lengthwise through a screw.

Crest—The surface of the thread corresponding to the major diameter of an external thread and the minor diameter of an internal thread.

Root—The surface of the thread corresponding to the minor diameter of an external thread and the major diameter of an internal thread.

Depth—The distance from the root of a thread to the crest, measured perpendicularly to the axis.

Pitch—The distance from a point on a screw thread to a corresponding point on the next thread, measured parallel to the axis.

Lead—The distance a screw thread advances in one turn, measured parallel to the axis. On a single-thread screw the lead and the pitch are identical; on a double-thread screw the lead is twice the pitch; on a triple-thread screw the lead is three times the pitch.

Gears

When gears are shown on machine drawings, the usual practice is to show only enough gear teeth to indicate necessary dimensions. Gear nomenclature is illustrated in Figure 14-15. The terms are described below.

Pitch diameter (PD)—Diameter of the pitch circle (or line) which equals the number of teeth divided by the diametral pitch.

Number of teeth (N)—The diametral pitch multiplied by the diameter of the pitch circle (DP × PD).

Diametral Pitch (DP)—The number of teeth to each item of the pitch diameter or the number of teeth divided by the pitch diameter. Diametral pitch is usually referred to simply as pitch.

Addendum circle (AC)—The circle over the tops of the teeth.

Outside diameter (OD)—Diameter of the addendum circle.

Circular pitch (CP)—Length of the arc of the pitch circle between the centers or corresponding points of adjacent teeth.

Addendum (A)—The height of the tooth above the pitch circle or the radial distance between the pitch circle and the top of the tooth.

Dedendum (D)—The length of the portion of the tooth from the pitch circle to the base of the tooth.

Chordal pitch—The distance from center to center of teeth measured along a straight line or chord of the pitch circle.

Root diameter (RD)—The diameter of the circle at the root of the teeth.

Clearance (C)—The distance between the bottom of a tooth and the top of a mating tooth.

Whole depth (WD)—The distance from the top of the tooth to the bottom, including the clearance.

Face—The working surface of the tooth above the pitch line.

Thickness—The width of the tooth, taken as a chord of the pitch circle.

Pitch circle—The circle having the pitch diameter.

Working depth—The greatest depth to which a tooth of one gear extends into the tooth space of another gear.

Rack teeth—A rack may be compared to a spur gear that has been straightened out. The linear pitch of the rack teeth must equal the circular pitch of the mating gear.

Helical Springs

There are three classifications of helical springs, compression, extension, and torsion. The true method of presentation, is seldom used. Springs are usually shown with straight lines; Figure 14-16 shows several methods of spring representation. Springs are sometimes shown as single line drawings as depicted in Figure 14-17.

Finish Marks

The military standards for finish marks are set forth in ANSIB 46.1-1962. Many metal surfaces must be finished with machine tools for various reasons. The acceptable roughness of a surface depends upon the use to which the part will be put. Sometimes only certain surfaces need to be finished. To indicate these surfaces and to specify the degree of finish desired, a modified V symbol is used with a number of numbers above it. On a drawing, this symbol touches the line which represents the surface to be finished. The proportions of the surface roughness symbol are shown in Figure 14-18. On small drawings the symbol is proportionately smaller.

The number in the angle of the check mark tells the machinist what degree of finish the surface should have. It indicates the root-mean-square value of the surface roughness height in millionths of an inch. In other words, it is a measure of the depth of the scratches made by the machining or abrading process.

Wherever possible, the surface roughness symbol is drawn touching the line representing the surface

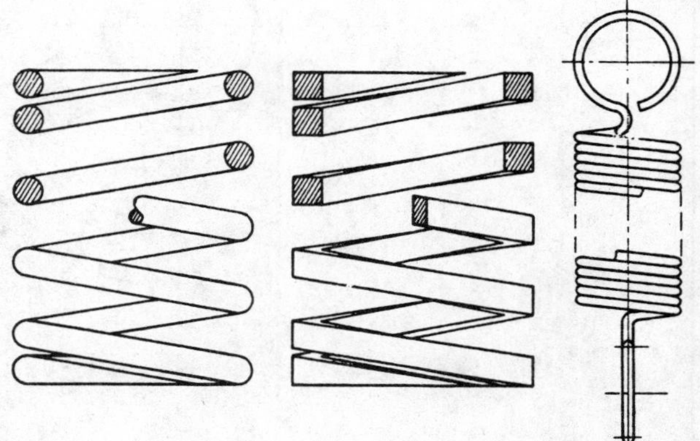

Figure 14-16. Representations of common types of helical springs.

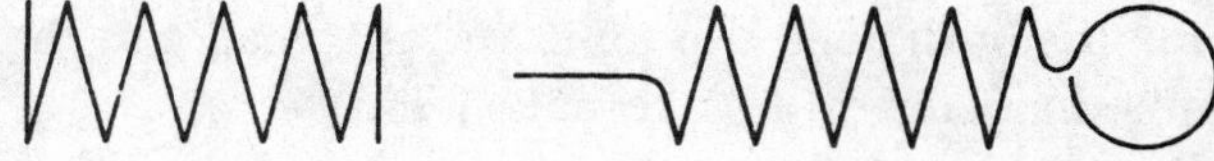

Figure 14-17. Single-line representations of springs.

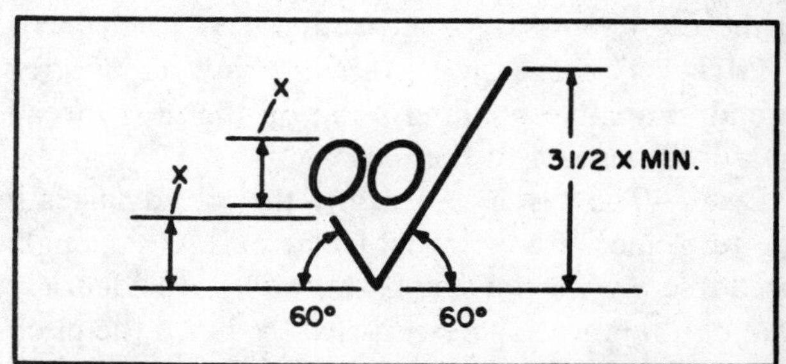

Figure 14-18. Proportions for basic symbol.

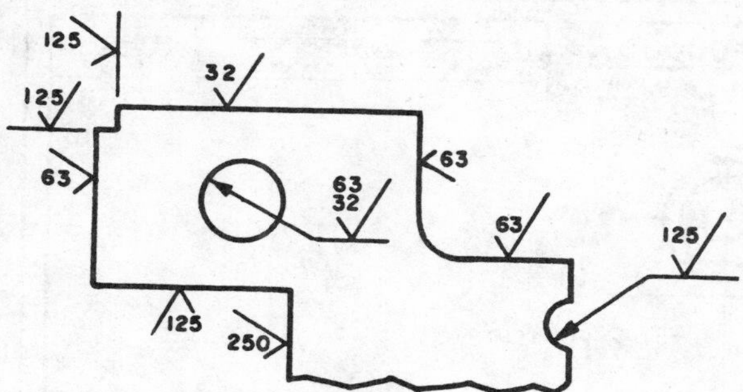

Figure 14-19. Methods of placing surface roughness symbols.

to which it refers. If space is limited, the symbol may be placed on an extension line from that surface or on the tail of a leader with an arrow touching that surface as shown in Figure 14-19.

When a part is to be finished to the same roughness all over, a note on the drawing will include the direction "finish all over," along with the finish mark and the proper number. For example, "finish all over^{32}." When a part is to be finished all over but a few surfaces vary in roughness, the surface roughness symbol and number or numbers are applied to the lines representing these surfaces and a note on the drawing will include the surface roughness symbol for the rest of the surfaces. For example, "^{32}all over except as noted" (Figure 14-20).

The following military standards contain most of the information on symbols, conventions, tolerances, and abbreviations used in shop or working drawings:

ANSI 14.5,	Dimensioning and Tolerancing
MIL-STD-9A,	Screw Thread Conventions and
ANSIB 46.1,	Methods of Specifying Surface Texture
MIL-STD-12, C,	Abbreviations for Use on Drawings and in Technical-Type Publications

Detail Drawings

A detail drawing is a print illustrating a single component or part, showing a complete and exact description of its shape, dimensions, and how it is made (construction). A complete detail drawing will show in a direct and simple manner the shape, exact size, type of material, type of finish for each part, tolerance, necessary shop operations, number of parts required, etc. A detail drawing is not to be confused with a detail view. A detail view shows a part of a drawing in the same plane and in the same arrangement, but in greater detail and to a larger scale than in the principal view.

Illustrated in Figure 14-21 is a relatively simple detail drawing of a clevis. Study this figure closely and apply the principles of reading two-view orthographic drawings. Figure 14-22 is an isometric (pictorial) view of the clevis shown in the detail drawing (Figure 14-21).

The dimensions on the detail drawing in Figure 14-21 are conventional, except the four toleranced dimensions given. On the top view, to the right of the part, is a hole requiring a diameter of .3125 + .0005, but no — (minus). This means that the diameter of the hole can be no less than .3125, but can be as large as .3130. On the bottom view, on the left, there is a diameter given as .665, $\mp$ .001.

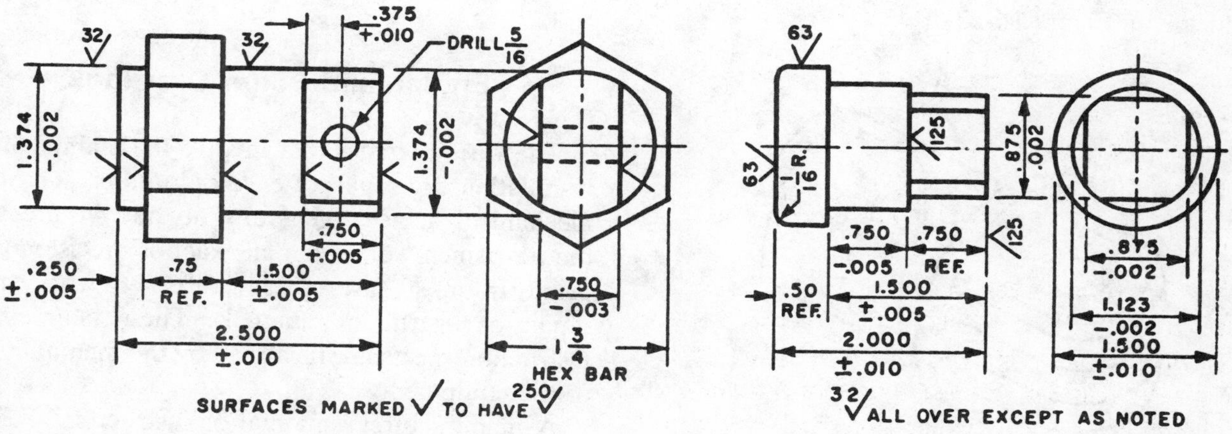

Figure 14-20. Typical examples of symbol use.

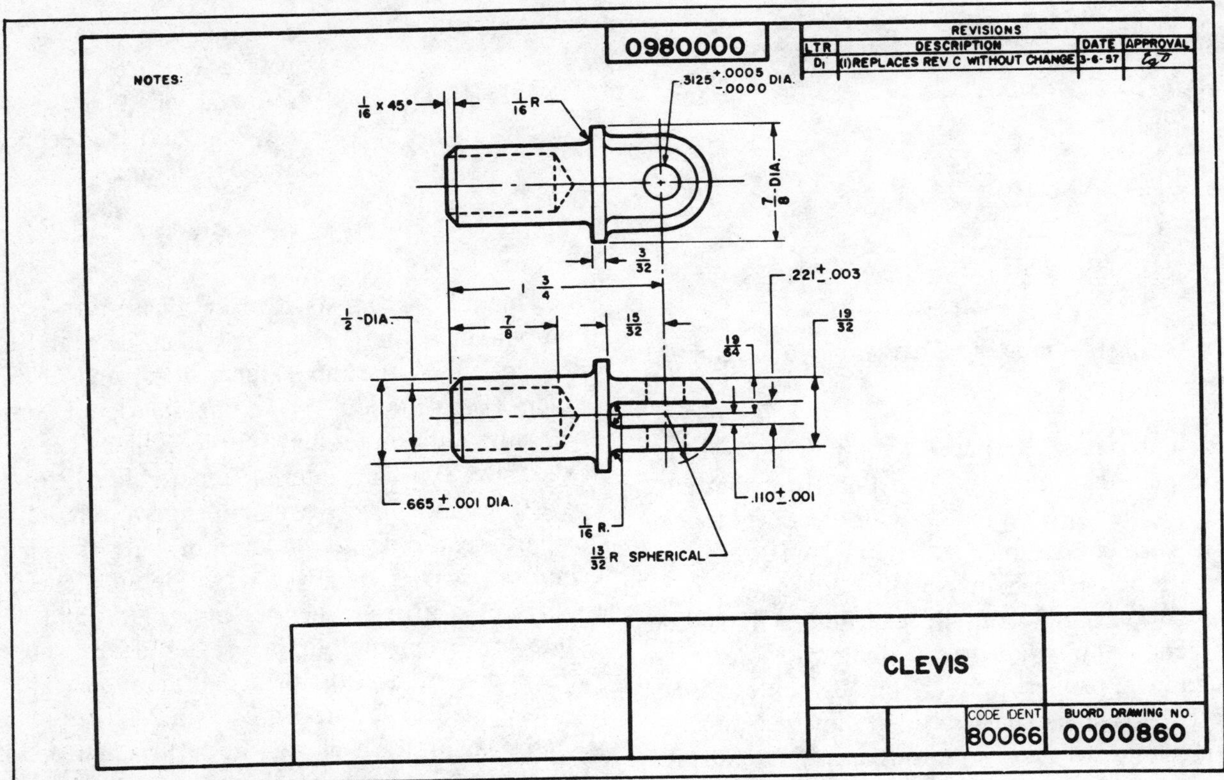

Figure 14-21. Detailed drawing of a clevis.

This means that this diameter can be a minimum of .664, and a maximum of .666. The other two toleranced dimension given are at the left of the bottom view.

Another detail drawing is shown in Figure 14-23. You may think the drawing is complicated, but in reality it is not. It does, however, have more symbols and abbreviations with which you were previously unfamiliar. Figure 14-24 is an isometric (pictorial) drawing of the base pivot shown orthographically in Figure 14-23.

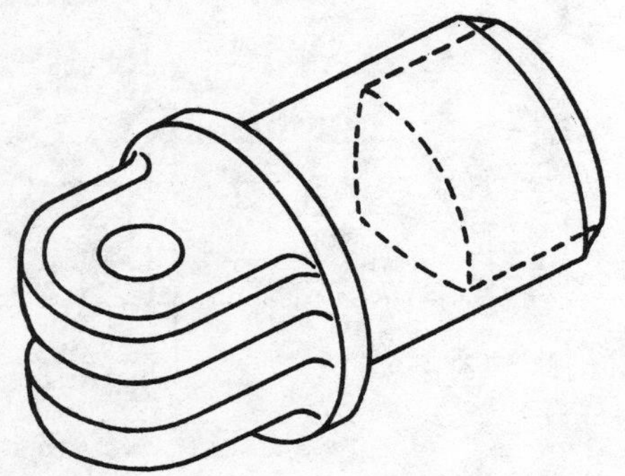

Figure 14-22. Isometric drawing of a clevis.

Section drawings, or views are often necessary in machine drawings due to the complicated parts or components; it is almost impossible to attempt to read the multiple hidden lines necessary to show the object in a regular orthographic print. For this reason in machine drawings one or more views are drawn to show the interior of the object by cutting away a portion of the part, as is shown in the upper portion of the view on the left of Figure 14-23. Section views were covered in an earlier chapter.

Engine and Boiler Drawings

The various trades that involve the maintenance, installation and repair of engines, motors, and other mechanical devices, generally do not require that the tradesmen work from an exact or precision type print. In most cases these craftsmen work from repair or instruction manuals. These manuals or instruction sheets are furnished by the manufacturer of the equipment.

A manufacturer's manual or instruction sheet is supplied to give necessary information on operating

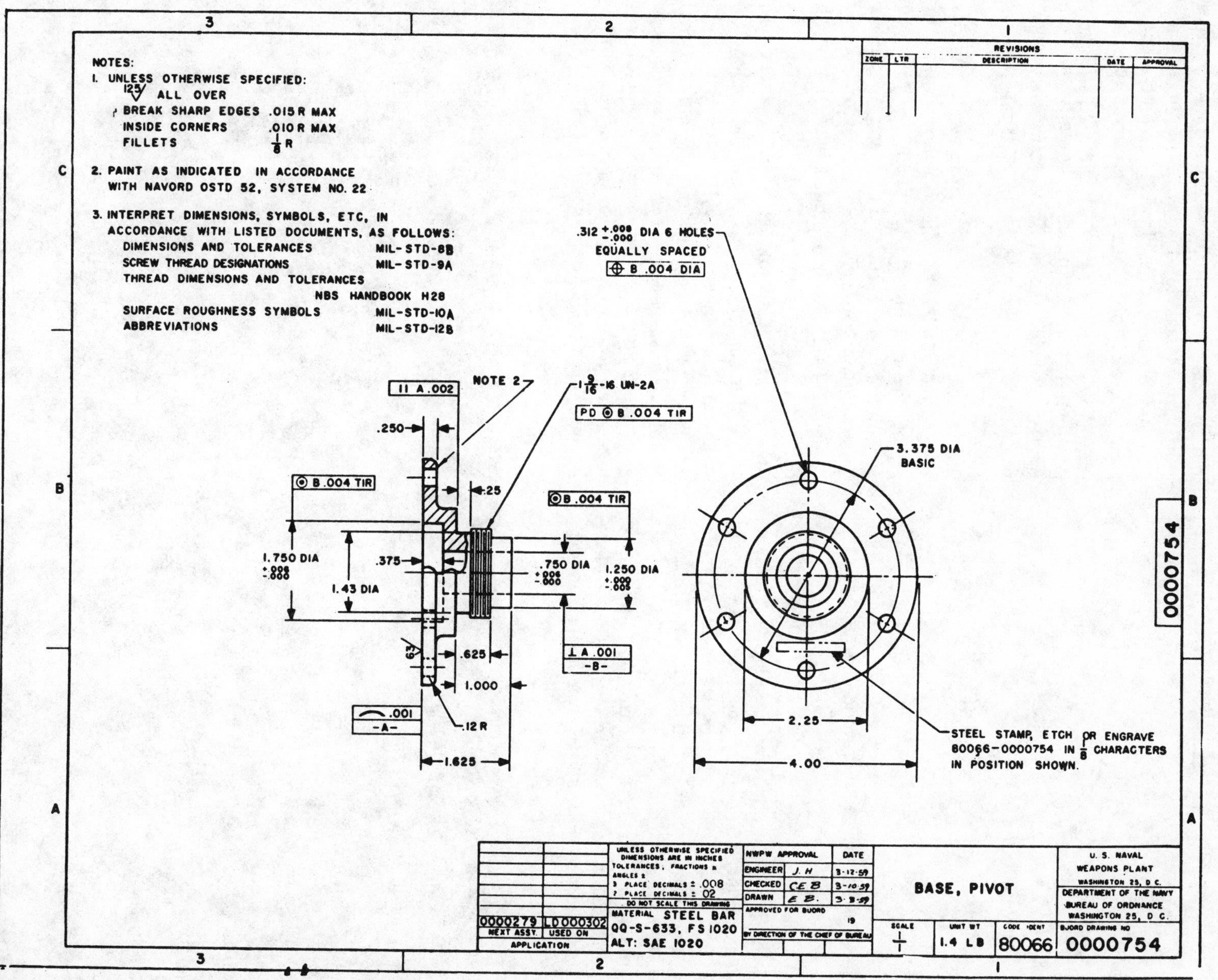

Figure 14-23. Detailed drawing of a base pivot.

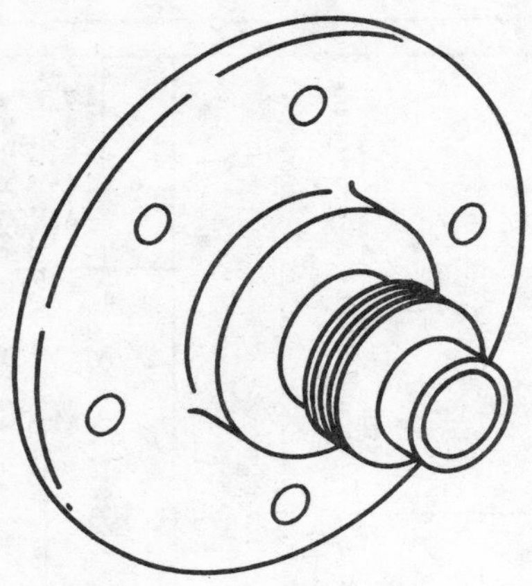

Figure 14-24. Isometric drawing of a base pivot.

principles, and procedures for installation, maintenance, and repair. To simplify this information, photographs, cutaway views, cross sections, exploded views, and other technical illustrations are used extensively.

Figure 14-25 can be read easily and understood as a photographic presentation of an exhaust manifold (attached to the internal combustion engine) and the intake manifold which is being held in a pair of hands, ready to be replaced on the engine.

Figure 14-26 can be classified as both a cutaway view, and a pictorial (drawn) presentation showing the operating principles of the 4-stroke cycle of a gasoline engine. If you carefully study each of the four views, you can easily visualize the operating principle of the 4 strokes—the intake, compression, power, and exhaust.

Figure 14-27 is a cross-section view of a cylinder (part cross-hatched) to show the bore and the limits of the piston stroke. TDC, means top dead center, top of the stroke; BDC, the bottom of the stroke (bottom dead center).

Figure 14-28 is an exploded view showing how the piston, connecting rod, and piston pin and bearings are assembled and attached to the crankpin of the crankshaft. On the right of the exploded view is a pictorial drawing of all the parts assembled. This type of print or illustration simplifies the assembly or disassembly of this particular piece of gear. Note also the cutaway portion of the piston that gives you an insight as to how the piston pin fits into the piston and connecting rod.

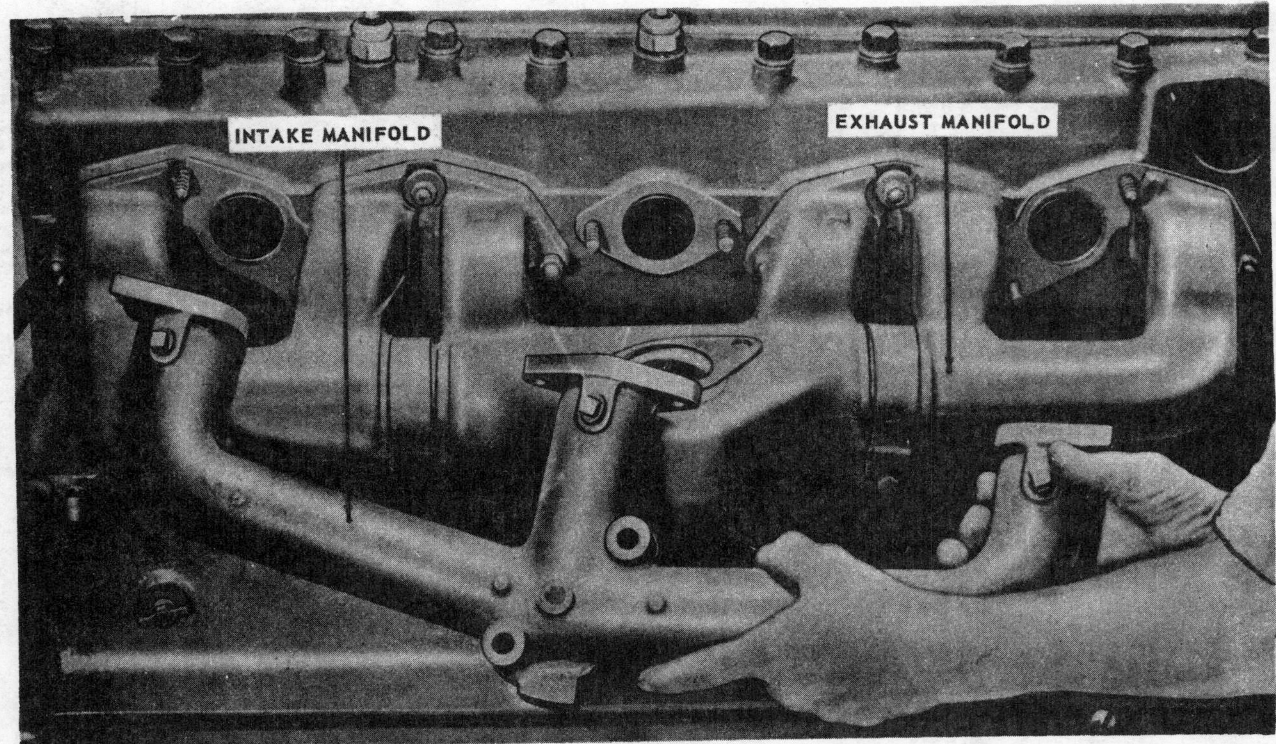

Figure 14-25. Intake and exhaust manifolds.

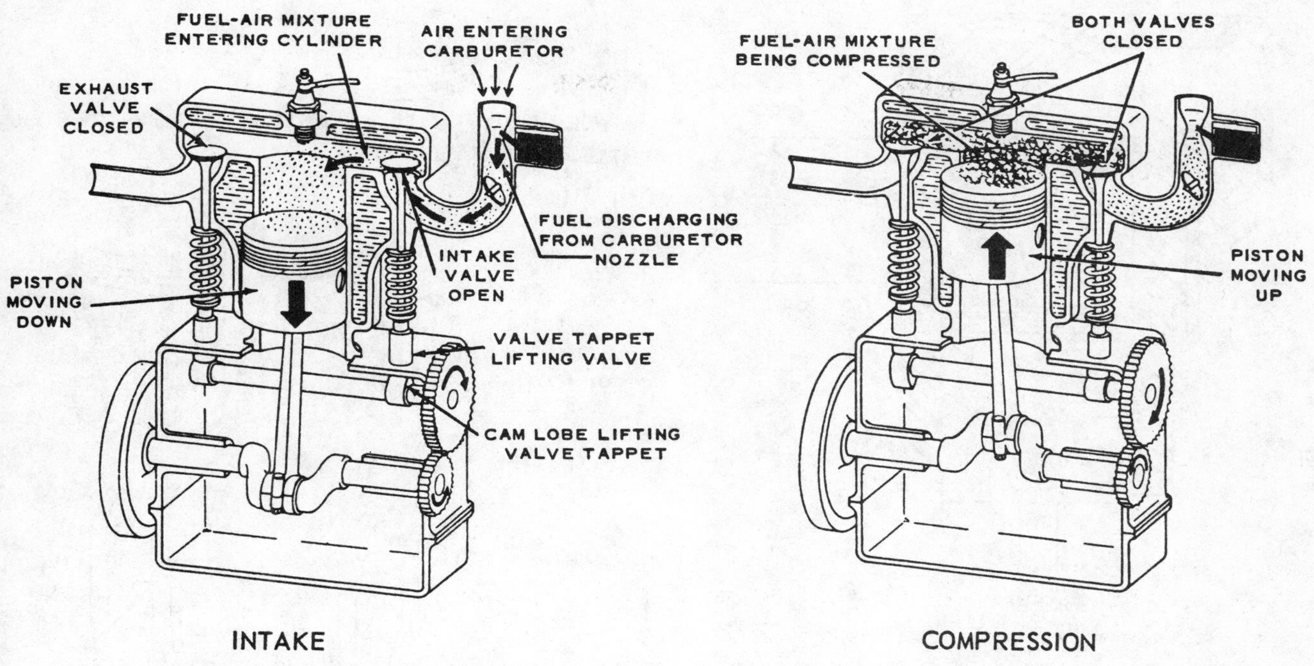

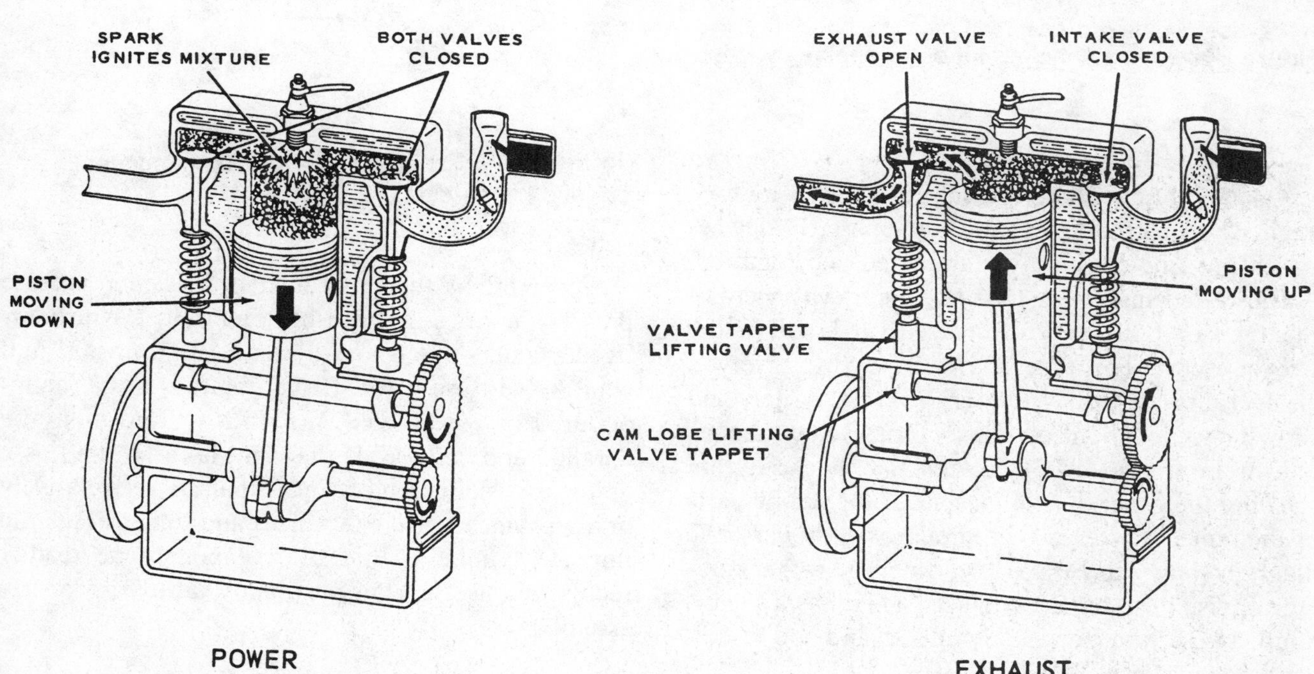

Figure 14-26. The four strokes of a four-cycle gasoline engine.

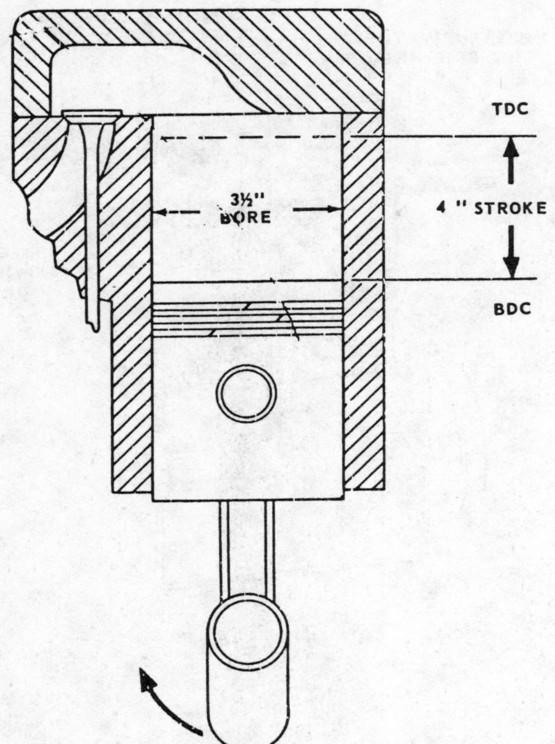

Figure 14-27. Bore and stroke of an engine cylinder.

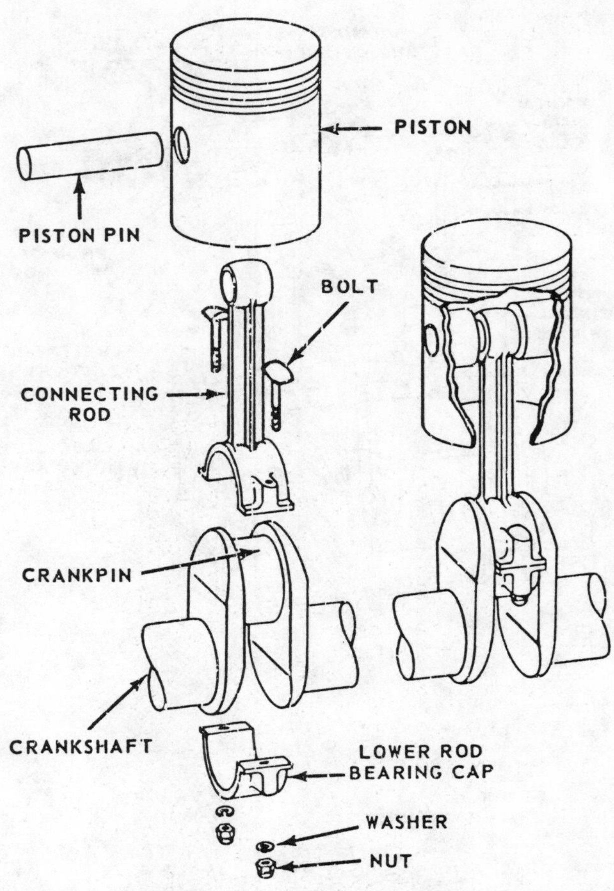

Figure 14-28. Exploded view of piston, connecting rod, and piston pin.

(text continued from page 258)

On board ship there are many types of motors, engines, and other mechanically operated devices which must be constantly maintained and repaired.

Illustrated in Figure 14-29 is a cutaway view of the type of boiler normally used on board a destroyer escort. This type of view is used to give you an overall picture of the interior of the boiler, and shows the location of the various components. Shown in Figure 14-30 is a single line diagram showing the general arrangement of the major parts of the boiler. If you are unfamiliar with boilers, the diagram shown in Figure 14-30 will be rather difficult to visualize, but having seen the cutaway view, figure 14-29, you can readily understand and visualize the diagram in Figure 14-30.

Figure 14-31 is a cutaway of the same engine showing the interior of the combustion section of the axial-flow engine. Figure 14-32 is an isometric diagram showing the left side of an axial-flow engine using a numbering system and key for the identification of the various parts.

The method the draftsman or designer uses to present the drawing or print of an object, is primarily dependent on who will read the drawing, and what he will use it for. If the object is to be manufactured or constructed, the print will normally be detailed and concise. If the print is to be used as a guide for maintenance and repair, it is relatively unimportant to give exact sizes and dimensions; the important point is that the component be readily recognized as the right part, and that it be properly assembled.

Patternmaker's Prints

Patternmakers must not only be able to read blueprints, they must also develop the skill to redraw the object (lay out a pattern) and then make the object, normally using wood (white pine).

text continued on page 263

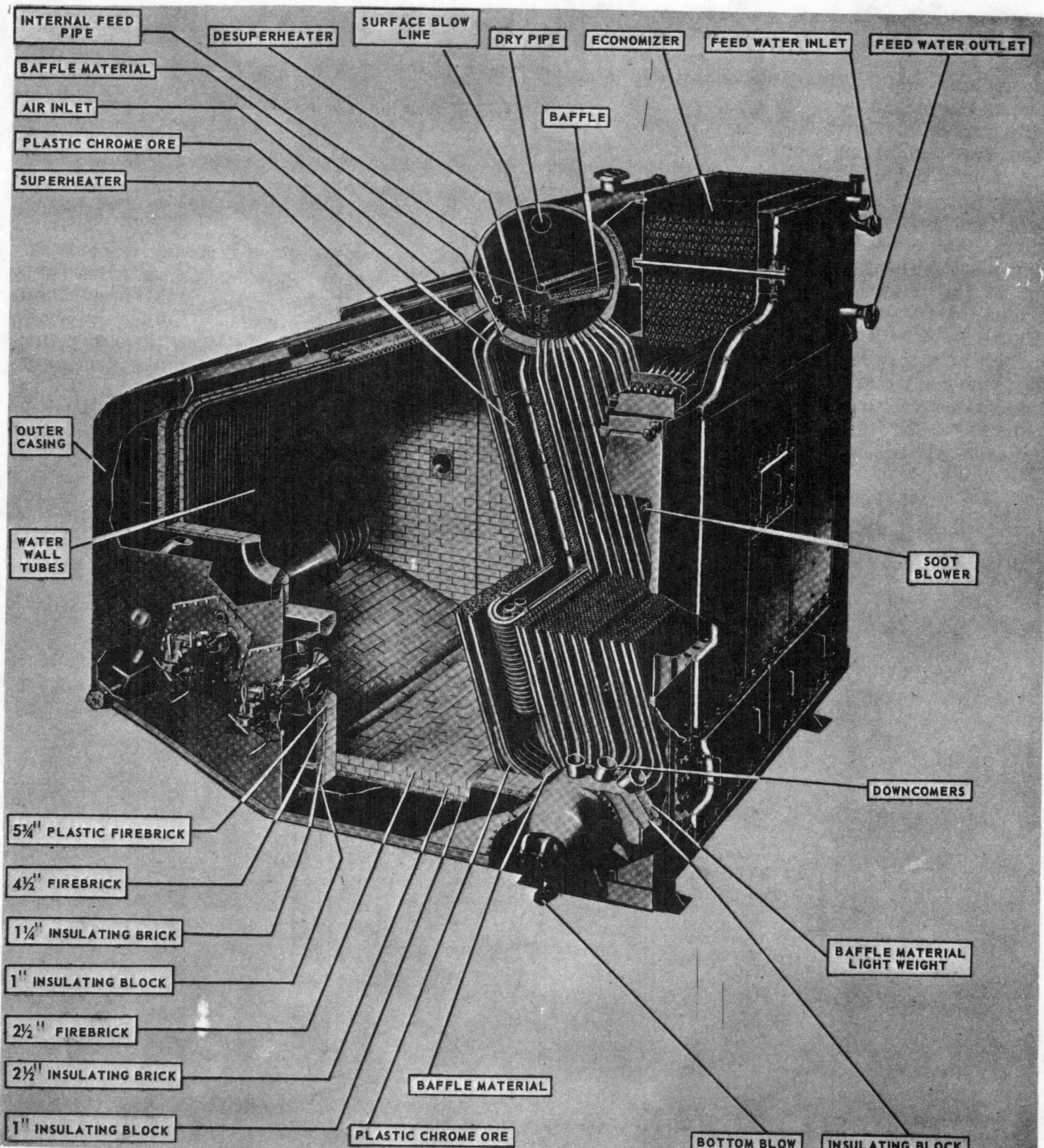

Figure 14-29. Cutaway view of destroyer escort boiler.

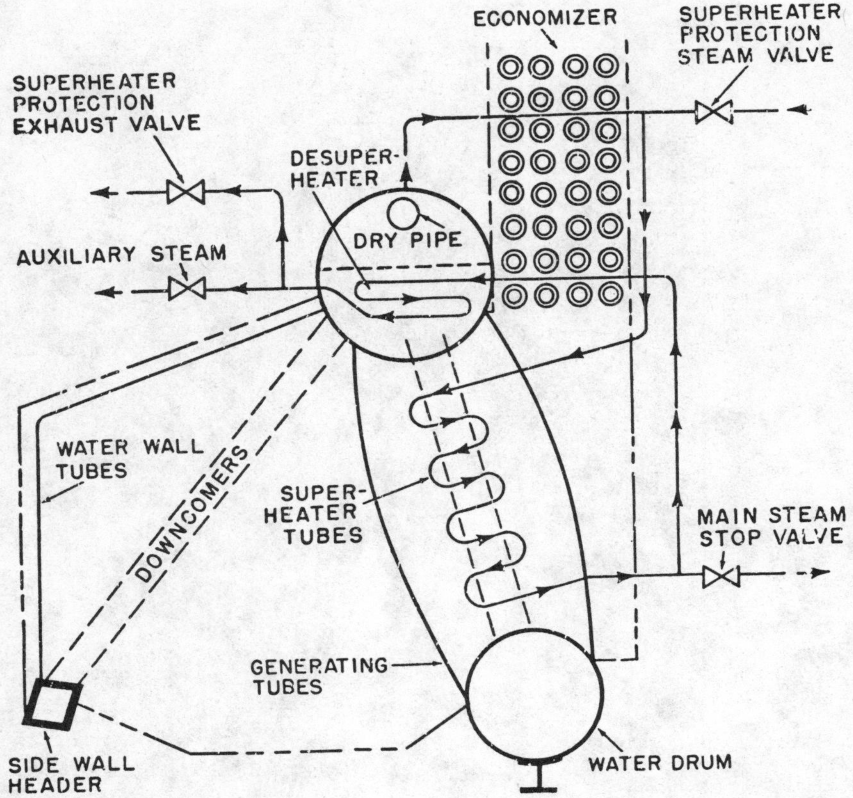

Figure 14-30. Line diagram showing the general arrangement of destroyer escort boiler.

Figure 14-31. A cutaway view of an axial-flow engine combustion section.

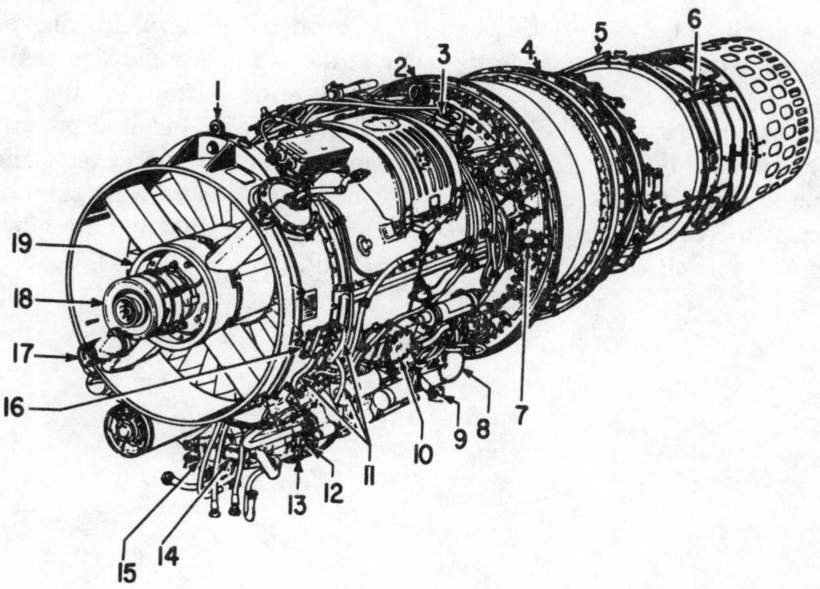

1. Forward engine mount link.
2. Engine fireseal.
3. Oil tank filler.
4. Afterburner sling aft support locating bracket.
5. Afterburner support linkage.
6. Afterburner hydraulic actuator shield.
7. Left main engine mount and mount fitting.
8. Oil vapor overboard duct.
9. Fuel strainer hose.
10. Flowmeter.
11. Engine pre-oiling hose.
12. Combined system hydraulic pump.
13. Engine drains.
14. Hydraulic pump cooling air diverging duct.
15. Flight system hydraulic pump.
16. Engine pre-oiling bracket.
17. Starter line and screen.
18. Starter.
19. Air inlet duct seal.

Figure 14-32. An isometric diagram illustrating the left side of an axial-flow engine.

Normally, the drawing required for a machined casting is dimensioned and noted for both the Patternmaker and the Machinist. An example is the base pivot shown in Figure 14-23. In some cases, however, particularly if the casting is difficult or complicated, two different drawings are made: one for the Machinist, and one for the Patternmaker. Figure 14-33 is a patternmaker or casting drawing of the base pivot shown in Figure 14-23. Make a close study and note the differences between the two drawings.

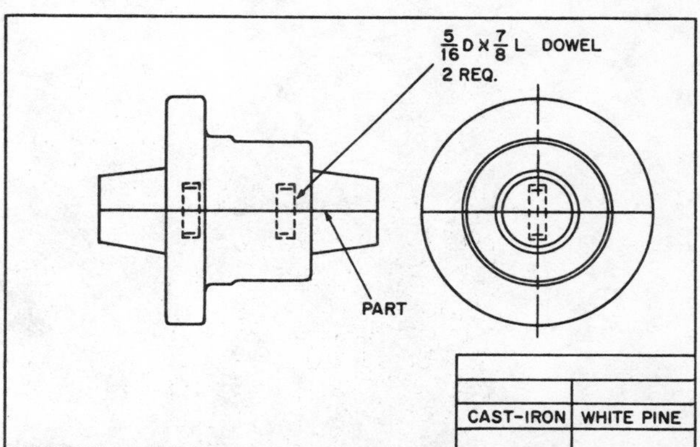

Figure 14-33. Casting drawing of a base pivot.

If the Patternmaker is to lay out a pattern from either a machine drawing or casting drawing, it is of the utmost importance that he use the proper shrink rule. A *shrink rule* is a ruler devised to compensate for the contraction or shrinkage of a casting while its metal is cooling and contracting. It is an expanded rule which has its graduation increased to allow for this shrinkage of the casting. Since all metals do not contract or shrink the same amount, it is necessary to have more than one size of shrink rule. A variety of shrink rule sizes are therefore available. The most common shrink rules are: $\frac{1}{10}$-$\frac{1}{8}$ inch per foot rule for cast iron, $\frac{3}{16}$-inch rule for brass and bronze, and $\frac{3}{32}$-$\frac{3}{16}$-inch rule for aluminum.

In Figure 14-33 note that most of the dimensions are larger than those given in the machine detail drawing shown in Figure 14-23. The reason for this is that the casting is to be machined, therefore extra material is required, or the part would be undersized after machining.

Looking again at the pattern drawing in Figure 14-33 note that each side of the object has a 1¼-inch "extra" round tapered part; this is called a "core print." A core print is used to anchor or support the core within a mold at each end.

Also on the pattern drawing are two notations, one at the top giving the size of two dowels and the other near the bottom on the right "part." These notations indicate that the pattern will be split, or cut in two, at the "part" line, and the two dowels will be placed as shown, extending ½ inch into each portion of the pattern after the pattern has been split.

APPENDIX A

Dimensions and Weights of Steel Pipe

Table A-1. Commercial Wrought Steel Pipe Data.

Note 1: The letters "s", "x", and "xx" in the column of Schedule Numbers indicate Standard, Extra Strong, and Double Extra Strong Pipe, respectively.

Note 2: The values shown in square feet for the Transverse Internal Area also represent the volume in cubic feet per foot of pipe length.

Nominal Pipe Size	Outside Diameter (D)	Schedule No.	Wall Thickness (t)	Inside Diameter (d)	Area of Metal (a)	Transverse Internal Area		Moment of Inertia (I)	Weight of Pipe	Weight of Water	External Surface	Section Modulus
		See Note 1					*See Note 2*					$\left(2\frac{I}{D}\right)$
Inches	Inches		Inches	Inches	Square Inches	Square Inches	Square Feet	Inches to 4th Power	Pounds per foot	Pounds per foot of pipe	Sq. Ft. per foot of pipe	
1/8	0.405	40s	.068	.269	.0720	.0568	.00040	.00106	.244	.025	.106	.00523
		80x	.095	.215	.0925	.0364	.00025	.00122	.314	.016	.106	.00602
1/4	0.540	40s	.088	.364	.1250	.1041	.00072	.00331	.424	.045	.141	.01227
		80x	.119	.302	.1574	.0716	.00050	.00377	.535	.031	.141	.01395
3/8	0.675	40s	.091	.493	.1670	.1910	.00133	.00729	.567	.083	.178	.02160
		80x	.126	.423	.2173	.1405	.00098	.00862	.738	.061	.178	.02554
1/2	0.840	40s	.109	.622	.2503	.3040	.00211	.01709	.850	.132	.220	.04069
		80x	.147	.546	.3200	.2340	.00163	.02008	1.087	.102	.220	.04780
		160	.187	.466	.3836	.1706	.00118	.02212	1.300	.074	.220	.05267
		...xx	.294	.252	.5043	.050	.00035	.02424	1.714	.022	.220	.05772
3/4	1.050	40s	.113	.824	.3326	.5330	.00371	.03704	1.130	.231	.275	.07055
		80x	.154	.742	.4335	.4330	.00300	.04479	1.473	.188	.275	.08531
		160	.218	.614	.5698	.2961	.00206	.05269	1.940	.128	.275	.10036
		...xx	.308	.434	.7180	.148	.00103	.05792	2.440	.064	.275	.11032
1	1.315	40s	.133	1.049	.4939	.8640	.00600	.08734	1.678	.375	.344	.1328
		80x	.179	.957	.6388	.7190	.00499	.1056	2.171	.312	.344	.1606
		160	.250	.815	.8365	.5217	.00362	.1251	2.840	.230	.344	.1903
		...xx	.358	.599	1.0760	.282	.00196	.1405	3.659	.122	.344	.2136
1 1/4	1.660	40s	.140	1.380	.6685	1.495	.01040	.1947	2.272	.649	.435	.2346
		80x	.191	1.278	.8815	1.283	.00891	.2418	2.996	.555	.435	.2913
		160	.250	1.160	1.1070	1.057	.00734	.2839	3.764	.458	.435	.3421
		...xx	.382	.896	1.534	.630	.00438	.3411	5.214	.273	.435	.4110
1 1/2	1.900	40s	.145	1.610	.7995	2.036	.01414	.3099	2.717	.882	.497	.3262
		80x	.200	1.500	1.068	1.767	.01225	.3912	3.631	.765	.497	.4118
		160	.281	1.338	1.429	1.406	.00976	.4824	4.862	.608	.497	.5078
		...xx	.400	1.100	1.885	.950	.00660	.5678	6.408	.42	.497	.5977
2	2.375	40s	.154	2.067	1.075	3.355	.02330	.6657	3.652	1.45	.622	.5606
		80x	.218	1.939	1.477	2.953	.02050	.8679	5.022	1.28	.622	.7309
		160	.343	1.689	2.190	2.241	.01556	1.162	7.440	.97	.622	.979
		...xx	.436	1.503	2.656	1.774	.01232	1.311	9.029	.77	.622	1.104
2 1/2	2.875	40s	.203	2.469	1.704	4.788	.03322	1.530	5.79	2.07	.753	1.064
		80x	.276	2.323	2.254	4.238	.02942	1.924	7.66	1.87	.753	1.339
		160	.375	2.125	2.945	3.546	.02463	2.353	10.01	1.54	.753	1.638
		...xx	.552	1.771	4.028	2.464	.01710	2.871	13.70	1.07	.753	1.997

(table continued on next page)

Table A-1 continued

Note 1: The letters "s", "x", and "xx" in the column of Schedule Numbers indicate Standard, Extra Strong, and Double Extra Strong Pipe, respectively.

Note 2: The values shown in square feet for the Transverse Internal Area also represent the volume in cubic feet per foot of pipe length.

Nominal Pipe Size Inches	Outside Diameter (D) Inches	Schedule No. *See Note 1*	Wall Thickness (t) Inches	Inside Diameter (d) Inches	Area of Metal (a) Square Inches	Transverse Internal Area Square Inches	Transverse Internal Area *See Note 2* Square Feet	Moment of Inertia (I) Inches to 4th Power	Weight of Pipe Pounds per foot	Weight of Water Pounds per foot of pipe	External Surface Sq. Ft. per foot of pipe	Section Modulus $(2\frac{I}{D})$
3	3.500	40s	.216	3.068	2.228	7.393	.05130	3.017	7.58	3.20	.916	1.724
		80x	.300	2.900	3.016	6.605	.04587	3.894	10.25	2.86	.916	2.225
		160	.438	2.624	4.205	5.408	.03755	5.032	14.32	2.35	.916	2.876
		...xx	.600	2.300	5.466	4.155	.02885	5.993	18.58	1.80	.916	3.424
3½	4.000	40s	.226	3.548	2.680	9.886	.06870	4.788	9.11	4.29	1.047	2.394
		80x	.318	3.364	3.678	8.888	.06170	6.280	12.51	3.84	1.047	3.140
4	4.500	40s	.237	4.026	3.174	12.73	.08840	7.233	10.79	5.50	1.178	3.214
		80x	.337	3.826	4.407	11.50	.07986	9.610	14.98	4.98	1.178	4.271
		120	.438	3.624	5.595	10.31	.0716	11.65	19.00	4.47	1.178	5.178
		160	.531	3.438	6.621	9.28	.0645	13.27	22.51	4.02	1.178	5.898
		...xx	.674	3.152	8.101	7.80	.0542	15.28	27.54	3.38	1.178	6.791
5	5.563	40s	.258	5.047	4.300	20.01	.1390	15.16	14.62	8.67	1.456	5.451
		80x	.375	4.813	6.112	18.19	.1263	20.67	20.78	7.88	1.456	7.431
		120	.500	4.563	7.953	16.35	.1136	25.73	27.10	7.09	1.456	9.250
		160	.625	4.313	9.696	14.61	.1015	30.03	32.96	6.33	1.456	10.796
		...xx	.750	4.063	11.340	12.97	.0901	33.63	38.55	5.61	1.456	12.090
6	6.625	40s	.280	6.065	5.581	28.89	.2006	28.14	18.97	12.51	1.734	8.50
		80x	.432	5.761	8.405	26.07	.1810	40.49	28.57	11.29	1.734	12.22
		120	.562	5.501	10.70	23.77	.1650	49.61	36.40	10.30	1.734	14.98
		160	.718	5.189	13.32	21.15	.1469	58.97	45.30	9.16	1.734	17.81
		...xx	.864	4.897	15.64	18.84	.1308	66.33	53.16	8.16	1.734	20.02
8	8.625	20	.250	8.125	6.57	51.85	.3601	57.72	22.36	22.47	2.258	13.39
		30	.277	8.071	7.26	51.16	.3553	63.35	24.70	22.17	2.258	14.69
		40s	.322	7.981	8.40	50.03	.3474	72.49	28.55	21.70	2.258	16.81
		60	.406	7.813	10.48	47.94	.3329	88.73	35.64	20.77	2.258	20.58
		80x	.500	7.625	12.76	45.66	.3171	105.7	43.39	19.78	2.258	24.51
		100	.593	7.439	14.96	43.46	.3018	121.3	50.87	18.83	2.258	28.14
		120	.718	7.189	17.84	40.59	.2819	140.5	60.63	17.59	2.258	32.58
		140	.812	7.001	19.93	38.50	.2673	153.7	67.76	16.68	2.258	35.65
		...xx	.875	6.875	21.30	37.12	.2578	162.0	72.42	16.10	2.258	37.56
		160	.906	6.813	21.97	36.46	.2532	165.9	74.69	15.80	2.258	38.48
10	10.750	20	.250	10.250	8.24	82.52	.5731	113.7	28.04	35.76	2.814	21.15
		30	.307	10.136	10.07	80.69	.5603	137.4	34.24	34.96	2.814	25.57
		40s	.365	10.020	11.90	78.86	.5475	160.7	40.48	34.20	2.814	29.90
		60x	.500	9.750	16.10	74.66	.5185	212.0	54.74	32.35	2.814	39.43
		80	.593	9.564	18.92	71.84	.4989	244.8	64.33	31.13	2.814	45.54
		100	.718	9.314	22.63	68.13	.4732	286.1	76.93	29.53	2.814	53.22
		120	.843	9.064	26.24	64.53	.4481	324.2	89.20	27.96	2.814	60.32
		140	1.000	8.750	30.63	60.13	.4176	367.8	104.13	26.06	2.814	68.43
		160	1.125	8.500	34.02	56.75	.3941	399.3	115.65	24.59	2.814	74.29
12	12.75	20	.250	12.250	9.82	117.86	.8185	191.8	33.38	51.07	3.338	30.2
		30	.330	12.090	12.87	114.80	.7972	248.4	43.77	49.74	3.338	39.0
		...s	.375	12.000	14.58	113.10	.7854	279.3	49.56	49.00	3.338	43.8
		40	.406	11.938	15.77	111.93	.7773	300.3	53.53	48.50	3.338	47.1
		..x	.500	11.750	19.24	108.43	.7528	361.5	65.42	46.92	3.338	56.7
		60	.562	11.626	21.52	106.16	.7372	400.4	73.16	46.00	3.338	62.8
		80	.687	11.376	26.03	101.64	.7058	475.1	88.51	44.04	3.338	74.6
		100	.843	11.064	31.53	96.14	.6677	561.6	107.20	41.66	3.338	88.1
		120	1.000	10.750	36.91	90.76	.6303	641.6	125.49	39.33	3.338	100.7
		140	1.125	10.500	41.08	86.59	.6013	700.5	133.68	37.52	3.338	109.9
		160	1.312	10.126	47.14	80.53	.5592	781.1	160.27	34.89	3.338	122.6
14	14.00	10	.250	13.500	10.80	143.14	.9940	255.3	36.71	62.03	3.665	36.6
		20	.312	13.376	13.42	140.52	.9758	314.4	45.68	60.89	3.665	45.0
		30s	.375	13.250	16.05	137.88	.9575	372.8	54.57	59.75	3.665	53.2
		40	.438	13.124	18.66	135.28	.9394	429.1	63.37	58.64	3.665	61.3
		..x	.500	13.000	21.21	132.73	.9217	483.8	72.09	57.46	3.665	69.1
		60	.593	12.814	24.98	128.96	.8956	562.3	84.91	55.86	3.665	80.3
		80	.750	12.500	31.22	122.72	.8522	687.3	106.13	53.18	3.665	98.2
		100	.937	12.126	38.45	115.49	.8020	824.4	130.73	50.04	3.665	117.8
		120	1.093	11.814	44.32	109.62	.7612	929.6	150.67	47.45	3.665	132.8
		140	1.250	11.500	50.07	103.87	.7213	1027.0	170.22	45.01	3.665	146.8
		160	1.406	11.188	55.63	98.31	.6827	1117.0	189.12	42.60	3.665	159.6

Table A-1 continued

Note 1: The letters **"s"**, **"x"**, and **"xx"** in the column of Schedule Numbers indicate Standard, Extra Strong, and Double Extra Strong Pipe, respectively.

Note 2: The values shown in square feet for the Transverse Internal Area also represent the volume in cubic feet per foot of pipe length.

Nominal Pipe Size	Outside Diameter (D)	Schedule No.	Wall Thickness (t)	Inside Diameter (d)	Area of Metal (a)	Transverse Internal Area		Moment of Inertia (I)	Weight of Pipe	Weight of Water	External Surface	Section Modulus
		See Note 1					*See Note 2*					$\left(2\frac{I}{D}\right)$
Inches	Inches		Inches	Inches	Square Inches	Square Inches	Square Feet	Inches to 4th Power	Pounds per foot	Pounds per foot of pipe	Sq. Ft. per foot of pipe	
16	16.00	10	.250	15.500	12.37	188.69	1.3103	383.7	42.05	81.74	4.189	48.0
		20	.312	15.376	15.38	185.69	1.2895	473.2	52.36	80.50	4.189	59.2
		30s	.375	15.250	18.41	182.65	1.2684	562.1	62.58	79.12	4.189	70.3
		40x	.500	15.000	24.35	176.72	1.2272	731.9	82.77	76.58	4.189	91.5
		60	.656	14.688	31.62	169.44	1.1766	932.4	107.50	73.42	4.189	116.6
		80	.843	14.314	40.14	160.92	1.1175	1155.8	136.46	69.73	4.189	144.5
		100	1.031	13.938	48.48	152.58	1.0596	1364.5	164.83	66.12	4.189	170.5
		120	1.218	13.564	56.56	144.50	1.0035	1555.8	192.29	62.62	4.189	194.5
		140	1.438	13.124	65.78	135.28	.9394	1760.3	223.64	58.64	4.189	220.0
		160	1.593	12.814	72.10	128.96	.8956	1893.5	245.11	55.83	4.189	236.7
18	18.00	10	.250	17.500	13.94	240.53	1.6703	549.1	47.39	104.21	4.712	61.1
		20	.312	17.376	17.34	237.13	1.6467	678.2	59.03	102.77	4.712	75.5
		..s	.375	17.250	20.76	233.71	1.6230	806.7	70.59	101.18	4.712	89.6
		30	.438	17.124	24.17	230.30	1.5990	930.3	82.06	99.84	4.712	103.4
		..x	.500	17.000	27.49	226.98	1.5763	1053.2	92.45	98.27	4.712	117.0
		40	.562	16.876	30.79	223.68	1.5533	1171.5	104.75	96.93	4.712	130.1
		60	.750	16.500	40.64	213.83	1.4849	1514.7	138.17	92.57	4.712	168.3
		80	.937	16.126	50.23	204.24	1.4183	1833.0	170.75	88.50	4.712	203.8
		100	1.156	15.688	61.17	193.30	1.3423	2180.0	207.96	83.76	4.712	242.3
		120	1.375	15.250	71.81	182.66	1.2684	2498.1	244.14	79.07	4.712	277.6
		140	1.562	14.876	80.66	173.80	1.2070	2749.0	274.23	75.32	4.712	305.5
		160	1.781	14.438	90.75	163.72	1.1369	3020.0	308.51	70.88	4.712	335.6
20	20.00	10	.250	19.500	15.51	298.65	2.0740	756.4	52.73	129.42	5.236	75.6
		20s	.375	19.250	23.12	290.04	2.0142	1113.0	78.60	125.67	5.236	111.3
		30x	.500	19.000	30.63	283.53	1.9690	1457.0	104.13	122.87	5.236	145.7
		40	.593	18.814	36.15	278.00	1.9305	1703.0	122.91	120.46	5.236	170.4
		60	.812	18.376	48.95	265.21	1.8417	2257.0	166.40	114.92	5.236	225.7
		80	1.031	17.938	61.44	252.72	1.7550	2772.0	208.87	109.51	5.236	277.1
		100	1.281	17.438	75.33	238.83	1.6585	3315.2	256.10	103.39	5.236	331.5
		120	1.500	17.000	87.18	226.98	1.5762	3754.0	296.37	98.35	5.236	375.5
		140	1.750	16.500	100.33	213.82	1.4849	4216.0	341.10	92.66	5.236	421.7
		160	1.968	16.064	111.49	202.67	1.4074	4585.5	379.01	87.74	5.236	458.5
24	24.00	10	.250	23.500	18.65	433.74	3.0121	1315.4	63.41	187.95	6.283	109.6
		20s	.375	23.250	27.83	424.56	2.9483	1942.0	94.62	183.95	6.283	161.9
		..x	.500	23.000	36.91	415.48	2.8853	2549.5	125.49	179.87	6.283	212.5
		30	.562	22.876	41.39	411.00	2.8542	2843.0	140.80	178.09	6.283	237.0
		40	.687	22.626	50.31	402.07	2.7921	3421.3	171.17	174.23	6.283	285.1
		60	.968	22.064	70.04	382.35	2.6552	4652.8	238.11	165.52	6.283	387.7
		80	1.218	21.564	87.17	365.22	2.5362	5672.0	296.36	158.26	6.283	472.8
		100	1.531	20.938	108.07	344.32	2.3911	6849.9	367.40	149.06	6.283	570.8
		120	1.812	20.376	126.31	326.08	2.2645	7825.0	429.39	141.17	6.283	652.1
		140	2.062	19.876	142.11	310.28	2.1547	8625.0	483.13	134.45	6.283	718.9
		160	2.343	19.314	159.41	292.98	2.0346	9455.9	541.94	126.84	6.283	787.9

APPENDIX B
Steel Flange Data

Table B-1. Flanging Processes. Welded Flanged Joints.

Welded flanged joints can be furnished in the types illustrated here. The Cranelap stub ends with Cranelap flange, also illustrated, afford an auxiliary flanged connection for welding.

Application: Any of the welded flanged joints shown at the right can be applied to straight pipe, pipe bends, the ends and nozzles of welded headers, and the flanged ends of welded assemblies. Special shop equipment assures the perfect alignment of flange faces on all Crane Welded Flanged Joints.

Welding: The shop welding of these flanged joints is performed by Crane welders working under approved procedure control.

Special piping materials: These types of welded flanged joints can be furnished on many special piping materials, including numerous alloy steels, with facilities for heat-treating after fabrication.

Complete information and prices will be furnished on application.

Forged Steel Screwed Flange, Seal-Welded

A Crane Forged Steel Screwed Flange is used in this joint. The pipe and the flange are accurately threaded; the flange is made up tight on the pipe, seal-welded, and then refaced. The joint is sealed by fillet-welding the back of the flange to the pipe, thus assuring no leakage through the threads.

The refacing assures perfect alignment of the flange faces, and that the end of the pipe is flush with the face of the flange. The threads retain the function of holding the flange securely on the pipe, hence there is no shearing action.

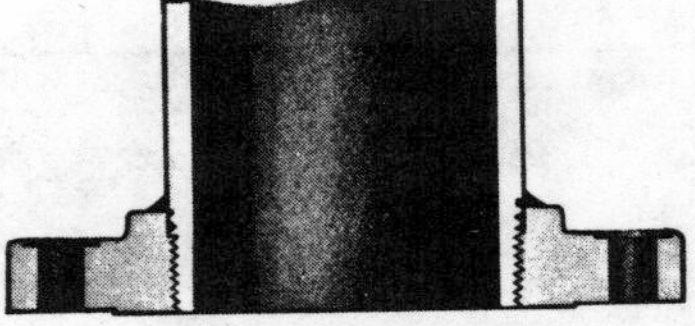

Screwed Flange
Seal-Welded and Refaced

Forged Steel Welding Neck Flange

Crane Welding Neck Flanges are of forged steel. They are machined with a beveled end and bored to match the inside diameter of the pipe to which they are applied. A butt-weld is used to attach the welding neck flange to the pipe, which is also machine beveled.

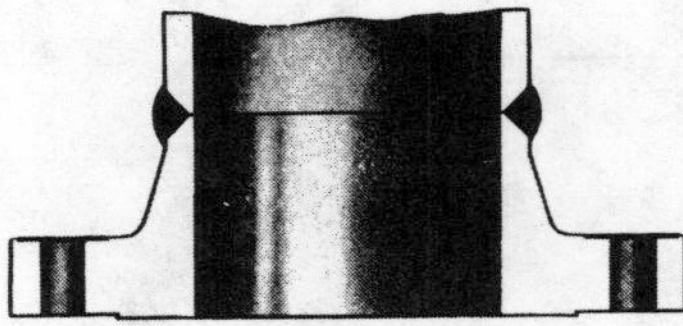

Welding Neck Flange
Butt-Welded to Pipe

Forged Steel Slip-On Welding Flange

Crane Forged Steel Slip-On Welding Flanges are bored for a snug fit on the pipe and, when applied to fabricated piping, are welded at the front and back through the two methods defined below and illustrated at the right.

Type No. 1: Type No. 1 is Crane standard for welded flanged joints using Forged Steel Slip-On Welding Flanges. Regular flanges are utilized with the end of the pipe set back from the face of the flange and the flange welded to the pipe both in front and back.

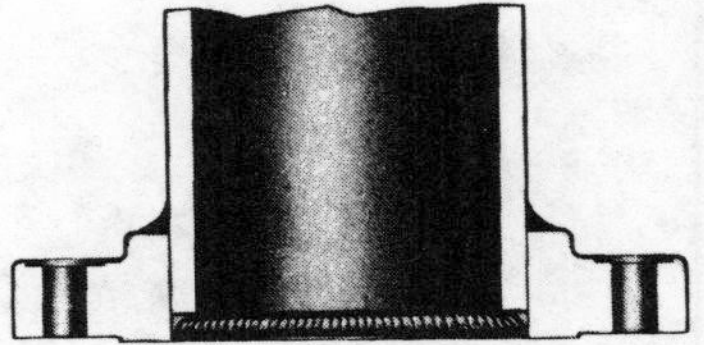

Type No. 1
Slip-On Welding Flange
Welded Front and Back

(table continued on next page)

Table B-1 continued

Type No. 2: Type No. 2 is furnished on special order only; slip-on flanges with a special front groove for welding are used. The pipe is flush with the flange face; this is accomplished by refacing, after both the front and back of the flange are welded to the pipe.

Code limitation: When piping must comply with the American Standard Code for Pressure Piping or the ASME Boiler and Pressure Vessel Code, the use of the slip-on flanged joint is permissible on all sizes of flanges listed under primary service pressure ratings up to and including the 900-pound class, and in sizes 2½-inch and smaller of the 1500-pound class, of the American Steel Flange Standard (ASA B16.5-1957).

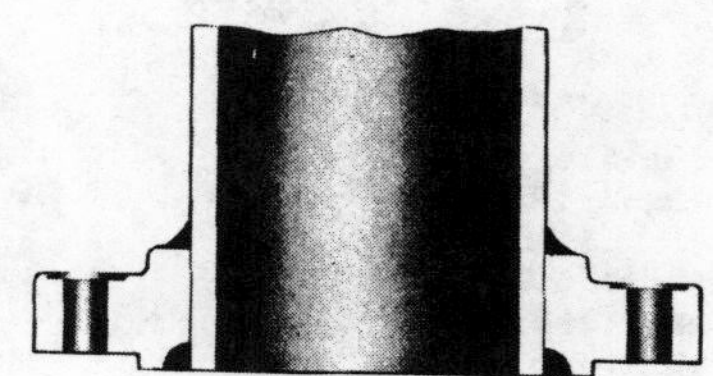

Type No. 2
Slip-On Welding Flange
Welded Front and Back and Refaced

Cranelap Stub Ends and Cranelap Flange

The Cranelap stub end with Cranelap flange can be applied to fabricated piping. Both the stub end and the pipe are machine beveled. A butt-weld is used to complete the joint.

This type of joint has all of the advantages of the regular Cranelap joint, as described on pages 363 to 365. In most cases, piping can be fabricated with Cranelap joints applied directly, which eliminates the weld necessary for the application of the Cranelap stub end with Cranelap flange.

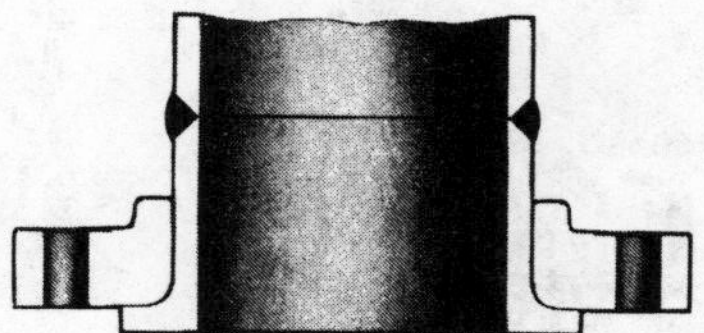

Cranelap Flange with
Cranelap Stub End
Butt-Welded to Pipe

Courtesy of Crane Co.

Table B-2 continued

Table B-2. Forged Steel Flanges.

Screwed Flange

No. 556, 150-Pound
No. 291 E, 300-Pound
No. 651 E, 400-Pound
No. 856 E, 600-Pound
No. 1266 E, 900-Pound
No. 1556 E, 1500-Pound

Slip-On Welding Flange

No. 554, 150-Pound
No. 294 E, 300-Pound
No. 694 E, 400-Pound
No. 854 E, 600-Pound
No. 1294 E, 900-Pound
No. 1594 E, 1500-Pound

Welding Neck Flange

No. 568, 150-Pound
No. 296 E, 300-Pound
No. 656 E, 400-Pound
No. 855 E, 600-Pound
No. 1265 E, 900-Pound
No. 1565 E, 1500-Pound

Cranelap Flange

No. 572, 150-Pound
No. 496 E, 300-Pound
No. 664 E, 400-Pound
No. 862 E, 600-Pound
No. 1262 E, 900-Pound
No. 1562 E, 1500-Pound

Blind Flange

No. 556½, 150-Pound
No. 297 E, 300-Pound
No. 657 E, 400-Pound
No. 858 E, 600-Pound
No. 1267 E, 900-Pound
No. 1557 E, 1500-Pound

A variety of types in seven different pressure classes . . . 150 to 2500-Pound

Description of materials...........page 8
Working pressures..............page 309
Sizes and weights.....pages 310 and 311
Dimensions...........pages 312 and 313
Prices........................on request

The Crane line of Forged Steel Flanges comprises the complete assortment of straight and reducing types illustrated on this page. Made in seven different pressure classes . . . . 150, 300, 400, 600, 900, 1500, and 2500-Pound . . . . they are available in a variety of materials and with various flange facings, providing a correct type for any service requirement.

Materials: Crane flanges are made of carbon steel forgings having a highly refined grain structure and generally excellent physical properties well in excess of recognized minimum requirements.

In the 150 and 300-pound pressure classes, the flanges are regularly made of carbon steel conforming to ASTM Specification A 181, Grade II; on special order, they can be furnished heat-treated (normalized or annealed) to conform to ASTM Specification A 105, Grade II.

In the 400-pound and higher pressure classes, the flanges are regularly made of carbon steel conforming to ASTM Specification A 105, Grade II.

In addition, flanges in 300-pound and higher pressure classes can be made to order of Crane No. 5 Chrome-Molybdenum Forged Steel (ASTM A 182, Grade F5a).

American Standard: The dimensions and drilling of all flanges conform to the American Steel Flange Standard B16.5-1957, for their respective pressure class.

This Standard does not include slip-on welding flanges of the 2500-pound class nor sizes 3-inch and larger of the 1500-pound class; in such classes and sizes, Crane slip-on welding flanges have the same dimensions as American Standard Steel Screwed Flanges, being bored instead of threaded.

Flange facings: The 150 and 300-Pound Screwed, Slip-On Welding, Welding Neck, and Blind Flanges are regularly furnished with an American Standard $\frac{1}{16}$-inch raised face.

The aforementioned flanges, in 400-pound and higher pressure classes, are regularly furnished with an American Standard ¼-inch male face (large male).

Reducing Screwed Flange

No. 558½, 150-Pound
No. 292 E, 300-Pound
No. 658 E, 400-Pound
No. 857 E, 600-Pound
No. 1263 E, 900-Pound
No. 1558 E, 1500-Pound

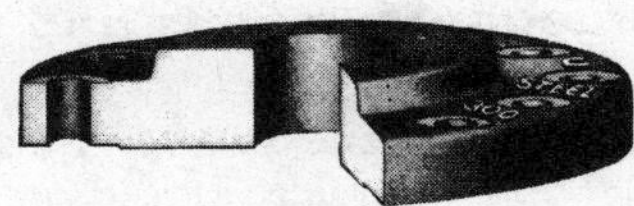

Reducing Slip-On Welding Flange

No. 554½, 150-Pound
No. 290 E, 300-Pound
No. 693 E, 400-Pound
No. 853 E, 600-Pound
No. 1295 E, 900-Pound
No. 1595 E, 1500-Pound

Other types of facings such as ring joint, female, tongue, groove, etc., can be furnished; see pages 332 to 335 for complete information.

In addition, flanges of any pressure class are available with a flat face (raised or male face removed); the flat face will have a spiral serrated finish.

Finish of flange faces: The $\frac{1}{16}$-inch raised faces and the ¼-inch large male faces are regularly furnished with a serrated finish. A smooth finish can be furnished when specified.

Drilling: The flanges are regularly furnished faced, drilled, and spot faced to the corresponding pressure class of the American Standard. They can be furnished faced only, when specified.

Reducing flanges: The Reducing Screwed and Reducing Slip-On Welding Flanges, illustrated above, are available in any size reduction; prices are based on the outside diameter of the flange. For ordering information, see page 311.

Reducing Welding Neck Flanges and Eccentric Reducing Screwed or Slip-On Welding Flanges can be made to order; information on request.

Reducing Cranelap Flanges are not recommended and, consequently, are not manufactured. Another type of flanged joint or connection should be used.

(table continued on next page)

Table B-2 continued

Forged Steel Flanges. Dimensions in Inches.

(table continued on next page)

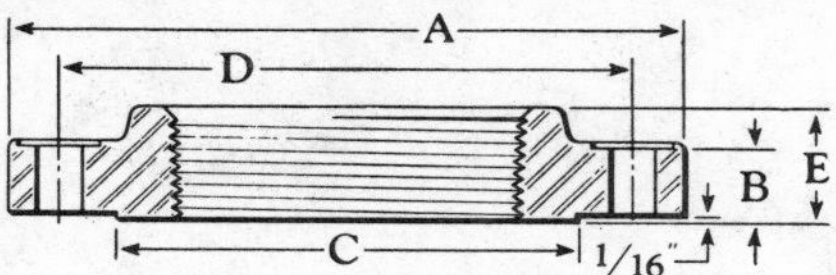

Screwed Flange
150 and 300-Pound

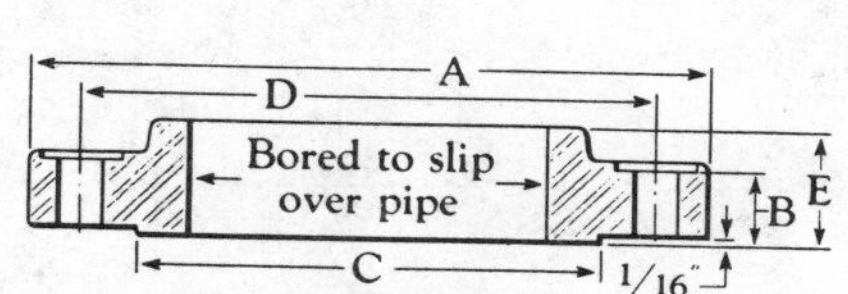

Slip-On Welding Flange
150 and 300-Pound

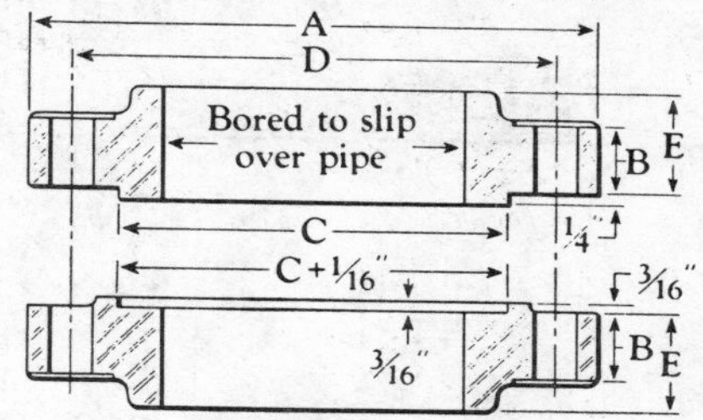

Slip-On Welding Flange
400, 600, 900, and 1500-Pound

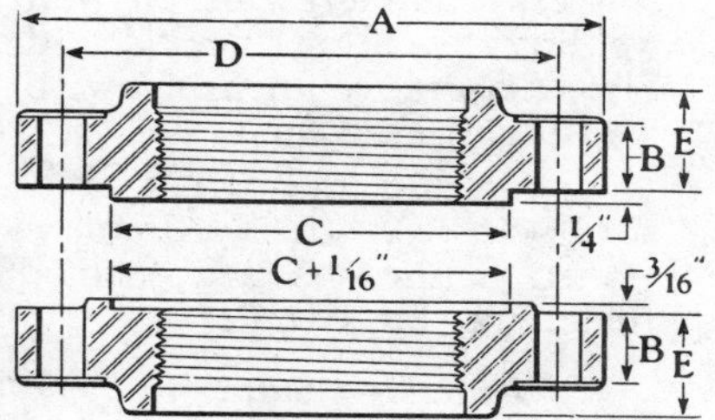

Screwed Flange
400, 600, 900, 1500, and 2500-Pound

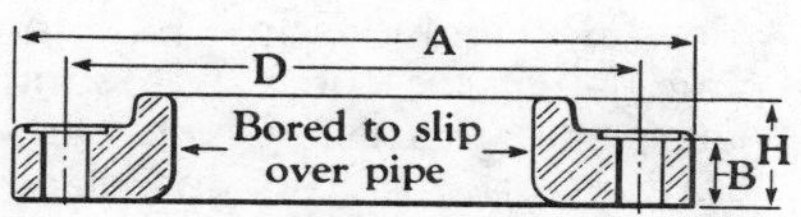

Cranelap Flange
150 and 300-Pound

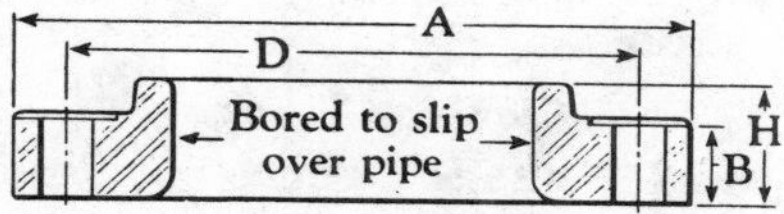

Cranelap Flange
400, 600, 900, 1500, and 2500-Pound

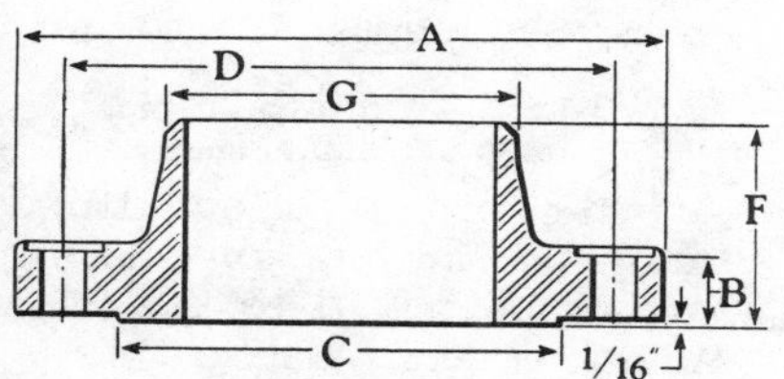

Welding Neck Flange
150 and 300-Pound

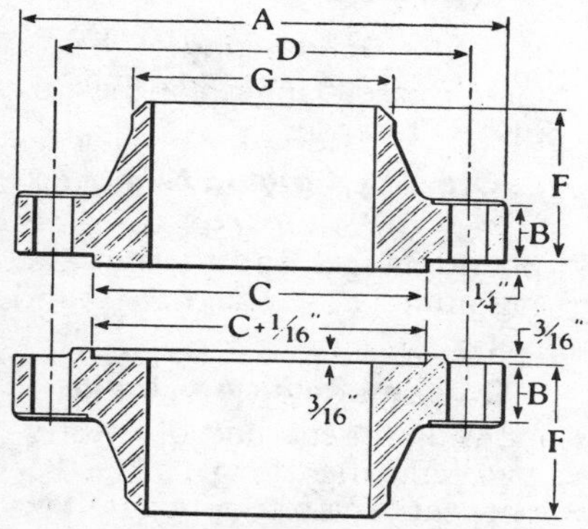

Welding Neck Flange
400, 600, 900, 1500, and 2500-Pound

Class	Pipe Size	A	B	C	D	Bolts No.	Bolts Dia.	E	F	G	H
150 Pound	1/2	3 1/2	7/16	1 3/8	2 3/8	4	1/2	5/8	1 7/8	0.84	5/8
	3/4	3 7/8	1/2	1 11/16	2 3/4	4	1/2	5/8	2 1/16	1.05	5/8
	1	4 1/4	9/16	2	3 1/8	4	1/2	11/16	2 3/16	1.32	11/16
	1 1/4	4 5/8	5/8	2 1/2	3 1/2	4	1/2	13/16	2 1/4	1.66	13/16
	1 1/2	5	11/16	2 7/8	3 7/8	4	1/2	7/8	2 7/16	1.90	7/8
	2	6	3/4	3 5/8	4 3/4	4	5/8	1	2 1/2	2.38	1
	2 1/2	7	7/8	4 1/8	5 1/2	4	5/8	1 1/8	2 3/4	2.88	1 1/8
	3	7 1/2	15/16	5	6	4	5/8	1 3/16	2 3/4	3.50	1 3/16
	3 1/2	8 1/2	15/16	5 1/2	7	8	5/8	1 1/4	2 13/16	4.00	1 1/4
	4	9	15/16	6 3/16	7 1/2	8	5/8	1 5/16	3	4.50	1 5/16
	5	10	15/16	7 5/16	8 1/2	8	3/4	1 7/16	3 1/2	5.56	1 7/16
	6	11	1	8 1/2	9 1/2	8	3/4	1 9/16	3 1/2	6.63	1 9/16
	8	13 1/2	1 1/8	10 5/8	11 3/4	8	3/4	1 3/4	4	8.63	1 3/4
	10	16	1 3/16	12 3/4	14 1/4	12	7/8	1 15/16	4	10.75	1 15/16
	12	19	1 1/4	15	17	12	7/8	2 3/16	4 1/2	12.75	2 3/16
	14	21	1 3/8	16 1/4	18 3/4	12	1	2 1/4	5	14.00	3 1/8
	16	23 1/2	1 7/16	18 1/2	21 1/4	16	1	2 1/2	5	16.00	3 7/16
	18	25	1 9/16	21	22 3/4	16	1 1/8	2 11/16	5 1/2	18.00	3 13/16
	20	27 1/2	1 11/16	23	25	20	1 1/8	2 7/8	5 11/16	20.00	4 1/16
	24	32	1 7/8	27 1/4	29 1/2	20	1 1/4	3 1/4	6	24.00	4 3/8
300 Pound	1/2	3 3/4	9/16	1 3/8	2 5/8	4	1/2	7/8	2 1/16	0.84	. . .
	3/4	4 5/8	5/8	1 11/16	3 1/4	4	5/8	1	2 1/4	1.05	1
	1	4 7/8	11/16	2	3 1/2	4	5/8	1 1/16	2 7/16	1.32	1 1/16
	1 1/4	5 1/4	3/4	2 1/2	3 7/8	4	5/8	1 1/16	2 9/16	1.66	1 1/16
	1 1/2	6 1/8	13/16	2 7/8	4 1/2	4	3/4	1 3/16	2 11/16	1.90	1 3/16
	2	6 1/2	7/8	3 5/8	5	8	5/8	1 5/16	2 3/4	2.38	1 5/16
	2 1/2	7 1/2	1	4 1/8	5 7/8	8	3/4	1 1/2	3	2.88	1 1/2
	3	8 1/4	1 1/8	5	6 5/8	8	3/4	1 11/16	3 1/8	3.50	1 11/16
	3 1/2	9	1 3/16	5 1/2	7 1/4	8	3/4	1 3/4	3 3/16	4.00	. . .
	4	10	1 1/4	6 3/16	7 7/8	8	3/4	1 7/8	3 3/8	4.50	1 7/8
	5	11	1 3/8	7 5/16	9 1/4	8	3/4	2	3 7/8	5.56	2
	6	12 1/2	1 7/16	8 1/2	10 5/8	12	3/4	2 1/16	3 7/8	6.63	2 1/16
	8	15	1 5/8	10 5/8	13	12	7/8	2 7/16	4 3/8	8.63	2 7/16
	10	17 1/2	1 7/8	12 3/4	15 1/4	16	1	2 5/8	4 5/8	10.75	3 3/4
	12	20 1/2	2	15	17 3/4	16	1 1/8	2 7/8	5 1/8	12.75	4
	14	23	2 1/8	16 1/4	20 1/4	20	1 1/8	3	5 5/8	14.00	4 3/8
	16	25 1/2	2 1/4	18 1/2	22 1/2	20	1 1/4	3 1/4	5 3/4	16.00	4 3/4
	18	28	2 3/8	21	24 3/4	24	1 1/4	3 1/2	6 1/4	18.00	5 1/8
	20	30 1/2	2 1/2	23	27	24	1 1/4	3 3/4	6 3/8	20.00	5 1/2
	24	36	2 3/4	27 1/4	32	24	1 1/2	4 3/16	6 5/8	24.00	6
400 Pound (For smaller sizes, use 600-Pound)	4	10	1 3/8	6 3/16	7 7/8	8	7/8	2	3 1/2	4.50	2
	5	11	1 1/2	7 5/16	9 1/4	8	7/8	2 1/8	4	5.56	2 1/8
	6	12 1/2	1 5/8	8 1/2	10 5/8	12	7/8	2 1/4	4 1/16	6.63	2 1/4
	8	15	1 7/8	10 5/8	13	12	1	2 11/16	4 5/8	8.63	2 11/16
	10	17 1/2	2 1/8	12 3/4	15 1/4	16	1 1/8	2 7/8	4 7/8	10.75	4
	12	20 1/2	2 1/4	15	17 3/4	16	1 1/4	3 1/8	5 3/8	12.75	4 1/4
	14	23	2 3/8	16 1/4	20 1/4	20	1 1/4	3 5/16	5 7/8	14.00	4 5/8
	16	25 1/2	2 1/2	18 1/2	22 1/2	20	1 3/8	3 11/16	6	16.00	5
	18	28	2 5/8	21	24 3/4	24	1 3/8	3 7/8	6 1/2	18.00	5 3/8
	20	30 1/2	2 3/4	23	27	24	1 1/2	4	6 5/8	20.00	5 3/4
	24	36	3	27 1/4	32	24	1 3/4	4 1/2	6 7/8	24.00	6 1/4

Table B-2 continued

Dimensions, in Inches — continued

Class	Pipe Size	A	B	C	D	Bolts		E	F	G	H
						No.	Dia.				
600 Pound	½	3¾	9/16	1⅜	2⅝	4	½	⅞	2 1/16	0.84	⅞
	¾	4⅝	⅝	1 11/16	3¼	4	⅝	1	2¼	1.05	1
	1	4⅞	11/16	2	3½	4	⅝	1 1/16	2 7/16	1.32	1 1/16
	1¼	5¼	13/16	2½	3⅞	4	⅝	1⅛	2⅝	1.66	1⅛
	1½	6⅛	⅞	2⅞	4½	4	¾	1¼	2¾	1.90	1¼
	2	6½	1	3⅝	5	8	⅝	1 7/16	2⅞	2.38	1 7/16
	2½	7½	1⅛	4⅛	5⅞	8	¾	1⅝	3⅛	2.88	1⅝
	3	8¼	1¼	5	6⅝	8	¾	1 13/16	3¼	3.50	1 13/16
	4	10¾	1½	6 3/16	8½	8	⅞	2⅛	4	4.50	2⅛
	5	13	1¾	7 5/16	10½	8	1	2⅜	4½	5.56	2⅜
	6	14	1⅞	8½	11½	12	1	2⅝	4⅝	6.63	2⅝
	8	16½	2 3/16	10⅝	13¾	12	1⅛	3	5¼	8.63	3
	10	20	2½	12¾	17	16	1¼	3⅜	6	10.75	4⅜
	12	22	2⅝	15	19¼	20	1¼	3⅝	6⅛	12.75	4⅝
	14	23¾	2¾	16¼	20¾	20	1⅜	3 11/16	6½	14.00	5
	16	27	3	18½	23¾	20	1½	4 3/16	7	16.00	5½
	18	29¼	3¼	21	25¾	20	1⅝	4⅝	7¼	18.00	6
	20	32	3½	23	28½	24	1⅝	5	7½	20.00	6½
	24	37	4	27¼	33	24	1⅞	5½	8	24.00	7¼
900 Pound (For smaller sizes, use 1500 Pound)	3	9½	1½	5	7½	8	⅞	2⅛	4	3.50	2⅛
	4	11½	1¾	6 3/16	9¼	8	1⅛	2¾	4½	4.50	2¾
	5	13¾	2	7 5/16	11	8	1¼	3⅛	5	5.56	3⅛
	6	15	2 3/16	8½	12½	12	1⅛	3⅜	5½	6.63	3⅜
	8	18½	2½	10⅝	15½	12	1⅜	4	6⅜	8.63	4½
	10	21½	2¾	12¾	18½	16	1⅜	4¼	7¼	10.75	5
	12	24	3⅛	15	21	20	1⅜	4⅝	7⅞	12.75	5⅝
	14	25¼	3⅜	16¼	22	20	1½	5⅛	8⅜	14.00	6⅛
	16	27¾	3½	18½	24¼	20	1⅝	5¼	8½	16.00	6½
	18	31	4	21	27	20	1⅞	6	9	18.00	7½
	20	33¾	4¼	23	29½	20	2	6¼	9¾	20.00	8¼
	24	41	5½	27¼	35½	20	2½	8	11½	24.00	10½
1500 Pound	½	4¾	⅞	1⅜	3¼	4	¾	1¼	2⅜	0.84	1¼
	¾	5⅛	1	1 11/16	3½	4	¾	1⅜	2¾	1.05	1⅜
	1	5⅞	1⅛	2	4	4	⅞	1⅝	2⅞	1.32	1⅝
	1¼	6¼	1⅛	2½	4⅜	4	⅞	1⅝	2⅞	1.66	1⅝
	1½	7	1¼	2⅞	4⅞	4	1	1¾	3¼	1.90	1¾
	2	8½	1½	3⅝	6½	8	⅞	2¼	4	2.38	2¼
	2½	9⅝	1⅝	4⅛	7½	8	1	2½	4⅛	2.88	2½
	3	10½	1⅞	5	8	8	1⅛	2⅞	4⅝	3.50	2⅞
	4	12¼	2⅛	6 3/16	9½	8	1¼	3 9/16	4⅞	4.50	3 9/16
	5	14¾	2⅞	7 5/16	11½	8	1½	4⅛	6⅛	5.56	4⅛
	6	15½	3¼	8½	12½	12	1⅜	4 11/16	6¾	6.63	4 11/16
	8	19	3⅝	10⅝	15½	12	1⅝	5⅝	8⅜	8.63	5⅝
	10	23	4¼	12¾	19	12	1⅞	6¼	10	10.75	7
	12	26½	4⅞	15	22½	16	2	7⅛	11⅛	12.75	8⅝
	14	29½	5¼	16¼	25	16	2¼	...	11¾	14.00	9½
2500 Pound	½	5¼	1 3/16	1⅜	3½	4	¾	1 9/16	2⅞	0.84	1 9/16
	¾	5½	1¼	1 11/16	3¾	4	¾	1 11/16	3⅛	1.05	1 11/16
	1	6¼	1⅜	2	4¼	4	⅞	1⅞	3½	1.32	1⅞
	1¼	7¼	1½	2½	5⅛	4	1	2 1/16	3¾	1.66	2 1/16
	1½	8	1¾	2⅞	5¾	4	1⅛	2⅜	4⅜	1.90	2⅜
	2	9¼	2	3⅝	6¾	8	1	2¾	5	2.38	2¾
	2½	10½	2¼	4⅛	7¾	8	1⅛	3⅛	5⅝	2.88	3⅛
	3	12	2⅝	5	9	8	1¼	3⅝	6⅝	3.50	3⅝
	4	14	3	6 3/16	10¾	8	1½	4¼	7½	4.50	4¼
	5	16½	3⅝	7 5/16	12¾	8	1¾	5⅛	9	5.56	5⅛
	6	19	4¼	8½	14½	8	2	6	10¾	6.63	6
	8	21¾	5	10⅝	17¼	12	2	7	12½	8.63	7
	10	26½	6½	12¾	21¼	12	2½	9	16½	10.75	9
	12	30	7¼	15	24⅜	12	2¾	10	18¼	12.75	10

Courtesy of Crane Co.

Regular Facings

In 150 and 300-pound pressure classes, the screwed, slip-on welding, welding neck, and blind flanges are furnished with a 1/16-inch raised face.

In 400-pound and higher pressure classes, the aforementioned flanges have a ¼-inch male face (large male).

American Standard

The dimensions and drilling of flanges conform to the American Steel Flange Standard, B16.5-1957, for their respective pressure class. This Standard does not include slip-on welding flanges in the 2500-pound class nor sizes 3-inch and larger of the 1500-pound class. Crane flanges of this type have the same dimensions as American Standard Steel Screwed Flanges, being bored instead of threaded.

Cranelap Flanges

For sizes, see pages 310 and 311. For information on complete Cranelap Joints, see page 363. Cranelap flanges also are recommended for use in combination with Cranelap stub ends; see pages 298 to 301.

3-inch Cranelap Joints
(300 and 600-Pound)

When 3-inch 300 or 600-pound flanges with ring joint facing are to be bolted to Cranelap joints, orders must so specify; they require a groove of special pitch diameter. See page 334 for dimensions.

Galvanizing

Galvanized flanges can be furnished to order.

Order by Catalog Number

When ordering 150 to 1500-pound flanges, specify the catalog number; see pages 310 and 311.

Ordering Reducing Flanges

For correct method of specifying reducing flange sizes, see note at bottom of page 311.

Table B-3. Ring Joint Facing and Rings (American Standard).

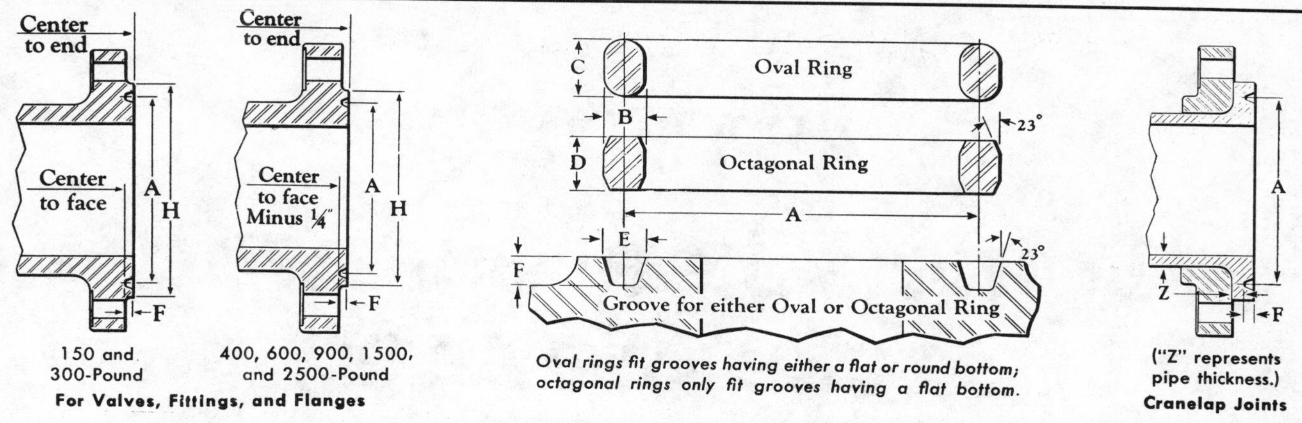

150 and 300-Pound — 400, 600, 900, 1500, and 2500-Pound

For Valves, Fittings, and Flanges

Oval rings fit grooves having either a flat or round bottom; octagonal rings only fit grooves having a flat bottom.

("Z" represents pipe thickness.)

Cranelap Joints

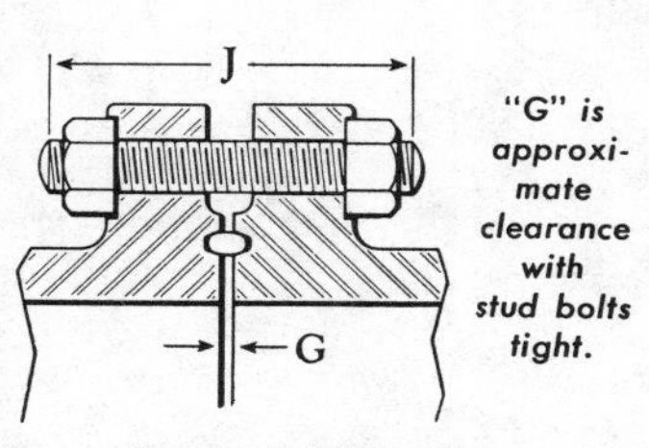

"G" is approximate clearance with stud bolts tight.

Assembled Ring Joint

*Dimension "J" does not apply to Cranelap Joints; see "Stud bolt lengths" on next page.

†Caution: 3-inch 300 and 600-Pound Cranelap Ring Joints use Ring No. R 30, having a pitch diameter of 4 5/8 inches. When 3-inch 300 or 600-Pound ring joint valves, fittings, or flanges are to be bolted to Cranelap joints, orders must specify; they will be machined special.

Class	Size	Ring No.	A	B	C	D	E	F	G	H	*J	Stud Bolts No.	Stud Bolts Dia.
150 Pound	1	R 15	1 7/8	5/16	9/16	1/2	11/32	1/4	5/32	2 1/2	3	4	1/2
	1 1/4	R 17	2 1/4	5/16	9/16	1/2	11/32	1/4	5/32	2 7/8	3	4	1/2
	1 1/2	R 19	2 9/16	5/16	9/16	1/2	11/32	1/4	5/32	3 1/4	3 1/4	4	1/2
	2	R 22	3 1/4	5/16	9/16	1/2	11/32	1/4	5/32	4	3 1/2	4	5/8
	2 1/2	R 25	4	5/16	9/16	1/2	11/32	1/4	5/32	4 3/4	3 3/4	4	5/8
	3	R 29	4 1/2	5/16	9/16	1/2	11/32	1/4	5/32	5 1/4	4	4	5/8
	3 1/2	R 33	5 3/16	5/16	9/16	1/2	11/32	1/4	5/32	6 1/16	4	8	5/8
	4	R 36	5 7/8	5/16	9/16	1/2	11/32	1/4	5/32	6 3/4	4	8	5/8
	5	R 40	6 3/4	5/16	9/16	1/2	11/32	1/4	5/32	7 5/8	4 1/4	8	3/4
	6	R 43	7 5/8	5/16	9/16	1/2	11/32	1/4	5/32	8 5/8	4 1/4	8	3/4
	8	R 48	9 3/4	5/16	9/16	1/2	11/32	1/4	5/32	10 3/4	4 1/2	8	3/4
	10	R 52	12	5/16	9/16	1/2	11/32	1/4	5/32	13	5	12	7/8
	12	R 56	15	5/16	9/16	1/2	11/32	1/4	5/32	16	5	12	7/8
	14	R 59	15 5/8	5/16	9/16	1/2	11/32	1/4	1/8	16 3/4	5 1/2	12	1
	16	R 64	17 7/8	5/16	9/16	1/2	11/32	1/4	1/8	19	5 3/4	16	1
	18	R 68	20 3/8	5/16	9/16	1/2	11/32	1/4	1/8	21 1/2	6 1/4	16	1 1/8
	20	R 72	22	5/16	9/16	1/2	11/32	1/4	1/8	23 1/2	6 1/2	20	1 1/8
	24	R 76	26 1/2	5/16	9/16	1/2	11/32	1/4	1/8	28	7 1/4	20	1 1/4

Class	Size	Ring No.	A	B	C	D	E	F	G 300 Lb.	G 400 Lb.	G 600 Lb.	H	*J 300 Lb.	*J 400 Lb.	*J 600 Lb.	No. of Stud Bolts 300 Lb.	No. of Stud Bolts 400 Lb.	No. of Stud Bolts 600 Lb.	Dia. of Stud Bolts 300 Lb.	Dia. of Stud Bolts 400 Lb.	Dia. of Stud Bolts 600 Lb.
†300, 400, and †600 Pound	1/2	R 11	1 11/32	1/4	7/16	3/8	9/32	7/32	1/8	...	1/8	2	3	....	3	4	...	4	1/2	...	1/2
	3/4	R 13	1 11/16	5/16	9/16	1/2	11/32	1/4	5/32	...	5/32	2 1/2	3 1/4	...	3 1/4	4	...	4	5/8	...	5/8
	1	R 16	2	5/16	9/16	1/2	11/32	1/4	5/32	...	5/32	2 3/4	3 1/2	...	3 1/2	4	...	4	5/8	...	5/8
	1 1/4	R 18	2 3/8	5/16	9/16	1/2	11/32	1/4	5/32	...	5/32	3 1/8	3 1/2	...	3 3/4	4	...	4	5/8	...	5/8
	1 1/2	R 20	2 11/16	5/16	9/16	1/2	11/32	1/4	5/32	...	5/32	3 9/16	4	...	4	4	...	4	3/4	...	3/4
	2	R 23	3 1/4	7/16	11/16	5/8	15/32	5/16	7/32	...	3/16	4 1/4	4	...	4 1/4	8	...	8	5/8	...	5/8
	2 1/2	R 26	4	7/16	11/16	5/8	15/32	5/16	7/32	...	3/16	5	4 1/2	...	4 3/4	8	...	8	3/4	...	3/4
	†3	†R 31	†4 7/8	7/16	11/16	5/8	15/32	5/16	7/32	...	3/16	5 3/4	4 3/4	...	5	8	...	8	3/4	...	3/4
	3 1/2	R 34	5 3/16	7/16	11/16	5/8	15/32	5/16	7/32	...	3/16	6 1/4	5	...	5 1/2	8	...	8	3/4	...	7/8
	4	R 37	5 7/8	7/16	11/16	5/8	15/32	5/16	7/32	7/32	3/16	6 7/8	5	5 1/2	5 3/4	8	8	8	3/4	7/8	7/8
	5	R 41	7 1/8	7/16	11/16	5/8	15/32	5/16	7/32	7/32	3/16	8 1/4	5 1/4	5 3/4	6 1/2	8	8	8	3/4	7/8	1
	6	R 45	8 5/16	7/16	11/16	5/8	15/32	5/16	7/32	7/32	3/16	9 1/2	5 1/2	6	6 3/4	12	12	12	3/4	7/8	1
	8	R 49	10 5/8	7/16	11/16	5/8	15/32	5/16	7/32	7/32	3/16	11 7/8	6	6 3/4	7 3/4	12	12	12	7/8	1	1 1/8
	10	R 53	12 3/4	7/16	11/16	5/8	15/32	5/16	7/32	7/32	3/16	14	6 3/4	7 1/2	8 1/2	16	16	16	1	1 1/8	1 1/4
	12	R 57	15	7/16	11/16	5/8	15/32	5/16	7/32	7/32	3/16	16 1/4	7 1/4	8	8 3/4	16	16	20	1 1/8	1 1/4	1 1/4
	14	R 61	16 1/2	7/16	11/16	5/8	15/32	5/16	7/32	7/32	3/16	18	7 1/2	8 1/4	9 1/4	20	20	20	1 1/8	1 1/4	1 3/8
	16	R 65	18 1/2	7/16	11/16	5/8	15/32	5/16	7/32	7/32	3/16	20	8	8 3/4	10	20	20	20	1 1/4	1 3/8	1 1/2
	18	R 69	21	7/16	11/16	5/8	15/32	5/16	7/32	7/32	3/16	22 5/8	8 1/4	9	10 3/4	24	24	20	1 1/4	1 3/8	1 5/8
	20	R 73	23	1/2	3/4	11/16	17/32	3/8	7/32	7/32	3/16	25	8 3/4	9 3/4	11 1/2	24	24	24	1 1/4	1 1/2	1 5/8
	24	R 77	27 1/4	5/8	7/8	13/16	21/32	7/16	1/4	1/4	7/32	29 1/2	10	11	13 1/4	24	24	24	1 1/2	1 3/4	1 7/8

Courtesy of Crane Co.

APPENDIX C

Forged Steel Screwed Socket-Welding Fitting Data

Table C-1. Forged Steel Screwed Fittings. 2,000, 3,000 and 6,000-Pound W.O.G.

90° Elbow
No. 240, 2000-Pound
No. 380, 3000-Pound
No. 660, 6000-Pound

Tee
No. 241, 2000-Pound
No. 381, 3000-Pound
No. 661, 6000-Pound

45° Elbow
No. 242, 2000-Pound
No. 382, 3000-Pound
No. 662, 6000-Pound

Cross
No. 243, 2000-Pound
No. 383, 3000-Pound
No. 663, 6000-Pound

90° Street Elbow
No. 384, 3000-Pound
No. 664, 6000-Pound

45° Y-Bend
No. 245, 2000-Pound
No. 665, 6000-Pound

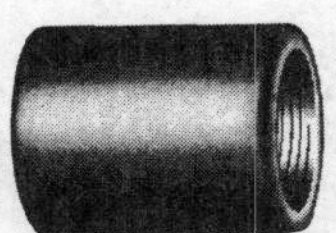

Coupling
No. 386, 3000-Pound
No. 666, 6000-Pound

Reducer
No. 387, 3000-Pound
No. 667, 6000-Pound

Half Coupling
No. 388, 3000-Pound
No. 668, 6000-Pound

Cap
No. 389, 3000-Pound

(table continued on next page)

Table C-1 continued

Round Head Plug
No. 308, 3000-Pound

Square Head Plug
No. 309, 3000-Pound

Hexagon Head Plug
No. 602, 6000-Pound

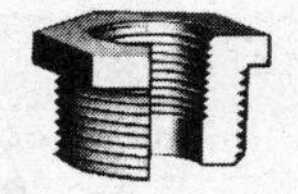

Hexagon Bushing
No. 600, 6000-Pound

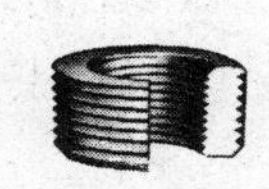

Face Bushing
No. 601, 6000-Pound

Courtesy of Crane Co.

Recommendations: These are unusually strong, rugged fittings. They are ideally suited for high pressure hydraulic lines and for high pressure-temperature service in oil refineries, oil and gas fields, central power stations, and industrial and chemical plants.

The 2000-Pound W.O.G. Fittings, exceptionally compact and light in weight, are intended for services beyond the temperature range of malleable iron fittings and for many relatively low pressure installations where the extra strength and safety afforded by steel fittings are desired.

Materials and design: Elbows, tees, crosses, and Y-bends are forged solid; the caps, couplings, reducers, plugs, and bushings are machined from solid steel. Carbon steel billets or bar stock used in the manufacturing process are subject to rigid specifications for strength, toughness, and resistance to temperature and shock.

The fittings feature liberal metal sections throughout and have an ample factor of safety over the recommended working pressures. All openings are drilled; on forged fittings, each opening is reinforced with a wide band which completely surrounds the thread chamber, extending beyond the last thread. The design provides the requisite strength, adds to the compact, neat appearance, and permits a sure wrench grip.

Threads: Threads are long and are accurately cut to gauge. All openings are in true alignment and chamfered to permit easy entrance of pipe.

MSS ratings: Working pressures agree with those in the MSS Standard for Forged Steel Screwed Fittings, No. SP-49-1956.

Working Pressures

Steam, Water, Oil, Oil Vapor, Gas, or Air

Temp.	Psi, Non-Shock Carbon Steel ASTM A105, Grade II		
Deg. Fahr.	2000 Pound W.O.G.	3000 Pound W.O.G.	6000 Pound W.O.G.
100°	2000	3000	6000
150	1970	2955	5915
200	1940	2915	5830
250	1915	2875	5750
300	1895	2845	5690
350	1875	2810	5625
400	1850	2775	5550
450	1810	2715	5430
500	1735	2605	5210
550	1640	2460	4925
600	1540	2310	4620
650	1430	2150	4300
700	1305	1960	3920
750	1180	1775	3550
800	1015	1525	3050
850[1]	830	1250	2500
875[1]	725	1090	2180
900[1]	615	925	1855
925[1,2]	520	785	1570
950[1,2]	425	640	1285
975[1,2]	330	500	1000
1000[1,2]	235	355	715

[1]Product used within the jurisdiction of Section 1, Power Boilers, of the ASME Boiler and Pressure Vessel Code is subject to the same maximum temperature limitations placed upon the material in Table P7, 1959 edition thereof.

[2]Product used within the jurisdiction of Section 1, Power Piping, of the ASA Code for Pressure Piping B31.1 is subject to the same maximum temperature limitations placed upon piping of the same general composition in Table 2a, 1955 edition thereof.

Table E-6 continued

Forged Steel Screwed Fittings. Dimensions in Inches.

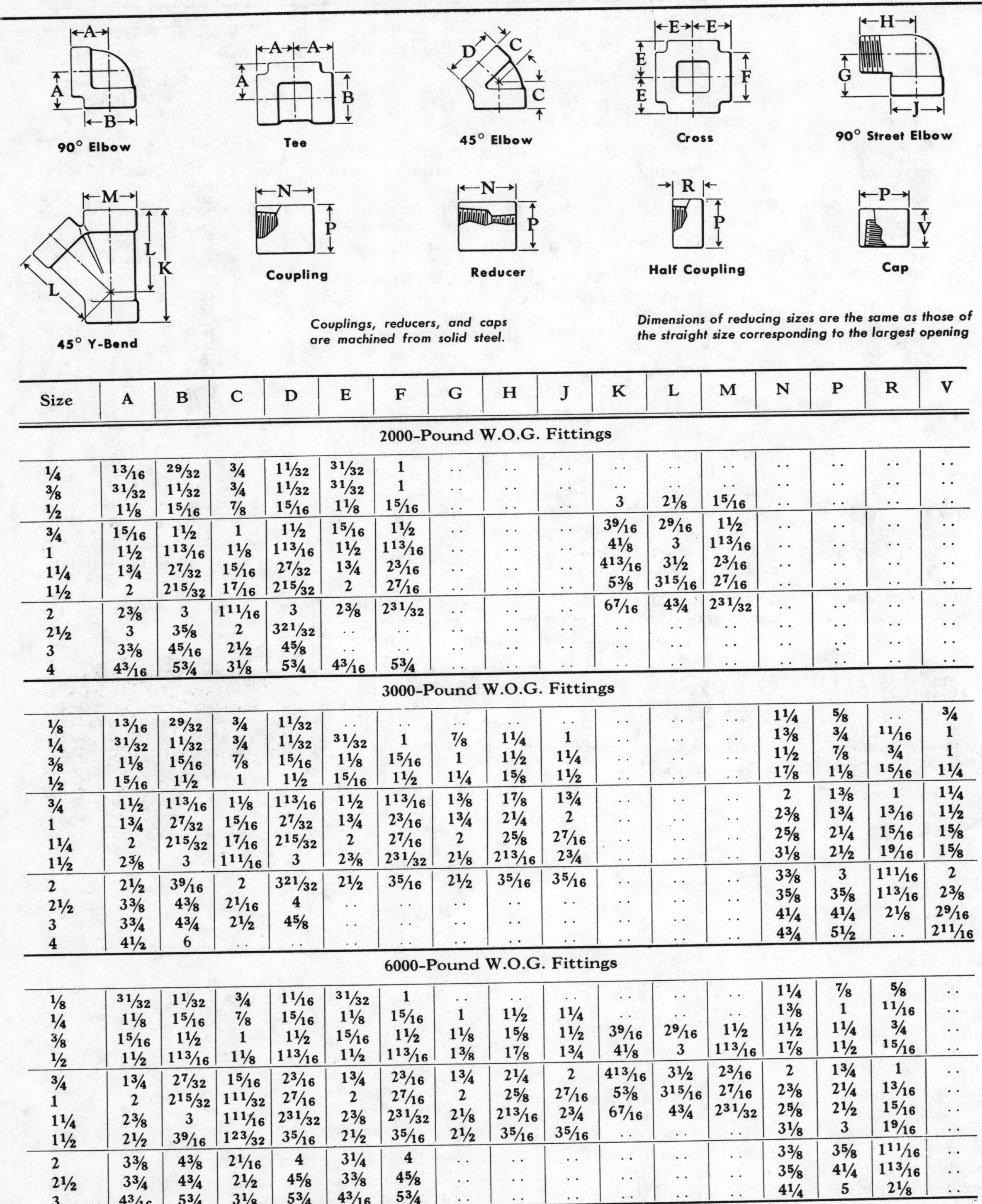

Size	A	B	C	D	E	F	G	H	J	K	L	M	N	P	R	V
2000-Pound W.O.G. Fittings																
1/4	13/16	29/32	3/4	1 1/32	31/32	1	..	..	..	..	..	..	..	..	..	..
3/8	31/32	1 1/32	3/4	1 1/32	31/32	1	..	..	..	..	..	..	..	..	..	..
1/2	1 1/8	1 5/16	7/8	1 5/16	1 1/8	1 5/16	..	..	..	3	2 1/8	1 5/16	..	..	..	..
3/4	1 5/16	1 1/2	1	1 1/2	1 5/16	1 1/2	..	..	..	3 9/16	2 9/16	1 1/2	..	..	..	..
1	1 1/2	1 13/16	1 1/8	1 13/16	1 1/2	1 13/16	..	..	..	4 1/8	3	1 13/16	..	..	..	..
1 1/4	1 3/4	2 7/32	1 5/16	2 7/32	1 3/4	2 3/16	..	..	..	4 13/16	3 1/2	2 3/16	..	..	..	..
1 1/2	2	2 15/32	1 7/16	2 15/32	2	2 7/16	..	..	..	5 3/8	3 15/16	2 7/16	..	..	..	..
2	2 3/8	3	1 11/16	3	2 3/8	2 31/32	..	..	..	6 7/16	4 3/4	2 31/32	..	..	..	..
2 1/2	3	3 5/8	2	3 21/32	..	..	..	..	..	..	..	..	..	..	..	..
3	3 3/8	4 5/16	2 1/2	4 5/8	..	..	..	..	..	..	..	..	..	..	..	..
4	4 3/16	5 3/4	3 1/8	5 3/4	4 3/16	5 3/4	..	..	..	..	..	..	..	..	..	..
3000-Pound W.O.G. Fittings																
1/8	13/16	29/32	3/4	1 1/32	..	..	..	..	..	..	..	..	1 1/4	5/8	..	3/4
1/4	31/32	1 1/32	3/4	1 1/32	31/32	1	7/8	1 1/4	1	..	..	..	1 3/8	3/4	11/16	1
3/8	1 1/8	1 5/16	7/8	1 5/16	1 1/8	1 5/16	1	1 1/2	1 1/4	..	..	..	1 1/2	7/8	3/4	1
1/2	1 5/16	1 1/2	1	1 1/2	1 5/16	1 1/2	1 1/4	1 5/8	1 1/2	..	..	..	1 7/8	1 1/8	15/16	1 1/4
3/4	1 1/2	1 13/16	1 1/8	1 13/16	1 1/2	1 13/16	1 3/8	1 7/8	1 3/4	..	..	..	2	1 3/8	1	1 1/4
1	1 3/4	2 7/32	1 5/16	2 7/32	1 3/4	2 3/16	1 3/4	2 1/4	2	..	..	..	2 3/8	1 3/4	1 3/16	1 1/2
1 1/4	2	2 15/32	1 7/16	2 15/32	2	2 7/16	2	2 5/8	2 7/16	..	..	..	2 5/8	2 1/4	1 5/16	1 5/8
1 1/2	2 3/8	3	1 11/16	3	2 3/8	2 31/32	2 1/8	2 13/16	2 3/4	..	..	..	3 1/8	2 1/2	1 9/16	1 5/8
2	2 1/2	3 9/16	2	3 21/32	2 1/2	3 5/16	2 1/2	3 5/16	3 5/16	..	..	..	3 3/8	3	1 11/16	2
2 1/2	3 3/8	4 3/8	2 1/16	4	..	..	..	..	..	..	..	..	3 5/8	3 5/8	1 13/16	2 3/8
3	3 3/4	4 3/4	2 1/2	4 5/8	..	..	..	..	..	..	..	..	4 1/4	4 1/4	2 1/8	2 9/16
4	4 1/2	6	..	..	..	..	..	..	..	..	..	..	4 3/4	5 1/2	..	2 11/16
6000-Pound W.O.G. Fittings																
1/8	31/32	1 1/32	3/4	1 1/16	31/32	1	..	..	..	..	..	..	1 1/4	7/8	5/8	..
1/4	1 1/8	1 5/16	7/8	1 5/16	1 1/8	1 5/16	1	1 1/2	1 1/4	..	..	..	1 3/8	1	11/16	..
3/8	1 5/16	1 1/2	1	1 1/2	1 5/16	1 1/2	1 1/8	1 5/8	1 1/2	3 9/16	2 9/16	1 1/2	1 1/2	1 1/4	3/4	..
1/2	1 1/2	1 13/16	1 1/8	1 13/16	1 1/2	1 13/16	1 3/8	1 7/8	1 3/4	4 1/8	3	1 13/16	1 7/8	1 1/2	15/16	..
3/4	1 3/4	2 7/32	1 5/16	2 3/16	1 3/4	2 3/16	1 3/4	2 1/4	2	4 13/16	3 1/2	2 3/16	2	1 3/4	1	..
1	2	2 15/32	1 11/32	2 7/16	2	2 7/16	2	2 5/8	2 7/16	5 3/8	3 15/16	2 7/16	2 3/8	2 1/4	1 3/16	..
1 1/4	2 3/8	3	1 11/16	2 31/32	2 3/8	2 31/32	2 1/8	2 13/16	2 3/4	6 7/16	4 3/4	2 31/32	2 5/8	2 1/2	1 5/16	..
1 1/2	2 1/2	3 9/16	1 23/32	3 5/16	2 1/2	3 5/16	2 1/2	3 5/16	3 5/16	..	..	..	3 1/8	3	1 9/16	..
2	3 3/8	4 3/8	2 1/16	4	3 1/4	4	..	..	..	..	..	..	3 3/8	3 5/8	1 11/16	..
2 1/2	3 3/4	4 3/4	2 1/2	4 5/8	3 3/8	4 5/8	..	..	..	..	..	..	3 5/8	4 1/4	1 13/16	..
3	4 3/16	5 3/4	3 1/8	5 3/4	4 3/16	5 3/4	..	..	..	..	..	..	4 1/4	5	2 1/8	..

Courtesy of Crane Co.

Table C-2. Steel Socket-Welding Reducer Inserts. Advantages of Steel Socket-Welding Fittings.

Reducer Insert

No. 1250
For use with Standard or Schedule 40 Pipe

No. 1390
For use with Extra Strong or Schedule 80 Pipe

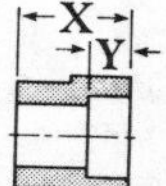

Type No. 1

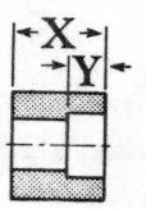

Type No. 2

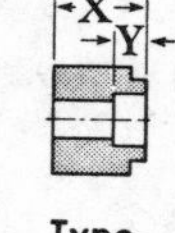

Type No. 3

Socket-Welding Reducer Inserts serve the same purpose as threaded bushings used with screwed fittings. The Inserts simplify the problem of making reductions in the sizes of pipe in socket-welded lines and avoid the delays and extra costs involved in procuring reducing fittings. By carrying an assorted stock of Reducer Inserts, the user can make up many combinations of reducing sizes in any type of fitting in the least amount of time and at low cost.

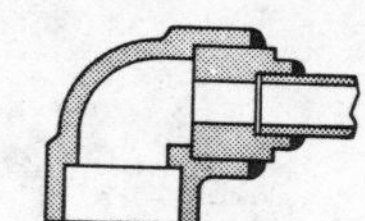

Socket-welding reducer insert applied to a socket - welding elbow.

The illustration at the right shows one of these fittings inserted in a straight size 90° elbow to make a reducing size. The Insert is placed in the end of the fitting and fillet welded; the smaller size pipe is then inserted in the smaller pipe end of the Reducer Insert and welded.

Reducer Inserts are available in 1, 1¼, 1½, and 2-inch sizes, the No. 1250 being made for use with Standard or Schedule 40 Pipe and the No. 1390 being made for use with Extra Strong or Schedule 80 Pipe. The fittings are made in three types, depending on the size, as illustrated above. They are made of carbon steel conforming to requirements of ASTM Specification A105, Grade II. Prices are furnished on request.

Weights and Dimensions

Size	Pounds, Each Carbon Steel		Dimensions, in Inches		
Inches	No. 1250	No. 1390	Type	X	Y
1 × ¾	.40	.44	No. 1	1½	9/16
× ½	.36	.44	No. 2	15/16	½
1¼ × 1	.6	.7	No. 1	1 11/16	⅝
× ¾	.6	.7	No. 2	1⅜	9/16
× ½	.6	.9	No. 3	1⅜	½
1½ × 1¼	.6	.9	No. 1	1⅞	11/16
× 1	.6	.9	No. 2	1 7/16	⅝
× ¾	.9	.9	No. 3	1 7/16	9/16
× ½	.9	.9	No. 3	1 7/16	½
2 × 1½	.9	1.2	No. 2	1 9/16	¾
× 1¼	1.1	1.2	No. 3	1 9/16	11/16
× 1	1.2	1.0	No. 3	1 9/16	⅝
× ¾	1.2	1.0	No. 3	1 9/16	9/16
× ½	1.2	1.0	No. 3	1 9/16	½

On installations where the pipe need not butt against the shoulder, the socket compensates for inaccuracies in measuring and cutting the pipe.

On installations where the pipe must butt against the shoulder, free, uninterrupted flow is assured. Fitting and pipe have the same inside diameter.

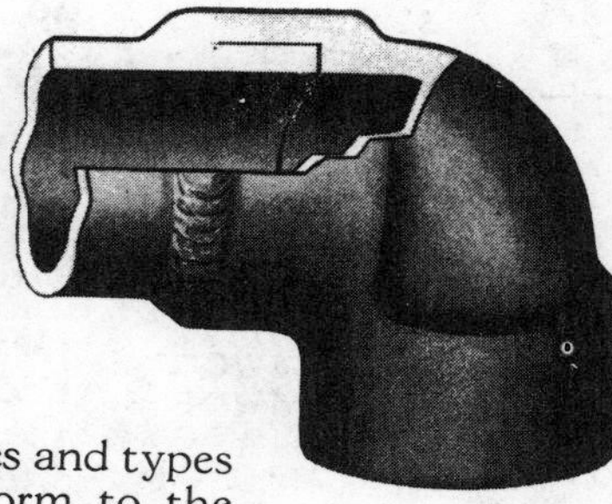

The socket type weld has advantages over the butt-weld that recommend it strongly for fittings for small size pipe lines. The pipe used with it does not require beveling.

Since the pipe end slips into, and is supported by the socket, the joint is self-aligning; tack welding and special clamps to line up and hold the joint are unnecessary. Because the weld metal is deposited on the outer surface of the pipe, there is no danger of forming welding icicles which would clog the line and restrict flow.

These fittings have the same inside diameter as Standard or Schedule 40, Extra Strong or Schedule 80, Schedule 160, or Double Extra Strong Pipe, depending upon the weight of fitting ordered. On installations where the pipe is butted against the shoulder at the back of the socket, the fittings permit free, uninterrupted flow of the fluid in the line.

The fittings, in the sizes and types covered therein, conform to the American Standard, ASA B16.11–1946, as explained on page 294. The bore diameter of the socket is the same for all fittings regardless of the pressure class.

Size of fitting	Bore diameter of socket (Minimum)
⅛"	0.420"
¼	0.555
⅜	0.690
½	0.855
¾	1.065
1	1.330
1¼	1.675
1½	1.915
2	2.406
2½	2.906
3	3.535
4	4.545

Forged Steel Socket-Welding Fittings have ample metal sections throughout; the long, low band forming the socket wall extends beyond the shoulder at the back of the bore, leaving no weak corner. Exhaustive tests show conclusively that the fittings are as strong as, or stronger than, pipe.

Being only slightly larger than pipe, the fittings, when welded in place, make an exceptionally neat appearing, workmanlike installation.

Courtesy of Crane Co.

Table C-3. Forged Steel Socket-Welding Fittings. 2,000, 3,000, 4,000, and 6,000-Pound W.O.G.

90° Elbow
No. 1240, 2000-Pound WOG
No. 1380, 3000-Pound WOG
No. 1460, 4000-Pound WOG
No. 1660, 6000-Pound WOG

Tee
No. 1241, 2000-Pound WOG
No. 1381, 3000-Pound WOG
No. 1461, 4000-Pound WOG
No. 1661, 6000-Pound WOG

45° Elbow
No. 1242, 2000-Pound WOG
No. 1382, 3000-Pound WOG
No. 1462, 4000-Pound WOG
No. 1662, 6000-Pound WOG

Cross
No. 1243, 2000-Pound WOG
No. 1383, 3000-Pound WOG
No. 1463, 4000-Pound WOG
No. 1663, 6000-Pound WOG

2000-Pound WOG Fittings are for use with Schedule 40 or Standard pipe
3000-Pound WOG Fittings are for use with Schedule 80 or Extra Strong pipe
4000-Pound WOG Fittings are for use with Schedule 160 pipe
6000-Pound WOG Fittings are for use with Double Extra Strong pipe

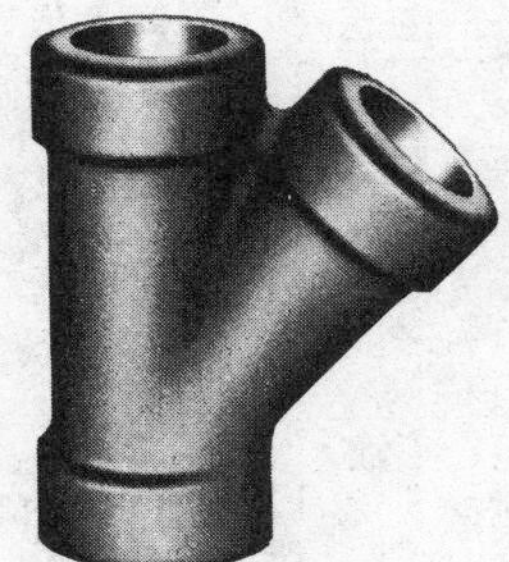

45° Y-Bend
No. 1245, 2000-Pound WOG
No. 1385, 3000-Pound WOG
No. 1465, 4000-Pound WOG
No. 1665, 6000-Pound WOG

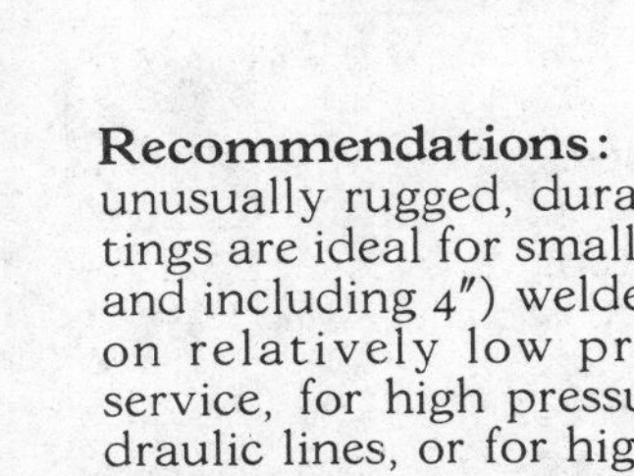

Coupling
No. 1246, 2000-Pound WOG
No. 1386, 3000-Pound WOG
No. 1466, 4000-Pound WOG
No. 1666, 6000-Pound WOG

Reducer
No. 1247, 2000-Pound WOG
No. 1387, 3000-Pound WOG
No. 1467, 4000-Pound WOG
No. 1667, 6000-Pound WOG

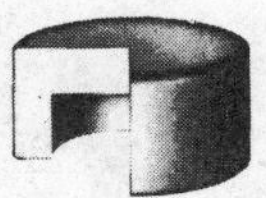

Cap
No. 1249, 2000-Pound WOG
No. 1389, 3000-Pound WOG
No. 1469, 4000-Pound WOG
No. 1669, 6000-Pound WOG

Recommendations: These unusually rugged, durable fittings are ideal for small (up to and including 4″) welded lines on relatively low pressure service, for high pressure hydraulic lines, or for high pressure-temperature service.

The 2000-Pound WOG Fittings are for use with Schedule 40 or Standard pipe . . . the 3000-Pound, with Schedule 80 or Extra Strong pipe . . . the 4000-Pound, with Schedule 160 pipe . . . and the 6000-Pound, with Double Extra Strong pipe.

Design: Elbows, tees, crosses, and Y-bends are forged solid; their openings are reinforced with a wide band which completely surrounds the socket chamber, extends well beyond the back of the socket, and meets recognized requirements for socket-weld dimensions. Reducer inserts (see page 296), couplings, reducers, and caps are machined from solid steel. Openings of all fittings are drilled and the ends are bored to slip over pipe.

Materials: The fittings are made from high grade carbon steel (ASTM A 105, Grade II) of unusual strength and toughness. It is particularly suitable for fusion welding.

American Standard: These fittings conform to the American Standard for Steel Socket-Welding Fittings (B16.11-1946). This Standard includes elbows, tees, crosses, and couplings in sizes 3-inch and smaller for use with Schedule 40, Schedule 80, and Schedule 160 pipe.

Working Pressures*
Steam, Water, Oil, Oil Vapor, Gas, or Air

Material	Temp. Deg. Fahr.	Pounds, Non-Shock: 2000 Pound WOG	3000 Pound WOG	4000 Pound WOG	6000 Pound WOG
Carbon Steel ASTM A 105 Grade II	100°	2000	3000	4000	6000
	150	1970	2955	3940	5915
	200	1940	2915	3885	5830
	250	1915	2875	3830	5750
	300	1895	2845	3790	5690
	350	1875	2810	3750	5625
	400	1850	2775	3700	5550
	450	1810	2715	3620	5430
	500	1735	2605	3470	5210
	550	1640	2460	3280	4925
	600	1540	2310	3080	4620
	650	1430	2150	2865	4300
	700	1305	1960	2610	3920
	750	1180	1775	2365	3550
	800	1015	1525	2030	3050
	850	830	1250	1665	2500
	875	725	1090	1450	2180
	900	615	925	1235	1855
	925	520	785	1045	1570
	950	425	640	855	1285
	975	330	500	665	1000
	1000	235	355	475	715

****Note:*** When pipe is rated in accordance with the Code for Pressure Piping or any other Code, these fittings may be used for the same pressures and temperatures as the pipe even though such ratings exceed those in the table above.

The fittings, of course, must be made of a material having chemical and physical properties comparable to the pipe, and must be of suitable weight, as indicated by the schedule numbers.

(table continued on next page)

Table C-3 continued

Forged Steel Socket-Welding Fittings. Dimensions in Inches.

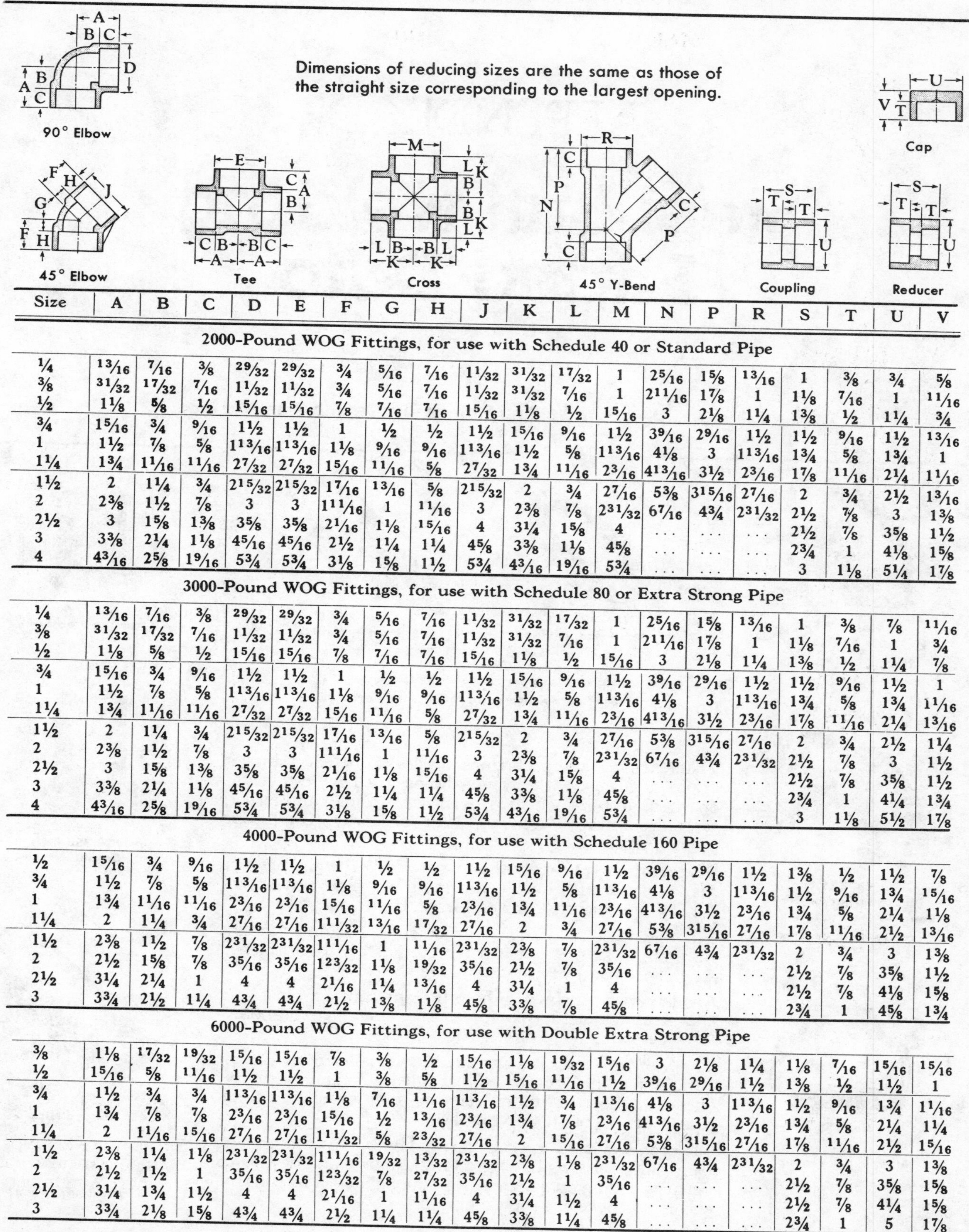

Size	A	B	C	D	E	F	G	H	J	K	L	M	N	P	R	S	T	U	V
2000-Pound WOG Fittings, for use with Schedule 40 or Standard Pipe																			
1/4	13/16	7/16	3/8	29/32	29/32	3/4	5/16	7/16	1 1/32	31/32	17/32	1	2 5/16	1 5/8	1 3/16	1	3/8	3/4	5/8
3/8	31/32	17/32	7/16	1 1/32	1 1/32	3/4	5/16	7/16	1 1/32	31/32	7/16	1	2 11/16	1 7/8	1	1 1/8	7/16	1	11/16
1/2	1 1/8	5/8	1/2	1 5/16	1 5/16	7/8	7/16	7/16	1 5/16	1 1/8	1/2	1 5/16	3	2 1/8	1 1/4	1 3/8	1/2	1 1/4	3/4
3/4	1 5/16	3/4	9/16	1 1/2	1 1/2	1	1/2	1/2	1 1/2	1 5/16	9/16	1 1/2	3 9/16	2 9/16	1 1/2	1 1/2	9/16	1 1/2	13/16
1	1 1/2	7/8	5/8	1 13/16	1 13/16	1 1/8	9/16	9/16	1 13/16	1 1/2	5/8	1 13/16	4 1/8	3	1 13/16	1 3/4	5/8	1 3/4	1
1 1/4	1 3/4	1 1/16	11/16	2 7/32	2 7/32	1 5/16	11/16	5/8	2 7/32	1 3/4	11/16	2 3/16	4 13/16	3 1/2	2 3/16	1 7/8	11/16	2 1/4	1 1/16
1 1/2	2	1 1/4	3/4	2 15/32	2 15/32	1 7/16	13/16	5/8	2 15/32	2	3/4	2 7/16	5 3/8	3 15/16	2 7/16	2	3/4	2 1/2	1 3/16
2	2 3/8	1 1/2	7/8	3	3	1 11/16	1	11/16	3	2 3/8	7/8	2 31/32	6 7/16	4 3/4	2 31/32	2 1/2	7/8	3	1 3/8
2 1/2	3	1 5/8	1 3/8	3 5/8	3 5/8	2 1/16	1 1/8	15/16	4	3 1/4	1 5/8	4	...	...	...	2 1/2	7/8	3 5/8	1 1/2
3	3 3/8	2 1/4	1 1/8	4 5/16	4 5/16	2 1/2	1 1/4	1 1/4	4 5/8	3 3/8	1 1/8	4 5/8	...	...	...	2 3/4	1	4 1/8	1 5/8
4	4 3/16	2 5/8	1 9/16	5 3/4	5 3/4	3 1/8	1 5/8	1 1/2	5 3/4	4 3/16	1 9/16	5 3/4	...	...	...	3	1 1/8	5 1/4	1 7/8
3000-Pound WOG Fittings, for use with Schedule 80 or Extra Strong Pipe																			
1/4	13/16	7/16	3/8	29/32	29/32	3/4	5/16	7/16	1 1/32	31/32	17/32	1	2 5/16	1 5/8	1 3/16	1	3/8	7/8	11/16
3/8	31/32	17/32	7/16	1 1/32	1 1/32	3/4	5/16	7/16	1 1/32	31/32	7/16	1	2 11/16	1 7/8	1	1 1/8	7/16	1	3/4
1/2	1 1/8	5/8	1/2	1 5/16	1 5/16	7/8	7/16	7/16	1 5/16	1 1/8	1/2	1 5/16	3	2 1/8	1 1/4	1 3/8	1/2	1 1/4	7/8
3/4	1 5/16	3/4	9/16	1 1/2	1 1/2	1	1/2	1/2	1 1/2	1 5/16	9/16	1 1/2	3 9/16	2 9/16	1 1/2	1 1/2	9/16	1 1/2	1
1	1 1/2	7/8	5/8	1 13/16	1 13/16	1 1/8	9/16	9/16	1 13/16	1 1/2	5/8	1 13/16	4 1/8	3	1 13/16	1 3/4	5/8	1 3/4	1 1/16
1 1/4	1 3/4	1 1/16	11/16	2 7/32	2 7/32	1 5/16	11/16	5/8	2 7/32	1 3/4	11/16	2 3/16	4 13/16	3 1/2	2 3/16	1 7/8	11/16	2 1/4	1 3/16
1 1/2	2	1 1/4	3/4	2 15/32	2 15/32	1 7/16	13/16	5/8	2 15/32	2	3/4	2 7/16	5 3/8	3 15/16	2 7/16	2	3/4	2 1/2	1 1/4
2	2 3/8	1 1/2	7/8	3	3	1 11/16	1	11/16	3	2 3/8	7/8	2 31/32	6 7/16	4 3/4	2 31/32	2 1/2	7/8	3	1 1/2
2 1/2	3	1 5/8	1 3/8	3 5/8	3 5/8	2 1/16	1 1/8	15/16	4	3 1/4	1 5/8	4	...	...	...	2 1/2	7/8	3 5/8	1 1/2
3	3 3/8	2 1/4	1 1/8	4 5/16	4 5/16	2 1/2	1 1/4	1 1/4	4 5/8	3 3/8	1 1/8	4 5/8	...	...	...	2 3/4	1	4 1/4	1 3/4
4	4 3/16	2 5/8	1 9/16	5 3/4	5 3/4	3 1/8	1 5/8	1 1/2	5 3/4	4 3/16	1 9/16	5 3/4	...	...	...	3	1 1/8	5 1/2	1 7/8
4000-Pound WOG Fittings, for use with Schedule 160 Pipe																			
1/2	1 5/16	3/4	9/16	1 1/2	1 1/2	1	1/2	1/2	1 1/2	1 5/16	9/16	1 1/2	3 9/16	2 9/16	1 1/2	1 3/8	1/2	1 1/2	7/8
3/4	1 1/2	7/8	5/8	1 13/16	1 13/16	1 1/8	9/16	9/16	1 13/16	1 1/2	5/8	1 13/16	4 1/8	3	1 13/16	1 1/2	9/16	1 3/4	15/16
1	1 3/4	1 1/16	11/16	2 3/16	2 3/16	1 5/16	11/16	5/8	2 3/16	1 3/4	11/16	2 3/16	4 13/16	3 1/2	2 3/16	1 3/4	5/8	2 1/4	1 1/8
1 1/4	2	1 1/4	3/4	2 7/16	2 7/16	1 11/32	13/16	17/32	2 7/16	2	3/4	2 7/16	5 3/8	3 15/16	2 7/16	1 7/8	11/16	2 1/2	1 3/16
1 1/2	2 3/8	1 1/2	7/8	2 31/32	2 31/32	1 11/16	1	11/16	2 31/32	2 3/8	7/8	2 31/32	6 7/16	4 3/4	2 31/32	2	3/4	3	1 3/8
2	2 1/2	1 5/8	7/8	3 5/16	3 5/16	1 23/32	1 1/8	19/32	3 5/16	2 1/2	7/8	3 5/16	...	...	...	2 1/2	7/8	3 5/8	1 1/2
2 1/2	3 1/4	2 1/4	1	4	4	2 1/16	1 1/4	13/16	4	3 1/4	1	4	...	...	...	2 1/2	7/8	4 1/8	1 5/8
3	3 3/4	2 1/2	1 1/4	4 3/4	4 3/4	2 1/2	1 3/8	1 1/8	4 5/8	3 3/8	7/8	4 5/8	...	...	...	2 3/4	1	4 5/8	1 3/4
6000-Pound WOG Fittings, for use with Double Extra Strong Pipe																			
3/8	1 1/8	17/32	19/32	1 5/16	1 5/16	7/8	3/8	1/2	1 5/16	1 1/8	19/32	1 5/16	3	2 1/8	1 1/4	1 1/8	7/16	1 5/16	15/16
1/2	1 5/16	5/8	11/16	1 1/2	1 1/2	1	3/8	5/8	1 1/2	1 5/16	11/16	1 1/2	3 9/16	2 9/16	1 1/2	1 3/8	1/2	1 1/2	1
3/4	1 1/2	3/4	3/4	1 13/16	1 13/16	1 1/8	7/16	11/16	1 13/16	1 1/2	3/4	1 13/16	4 1/8	3	1 13/16	1 1/2	9/16	1 3/4	1 1/16
1	1 3/4	7/8	7/8	2 3/16	2 3/16	1 5/16	1/2	13/16	2 3/16	1 3/4	7/8	2 3/16	4 13/16	3 1/2	2 3/16	1 3/4	5/8	2 1/4	1 1/4
1 1/4	2	1 1/16	15/16	2 7/16	2 7/16	1 11/32	5/8	23/32	2 7/16	2	15/16	2 7/16	5 3/8	3 15/16	2 7/16	1 7/8	11/16	2 1/2	1 5/16
1 1/2	2 3/8	1 1/4	1 1/8	2 31/32	2 31/32	1 11/16	19/32	1 3/32	2 31/32	2 3/8	1 1/8	2 31/32	6 7/16	4 3/4	2 31/32	2	3/4	3	1 3/8
2	2 1/2	1 1/2	1	3 5/16	3 5/16	1 23/32	7/8	27/32	3 5/16	2 1/2	1	3 5/16	...	...	...	2 1/2	7/8	3 5/8	1 5/8
2 1/2	3 1/4	1 3/4	1 1/2	4	4	2 1/16	1	1 1/16	4	3 1/4	1 1/2	4	...	...	...	2 1/2	7/8	4 1/4	1 5/8
3	3 3/4	2 1/8	1 5/8	4 3/4	4 3/4	2 1/2	1 1/4	1 1/4	4 5/8	3 3/8	1 1/4	4 5/8	...	...	...	2 3/4	1	5	1 7/8

Courtesy of Crane Co.

APPENDIX D

Steel Butt-welding Fitting Data

Table D-1. Steel Butt-Welding Fittings for Use with Standard Pipe.

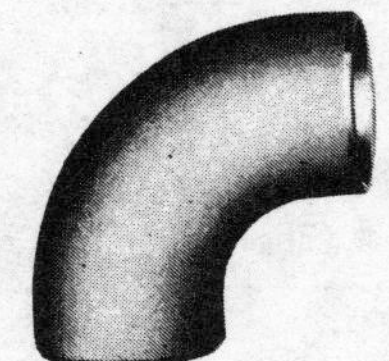

No. 352 E
90° Long Radius Elbow
Straight and Reducing

No. 331 E
90° Short Radius Elbow

No. 354 E
45° Long Radius Elbow

No. 574
Cranelap Stub End

For sizes and weights, see the facing page.
For pressure-temperature ratings, see page 305.

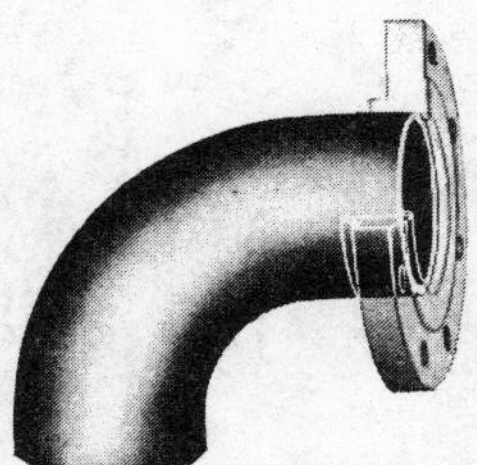

No. 335 E
90° Long Radius Elbow
Long Tangent on One End
(flange is not included)

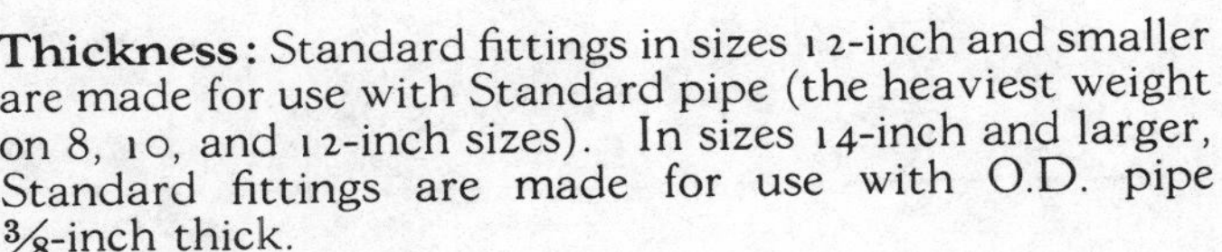

Thickness: Standard fittings in sizes 12-inch and smaller are made for use with Standard pipe (the heaviest weight on 8, 10, and 12-inch sizes). In sizes 14-inch and larger, Standard fittings are made for use with O.D. pipe ⅜-inch thick.

Materials: Unless otherwise specified, the fittings are made of carbon steel conforming to requirements of ASTM Specification A 234, Grade B.

Fittings made of Grade A carbon steel, genuine wrought iron, stainless steel, or other materials can be furnished when specified; information on request.

American Standard: These fittings conform, in types and sizes included therein, to the American Standard for Steel Butt-Welding Fittings, B16.9-1958.

The Standard does not include sizes smaller than 1-inch, nor does it include 90° elbows with a long tangent on one end, short radius 90° elbows, crosses, short radius return bends, or shaped nipples.

90° elbow with long tangent: The No. 335 E are 90° long radius elbows having a long tangent on one end to permit welding on a slip-on welding flange. The tangent end is not beveled; the other end is beveled.

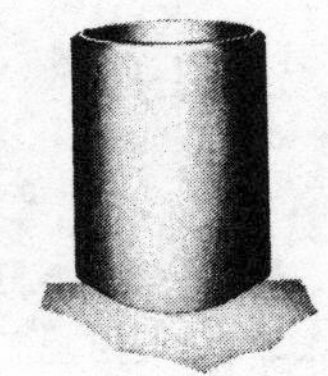

No. 350 E
90° Shaped Nipple

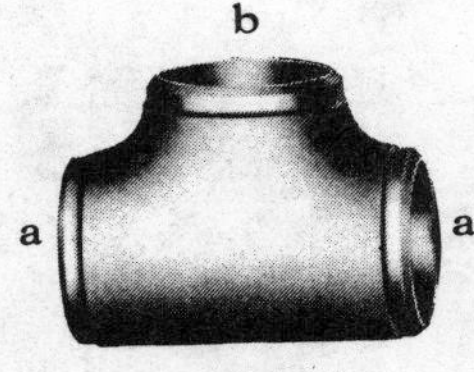

No. 353 E
Tee
Straight and Reducing

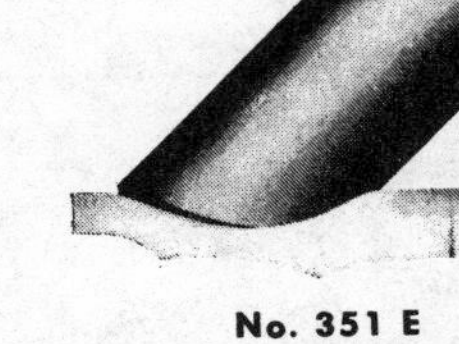

No. 351 E
45° Shaped Nipple

(table continued on next page)

Table D-1 continued

Steel Butt-Welding Fittings for Use with Standard Pipe.

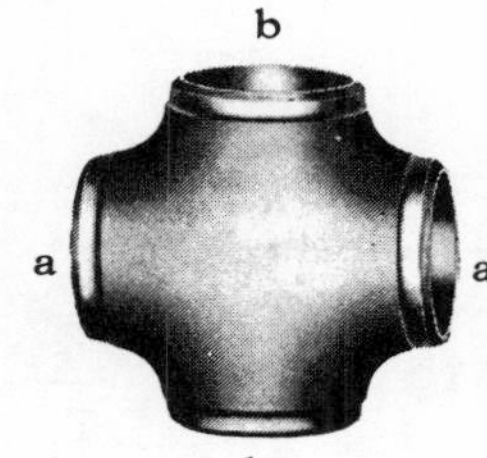

No. 336 E
Cross
Straight and Reducing

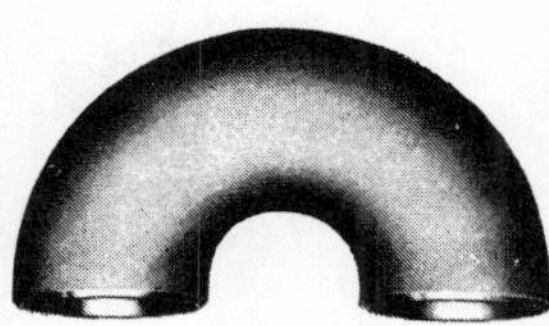

Return Bend
No. 372 E, Short Radius
No. 373 E, Long Radius

Courtesy of Crane Co.

Cranelap stub ends: Cranelap stub ends, made of Grade B seamless steel pipe lapped to the full thickness of the pipe wall, and Cranelap flanges (see page 308) afford an ideal method of installing flanged equipment in a welded line. The swivel flange eliminates the difficulty of aligning bolt holes and permits installing the equipment at any angle.

Shaped nipples: Shaped nipples eliminate the use of templates when saddling one pipe upon another; they save erection time and assure an accurate fit. Both ends are beveled for welding. When ordering, be sure to specify both the pipe size and the nominal size of the header on which the nipple will be used; header sizes which the nipples are shaped to fit are included in the upper table on the facing page.

Prices: Prices are furnished on request.

Ordering reducing tees and crosses: When ordering reducing tees and crosses, specify the size of openings in the sequence of the lower case letters (a and b) shown on their illustrations at the left.

No. 357 E
Cap

Concentric Reducer

Eccentric Reducer

Table D-1 continued

Steel Butt-Welding Fittings. Dimensions in Inches.

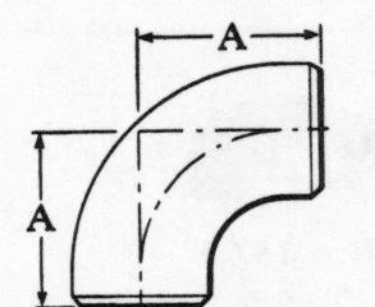

90° Long Radius Elbow
Straight or Reducing

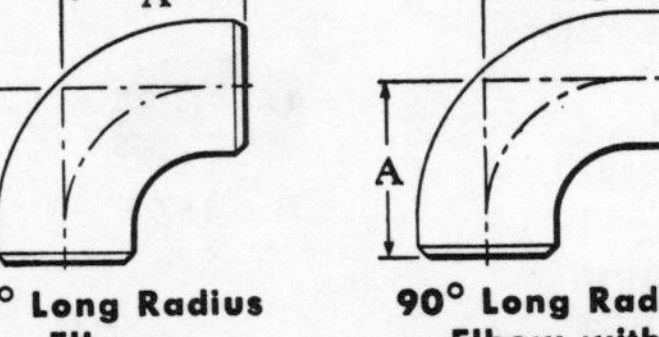

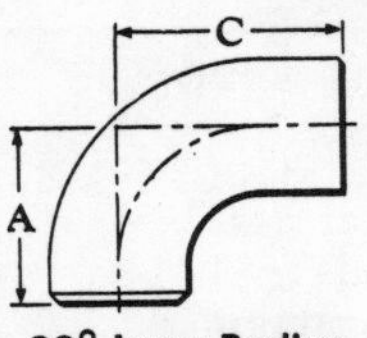

90° Long Radius Elbow with Long Tangent on One End

90° Short Radius Elbow

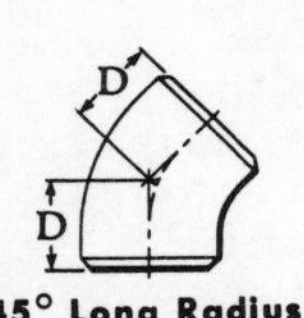

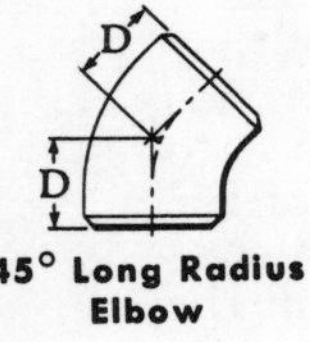

45° Long Radius Elbow

Straight Tee

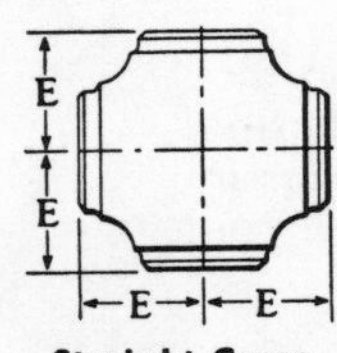

Straight Cross

Standard, Extra Strong, Schedule 160, and Double Extra Strong Fittings have the same outside dimensions.

American Standard: These fittings conform, in sizes and types included therein, to the American Standard, B16.9-1958; see page 293.

Thickness: Standard Fittings 12-inch and smaller are made for use with Standard pipe (heaviest weight on 8, 10, and 12-inch sizes); sizes 14-inch and larger are made for use with O.D. pipe ⅜-inch thick.

Extra Strong Fittings 12-inch and smaller are made for use with Extra Strong pipe; larger sizes are made for use with O.D. pipe ½-inch thick.

Schedule 160 Fittings are made for use with Schedule 160 pipe.

Double Extra Strong Fittings are made for use with Double Extra Strong pipe.

Cap

Short Radius Return Bend

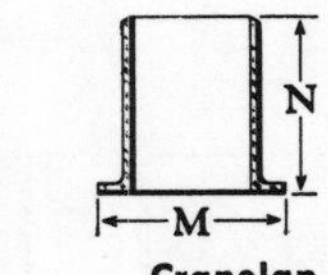

Cranelap Stub End

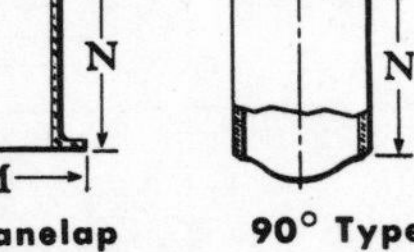

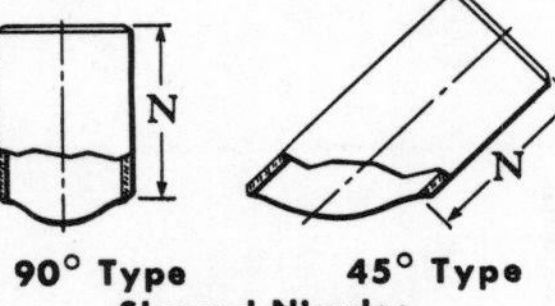

90° Type 45° Type
Shaped Nipples

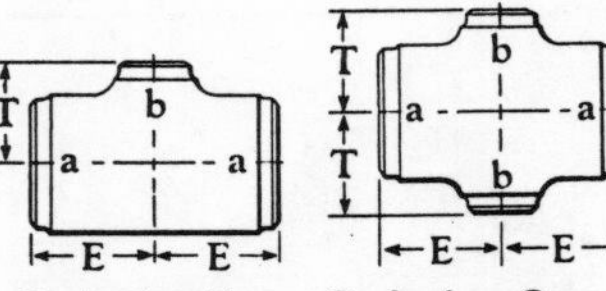

Reducing Tee **Reducing Cross**

Dimension "T" is shown in table below; refer to large table for dimension "E".

Size a × b	T
1 ×¾ *	1½
1¼×1 *	1⅞
×¾ *	1⅞
1½×1¼*	2¼
×1 *	2¼
×¾ *	2¼
2 ×1½	2⅜
×1¼	2¼
×1	2
×¾	1¾
2½×2	2¾
×1½	2⅝
×1¼	2½
×1	2¼
3 ×2½	3¼
×2	3
×1½	2⅞
×1¼	2¾
3½×3	3⅝
×2½	3½
×2	3¼
×1½	3⅛
4 ×3½	4
×3	3⅞
×2½	3¾
×2	3½
×1½	3⅜
5 ×4	4⅝
×3½	4½
×3	4⅜
×2½	4¼
×2	4⅛
6 ×5	5⅜
×4	5⅛

Size a × b	T
6×3½	5
×3	4⅞
×2½	4¾
8×6	6⅝
×5	6⅜
×4	6⅛
×3½	6
10×8	8
×6	7⅝
×5	7½
×4	7¼
12×10	9½
×8	9
×6	8⅝
×5	8½
14×12*	10⅝
×10*	10⅛
×8 *	9⅝
×6 *	9¼
16×14*	12
×12*	11⅝
×10*	11⅛
×8 *	10⅝
18×16*	13
×14*	13
×12*	12⅝
×10*	12⅛
20×18*	14½
×16*	14
×14*	14
×12*	13⅝
24×20*	17
×18*	16½
×16*	16

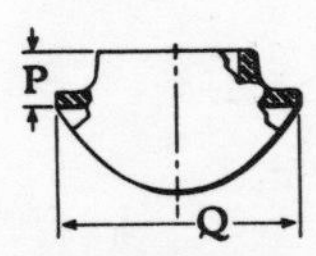

Reinforcing Welding Saddle

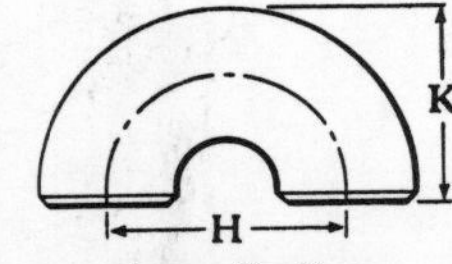

Long Radius Return Bend

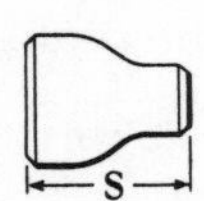

Concentric Reducer

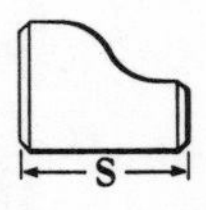

Eccentric Reducer

Size	A	B	C	D	E	F	G	H	J	K	M	N	P	Q	S	Pipe Schedule Numbers for: Std. Ftgs.	Extra Strong
½	1½	..	..	⅝	..	..	..	3	..	1¹⁵⁄₁₆	..	..	..	..	..	40	80
¾	1⅛	..	..	⁷⁄₁₆	1⅛*	..	..	2¼	..	1¹¹⁄₁₆	1¹¹⁄₁₆	4	..	..	2	40	80
1	1½	1	..	⅞	1½*	1½	2	3	1⅝	2³⁄₁₆	2	4	1	4⅛	2	40	80
1¼	1⅞	1¼	..	1	1⅞*	1½	2½	3¾	2¹⁄₁₆	2¾	2½	4	1⅛	4¾	2	40	80
1½	2¼	1½	3¼	1⅛	2¼*	1½	3	4½	2⁷⁄₁₆	3¼	2⅞	4	1⅛	5¼	2½	40	80
2	3	2	4¼	1⅜	2½	1½	4	6	3³⁄₁₆	4³⁄₁₆	3⅝	6	1⅛	5¹⁵⁄₁₆	3	40	80
2½	3¾	2½	5	1¾	3	1½	5	7½	3¹⁵⁄₁₆	5³⁄₁₆	4⅛	6	1⅛	6⁷⁄₁₆	3½	40	80
3	4½	3	5¾	2	3⅜	2	6	9	4¾	6¼	5	6	1⅛	7⁹⁄₁₆	3½	40	80
3½	5¼	3½	6¾	2¼	3¾	2½	..	..	..	..	5½	6	1¼	8¹⁄₁₆	4	40	80
4	6	4	7½	2½	4⅛	2½	8	12	6¼	8¼	6³⁄₁₆	6	1⅜	8⁹⁄₁₆	4	40	80
5	7½	5	9	3⅛	4⅞	3	10	15	7¾	10⁵⁄₁₆	7⁵⁄₁₆	8	1⅜	9⅝	5	40	80
6	9	6	10¾	3¾	5⅝	3½	12	18	9⁵⁄₁₆	12⁵⁄₁₆	8½	8	1½	11¾	5½	40	80
8	12	8	13¾	5	7	4	16	24	12⁵⁄₁₆	16⁵⁄₁₆	10⅝	8	2	14¾	6	40	80
10	15	10	17	6¼	8½	5	20	30	15⅜	20⅜	12¾	10	2½	17⅞	7	40	60
12	18	12	20½	7½	10	6	24	36	18⅜	24⅜	15	10	2¾	20⅞	8	..	..
14	21	14	..	8¾	11*	6½	28	42	21	28	16¼	12	3¼	22⅛	13	30	..
16	24	16	..	10	12*	7	32	48	24	32	18½	12	3½	24⅛	14	30	40
18	27	18	..	11¼	13½*	8	36	54	27	36	21	12	4	26⅛	15	..	..
20	30	20	..	12½	15*	9	40	60	30	40	23	12	4	30⅛	20	20	30
24	36	24	..	15	17*	10½	48	72	36	48	27¼	12	4	34⅛	20	20	..

Courtesy of Crane Co.

APPENDIX E
Valve Data

**Table E-1. Bronze Gate Valves.
Names of Parts.**

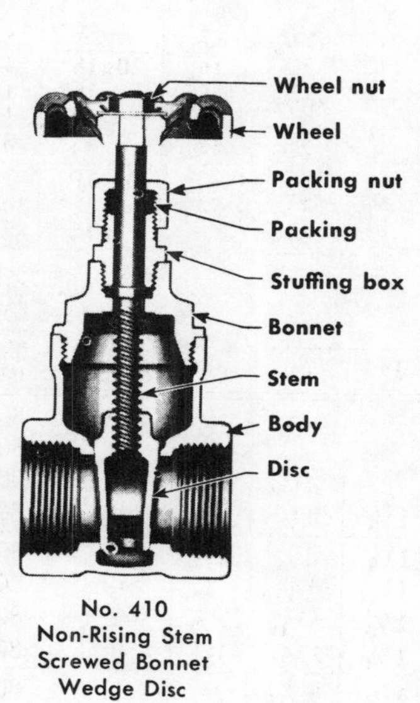

No. 410
Non-Rising Stem
Screwed Bonnet
Wedge Disc

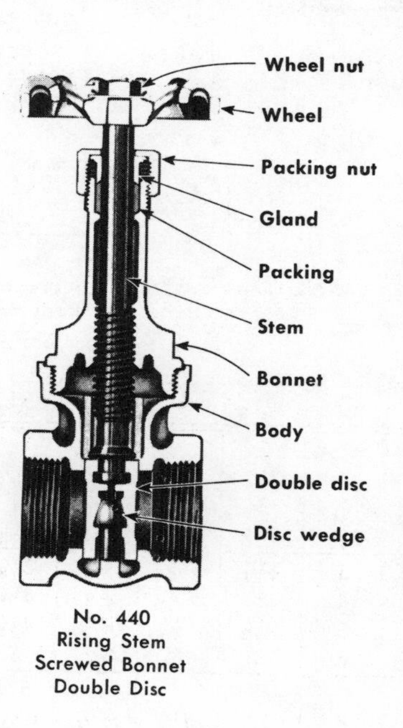

No. 440
Rising Stem
Screwed Bonnet
Double Disc

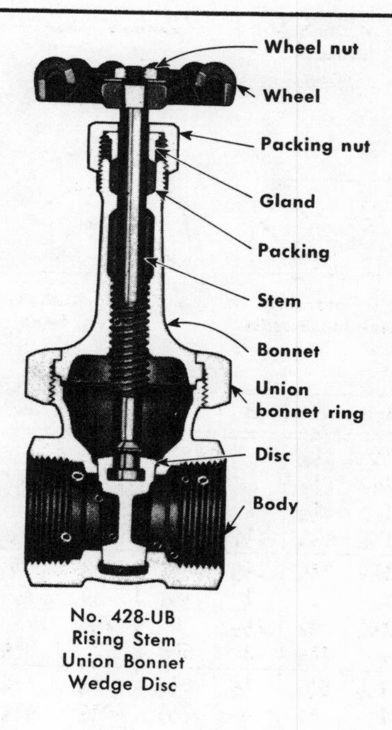

No. 428-UB
Rising Stem
Union Bonnet
Wedge Disc

(table continued on next page)

Table E-1 continued

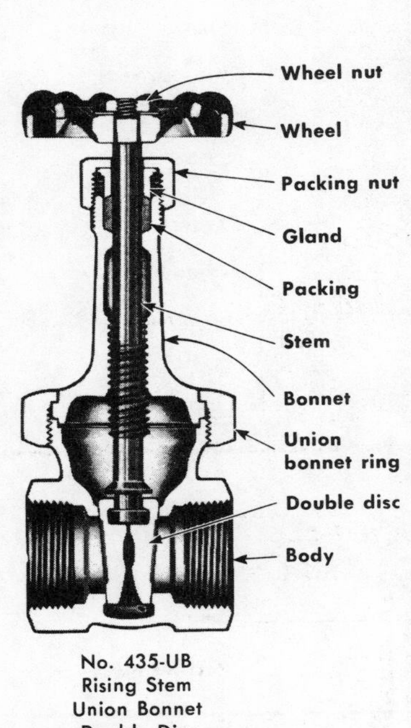

No. 435-UB
Rising Stem
Union Bonnet
Double Disc

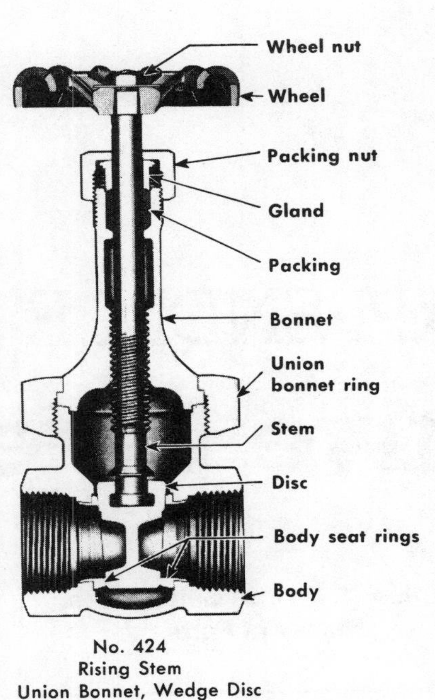

No. 424
Rising Stem
Union Bonnet, Wedge Disc
Expanded Seats

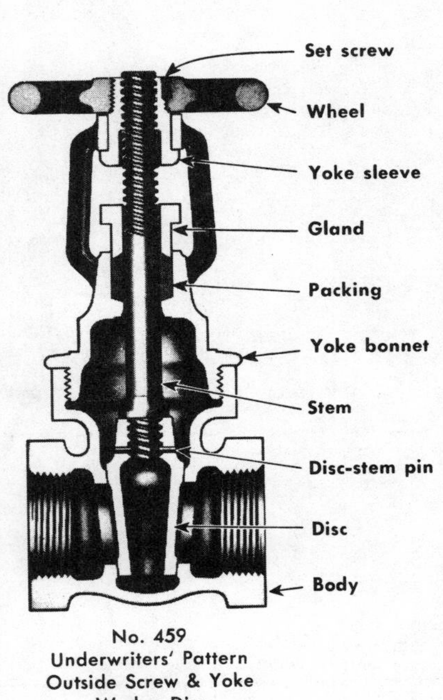

No. 459
Underwriters' Pattern
Outside Screw & Yoke
Wedge Disc

Courtesy of Crane Co.

Table E-2. Bronze Globe and Angle Valves. Names of Parts.

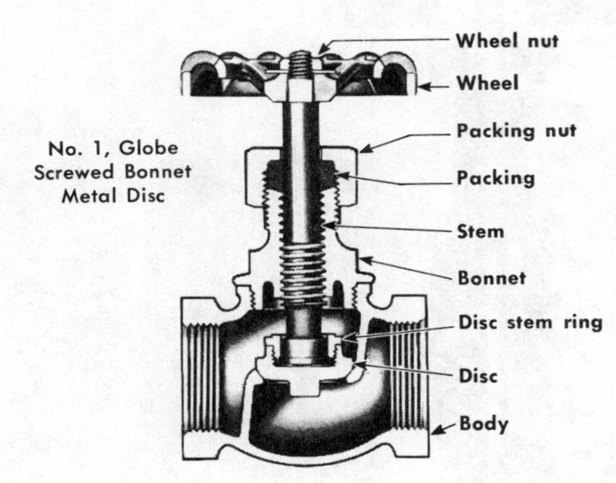

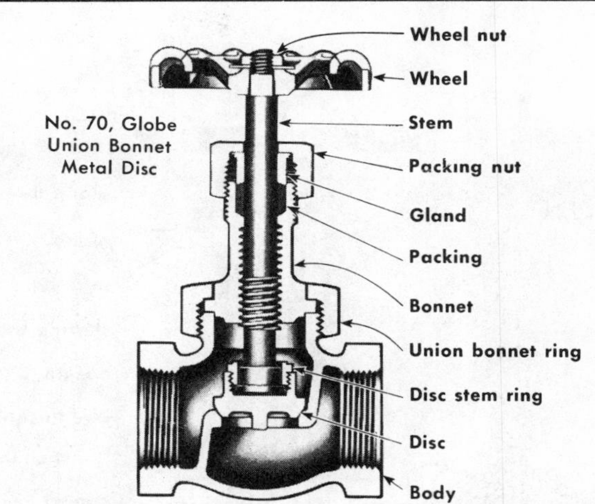

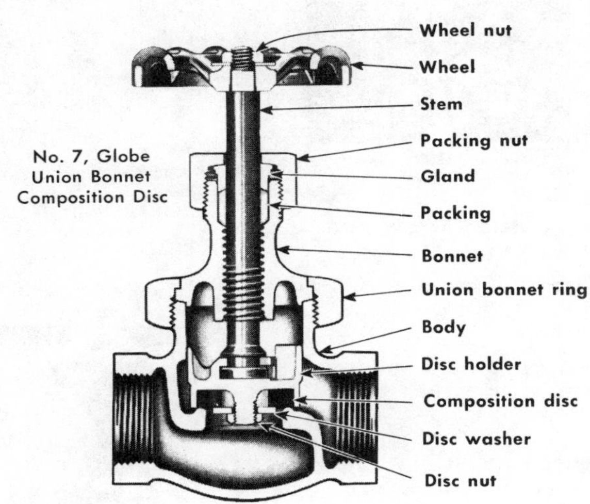

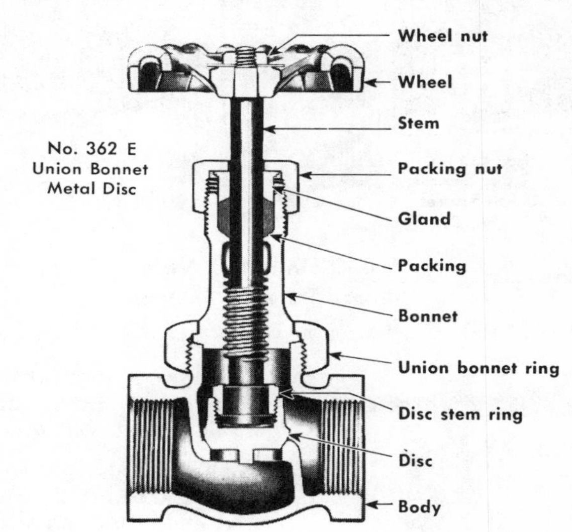

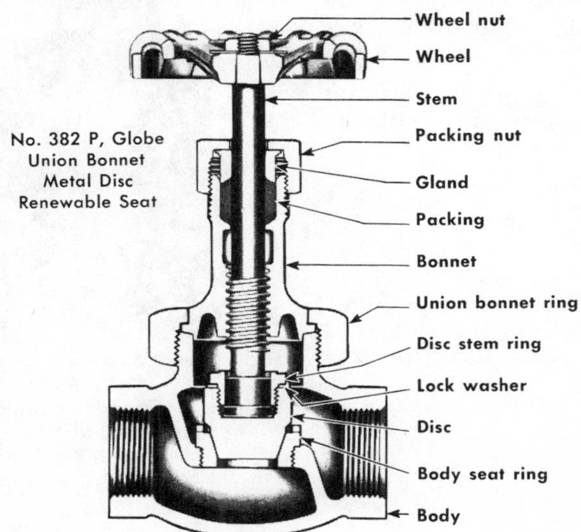

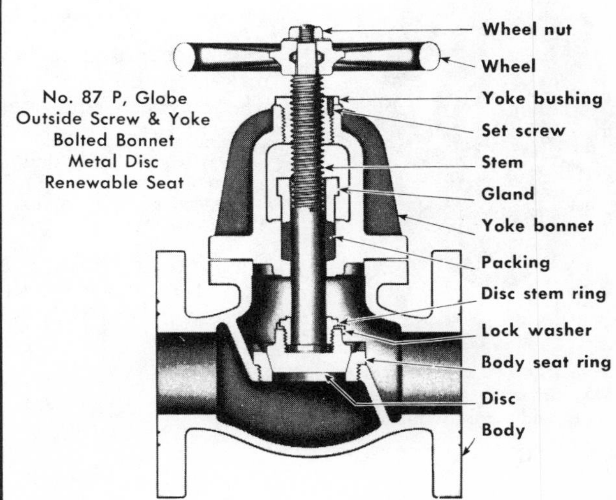

Courtesy of Crane Co.

Table E-3. 125-Pound Ferrosteel Wedge Gate Valves. Name of Parts.

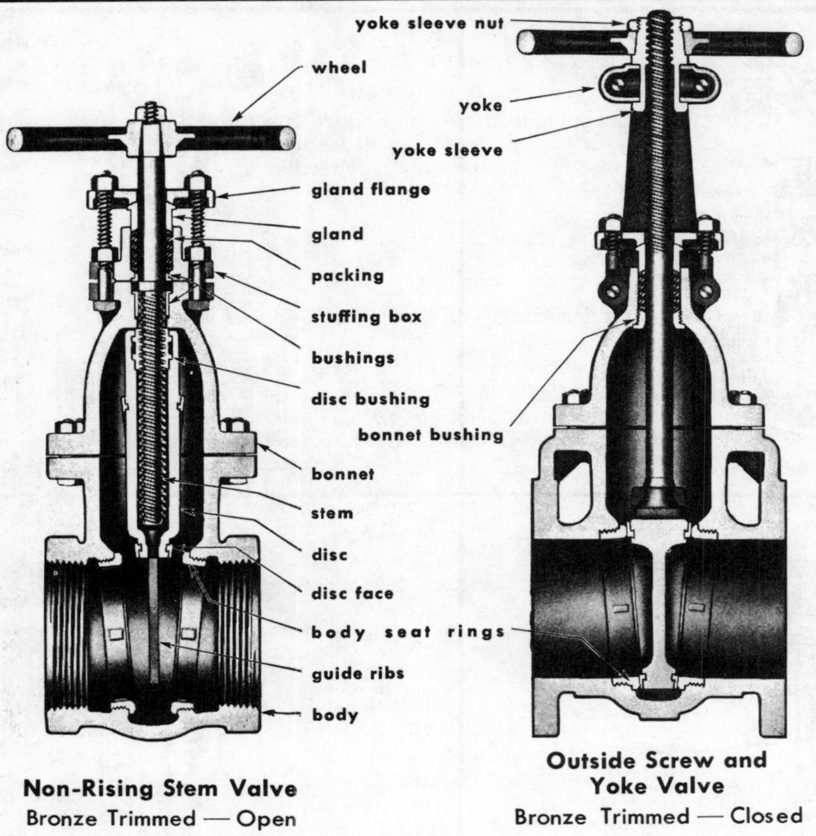

Non-Rising Stem Valve
Bronze Trimmed — Open

Outside Screw and Yoke Valve
Bronze Trimmed — Closed

Non-Rising Stem Flanged
Bronze trimmed and all-iron valves with screwed, flanged, or hub ends are available.

Crane 125-Pound Ferrosteel Wedge Gate Valves are described in detail on this and the seven pages that follow.

The valves have proved their versatility and dependability in practically every industry. Sizes 16-inch and smaller, particularly, with many features of unusual merit, set a new peak for quality in iron body wedge gate valves; unusual strength, long service life, and fine all-around adaptability in all types of installations are a few of the advantages of the Crane line.

Bronze trimmed and all-iron non-rising stem and outside screw and yoke valves are available; refer to the listings below:

The following alloy iron valves, made to the same high standard of quality and exacting design as the 125-pound valves, are also included in this section:

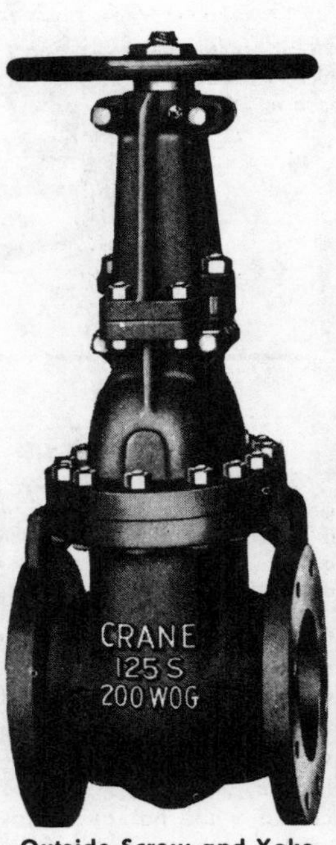

Outside Screw and Yoke Flanged
Bronze trimmed screwed and flanged, or all-iron flanged valves, are available.

Courtesy of Crane Co.

Table E-4. Cast Steel Wedge Gate Valves. 150 to 300-Pound.

"Flex Gate" Valves

(Sizes 2 to 12-inch regular; sizes 14 to 36-inch on order)

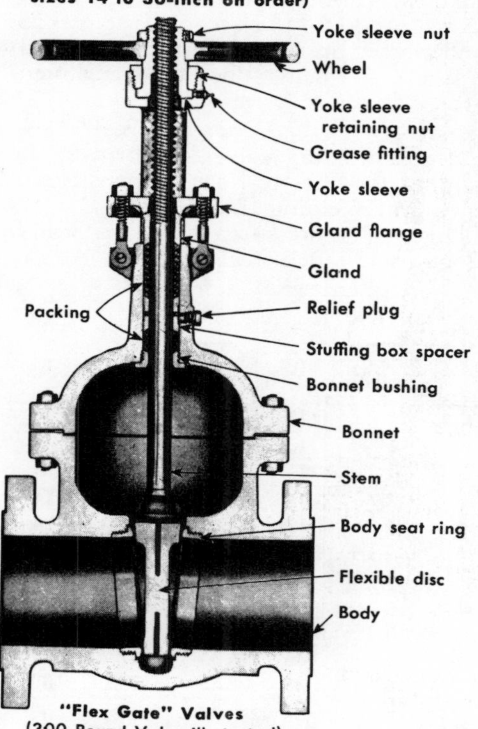

"Flex Gate" Valves
(300-Pound Valve illustrated)

Regularly furnished in sizes 2 to 12-inch; available on order in sizes 14-inch tc 36-inch.

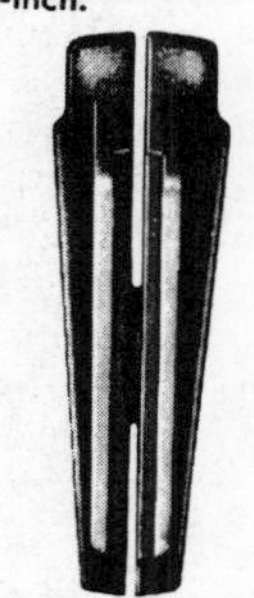

Crane's patented one-piece flexible disc . . . solid through the center only . . . permits each disc face to move independently of the other.

Crane "Flex Gate" Valves offer a host of benefits. The design effects easy operation; less torque is required to seat and unseat . . . . the disc will not stick in the closed position, even if closed while hot and allowed to cool . . . . the resiliency of the construction compensates for minor misalignment of seats due to pipe-line deflection . . . . and the valves are tight over a wide range of pressures on both the inlet seat and the outlet seat.

A complete line . . . featuring Crane's patented "Flex Gate" design in sizes 2 to 12-inch

The term "Flex Gate" is a Crane Co. trademark; registration pending in the U.S. Patent Office.

Crane 150 and 300-Pound Cast Steel Wedge Gate Valves offer dependable service in steam, water, oil, and oil vapor lines. Quality materials and fine workmanship combined with tested designs assure high utility in severe service. A variety of trim materials are furnished.

The line, in the popular 2 to 12-inch size range, introduces Crane's "Flex Gate" Valves . . . . with patented one-piece flexible wedge disc . . . . a major step forward in fine valve construction.

"Flex Gate" Valves: Crane "Flex Gate" Valves feature a new concept in valve design . . . . a flexible wedge disc. Instead of being made solid with both seating faces maintained in the same rigid position, flexibility . . . . or resiliency . . . is attained by having the two faces separated from each other except fo a small section at the center. See the two illustrations at the left.

The shape of the flexible disc can be likened to two wheels on a very short axle. The "axle" or spud at the center of the disc is amply strong to carry the two halves of the disc together at all times . . . . and yet, it permits a degree of action between them. It is this "flexibility" that makes the disc tight on both faces over a wide range of pressures . . . . prevents sticking during temperature changes, and assures minimum operating torque.

Although each disc face can move independently of the other . . . up to two full degrees . . . . the construction is one-piece. There are no loose parts to cause harmful vibration.

Solid Disc Valves: Crane Solid Wedge Disc Valves, illustrated at the upper right, are regularly furnished in the 1½-inch size and in sizes 14 to 24-inch; they are optional in sizes 2 to 12-inch. As in the "Flex Gate" design, careful engineering and workmanship are combined to produce a quality product highly dependable in severe service.

The disc is the solid web type. The facings are smoothly and accurately machined, and are then ground to a mirror-like finish. The disc is carefully fitted into the valve so that an even, wide, and true contact is made with the corresponding faces of the body seat rings.

Disc guides; stem connection: Both the flexible disc and the solid disc have long, machined guide slots which engage the guide ribs in the body to maintain true alignment of the disc throughout its travel. The seating faces do not contact each other until the valve is virtually closed. A tee-head disc-stem connection prevents lateral strains on the stem.

Solid Disc Valves

(Sizes 1½ to 24-inch regular; sizes 30 and 36-inch on order)

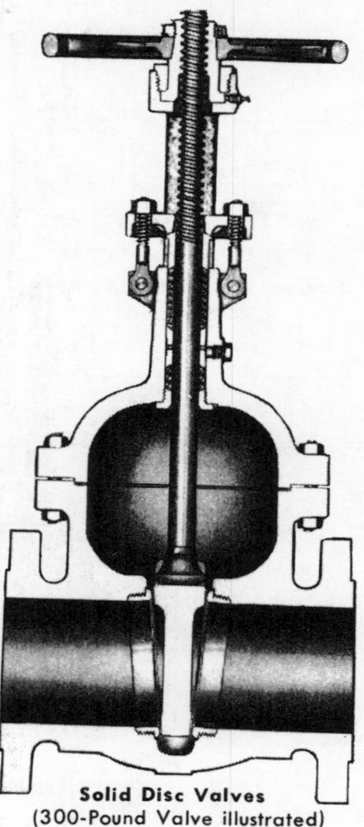

Solid Disc Valves
(300-Pound Valve illustrated)

Solid wedge disc valves, except for the disc, are the same design as the "Flex Gate" Valves.

(table continued on next page)

Table E-4 continued

Cast Steel Wedge Gate Valves. 150 to 1500-Pound Dimensions.

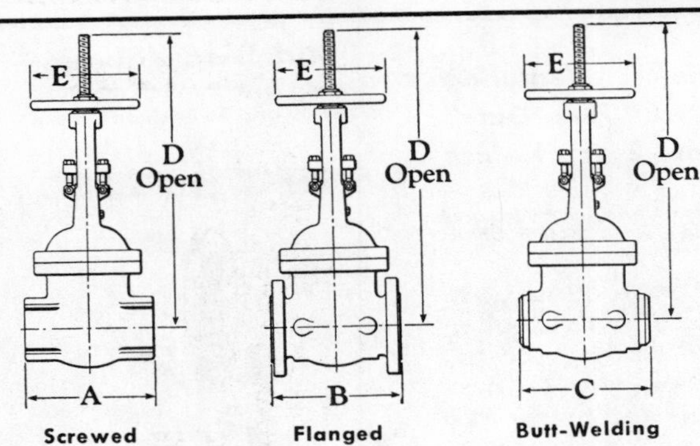

Face to face: Flanged valves of the 150 and 300-pound pressure classes are regularly furnished with a 1/16-inch raised face; those of the 400, 600, 900, 1500, and 2500-pound classes are regularly furnished with ¼-inch high large male facing; face to face dimensions include these facings.

All flanged and butt-welding valves conform to the American Standard for Face-to-Face Dimensions of Ferrous Flanged and Welding End Valves, B16.10-1957, for their respective pressure class. This Standard does not include 3½-inch steel valves.

Dimensions, in Inches

Class	Size of Valve	A	B	C	D	E
150-Pound	2	6¼	7	8½	15¾	8
	2½	7	7½	9½	16½	8
	3	7⅜	8	11⅛	20¾	9
	3½		8½	...	23	9
	4	8	9	12	25¾	10
	5	...	10	15	30½	12
	6	...	10½	15⅞	35¼	14
	8	...	11½	16½	44	16
	10	...	13	18	52½	18
	12	...	14	19¾	60½	18
	14	...	15	22½	70¼	22
	16	...	16	24	79¾	24
	18	...	17	26	89	27
	20	...	18	28	97¼	30
	24	...	20	32	112¾	30
300-Pound	1½	...	7½	...	16¾	8
	2	7	8½	8½	18	8
	2½	8	9½	9½	19	8
	3	9	11⅛	11⅛	23¼	9
	4	11	12	12	28¼	10
	5	...	15	15	33½	12
	6	...	15⅞	15⅞	38½	14
	8	...	16½	16½	47	16
	10	...	18	18	56½	20
	12	...	19¾	19¾	64¼	20
	14	...	30	30	75¼	27
	16	...	33	33	81	27
	18	...	36	36	91½	30
	20	...	39	39	99¾	36
	24	...	45	45	120½	36
400-Pound	4	...	16	16	30¾	12
	5	...	18	18	35	14
	6	...	19½	19½	40¼	16
	8	...	23½	23½	50½	20
	10	...	26½	26½	59¾	24
	12	...	30	30	67¾	24
	14	...	32½	32½	74¾	27
	16	...	35½	35½	80¾	27

Class	Size of Valve	B	C	D	E
600-Pound	2	11½	11½	18¼	8
	2½	13	13	22¼	9
	3	14	14	25¾	10
	3½	15	...	32	14
	4	17	17	31½	14
	5	20	20	36¾	16
	6	22	22	42¾	20
	8	26	26	52¼	24
	10	31	31	62¼	27
	12	33	33	70	27
	14	35	35	77¼	30
	16	39	39	83¾	30
	18	43	43	93¾	36
	20	47	47	104½	36
	24	55	55	126	42
900-Pound	3	15	15	27¼	12
	4	18	18	31½	14
	5	22	22	36¾	16
	6	24	24	42¾	20
	8	29	29	52½	24
	10	33	33	62¼	27
	12	38	38	73½	30
	14	40½	40½	77¼	30
	16	44½	44½	85¾	36
1500-Pound	1	10	10	16	8
	1¼	11	11	16½	8
	1½	12	12	20	9
	2	14½	14½	22⅛	10
	2½	16½	16½	26⅜	12
	3	18½	18½	28	14
	4	21½	21½	33	16
	5	26½	26½	38¾	20
	6	27¾	27¾	47	24
	8	32¾	32¾	55	27
2500-Pound	Dimensions of 2500-Pound Valves are furnished on application.				

Courtesy of Crane Co.

Table E-5. Cast Steel Globe and Angle Valves.

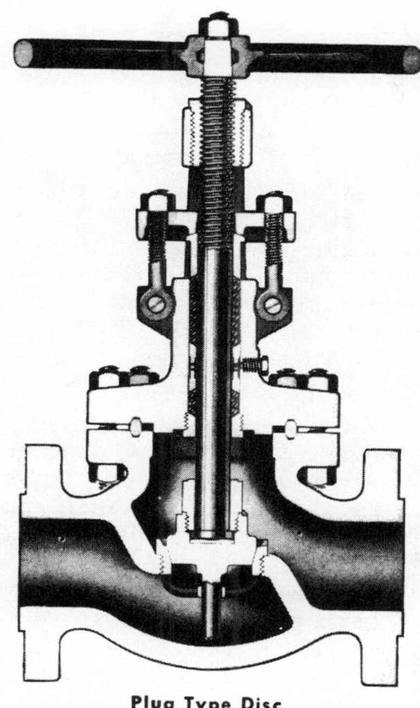

Plug Type Disc

Crane Steel Globe and Angle Valves embody many refinements in design and materials.

Disc and seat: The "XR" trimmed valves (for steam, water, or general service) and "U" trimmed valves (for steam, water, oil, or oil vapor service) in sizes 6-inch and smaller have a plug type disc and seat (illustrated at left); the 8-inch size has a flat disc and seat (not illustrated). The 2-inch valves do not have a disc stem guide.

All sizes of "X" trimmed valves (for oil or oil vapor service) are furnished with a 35° taper seat and a ball shaped seating face on the disc (illustrated at right).

Body seat ring: All valves have the shoulder-type screwed-in body seat ring for utmost tightness and security; in "U" trimmed valves, the rings are also seal brazed or seal welded.

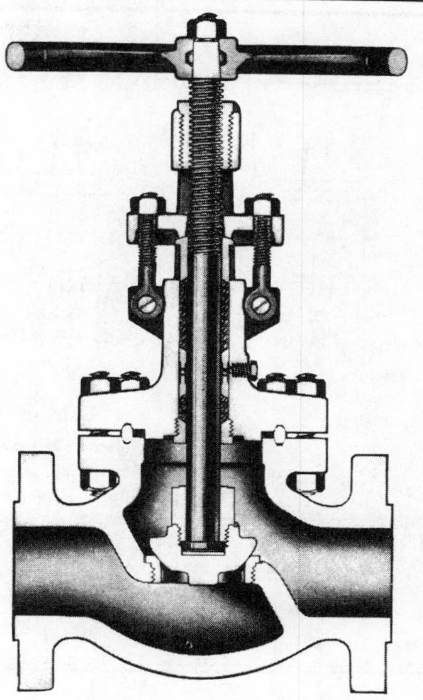

Ball Type Disc

Body and bonnet: The body and bonnet have heavy metal sections with liberal reinforcement at points subjected to greatest stress. The bonnet is fitted with a stem hole bushing.

Bonnet joint: A ring-type bonnet joint holds pressure easily on the 400, 600, 900, 1500, and 2500-pound valves, assuring tightness and maximum strength. On 150 and 300-pound valves, a close-fitting male and female bonnet joint retains the gasket and accurately centers the working parts.

The 300-pound and higher pressure valves have through stud bolts in the bonnet joint. The 150-pound valves employ studs, threaded into the bonnet flange on the body.

Stuffing box: The stuffing box on all valves is deep, assuring tightness and long packing life. The stuffing box is the lantern-type on all except the 150-pound valves. When wide open, the valves can be repacked while under pressure.

Gland: A two-piece ball-type gland and gland flange assure even pressure on the packing without binding on the stem. The gland flange is held in place by swinging eye bolts; the bolts will not loosen in service.

Stem: The stem is of liberal diameter and has unusual strength. Threads are clean and accurately cut and have long engagement with the yoke bushing. The stem and disc are held together by a disc stem ring, which permits the disc to swivel.

Drilling: Flanged valves of each pressure class are furnished with the end flanges faced, drilled, and spot faced (FD & SF) unless otherwise ordered.

When orders so specify, flanged valves can be furnished faced only.

Flange facings: The 150 and 300-pound flanged valves are regularly furnished with an American Standard 1/16-inch raised face on the end flanges; the 400, 600, 900, 1500, and 2500-pound flanged valves regularly have a 1/4-inch male face (large male).

When so ordered, valves can be furnished with other types of facings, such as ring joint, female, tongue, groove, etc.; see pages 332 to 335.

Finish of flange faces: The 1/16-inch raised faces of the 150 and 300-pound valves and the 1/4-inch male faces of the 400-pound and higher pressure class valves are regularly furnished with a serrated finish.

A smooth finish can be furnished on the raised or male faces, when specified.

American Standard: In design and materials, Crane Cast Steel Globe and Angle Valves exceed the requirements of Standards issued by the American Standards Association.

The butt-welding valve ends and the dimensions and drilling of end flanges on flanged valves conform to the American Steel Flange Standard, B16.5-1957, for their respective pressure class.

Flanged and butt-welding valves conform to the American Standard for Face-to-Face and End-to-End Dimensions of Ferrous Flanged and Welding End Valves, B16.10-1957, for their respective pressure class. This Standard does not include 3½-inch steel valves.

(table continued on next page)

Table E-5 continued

Cast Steel Globe and Angle Valves. Dimensions in Inches.

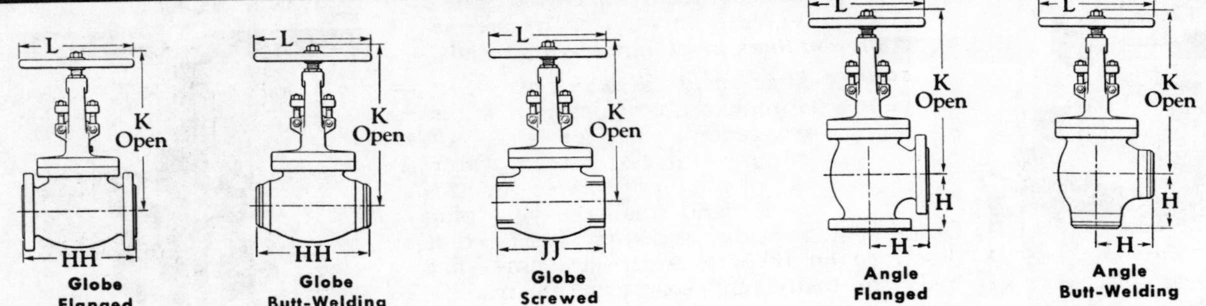

Globe Flanged · Globe Butt-Welding · Globe Screwed · Angle Flanged · Angle Butt-Welding

Dimensions, in Inches

All dimensions shown below apply to valves without gears; dimensions "HH" and "H" apply also to valves with gears. For sizes regularly furnished with gears, see asterisked (*) note at right.

Class	Size	Globe Valves						Angle Valves		All Valves
		Flanged		Butt-Welding		Screwed		Flanged or Butt-Welding†		
		HH	K	HH	K	JJ	K	H	K	L
150 Pound	2	8	13¾	8	13¾	8	13¾	4	12½	8
	2½	8½	14½	8½	14½	...	...	4¼	13	8
	3	9½	16½	9½	16½	...	...	4¾	15	9
	3½	10½	17¼	...	...	...	...	...	...	9
	4	11½	19¾	11½	19¾	...	...	5¾	17¾	10
	5	14	23	14	23	...	...	7	20¾	10
	6	16	24½	16	24½	...	...	8	21¾	12
	8	19½	26	19½	26	...	...	9¾	23½	16
300 Pound	2	10½	17¾	10½	17¾	...	...	5¼	17¾	9
	2½	11½	19	11½	19	...	...	5¾	19	10
	‡3	12½	20½	12½	20½	...	...	6¼	20½	10
	3½	13¼	22½	...	...	...	...	6⅝	22½	12
	4	14	24¾	14	24¾	...	...	7	24¾	14
	5	15¾	26½	...	...	...	...	7⅞	26½	16
	6	17½	29¾	17½	29¾	...	...	8¾	29¾	18
	8	22	36½	22	36½	...	...	11	36½	24
400 Pound	4	16	25¼	16	25¼	...	...	8	25¼	14
	5	18	28½	18	28½	...	...	9	28½	18
	6	19½	31¼	19½	31¼	...	...	9¾	31¼	20
	8	23½	38¼	23½	38¼	...	...	11¾	38¼	27
600 Pound	2	11½	19	11½	19	...	...	...	...	10
	2½	13	21¼	13	21¼	...	...	6½	21¼	12
	‡3	14	23½	14	23½	...	...	7	23½	12
	3½	15	25	...	...	...	...	7½	25	14
	4	17	27½	17	27½	...	...	8½	27½	18
	5	20	30¾	20	30¾	...	...	10	30¾	20
	6	22	35	22	35	...	...	11	35	24
900 Pound	3	15	24	15	24	...	...	7½	24	12
	4	18	29½	18	29½	...	...	9	29½	20
	6	24	37¾	24	37¾	...	...	12	37¾	27
1500 Pound	2	14½	25⅛	14½	25⅛	...	...	...	...	14
	2½	16½	28⅛	16½	28⅛	...	...	...	...	18
	3	18½	33½	18½	33½	...	...	...	...	24

†Angle butt-welding valves are made only in the 600-pound class in sizes 2½, 3, 4, 5, and 6-inch.

‡When 3-inch 300 and 600-pound flanged valves with ring joint facing are to be bolted to Cranelap Joints, orders must so specify; a groove of special pitch diameter is required; see page 334 for dimensions.

***Ball-bearing yoke; gearing:** Crane Cast Steel Globe and Angle Valves, in the larger sizes of the 300-pound and higher pressure classes, are regularly furnished with a ball-bearing yoke and spur or bevel gears, as follows:

300-Pound 8-inch
400-Pound ...... 6 and 8-inch
600-Pound ...... 5 and 6-inch
900-Pound ..... 4 and 6-inch
1500-Pound 3-inch

Orders must state whether spur or bevel gears are wanted; see page 149 for description.

When specified, the above valves can be furnished without gears (plain bearing yoke).

Note: All dimensions apply to valves without gears. Face to face (HH) and center to face (H) dimensions also apply to valves with gears; for additional dimensions of geared valves; see page 149.

Face to face: The 150 and 300-pound flanged valves are regularly furnished with a 1/16-inch raised face; valves of the 400-pound and higher pressure classes have a ¼-inch high large male face. The face to face (HH) and center to face (H) dimensions include this facing.

Flanged and butt-welding valves conform to the American Standard for Face-to-Face Dimensions of Ferrous Flanged and Welding End Valves, B16.10-1957. This Standard does not include steel valves in the 3½-inch size.

Ordering: When ordering, specify catalog number and suffix; see preceding page.

Table E-6. Cast Steel Swing Check Valves

Screwed

For Oil, Oil Vapor, Steam, or Water

No. 148 X, 150-Pound
No. 158 X, 300-Pound

Flanged

For Oil, Oil Vapor, Steam, or Water

No. 147 X, 150-Pound
No. 159 X, 300-Pound
No. 169 X, 400-Pound
No. 175 X, 600-Pound
No. 187 X, 900-Pound
No. 199 X, 1500-Pound

Butt-Welding

For Oil, Oil Vapor, Steam, or Water

No. 147½ X, 150-Pound
No. 159½ X, 300-Pound
No. 169½ X, 400-Pound
No. 175½ X, 600-Pound
No. 187½ X, 900-Pound
No. 199½ X, 1500-Pound

A rugged line . . . designed for severe service on oil, oil vapor, steam, and water lines.

Crane Cast Steel Swing Check Valves, described on these facing pages, embody the many refinements in design and materials necessary to withstand severe service.

For working pressures, test pressures, service recommendations, and specification of materials, see pages 156 and 157. For weights and dimensions, see the facing page.

Materials: These valves, in all pressure classes, are regularly furnished with a body and cap made of Crane Carbon Steel conforming to requirements of ASTM A 216, Grade WCB.

Seating materials are Exelloy to Exelloy (Class "X" trim), suitable for steam, water, oil, oil vapor, air, or gas.

Design: On flanged and butt-welding valves the full port area is maintained without pockets, from the inlet port to the valve seat, to avoid turbulence. On the outlet side of the valve seat, the body is of generous proportions, allowing full swing of the disc and minimizing erosion and flow resistance.

Body seat ring: A shoulder-type screwed-in body seat ring provides maximum tightness and security.

Cap joint: Valves of the 150 and 300-pound pressure classes have a male and female type cap joint.

Valves of the 400, 600, 900, 1500, and 2500-pound pressure classes have a ring type cap joint.

Crane Triplex Steel studs and stud bolts assure an unusually strong and tight joint. The 150-pound valves are equipped with studs; all other valves have through stud bolts.

Flange facings: The 150 and 300-pound flanged valves are regularly furnished with an American Standard 1/16-inch raised face on the end flanges.

The 400, 600, 900, 1500, and 2500-pound flanged valves are regularly furnished with a ¼-inch male face (large male).

Cross Section of Cast Steel Swing Check Valve

Note: The 150-pound valves in sizes 14 and 16-inch (not Illustrated) have a bottom seated body seat ring . . . . and the complete disc, hinge, and hinge pin assembly is suspended from a hinge bracket; the bracket is securely fastened to a pad which is cast integral with the body.

When so ordered, flanged valves can be furnished with other types of facings, such as ring joint, female, tongue, groove, etc.

Finish of flange faces: The 1/16-inch raised faces and the ¼-inch male faces are regularly furnished with a serrated finish.

A smooth finish can be furnished on raised or male faces, when specified.

Standards: In design and materials, Crane Cast Steel Swing Check Valves exceed the requirements of Standards issued by the American Standards Association and the American Petroleum Institute.

The end flanges on flanged valves as well as the dimensions of butt-welding valve ends conform to the American Steel Standard, B16.5-1957, for their respective pressure class.

Flanged and butt-welding valves of all classes, in sizes 12-inch and smaller, conform to the American Standard for Face-to-Face and End-to-End Dimensions of Ferrous Flanged and Welding End Valves, B16.10-1957, for their respective pressure class. This Standard does not include 3½-inch steel valves.

Flanged and butt-welding valves of all classes conform also to the API Standard for Pipe Line Valves, No, 6-D, Ninth Edition, April, 1960. This Standard does not include a 3½ or 5-inch size.

Prices: Prices of all valves are furnished on request.

Courtesy of Crane Co.

Table E-6 continued

Cast Steel Swing Check Valves. Weights and Dimensions.

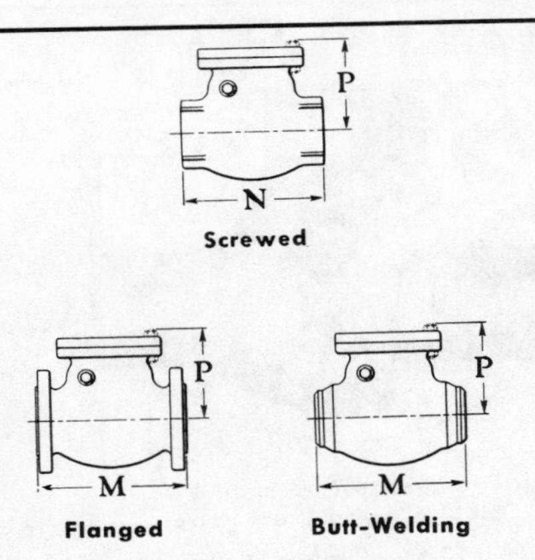

When ordering, specify catalog number and suffix; see the preceding page.

Drilling: Flanged valves are regularly furnished with end flanges faced, drilled and spot faced (FD & SF); they are drilled to the corresponding pressure class of the American Standard; they can be furnished faced only, when specified.

Face to face: Face to face dimensions (M) of flanged valves include the 1/16-inch raised face on the 150 and 300-pound pressure classes and the ¼-inch high large male face on the 400-pound and higher pressure classes.

Butt-welding valves: Unless otherwise ordered, 150 and 300-pound butt-welding valves are bored to match the inside diameter of standard pipe (heaviest weight on the 8, 10, and 12-inch sizes). For all other pressure classes, orders must specify the diameter of the bore (I.D. of pipe).

Smaller size 400 and 900-pound valves: For smaller size 400-pound valves, use the 600-pound valves. For smaller size 900-pound valves, use the 1500-pound valves.

2500-pound valves: Prices, weights, and dimensions of 2500-pound valves are furnished on request. For sizes and general description, see page 160.

Weights and Dimensions — Prices on Request

Pressure Class	Size	Pounds, Each			Dimensions, in Inches			
		Screwed Valves	Flanged Valves	Butt-Welding Valves	Screwed		Flanged or Butt-Welding	
	Inches		FD & SF		N	P	M	P
150 Pound	2	27	34	25	8	5	8	5
	2½	40	50	30	8½	5½	8½	5½
	3	50	65	50	9½	6	9½	6
	3½	...	94	...	...	...	10½	6½
	4	96	100	100	11½	7	11½	7
	5	...	140	120	...	...	13	8
	6	...	200	160	...	...	14	9
	8	...	390	360	...	...	19½	10¼
	10	...	510	...	...	...	24½	12⅛
	12	...	775	...	...	...	27½	13¾
	14	...	1200	...	...	...	35	on
	16	...	1450	...	...	...	39	request
300 Pound	2	40	62	47	9½	6¾	10½	6¾
	2½	70	80	60	10¾	8	11½	8
	*3	100	120	80	11¾	8½	12½	8½
	4	...	180	130	...	...	14	9¾
	5	...	250	240	...	...	15¾	10¾
	6	...	330	260	...	...	17½	11¾
	8	...	620	510	...	...	21	14
	10	...	920	760	...	...	24½	15
	12	...	1290	1015	...	...	28	16¾
400 Pound	4	...	200	190	...	...	16	10
	5	...	270	265	...	...	18	12
	6	...	395	310	...	...	19½	12½
	8	...	680	580	...	...	23½	14½
	10	...	900	820	...	...	26½	15¼
	12	...	1250	1150	...	...	30	16⅞
600 Pound	1¼	...	38	32	...	...	9	6¼
	1½	...	58	40	...	...	9½	6¾
	2	...	70	55	...	...	11½	7
	2½	...	105	70	...	...	13	8¼
	*3	...	140	100	...	...	14	9
	4	...	260	170	...	...	17	10¼
	5	...	400	300	...	...	20	12¾
	6	...	530	420	...	...	22	13½
	8	...	900	740	...	...	26	15¼
	10	...	1440	880	...	...	31	18¾
	12	...	1970	1200	...	...	33	21½
900 Pound	3	...	180	155	...	...	15	9½
	4	...	340	240	...	...	18	11
	6	...	640	500	...	...	24	13¾
	8	...	1180	890	...	...	29	16½
1500 Pound	1½	...	110	80	...	...	12	8¼
	2	...	160	130	...	...	14½	9¾
	2½	...	245	170	...	...	16½	10½
	3	...	280	210	...	...	18½	11¼
	4	...	630	390	...	...	21½	13¼
	5	...	950	480	...	...	26½	15¼
	6	...	1360	780	...	...	27¾	15¾
	8	...	2100	1320	...	...	32¾	18¼

***3-inch Cranelap Joints:** When 3-inch 300 and 600-pound flanged valves with ring joint facing are to be bolted to Cranelap Joints, orders must so specify. A groove of special pitch diameter is required; see page 334 for dimensions.

Courtesy of Crane Co.

Index